WHO'S WHO IN THE COSMIC ZOO?

An End Times Guide to ETs, Aliens, Exoplanets & Space Controversies

BOOK FIVE
THE HEAVENS

Ella Le Bain

Time Jumping During Appointed Times 308
Time As A Circle 310
The Shape Of Time 311
Time Zones And Parallel Universes 312

CHAPTER TWENTY-FOUR 314
THE BATTLE FOR TIME 314
Time Is Money 314
Secret Time Travel Projects 316
Project Pegasus 317
Times To Pray 318

CHAPTER TWENTY-FIVE 321
FROM SECRET SPACE PROGRAM TO SPACE FORCE 321
The 20 And Back Program 321

CHAPTER TWENTY-SIX 334
WHO OWNS SPACE? 334
Where Is Outer Space? 336
Mining The Moon 336
The Real Star Wars 337
Space Force Just Received Its First New Offensive Weapon 340
USAF Secy Warns Force Justified To Deter Space Attacks 342
USAF 343
US Army 345
Weapons In Space 345

CHAPTER TWENTY-SEVEN 347
THE NEW SPACE RACE 347
Private Aerospace Partners With NASA 348
SSP, Space Force & Private Industry 353
Commercial Aerospace 353

CHAPTER TWENTY-EIGHT 356
NASA'S BOMBSHELL ON THE MOON 356
Space Policies 357
Mining The Moon & Asteroids 361
Human Space Exploration Programs 363

CHAPTER TWENTY-NINE 368
PRESIDENTIAL SPACE POLICIES 368
   Space Commanders In Chief: From Ike To Trump 368
CHAPTER THIRTY 374
THE GRAYS & THE U.S. GOVERNMENT 374
   Tau IX Treaty For The Preservation Of Humanity 375
   US Presidents & Aliens 379
CHAPTER THIRTY-ONE 381
SECRET GOVERNMENT SPACE PROGRAMS 381
   Declassified: America's Secret Flying Saucer 381
   Undersea Military Bases 383
   TR-3B 387
   The New Space Race 388
   Space Force: 21ST Century Warfare 390
   Pilots Die Chasing UFOs 391
   Shoot Them Down 392
   Solar Warden AKA Secret Space Program 393
   Solar Warden Spaceships 394
   Controlling The High Ground 395
CHAPTER THIRTY-TWO 398
REPTILIANS HELPED NAZIS BUILD A SECRET SPACE PROGRAM IN ANTARCTICA 398
   Whistleblower Who Worked For Extraterrestrials 398
CHAPTER THIRTY-THREE 403
SPACE FORCE 403
   From Space Command To The New Space Force 403
   Space Force Logo V. Star Trek's Starfleet Command Logo 408
   Meaning Behind The Logo 410
   Has Space Force Taken Over The Secret Space Program? 411
   Space Warfare – Legal Ground 414
   Space Race Of The Roaring 20S 415
   To The Moon, Mars, Asteroids & Beyond 415
   Sci-Fi Movies Prep Us For Reality 417
CHAPTER THIRTY-FOUR 423

BIBLE DNA & TODAY'S GENETICISTS 423
   Sumerian Alien Connection 426
   Anunnaki And Racism 426
   Anunnaki-Nibiruans 427
   Dead Sea Scrolls 428
   Nibiru's Population Of Giant Aliens 432
   Survey Nibiru – New Space Force Mission? 432
   Annunaki Alien Threat – New Space Force Mission? 434

CHAPTER THIRTY-FIVE 436
CLOUDSHIPS, CHARIOTS & UFOS 436
   The 1ST Stargate Mentioned In The Bible 438
   Biblical Chariots Are Starships 442
   Armada Of Chariots Of God 442
   Horses And Chariots Of Fire 443
   Clouds As Chariots 443
   The Chariots Of God Spin And Glow: 444
   Ancient Star Wars Culminates In Divine End Times Appointment 444
   Christ Returns With Blazing Fire 446

CHAPTER THIRTY-SIX 449
THE CASE AGAINST FLAT EARTHERS 449
   Flat Earth Theory Fabricated To Distract From Planet X 450
   Learn To Discern The Spirits For God's Sake! 452
   Debunking Flat Earthers 454
   What Is The Firmanent? 464
   The End Times Antisemitic Flat Earth Cult 470
   Earth's Ancient Dome Of Water 476

CHAPTER THIRTY-SEVEN 482
THE REAL STAR WARS 482
   Connecting The Dots From Jediism To Judaism: Star Wars As Jewish Allegory 484
   Hannukah And Star Wars – The Uncanny Parallel 484
   In A Galaxy Of Hebrew Names 486
   The Force 487
   The Last Jedi And The Force Of Family 489

    Movies And Subliminal Messages 491
    Star Trek And God 491
    Battlestar Galactica 493
    Knights Templar And Star Wars 494
    The Tree Of Life In Kabbalist Thought 496
CHAPTER THIRTY-EIGHT 499
SIGNS AND WONDERS IN THE HEAVENS 499
    Indications Planet X Is Here 500
    Our Citizenship Is In Heaven 502
    Exopolitical Evangelism 505
    Heaven Is Inside The Earth & Above Us 506
CHAPTER THIRTY-NINE 509
SUN OR SON? 509
    Who Is The Greater Light? 509
    The Sheep And Goat Judgment 513
    Ancient Humans Stargazing 515
    Our Story Begins And Ends In Heaven 516
    The Ecliptic Cross 517
    The Solstice Cross 517
    The Throne Of David 518
    Nathan's Prophecy 519
    The Prophecy Of The Messiah 524
CONCLUDING WORDS 533
    The Metaphor Of Movies 533
    End Times Disclosure 535
    Making Heaven More Crowded 543
    Prepping Your Soul 546
    Faith In God's Faithfulness 549
    The Kingdom Of Heaven 551
NOTES AND BIBLIOGRAPHY 553
ABOUT THE AUTHOR 575
    Who Is Ella LeBain? 575

DEDICATION

This book is dedicated to the Lord of the Cosmos, the Lord of Heaven and Earth, without whose loving support, guidance and protection, this manuscript would not have been possible. I am forever grateful.

In addition, I am dedicating this book to all those who have been wounded by religion, religious spirits, all atheists, agnostics, and especially to those who consider themselves religious. May you all find something within these pages to heal your soul.

And finally, a special dedication to all who were told in one way or another that they just do not fit in. Heaven has a place for you. May you be inspired to find your place and reconcile with your Creator as you journey through the pages of this book.

I was told by some who consider themselves Christians that I can't mix God with astrology, then I was told by astrologers that I can't combine ufology with astrology, then I was told by astronomers that there's no place for astrology in astronomy, and finally I was told by my Jewish family and friends that you can't be Jewish and believe in Jesus. This is the book that contracts all those false statements! Let us connect all the dots and prove them all wrong.

When you set out to research the stars, the physics, the meanings, you must be inclusive, and understand that they are all connected. Just like in life, there is science put into language. Together they bring understanding. Understanding can bring Wisdom.

For this reason, this book is dedicated to all the Cosmic Misfits, Black Sheep and Orphans. There are no orphans in the Kingdom of Heaven – we are all adopted daughters and sons of the Most High God. May the Almighty Lord of the Cosmos bless you with the revelation of His goodness, understanding and everlasting love.

Acknowledgements

My deepest debt of gratitude goes to my loving and devoted husband and daughter, for their love and support. They have stuck with me through all the spiritual battles and never stopped believing in my vision and goal. They are my true loves and soulmates whom I am blessed to be journeying with through this Earth experience.

INTRODUCTION

"The planets are God's punctuation marks pointing the sentences of human fate, written in the constellations."
~James Lendall Basford

When you set out to research the Stars, their Physics, their meanings, you run the risk of being myopic if you choose one over the other.
To be succinct, you must be inclusive, and understand that they are connected. Just like in life, there's Science and Language. Together they bring understanding. Understanding brings Wisdom.
~Anonymous

"There are more things in Heaven and Earth, Horatio, than are dreamt of in your philosophy." ~ Hamlet
~William Shakespeare

The points discerned and discussed in the first half of this book will touch on these facts: NASA is calling Nibiru, also cited as Planet X, Planet 9. They are one and the same. NASA announced the discovery of 7 Earth-sized planets clustered around a dwarf star close to Earth. This is the Nemesis/Nibiru system of 7 planets which is intertwining with our Solar System. Please refer to the YouTube videos in the comments for further information and clarification.

In 2017, NASA disclosed its discovery of 7 Earth-sized planets orbiting a brown dwarf star and named it the TRAPPIST-1 system. It was named after the Transiting Planets and Planetesimals Small Telescope in Chile, which is abbreviated – TRAPPIST. Many of us have connected the dots to the Nemesis system which is our Sun's Twin Brown Dwarf that has 7 planets, in its orbit, one of which is Nibiru, which is approaching our Sun. Confusion exists over the identification of Nibiru and her giant sister planets, and all the name changes do not help. The newly discovered orbs have been dubbed Planet X, Planet Nine, Blue Kachina, Red Kachina, Red Dragon, Hercolubus, and more!

I continue to uphold with proofs that Nibiru, using the ancient Sumerian name, was always depicted by the ancients with wings and is red/gold. There are several researchers who are calling the Blue Planet, Nibiru, which is not in accordance with the Sumerian, Biblical, Greek, and Chinese accounts. The Chinese called it the Red Dragon because of its red fiery tail. The Greeks called it Hercolubus which was described as a giant red planet that caused Earth changes.

I define the differences in my third Book, *Who Are the Angels?* Chapter Sixteen: *The Second Coming and Nibiru*,[1] where I prove through documented research that it was the Blue Planet that eclipsed our Sun during the crucifixion of Jesus Christ and caused three hours of darkness in the biblical account.

Nibiru comes around our Sun every 3,600 years, causing Pole Shifts and cataclysmic Earth changes. Immanuel Velikovsky, author of *Worlds in Collision*, close friend, and colleague of Albert Einstein, also tracked these cataclysms on a 3,600-year cycle. His work in the 1950s was based entirely on the science of physics, confirmed by NASA technology and Hubble photos.

The fact that this binary system intersects with our own, sometimes colliding, during these cycles, is the reason the Lord promised to recreate the Heavens and Earth in Isaiah 65:17; Revelations 21:1. "For behold, I create new Heavens and a new Earth; And the former things will not be remembered or come to mind."

There are several more mentions of the Lord recreating the Heavens and Earth:

> "For behold, I create new Heavens and a new Earth; And the former things will not, and there will be no more sea."
> (Revelations 21:1)

> "For just as the new Heavens and the new Earth, which I will make, will endure before Me," declares the LORD, "so your descendants and your name will endure."
> (Isaiah 66:22)

> "…that the creation itself will be set free from its bondage to decay and brought into the glorious freedom of the children of God."
> (Romans 8:21)

> "But in keeping with God's promise, we are looking forward to a new Heaven and a new Earth, where righteousness dwells."
> (2 Peter 3:13)

> "At the time His voice shook the Earth, but now He has promised, 'Once more I will shake not only the Earth but also the heavens. The words 'once more' indicate the removing of what can be shaken- that is, created things- so that what cannot be shaken can remain."
> (Hebrews 12:26 - 27)

> "The sky (the Heavens) receded like a scroll being rolled up, and every mountain and island was moved from its place."
> (Revelation 6:14)

The Zohar, the foundational work in the literature of the Jewish Torah and mystical thought, states explicitly that the Messianic process will be accompanied by "several stars appearing" and one Red Star in particular as the harbinger of the Coming of the Messiah. The Zohar goes into great depth describing the number and various colors of these stars.

The Zohar coincides with the Torah verse, Numbers 24:17: "I see him, but not now; I behold him, but not nigh; there shall step forth a star out of Yaakov, and a scepter shall rise out of Yisrael, and shall smite through the corners of Moab, and break down all the sons of Seth."

I agree with multiple astronomers, sky watchers and observers who have connected the discovery of seven Earth-sized planets orbiting around a dwarf star to be no coincidence. This describes the Nemesis-Nibiru system exactly, and now it is actually here in our own Solar System! Many of us believe this was NASA's way of diverting attention from what is taking place in our own Solar System. Nemesis is a brown dwarf star that is possibly our Sun's evil twin because it is dark and seems to create cataclysmic events in 3,600 cycles. It has been observed to have seven planets with Moons, twelve objects in total.

LIFE IS OUT THERE – CLOSER THAN WE THINK

One of the first scientists to model the potential climate for habitable worlds in depth was Eric Wolf, a researcher at the University of Colorado, Boulder.[2] The astrobiologists are now claiming that the conditions for life do indeed exist in the TRAPPIST-1 System. The ancients recorded this same information on the Sumerian stone cuneiform tablets, in the Bible scriptures, in some extrabiblical scriptures, as well as repeated orally in Native American histories and prophesies. I for one do not believe in coincidences. It is all intelligent design.

July 17, 2020, Universe Today announced that astronomers confirmed through a collaborative study that TRAPPIST-1 planets have an atmosphere that would support life. The study, which recently appeared in the journal *Astrobiology*, was conducted by an international team of researchers from the Geneva Astronomical Observatory (GAO), the University of Bern, the Laboratoire d'astrophysique de Bordeaux (LAB), the Astrophysics Research Group at Imperial College London, and the Laboratory for Atmospheric and Space Physics (LASP) at the University of Colorado.[3]

Martin Turbet, who was a postdoctoral researcher at the GAO, and lead author on the study, explained to Universe Today via email:

> "We reviewed all existing works on the topic, ranging from observations with the best telescopes available (Hubble Space Telescope, Spitzer Space Telescope, Very Large Telescope, etc.) to the most sophisticated theoretical models

such as three-dimensional numerical climate models," said Turbet.

What they found was rather encouraging. For starters, they were able to determine that most of the TRAPPIST-1 planets had cloud-free, low-molecular-weight atmospheres, similar to what Earth's primordial atmosphere was like. Second, they found compelling evidence that those planets that did have atmospheres were likely composed of elements that have higher atomic weights. Turbet said, "We determined that the seven TRAPPIST-1 planets are unlikely to have hydrogen-dominated atmospheres. We also suggested that the atmospheres (if present) of the TRAPPIST-1 planets are most likely to be carbon dioxide-dominated, oxygen-dominated or water-dominated." [3]

In other words, of the seven TRAPPIST-1 planets, those that have atmospheres are likely to have the kind that are favorable to life (at least as we know it). That means carbon dioxide, an essential climate stabilizer necessary for photosynthetic organisms, oxygen gas, nitrogen, and volatile elements like water. It also includes cloud cover, which is not only an indication of water, but provides protection against stellar radiation.

Turbet said, "Next-generation missions—in particular the James Webb Space Telescope and the near-infrared ground-based spectrographs—will have the power to detect 'heavy' molecules such as carbon dioxide, oxygen, methane, etc. and thus they may have the potential to determine whether or not the TRAPPIST-1 planets have atmospheres, and if so, what they are made of."

The James Webb Space Telescope is scheduled to launch in 2021, whereas ground-based telescopes equipped with next-generation spectrographs are expected to come online throughout this decade. With these and even more powerful instruments planned for the future, astronomers expect to know with certainty that there is indeed life beyond Earth in our corner of the galaxy. [3]

A planet that goes by many names: Planet X, Nibiru, Hercolubus, Nemesis, Red Kachina, Red Dragon, Wormwood, the Destroyer, Marduk, Rahab, are but a handful of names this planet is known. It is on its way to pass Earth and, in its passing, will complete the pole shift already in progress.

The effects of this planet passing Earth include climate change, the increase in fireballs, meteorites, and mental, emotional, and spiritual meltdowns. Its energy is said to cause mental illness, and its approaching transit, major Earth changes.

The other factors being discerned and observed by astronomers and sky watchers is the fact that chemtrails appear just before Sunset, observed to occur only on the western sky. Astronomers suspect this is done deliberately to obscure the actual Sunset and the second Sun or its planets.

Two other planets have been observed and photographed: the white planet with a black hole, similar in appearance to the *Star Wars Death Star*, and the blue-green planet observed and recorded for years on our Neumayer Observatory on a 24/7 loop.

For the past few years, it has been blocked out of Google Sky, yet recently has shown up for the public to see. Recent photos taken through telescopes by various astronomers are of incoming Nibiru - Nemesis system. This is the only section blocked out on Google Sky. Its tails are iron oxide full of comets, meteorites, and fireballs. Revelation 12 is the prophecy of Nibiru and is written in astronomical language. On September 23, 2017, Nibiru and Jupiter were transiting through Virgo, the constellation of The Virgin. It was the time when astronomers said we would begin to see the red-winged planet, the planet of Crossing, with our naked eyes, just as we can view Jupiter in the night sky without binoculars.

> "And I shall give signs in the Heavens, miracles in the Earth:
> blood, fire and clouds of smoke."
> (Acts 2:19)

I wrote a rather lengthy chapter on Nibiru and all its many names, noting astronomical viewpoints that connect the dots of Bible prophesies and historical accounts of previous Nibiru transits.[1] See, *Who Are the Angels?* Chapter Sixteen: *The Second Coming and Nibiru* pp.338-363

In my book, I postulate that it could take a Shemitah, a 7-year cycle, from the time it passes the Earth on its way around the Sun, till it returns with unspeakable velocity to pass the Earth for a second time. I find it highly coincidental that almost all the astronomers, including those at NASA, are predicting it to take 24 to 36 hours to pass the Earth, whereas I pulled multiple scriptures throughout the End Times Bible Prophesies that refer to it as a 'day,' the Terrible 'Day' of the Lord.

SPACE FORCE, SOLAR WARDEN & DISCLOSURE

The second half of this book will focus on shedding some light on and understanding of UFO Disclosure, what has been being covered up for 70 years, and why we're in a Disclosure era, which is the real reason for instituting the Space Force. I laid the groundwork for the reality of the alien presence in Book One of *Who's Who in the Cosmic Zoo?* This connects the dots of Nibiru and UFO Disclosure points to a time that fulfills End Times Prophecies that were written down several millennia ago.

There are so many different layers to these controversies! One thing is clear, the level of human interest is greater than ever before, as the latest statistics show that more people have had experiences and want to know what is going. UFOs have been in the news, more so during period of the 21st Century than ever before. Under the Trump Administration, Americans have had 3 disclosures that came from government sources. The Department of Defense and the Pentagon announced respectively that were studying UFOs, taking them seriously as a threat to National

Security. In 2019 came the release of protocols on how pilots may report UFOs. These disclosures were huge. The U.S. government has never before admitted to the reality of UFOs, much less admit they were studying them, taking them seriously and have created protocols for accurate reporting. Eschewed the last sentence which contradicted the above.

We are seeing videos from the ISS Live feed that possibly reveals new high-flying Space Force assets. Unidentified spacecraft shows up on the ISS Live feed often, but we do not know if these are alien craft or from our own Secret Space Program. I watched an uninterrupted live video feed for 22 minutes. Nobody cut the feed, in fact, you can see this unidentified craft ascend into deep space, and the ISS camera is focused on its ascent as it appears to go INTERSTELLAR and turns green. This is highly unusual, and of course, NASA has made no comment.

The debate over the future of Space Force has now gone mainstream. Academics will be shocked once Space Force roles out the antigravity craft it is inheriting from the USAF SSP: "Space Force: What Will the Newest Military Branch Actually Do?" [7]

What is new is that Space Force will be an open, transparent, and accountable space program, unlike the USAF SSP that has been operating since the 1970s as a black op. This is a big difference and worth emphasizing as all the ultra-secret technology is transferred and unveiled as new.

Besides Aliens, Space Force is in charge of interstellar travel, patrolling the Earth and our solar system. That was the purpose behind the Secret Space Program Earth's Solar Warden. Reminds me of the scripture in Zechariah when the Lord of Heaven and Earth, ordered His horses to patrol the Earth. We all know horses do not fly. Horses are ancient vernacular for space chariots, or spaceships.

> "When the *powerful horses came out*, they were impatient to go and patrol the Earth. And he said, "Go, patrol the Earth." So, they patrolled the Earth."
> (Zechariah 6:7)

If there is another system intersecting with ours, Space Force is going to take over what Solar Warden has been doing. It will also take charge of all technology around the Sun. There are multiple objects around the Sun, some are ours, some are not.

I once had a Fundamentalist unfriend and block me for reporting there are two Suns. She said "God made the Sun, the Moon and stars. There's only one Sun." When I pointed out to her that stars are Suns and our Sun is a star, she unfriended me. Cognitive Dissonance is heavy with my generation. I explained that most star systems are binary, some are even more complex, with three or more Suns. Now I understand what Giordano Bruno went through! The poor man was burned alive for daring to suggest that we are not the center of the universe.

INTRODUCTION

> "Woe to those who call evil good and good evil,
> who put darkness for light and light for darkness,
> who put bitter for sweet and sweet for bitter."
> (Isaiah 5:20)

We need to get comfortable with the idea that we are not alone, not alone in terms of extraterrestrial or alien life, and not alone as a singular star system. Ours is a binary system, our Sun being a twin, an evil twin, the dark brown dwarf star called Nemesis.

RECENT UFO GALLUP POLL

Gallup polled Americans about what they then called flying saucers for the first time in 1966. Only 5 percent had seen one, whereas 91 percent of those they polled said they had never seen one, 4 percent never heard the term "flying saucer." Then in 1996, the company tried again: "Have you, yourself, ever seen anything you thought was a UFO (unidentified flying object)?" Disappointingly, 87 percent of participants said no, while a more interesting 12 percent said yes, and 1 percent, intriguingly, said they were not sure or refused to answer. [4]

The pollsters put the 'flying saucers' to rest for the time. Over 40 years passed, and then things got publicly, visibly weird, to the point where the Navy created a new protocol for pilots to report sightings, and the Pentagon has had to acknowledge they have top-secret video of a particularly infamous UFO incident that occurred off the coast of San Diego in 2004, between Navy pilots aboard the USS Nimitz and something that looked like an enormous Tic Tac, moving in baffling ways.

The June 2019 Gallup UFO poll did not ask if people who have seen a UFO believed it was piloted by an alien, only if they have ever seen one. Gallup realized in their wording, people might think the UFO they saw might be a drone or a military jet. So, the company conducted a follow up poll in August 2019, to clarify what people believed the craft might be; at that point, they discovered 33 percent of those polled thought some of the UFOs might be the result of alien visitation. [3]

Other Gallup poll findings include:
• The 68% today who believe the government is withholding information about UFOs is comparable to the 71% found in 1996. Both times, the results were similar among all main demographic groups, including by age, education, and party identification.

However, it appears that only about half the number who think the government is hiding something about UFOs think it is covering up information about alien space landings, specifically. This conclusion is based on the finding that far fewer people give credence to UFO sightings or have witnessed them, personally, than think the government knows more than its telling.

- 33% of U.S. adults believe that some UFO sightings over the years have in fact been alien spacecraft visiting Earth from other planets or galaxies.
- The majority, 60%, are skeptical, saying that all UFO sightings can be explained by human activity or natural phenomenon, while another 7% are unsure.
- Separate from possible space landings on Earth, about half of Americans, 49% think that people somewhat like ourselves exist elsewhere in the universe. Even more, 75%, believe that some form of life exists on other planets.

UFO believers are found in all demographic and political segments of society but are particularly likely to be residents of the West. That is the home of Area 51, as well as nearby Roswell, New Mexico — where a crash of some kind in 1947 sparked the Area 51 conspiracy theories that persist to this day.

- Forty percent of residents of the West believe some UFOs can be attributed to alien visitors. This compares with 32% of residents in the East and South and 27% in the Midwest. [5]

We are living through what the Book of Revelation calls Apocalypse, which means *unveiling, disclosure, revelation, an unfolding of things not previously known, and which could not be known apart from the unveiling*. UFO Disclosure is most certainly part of the End Times Prophecies. In my previous books, I laid bare the facts from history, ancient scriptures and connected some of those dots to our present reality. I have already established the dichotomy in perspectives, philosophies, and belief systems, as to how people are groomed to think about UFOs, extraterrestrial life, and the alien presence on Earth. It seems there are essentially two camps of people who believe that UFOs are real: those who believe they're here to help us, save us, guide us towards a higher evolution, and even ascension; and those who believe they are all demons, just the fallen angels laid out in the Apocryphal books of Enoch, Jubilees, Jasher. They both cannot be right. What I am going to suss out here is my own unique perspective, which is why I am writing these books, for the edification of those who are seeking truth, knowledge that leads to understanding.

Firstly, I am an Experiencer. My earliest memories are from age 2. I will be turning 60 in January of 2021, so you can do the math. I was orphaned at the age of 6 and then again at 15. I have been on my own in this world, or so it would seem to those in the physical realms. However, I have never really been alone. I have however, been fought over, but that story is for another book, which includes my detailed testimony, *Cinder-Ella's Shadow*.

Secondly, I come from a biracial family, I am a hybrid of sorts, half Jewish and half Italian. I am an only child, so I was curious about my ancestry, got a DNA test, and found out where my roots were. My Jewish roots come from Israel, Turkey, and the Middle East (a general term), yet my Italian mother had actually passed to me 46 per cent Greek DNA, and only 11 percent Italian. After doing some digging, I found that made perfect sense, as the Greeks took over most of Italy prior to the

Roman conquest, and had settled in and around Naples, the seat of my Mother's family. When I learned this piece of my ancestry, it made sense to me, as to why I collect books, sleep on books, and have always been hungry for knowledge. I am also the only one in my family that was born with blond hair. In fact, I was born bald, and only started to grow hair when I turned 2. This was evidenced by photos my mother took of me, always putting hats and wigs on me as a baby to cover my baldness. Both my parents have dark hair and brown eyes. I am the oddball in the family if there was a family.

Thirdly, I am a Jew for Jesus – as if things could not get more complicated! I have had 6 Close Encounters with Yeshua (Jesus Christ), one of which was a Near Death Experience in 2010, after suffering a heart attack. I saw both my deceased parents, and then I saw Jesus. I have seen Him before, 5 times in my life previously, since the age of 18, when He first appeared to me in the Negev desert, after I graduated and matriculated from Sde Boker, in 1979. He appeared in a live vision, while I was awake, He said, "I am the Messiah, follow me." So, I did. I have been excommunicated from my Jewish families and friends for my decision to believe in Christ and what I saw and heard very clearly. There was no ambiguity there, it was Jesus.

I have also been shunned, mocked, and gaslighted by my Italian Roman Catholic family members for knowing more scripture than they, and not following the ways of Babylon, which has made me invisible to them. Suffice it to say, I have lived and survived the past 45 years without the Tribe. This has made me vulnerable and susceptible to outside influences. And while I have had 6 memorable and very distinct close encounters with Jesus (Yeshua), I have also been repeatedly abducted by aliens since the age of 2.

I have vivid memories of my mother coming into my crib and finding me upside down with nose bleeds. They talked about it a lot too, so it stuck in my memory as these were not isolated incidents but there seemed to have been a bout of them. Then my mother was taken from me at the age of 6 after dying of bone marrow cancer at age 45.

When I got older, I had vivid memories of being used as an incubator for the alien hybridization program, until I realized how to end these contracts and agreements, in what I am calling demonic assignments. I have seen Grays, Reptilian Lizard Men, both dark and the luminous ones, the short Blue Frog men, the human looking ETs, and I have seen giant 35-foot beings that I believe were Guardians. I have memories of being taken on a spaceship while seeing the Earth from Space as a child multiple times, and have often had dreams of flying, not with wings, but just the ability to lift myself up and fly above the ground.

My experiences are real, and they are part of my conscious memories, which has shaped the writing of these books. I know what I know, and what I don't know, I don't know, you know? So, this installment connects dots for how and why this is all connected to the End of this Age, the Prophesies, and the Divine Grand Plan, and what we can we do about this era of Disclosure we are in? If this book helps just one person on their journey to Heaven, it will have all been worth the effort.

DISCERNING DISCLOSURE

> "The mind, once stretched by a new idea,
> never returns to its original dimensions."
> ~Ralph Waldo Emerson

The reason I put my Book Series together is to teach spiritual, scientific, and scriptural discernment. Most Christians seem to believe all aliens are demons. Yes, all demons happen to be aliens but not all aliens are demons. Yes, fallen angels became demonic but not all angels are aliens. Just do the math: Only one third of Heavens angels rebelled and fell, while two thirds remain faithful to the Creator. These are the good ETs, or good aliens, also called angels, a word which simply means *messenger*. God created them. God also created many other types of creatures, also mentioned in Scriptures that are simply out of this world and classified as 'alien.' Not all are evil. Book One of the *Who's Who in the Cosmic Zoo* series is written as an A-Z Compendium [6] which is being republished as an updated revised 4th Edition soon. In it, I identify the evil demonic aliens and those who are simply alien creatures mentioned in the Bible that serve the Kingdom of Heaven and the Lord Himself.

Later in this book, I will be sharing what started my journey of researching and writing, which began in 1990. I taught classes in Florida, called, *ETs, Aliens or Angels?* Which was a weekly gathering at a local metaphysical bookstore, about who were the ETs and Aliens. In 1995 I was invited to attend a one of kind Conference in Washington, DC called, *When Cosmic Cultures Meet*, where I met and interviewed the late great Archeoastronomer Zecharia Sitchin. We shared some common heritage, as Jews, and both lived in Israel. I was educated with Biblical Hebrew, and I became intrigued by Sitchin's interpretations, which were quite different from what I was taught in Hebrew School and Tanach classes. But then again, being a Messianic Jew, I hold a different perspective than most other Jews.

Sitchin taught me a lot. I was deeply inspired by his perspective on the Bible. I belonged to his international bible study group and got to attend his lectures each time he visited Colorado in the 1990s and early 2000s. I spent about 17 years compiling research on ETs and Aliens and was getting ready to publish my first book, when my life got rocked with a heart attack on February 7, 2010.

During my Near-Death Experience, the Lord Jesus took my hand and walked me through this bright green field, against this bright clear bluebird sky, it was just the two of us. Everything was bright. He told me I was going to be all right, that He wanted me to rewrite my book on ETs and Aliens with Him and put Him and His message in it. There is an old saying, *we teach what we need to learn*. He taught me discernment, where the good aliens are and where the evil aliens are in the original Hebrew Scriptures which I share in my Book Series. If you're starting with Book

INTRODUCTION

Five, you may want to go back and review all the Discernments, identifying the different groups of extraterrestrials and aliens, I shared on *Who* belongs to the Draconian Kingdom of Darkness and *Who* belongs to the Kingdom of Heaven, which is governed by the Kingdom of God.

In Book Five and Book Six of this book series, I'm going to show you how by understanding the Creator's signature in the stars, that there is a Clash of Two Kingdoms in the Heavens, and on Earth, as well as when and how this ongoing cosmic battle will come to an End. You get to choose which Kingdom you will serve. Book Six, *Heavens Witness* is devoted to the most ancient astronomy-astrology I could find, which coincidentally tells the same story of God's Divine Plan of Salvation, which is embedded within the Scriptures, which is literally Jesus in the Stars.

But, before anyone can make a choice, they first must gain knowledge through information, so they can make an educated choice.

This Book Series is about teaching Discernment, herein, allow me to take you on a journey in how to discern the truth, in God's first Creation, the Heavens. As you read, keep in mind, *The Truth is Stranger than Fiction!*

Thanks for your patience for hanging in there with me, as I know many of you have been waiting anxiously for Book Five. I am grateful and humbled for your kind support of my work.

Notes:
1. Ella LeBain, *Who Are the Angels?* Chapter Sixteen: *The Second Coming and Nibiru* pp.338-363, Skypath Books, 2015
2. Sarah Lewin, *New Model: Nearby Exoplanet TRAPPIST-1e May Be Just Right for Life*, https://www.space.com/36349-trappist-1e-just-right-for-life.html?utm_source=sp-newsletter&utm_medium=email&utm_campaign=20170405-sdc, April 05, 2017.
3. Universe Today, *Do the TRAPPIST-1 planets have atmospheres?*, https://www.universetoday.com/146958/do-the-trappist-1-planets-have-atmospheres/#more-146958
4. Anna Merlan, "Here's Why Gallup Polled Americans About UFOs for the First Time in Decades, if they're asking about aliens, they must be real." (February 25, 2020) https://www.vice.com/en_us/article/n7j4bb/why-gallup-polls-americans-about-ufo-aliens
5. Lydia Saad, *Americans Skeptical of UFOs, but Say Government Knows More*, https://news.gallup.com/poll/266441/americans-skeptical-ufos-say-government-knows.aspx, September 6, 2019.
6. Ella LeBain, *Who's Who in the Cosmic Zoo?* Book One, Trafford, 2012.
7. Leonard David, Space Force: What Will the Newest Military Branch Actually Do? https://exonews.org/space-force-what-will-the-newest-military-branch-actually-do/, February 9, 2020
8. Third Phase of the Moon, *Space Force New Asset? An Investigative report in regards to the ISS and a UFO close encounter*, February 22, 2020, https://youtu.be/xhCNgVg7enY

CHAPTER ONE

SIGNS IN THE HEAVENS

"There will be signs in the Sun, Moon and stars.
On Earth, nations will be in anguish and perplexity at the roaring and tossing of the seas. (Tsunamis) Men will faint from terror, apprehensive of what is coming on the world, *for the heavenly bodies will be shaken."*
(Luke 21:25-26)

PLANET X, NIBIRU, AND THE NEMESIS SYSTEM

The Terrible Day of the Lord that is mentioned in the books of Isaiah, Zephaniah, Joel, Malachi, and Revelation coincides with the passing of the planet Nibiru, the Planet of Crossing, close to Earth. Only the passage of another celestial object could produce such havoc upon the Earth that will cause men's hearts to fail them. As the scriptures say, "men's hearts failing them for fear, and for looking after those things which are coming on the Earth: for the powers of heaven shall be shaken." (Luke 21:26) In Revelation 8:11, Wormwood comes out of Nibiru, which is a comet that will destroy one third of the fresh water and ocean life, making them bitter.

The Book of Revelations Chapter 12 is a combination of astronomical markers combined with Christian symbolism. The event described in Chapter 12 depicts the alignment and transit of the planets Jupiter and Nibiru, the Great Fiery Red Dragon, passing through the belly of the woman which is the constellation of Virgo. All the hype for September 23 is based solely on the first three verses of Revelation 12:

> "And a great sign was seen in heaven; a Woman invested with the Sun and Moon under her feet. And on her head a crown of twelve stars, and being pregnant, she cried forth travailing and being pained to bring forth . . . And another sign was seen in heaven and behold! A great fiery red dragon, having seven heads and ten horns and on his heads, seven diadems."

Who or what are the seven heads and ten horns of the Red Dragon? They may be the seven exoplanets of the Nibiru/Nemesis system. Horns, as we have already established in this book series, are Biblical symbols of power, and represent authorities. We know from the Sumerian cuneiform tablets that the planet Nibiru, the planet of Crossing, was inhabited by giant humanoids, some reptilian looking. I realize many Christians believe this passage to be purely astronomical due to the description of the Sun, Moon and stars, and the Woman representing the constellation Virgo.[1] See, Book Three: *Who Are the Angels?* Chapter: *The Second Coming and Nibiru*, pp. 338-363.

There are several problems with everyone saying this is the great sign in the heaven that fulfills Revelation 12:1. Firstly, there aren't twelve bright stars in the constellation of Leo. If you look at all these pictures, they are using the planets Venus, Mars, and Mercury to add up to the nine bright stars. Venus, Mars, and Mercury are not stars, they are planets, despite looking like stars to us from Earth. Secondly, the Sun is NOT under the feet of the Woman at this time. The Woman is the Virgin of the constellation Virgo, and the only way the Sun could possibly be under the feet of this entire constellation, is for our own planet to be removed from the ecliptic plane and be placed far above the ecliptic plane as to as to make the Sun below Virgo.

Here is why this scripture was not fulfilled on September 23, 2017, in astronomical language: According to Nibiru researcher, Jason Breshears:

> "what this implies, sequentially, is that Nibiru will ascend into the inner system from directly underneath Earth, push our world onto the Dark Star's ecliptic plane with it, and with Earth shoved upward and out of the way, we will be able to first view Virgo before seeing Nibiru as it passes our planet on its 60-year journey over our Sun."[2]

According to the Maya, time will collapse into a newer more accelerated system. The Mayan Prophesies are reflected and confirmed in the prophesies of Revelation if we look at these scriptures through the eyes of science along with the awareness of the presence of the Nibiru system. Breshears concluded that Earth's new position being closer to the Sun will increase its rotational spin-rate by a third, by going from a 24 hour to a 16-hour day, and the year will be abbreviated from original year of 360 days to 240 days. The orbit is quickened, because the planets distance from the Sun is significantly shorter, and the rapid rotation increases our ability to see the Moon more often than is presently possible. This is how both day and night together can be reduced to a third without lengthening one or the other.

This is a deliberate act of God, for Christ said, ". . . unless those days be shortened, no flesh would be saved." (Matthew 24:22) This is proof of the Last Days Quickening. It is an astronomical event, caused by the incoming Nibiru system that will throw the Earth into a new orbit around the Sun. It is certainly not the first time this has happened which justifies the prophetic promises of Jesus to create a New Heaven and New Earth (see, Isaiah 65:17; Revelation 21:1) because the proximity of these two solar systems seem to collide with each other in one way or another every 3,600-years causing Earth changes, pole shifts and extinctions.[3] (See, *Worlds in Collision* by Immanuel Velikovsky.)

For those who doubt that this can happen, consider recent events of the Earths' shifting axis. The Inuit Elders, who base their very existence on the observation of the positions of the Sun, Moon, and stars to survive in the Arctic, have confirmed that the Earth has tilted and shifted. The markers and directions of rising and setting Sun and stars are not where they used to be observed. Many people I communicate with daily via social media have shared their observations of Sunrises and Sunsets being ten to twenty degrees off, based on common landmarks such as mountain peaks. This is backed up through a plethora of photos of exoplanets seen around the Sun during Sunrises and Sunsets all over the planet.

There is no question that Earths' axis has shifted. As Nibiru approaches perihelion with Earth, we have been experiencing major solar flares. X-class coronal mass ejections result in geomagnetic storms, some of which have knocked out a few satellites, triggered Earthquakes, volcanic eruptions, and extreme weather. Something is certainly disturbing our Sun outside of its normal 11-year Sunspot cycles. The entire solar system is perturbed, affecting the equilibrium of planet Earth.

This transit of Nibiru passing through the Virgo/Libra portion of the Milky Way, when it happens, will set off an electrical charge not just to Earth, but to all the other planets being affected by this massive heavenly body. The only real 'sign in the heavens' this transit represents, is that the Nibiru system will begin to appear closer in Earths skies, becoming more and more visible to the naked eye as it continues on its orbit around our Sun, which it does every 3,600 years. Because this event is unprecedented in modern times, we have only ancient records to draw upon its length of orbit. Up until now, many have relied on photos and videos taken all over the planet to see the Nibiru/Nemesis system. Soon, everyone will be able to see it with the naked eye because it is getting closer to our Sun as it passes Earth, which will eventually put all the doubters to rest.

The other issue and controversy amongst modern astronomers is the rate and speed of which this body is moving, which is erratic and travels in a rebellious corkscrew orbit through space. The said orbital pattern moves in an elliptical counterclockwise orbit around our solar system. The closer it gets to our Sun, the faster it moves. When it finally goes around our Sun, the speed will accelerate as the Sun will catapult it back out into its orbit through space, at which time it is expected to pass Earth for round two, the worse of the two passings, mentioned in the Jewish Bible as the Terrible Day of the Lord. This will happen in a single day, which is how fast it will be travelling.

It is also important to note that Revelation 12 cannot happen out of chronological order. There are many signs that still have not been fulfilled, all the way up to Revelation Chapters 8 and 11, along with the two witnesses in addition to the Lord's promise in Malachi 4:5, "See, I will send you Elijah the prophet before the coming of the great and dreadful day of the Lord." As far as I know, Elijah has not yet returned, meaning that the tribulation has not begun, nor can the Rapture happen.

There is a lot more to Revelation 12! We simply cannot cherry pick two verses out of context and discard the rest to make a prophetic doctrine, ever!

We are explicitly forbidden to set dates for the Lord's return. Even He did not know the date of His return, as we read in Matthew 24:36. If He did not know, how can we?? Further, it is dangerous to set dates, as it sets up expectations, and when that 'prophetic event' fails to happen, people lose faith, fall away, and begin to doubt not only the author of the prophesy, but the Rapture and Return of Christ as well. When we set dates, we stumble on our fellow believers. I remember in 2015 when many eagerly following the false predictions of Blood Moons,[4] (see my blog *The Hype and Anticlimax of Blood Moons*, http://spirituallydiscerning.com/?cat=1) how many Christians went into depression, and were subsequently vulnerable to attacks by the enemy because they had put their faith in astrological dates. There are consequences to following false prophets. One would think the church would have learned an important lesson back then, but like life, lessons repeat themselves until they are learned and properly integrated.

It is clear that no one knows the date or time of the Rapture, or the Second Coming of Jesus Christ. As scripture says: "But about that day or hour no one knows, not even the angels in heaven, nor the Son, but only the Father." (Matthew 24:36)

There have been so many false predictions that it boggles the mind that we have not gotten wise to it all yet! Here are just a few to jog your memories: the Rapture was to happen in 1988, and the 88 reasons why… the New World Order would overtake us in 1984, based on George Orwell's book of the same name… the Y2K computer bug that was going to end the world… the 12.21.2012 end date of the Mayan Calendar, and our Sun's eclipse of the Galactic Center, were all supposed to bring the end of the world as we know it. I think not!!! As of 2020, we are alive and well and living on the great planet Earth!

To be fair, New Agers have also had their fair share of false prophets and erroneous date setters whipping people up into a frenzy. The painful events of the Heavens Gate suicides led by Marshall Applewhite, who believed the deaths of his followers would land their souls in spaceships hidden by comet Hale-Bopp, is fresh evidence of delusion. Before this, the deaths of those sad misguided followers of Jim Jones who drank the Kool-Aid laced with poison, along with their innocent children, in the hopes that their suicides would allow them to escape Earth and be at peace in Heaven. Sheldan Nidle, author of *You Are Becoming a Galactic Human*, is known for his prediction that the world would end on December 17, 1996. Nidle predicted that it would happen with the arrival of 16 million spaceships and a host of angels from the photon belt. His undiscerning followers waited endlessly on the beaches of Hawaii and California only to be faced with the reality that they were misled. When this did not occur, Nidle claimed the angels had transferred humanity into a holographic projection to give it a second chance. Nidle was given the Pigasus award, designed to expose psychic frauds, for his failed prediction.[5]

The 12.21.2012 date, the end of the Mayan Calendar, saw thousands of believers travel to the Bugarach, France, considered to be a holy mountain based on a prediction of the apocalypse on that date. The Bugarach mountain in France is known for its UFO activity, as thousands came with the hopes that extraterrestrials would come to rescue them. The thousands who pilgrimage to Bugarach all left without meeting ET.

Can it be that all people, regardless of religious persuasion, seem to be implanted with an innate belief that God or the gods, angels or ETs from heaven will rescue us from Earth? Christians who believe in the Rapture are not much different than New Agers who believe ETs will save them from the miseries of planet Earth. The belief in being 'taken' is the same, regardless of what god or gods are believed to be the takers. Book One of *Who's Who in the Cosmic Zoo? A Spiritual Guide to ETs, Aliens, Gods & Angels* teaches the discernment of ETs and aliens based on history, scriptures, actual whistleblowers, and experiencers. Discernment is an important lesson for both Christians and New Agers to learn at this time.[1]

Most recently, the whole Blood Moon craze, begun by Hebrew Roots Movement Pastor Mark Biltz, has driven the ink and presses overtime. John Hagee plagiarized Biltzs' work by writing a book and producing a movie of the same name, *The Four Blood Moons*. They predicted the Rapture and the Return of Christ 2014-2016. You will remember that these lunar eclipses were just like all the other lunar eclipses that happen every six months. Again, as I've pointed out in previous blogs, this is due to the misuse, abuse and gross misunderstanding of the end times prophecy of Joel 2:31, "The Sun shall be turned into darkness, and the Moon into blood, before the great and the terrible day of the LORD come."

As stated in my blogs and books, the only way the Sun can go dark and the Moon red at the same time is due to an extraordinary astronomical event. It cannot happen with a regular lunar or solar eclipse, but it can be created through the presence of the passing of a planetary body four times the size of Earth, with a tail of comets trailing its orbit. That could cause the lights to go out on Earth and cover the Sun causing three days of darkness. This is the corresponding scripture in the New Testament to Joel 2:31 in the Old Testament: "I watched as he opened the sixth seal. There was a great Earthquake. The Sun turned black like sackcloth made of goat hair, the whole Moon turned blood red, and the stars in the sky fell to Earth, as figs drop from a fig tree when shaken by a strong wind. The heavens receded like a scroll being rolled up, and every mountain and island was removed from its place." (Revelation 6:12-14)

We know that this is no ordinary celestial event. The Sun goes dark but for a moment or two during a solar eclipse. During a solar eclipse, the Moon is in its dark phase. Blood Moon Lunar Eclipses happen on Full Moons, two weeks after solar eclipses. We know these events cannot take place simultaneously. The End Times Prophesy is an extraordinary astronomical event, which will be caused by the passing of Nibiru, Planet X, Hercolubus, Red Dragon, which ever name we fear most, will precede the return of the Lord to Earth.

CHAPTER ONE: SIGNS IN THE HEAVENS

Revelation 12:4 "Its tail swept a third of the stars out of the sky and flung them to the Earth." That is the tail of comets from the Nibiru system, which is called Wormwood in Revelation 8:11. Nibiru appears to have wings, but these are plasma trails full of debris, comets, meteorites, and fireballs. When this planetary body passes Earth on its trip around our Sun, it will bring about the Terrible Day of the Lord. Before that happens, it needs to get closer and pass through the Kuiper Belt, which is already full of asteroids, the remnants of a broken planet that was smashed by Nibiru's passing in the distant past.

Remember Harold Camping? Camping predicted that Jesus Christ would return to Earth on May 21, 2011, whereupon the saved would be taken up to heaven in the rapture. He said that there would follow five months of fire, brimstone, and plagues on Earth, with millions of people dying each day, culminating on October 21, 2011, with the final destruction of the world. He had previously predicted that Judgment Day would occur on or about September 6, 1994. [6,7,8]

After October 21, 2011, passed without the predicted apocalypse, the mainstream media labeled Camping a false prophet and commented that his ministry would collapse after the failed Doomsday prediction.[9] Not too long thereafter, Harold Camping died on December 15, 2013, when the world definitely ended for him. [10]

As I have said multiple times in my blogs and books, and is worth repeating here now, the signs in the Heavens about which Jesus spoke are not mere solar and lunar eclipses, which happen like clockwork every six months. The signs in the heavens, which in Hebrew is *shamayim*, are the presence of the exoplanets in the Nemesis system, which will herald the passing of the planetary comet Nibiru, or Wormwood. It behaves like a comet because of its corkscrew, rebellious orbit. It is known by the ancients as the planet of crossing, because it forms a cross in the sky between its deep red body surrounded by four Moons. Nibiru and her sister planets have been photographed by astronomers and sky watchers all over the planet. Just google it and you will see a plethora of photos proving that its approach is getting closer and closer in its orbit around our Sun. It will pass the Earth, not once, but twice, which will bring about the end of this age and the return of Christ. The Earth will be turned upside down, as predicted over 3,500 years ago by the Old Testament Prophet Isaiah.

> "Look, the LORD makes the Earth empty, makes it waste,
> turns it upside down, and scatters its inhabitants."
> (Isaiah 24:1)

The most devout and charitable Christians incessantly misuse Joel 2:23 every time we have a solar or lunar eclipse. "The Sun will be turned to darkness and the Moon to blood before the coming of the great and dreadful day of the LORD." This happens at least four times every two years and has happened since time immemorial. Solar eclipses are caused by a new Moon which happen two weeks apart from lunar eclipses which are full Moons. Joel 2:31 is neither. It is an extraordinary celestial event caused by the passing of the planet Nibiru, also known as Planet X or Wormwood, which will block out the Sun and turn the Moon red due its iron oxide environment. Nibiru is a gigantic corkscrew comet-like gas giant planet that has tails. Revelation 12:4 is an astronomical event, the dragon's tail being tails of comets, and Nibiru looks like it has wings, that orbits Nemesis, the brown dwarf twin of our Sun that appears around the Sun from time to time. Nibiru is a part of its system, but its shape is only semispherical as it appears to have wings.

There is so much controversy today over the appearance, reality, and timing of the passing of Nibiru. No one really knows the exact timing of when it will pass the Earth, but when it does, it will be visible in the heavens to the naked eye. Recently it has been observed through the Hubble telescope. The Rapture will happen when it happens in preparation for all of these end times events, including a spiritual revival which will overshadow wars and rumors of wars, Earthquakes, volcanic eruptions, and climate changes like birth pains, just as Christ predicted. All these events signal His Second Coming.

Historically speaking, the Lord has always poured out spiritual revival and dispatched His prophets (messengers) to get people ready when judgment and destruction are about to occur. This is always a great idea, regardless of time period. We can prepare our hearts, minds, and souls by getting right with God. Live each day as if it is your last day on Earth, and one day you are sure to be right.

I get that everyone is impatient and wants to just get on with the end already, but the end of this age means the end of the age of grace. Those who answer the Great Call for redemption get saved, and those who do not, will be claimed by Satan. Why would good Christians want that time to be shortened or abbreviated???

I find it deeply troubling that my fellow Christians are more focused on judgement than grace. Doesn't the scripture say, mercy triumphs over judgment? Is that not the redemptive power that emanates from Christ? James 2:13, "For judgment without mercy will be shown to anyone who has not been merciful. Mercy triumphs over judgment."

New Agers always ask me whether the return of Christ is physical or the manifestation of Christ Consciousness throughout the world. And my answer is always, both! The physical return of Yeshua Jesus Christ will inevitably create the mind of Christ in the world, during His Millennial Reign, which begins the precessional Age of Aquarius, the Age of Unity of Brotherhood, Peace and Understanding. To have Christ consciousness is to embody the mind of Christ (1 Corinthians 2:16) which comes from being born again through His Spirit and knowing His words and laws written into our heart and mind, just as the prophecies say in Joel 2:28-29, that in the last days, the Lord will pour out His Spirit into our hearts:

> "It will come about after this that I will pour out My Spirit on all humankind; And your sons and daughters will

> prophesy, your old men will dream dreams, your young men will see visions. "Even on the male and female servants I will pour out My Spirit in those days."
>
> (Joel 2:28-29)

> "This is the covenant I will make with the people of Israel after that time," declares the LORD. "I will put my law in their minds and write it on their hearts. I will be their God, and they will be my people."
>
> (Jeremiah 31:33)

The scene in Revelation 12 suggests an eclipse in the last and master degree of Leo, which rules the heart, truly making this an Eclipse of the Heart, as the Bonnie Tyler song goes. The dark side of Leo is pride, and God hates pride. Pride was the original sin that caused Lucifer to devolve into Satan and get kicked out of heaven. Pride is arrogance, narcissism, and can induce a psychological projection due to its stubborn denial. Pride will project its sins onto a scapegoat, it will point fingers at the righteous and pin its sin upon them, through psychological manipulation. It will deny its own condition by projecting it onto others. It will turn good into evil and evil into good (Isaiah 5:20). This is known as the Religious Spirit. Anyone familiar with my books series knows I discern and expose this demonic Spirit in each and every book! It is the cause of the great apostasy, which instead of bearing fruits of love, peace, and joy, it brings forth the bitter fruit of judgment, accusations and a critical spirit that persecutes, rejects, and excommunicates.

WARNINGS AGAINST FOOLISHNESS

> "Therefore, his calamity will come suddenly; Instantly he will be broken and there will be no healing. There are six things which the LORD hates, Yes, seven which are an abomination to Him: Haughty eyes (pride), a lying tongue, and hands that shed innocent blood, a heart that devises wicked plans, feet that run rapidly to evil, a false witness who utters lies, and one who spreads strife and division among brothers."
>
> (Proverbs 6:15-19)

Too many people leave churches with these deep psycho-spiritual wounds that lead them to reject God. As I always say, you can't throw the baby out with the bath water, meaning that just because broken, imperfect, and fractured Christians and other so called religious people, who are full of themselves instead of full of Christ, also fill the church, doesn't mean that God is far away, or thinks of you as less-than. God still loves you and wants a relationship with you. The problem is that too many of us look to other people for a relationship with God when God is as close as your very breath. In fact, it is He who gives you breath and life and the electricity that causes your heart to beat. It is His Spirit that one should seek, yes, as Christ said, 'by their fruits you will know them', *who* belongs to Him as His disciple. There is no better fruit than the love of God and love of others to identify *who* belongs to God and *who* does not. Those who do not exhibit this fruit should be cultivated for God's salvation, not rejected from the fold, for they are wounded, and their wounds have turned them to the dark side. But nothing is impossible with God.

Christ came to save those who are lost. He is in the business of taking souls through profound transformations. Only the maker of the heart can heal the broken-hearted (Psalm 147:3). It is not the job of Christians to judge the world, as that job rests solely on the shoulders of the Lord Himself. It is, however, the job of Christians to project the Spirit of Christ to others and be ambassadors of light for Heaven's Kingdom.

This is a time to choose which side you are going to take, that of God or of Lucifer? It is a time to repent of arrogance, the evil Religious Spirit that mistakenly thinks it is superior to everyone else and thinks that works can earn you salvation. There is only one way to obtain salvation and that is through faith in the name and person of Yeshua Jesus Christ, to become *alive again* with and through Christ: "But because of His great love for us, God, who is rich in mercy, made us alive with Christ, even when we were dead in our trespasses. It is by grace you have been saved! And God raised us up with Christ and seated us with Him in the heavenly realms in Christ Jesus. (Ephesians 2:4-6)

While we are all anxious for the coming of the Kingdom to Earth, we must busy ourselves to be about our Father's business, which is that no one should perish but have everlasting life through the redemptive power of Christ. This is not about religion but all about relationship through reconciliation. This is what true spiritual revival is about, bringing souls back into right relationship with the Creator through the light of Christ, who is the Son of light. While people look to the Sun for light and energy, in the Age to Come, those who are redeemed for Heaven's Kingdom on Earth will look to the Son for all light and energy, as this Sun will eventually be eclipsed by the Son of Heaven. "The Sun will no more be your light by day, nor will the brightness of the Moon shine on you, for the LORD will be your everlasting light, and your God will be your glory." (Isaiah 60:9)

> "When His parents saw the young Jesus, they were astonished. "Child, why have You done this to us?" His mother asked. "Your father and I have been anxiously searching for You." "Why were you looking for Me? He asked, "Did you not know that I had to be about My Father's business?"
>
> (Luke 2:48-49)

Jesus, He who is wise, is the only One who can save souls. Who will you welcome at the End Times Harvest? The signs in the heavens are pointing to the final phases of this grand experiment, perhaps it is time you get ready, for your redemption draws near.

> "Those who are wise will shine like the brightness of the heavens, and those who lead many to righteousness, like the stars for ever and ever."
> (Daniel 12:3)

> "The fruit of the righteous is a tree of life, and he who is wise wins souls."
> (Proverbs 11:30)

It blows my mind how popular astrology has become amongst Christians these days. I was an astrological counselor for 25 years and was excommunicated by my Christian community and rejected by my oldest best friend for even looking at astrology. I told them God works through the stars. Now Christians follow astrology. Hmmm! The Lord gave me this message for Christians which is in my books and my blog:

When you take astronomical data, and apply it to human events, dates, human interactive relationships, and the human psyche, that is the very definition of the language of astrology. Can we please just call it for what it is? And stop trying to pawn it off as astronomy?

We are not to worship the starry hosts. Nor are we to hang our hat on dates, for the Lord, Creator of the Stars, who named the stars, who enjoys their unquestioning obedience to Him, can at any time override any astrological influence according to His Will and Glory. I have seen it happen many times. He wants us to trust in Him alone, not trust in starry influences which can all be manipulated. There is a positive and negative to all these influences and when you lean on the Creator of the stars, you can reap better results. I have issues with all the date setting which Mark Biltz, John Hagee, and others have done about the blood Moons, which were inaccurate and deceptive. I think when they create an expectation and attach a biblically prophetic event like the Rapture or the Tribulation, it sets them up for failure and creates disappointment in those who trust their leadership. Those weak in the faith get discouraged and fall away. What does Scripture tell us about stumbling our brethren?? Making predictions is a risky sport.

The Lord has the power to override any astronomical event. He has historically not returned when people expected Him, in order to tarry and expand His Kingdom on Earth, as well as make fools of men for believing in astronomical dates and attaching expectations to them. I have wrestled with these false narratives among my fellow Christians for a long time. *I openly forgive them* for all the judgments and rejection I received for even studying astrology with astronomy. They never stopped me, but now I realize I was not wrong. Plus, at one time in our ancient history, astrology and astronomy were one science. Astrology birthed astronomy.

Nevertheless, if Christians really want to understand the prophesied signs in the heavens that Jesus warned us to watch and await or demystify the cryptic language of the Book of Revelation, then perhaps learning astronomy without the incessant denials that come from Christian mainstream is an avenue. Pole shifts happen, as a scientifically verified fact. Denying there is a long record of them ever happening is as ignorant as those who purport a flat Earth. Isaiah 24 is clear about the Earth turning upside down and moving to and fro like a drunkard. This is a magnetic polar shift, and according to NASA, it happens all the geologic time.

> "Behold, the LORD makes the Earth empty, and make it waste, and turns it upside down, and scatters abroad the inhabitants thereof."
> (Isaiah 24:1)

> "The Earth shall reel to and fro like a drunkard and shall be removed like a cottage; and the transgression thereof shall be heavy upon it; and it shall fall, and not rise again."
> (Isaiah 24:20)

The Harvest is at the end of the age, and the Harvesters are the Angels, (Matthew 13:39). Nowhere did Christ tell us to make dates about His Return. We attract more bees with honey than with vinegar. We cannot extend grace to those on whom we stand in judgment, condemnation, who we reject and excommunicate. Grace does not work like that and neither does the frequency of the Kingdom of Heaven. This information is relevant today as never before to encourage the churches to teach Revelation Prophesies. This cannot be done through the spiritual and mental blocks of antisemitism, funda-mentalism, and a general impatience and anxiousness for Christ's return at the expense of Israel, Messianics and the world's unsaved Jews, New Agers, Muslims, Gays, and so many others.

The painful reason why so many Jews reject Christ today has to do with the rude behavior of Christians towards them, for which we all should be deeply grieved. Jesus Himself is a Jew! The Church of Rome stole the Jewish scriptures and rewrote them into a blend of paganistic Christianity called Catholicism. [11] (See, *Who Is God? Book Two*). Jesus is returning to judge the Church! Matthew 25 is all about what happens when He returns to separate the sheep from the goats. Do not be a goat.

The End Times Prophecies are focused on Israel. Those Christians who are awake to this will be up and about God's business in these last days. Those who are still stuck in Replacement Theology, antisemitism, or harbor a deep-rooted jealousy and misunderstanding of Israel and Jews, will be blinded by the god of this world, and will miss the true signs in the heavens. Being distracted by astrology, numerology, and a denial of the true nature of the Earth and what is taking place in this solar system will avail nothing.

God wants His grace to reach all those who are wounded by the Church, all those who doubt who Christ is, all those who misunderstand scriptures, all those who place their faith in astrology instead of the Creator of Heaven and Earth. Christians are called to express Christ to the rest of the world, not create more cults and false expectations through theological date setting. We have seen that the Creator of the stars can override at any time our careful mathematical workings, just to make fools of prognosticators. I have watched Him do it many times.

"Be anxious about nothing, but through prayer and supplication, let your requests be made known to God with thanksgiving, and the supernatural peace of Christ will rule over you." (Philippians 4:6-7) A major discernment of 'who' walks with the King of the Kingdom and 'who' does not.

Be prepared, for death or Rapture, as I tell everyone, live each day as if it is your last day on Planet Earth, and one day, you are sure to be right.

Notes:
1. Ella LeBain, *Who Are the Angels?* Book Three of *Who's Who in the Cosmic Zoo? A Guide to ETs, Aliens, Gods & Angels*, Chapter: *The Second Coming and Nibiru*, pp. 338-363. Skypath Books, CO. 2016
2. Jason M. Breshears, Annunaki Homeworld, Orbital History and 2046AD Return of Planet Nibiru, The Book Tree, San Diego, CA 2011.
3. Immanuel Velikovsky, Worlds in Collision, Paradigma Ltd; Later Printing edition (October 1, 2009)
4. Ella LeBain, *The Hype and Anticlimax of Blood Moons*, http://spirituallydiscerning.com/?cat=1, 2014.
5. Sheldan Nidle, *You Are Becoming a Galactic Human*, Spiritual Education Endeavors; 1st edition (April 1, 1994).
6. Elizabeth Tenety (January 3, 2011). "May 21, 2011: Harold Camping says the end is near". Washington Post. Kimberly Winston (March 23, 2011). "Judgment Day: May 21, 2011". Washington Post.
7. Nelson, Chris (June 18, 2002). "A Brief History of the Apocalypse; 1971 – 1997: Millennial Madness". Retrieved June 23, 2007.
8. "Harold Camping Says End did come May 21, spiritually; Predicts New Date: October 21". International Business Times. Retrieved May 23, 2011.
9. "Did Harold Camping Ever Teach the End Was Coming In 1994?" on YouTube. July 30, 2009. Retrieved December 16, 2012.
10. "Harold Camping False Prophet: Ministry Probably Doomed". International Business Times. October 21, 2011.
11. Ella LeBain, *Who is God?* Book Two of *Who's Who in the Cosmic Zoo? A Guide to ETs, Aliens, Gods & Angels*, Skypath Books, CO. 2015

CHAPTER TWO

LIFE ON OTHER PLANETS?

Before the truth can set you free,
You need to recognize which false belief is holding you hostage.

The Bible tells us that there is extraterrestrial life that Jesus needs to join together under His Lordship, as One Shepherd, to dwell together in One Kingdom, that according to John 10:17 needs to be gathered together, to dwell as One. Jesus said,

> "I have other sheep that are not of this sheep pen. I must bring them also. They too will listen to my voice, and there shall be one flock and one shepherd."
> (John 10:17)

But there is also the mention of other worlds, indicating ETs in the Bible on the basis of a word in Hebrews 11:3: 'Through faith we understand that the *worlds* were framed by the word of God, so that things which are seen were not made of things which do appear.' The word *worlds* refer to other inhabitable planets. In the Greek, the word is αἰών (aiōn), where we derive the word *eons*, that modern translations render to the word *universe* (entire space-time continuum) because it correctly describes 'everything that exists in time and space, visible and invisible, present and eternal'. 2

I don't know why Christians think there is no Biblical Hebrew word for *the universe* where Jews pray every holiday, the blessing, 'Blessed are thou, O Lord, our God, King of the Universe (World)', which begins all the blessings over Shabbat candles, Hanukkah candles, bread, wine, Passover, Sukkot, etc. The Hebrew word *olam* can be translated as *world* or *universe*. The term *All of creation*, can also be translated as *universe* or even *multiverse*. The Bible starts out in Genesis with the merism *the heavens and the Earth* which describes the universe, the cosmos, in its entirety. It is clear that New Testament passages like the aforementioned Romans 8:18–22 and Hebrews 11:3 are pointing back to the Genesis (heavens and Earth) creation, and thus, everything that God made and when time as we know it began.

Jesus' teaching was causing division among the Jews because they always believed that salvation from the Lord was for them alone despite the fact that He reaffirmed that He would be the Messiah for all of humankind. Their self-righteous religious spirit was the cause of their spiritual blindness. They completely missed who Yeshua was: He was their Shaliach, in Hebrew, *the Sent One.*

The Bible indicates that the whole creation groans and travails under the weight of sin (Romans 8:18–22). Sin is the effect of the Curse following Adam's Fall as universal. But what if the Curse was not limited to Earth? What if it was on a species, a bloodline and perhaps extrapolated out into other realms? If our battle is not against flesh and blood, but against powers, principalities, and rulers of the darkness of this present world, who may not actually rule from this world, but perhaps on another world close to our world, like, let's say the Moon? Or Saturn? Or some other planet or Moon in our Solar System? As Ephesians 6:12 describes that our battle is against spiritual wickedness in the Heavens, then that means the Curse is not limited to Earth but extends out to the Heavens. Otherwise, what would be the point of God destroying this whole creation to make way for a new Heavens and Earth? (2 Peter 3:13, Revelation 21:1) [2]

Therefore, any extraterrestrials living elsewhere would also be affected by the Curse. Or perhaps we can call it, the battle of dualities, or the battle of the Titans?

When Christ (God) appeared in the flesh, He came to Earth not only to redeem humankind but ultimately the whole creation back to Himself (Romans 8:21, Colossians 1:20), that includes extraterrestrials and aliens. Being human happens to be a hot commodity in the alien and ET realms, which is why alien races abduct humans, to extract and use human DNA in their hybridization programs. This would qualify those hybrids for salvation under Christ's atoning death at Calvary because one needs to be a physical descendant of Adam for Christ to be our "kinsman-redeemer" (Isaiah 59:20). If the aliens are stealing human DNA through abductions, then that grafts them into the Evadamic race. See, Book One of *Who's Who in the Cosmic Zoo?* [2] Jesus was called the last Adam because there was a real first man, Adam [3] (1 Corinthians 15:22,45) — not a first Vulcan, Klingon, or Jedi. This is so One sinless human could take the punishment with which all humans were cursed, the curse of the genetic downgrading of human DNA by the Anunnaki, which creates the state that the Bible calls being born into sin [4] (Isaiah 53:6,10; Matthew 20:28; 1 John 2:2, 4:10). This One Human *kinsman-redeemer* had to be divine as well, in order to handle and overcome the task of being a perfect human without the inclination to fall into the sinful genetic patterns of the curse of humankind, without the need to atone for any (non-existent) sin of his own (Hebrews 7:27). [3,5]

Some wonder whether Christ's sacrifice might be repeated elsewhere for other beings? I don't think it's a coincidence that the Mazzaroth, the constellations in our Milky Way galaxy, tell the original story of the Good news of Redemption from a Son of Heaven, who is represented as a Hero/Kinsmen/Redeemer, was written just for Planet Earth. Unfortunately, due to space, I'm extending this topic into the next and final book of this series, *Heaven's Gospel – Book Six*, in which I detail the original meaning of all the main stars that make up the Milky Way, the zodiacal path of the Stars, which tells the story of Jesus Christ redeeming humankind, animal kind and all creation, from the death star and its curse. After spending most of my life fascinated with the stars, as a student of both astronomy and astrology, this coincidence and synchronicity cannot be ignored, in the face of End Times Prophecies.

ARE ALIENS NEW AGE?

"To be great in the kingdom of God you'll have to deny yourself and be willing to be mocked, hated by unbelievers and misunderstood by religious folks."
~ Steven Dutes, New York City Evangelist

Aliens are NOT New Age, in fact they are age-old, having been around for millennia. Christians really need to practice discernment here! Aliens and ETs are in the Bible. Some are demonic, some are creatures *created* to serve the Throne of God. My books carefully provide all those Scriptures proving that there is an ancient conflict with multiple groups of ETs and aliens, that according to Bible Prophecies are in a Divine appointment scheduled to fight against the Creator Lord over Earth in a real-life Star Wars. So, if all this is in the Bible, why do most Christians today mistakenly believe only one side of the story? What about the purpose of the Lord which is for His Gospel of Truth to be given to ALL Creation? Clearly the Gospel Word is NOT limited to Earth and humans.

The equivalent opposite viewpoint is for New Agers to believe that ETs and Aliens are here to save us, and we should all blindly get on the first spaceship out of here. Two opposite viewpoints, however, and both sides cannot be right.

The Truth is often found in the middle ground. The midpoint between the two opposites is balance.

My work proves that God created good alien beings and good ETs to serve Him. If you believe in the prayers of the Bible, the Torah, which gives praise to Avinu Malchenu, our Father and our King of the Universe of the HEAVENS, why would anyone who believes in the Sovereignty of God's Kingdom of Heaven think for a moment that He doesn't have ETs and aliens that serve Him and His Thrones? I prove in my books from Biblical Hebrew that those things which are called flying thrones that show up cloaked in clouds, are in fact, Spaceships.

I was gaslighted, verbally and emotionally abused, by a so-called Christian filled with hatred and prejudice, who insisted that anyone who writes about aliens or wellness and DELIVERANCE is New Age. I have got Good News (Gospel) for anyone reading this, only by the authority of JESUS Yeshua the Christ can you be set FREE from the negative Aliens, alien abductions, and all the related demonic strongholds and assignments from alien abductions.

You would have to get into my research presented in my books in greater detail to learn why I conclude that most people on Earth have been abducted, implanted, and used in one way or another in the alien human hybridization program, if you have the curiosity or desire. It is all there!!

For Christians to ignore this very real phenomenon in DELIVERANCE ministry and in the breaking of generational sins and curses is a sad case of missing a huge piece of Jesus' ministry. He came to set the captives FREE. That would be us Humans! Also, those creatures and ETs who were not imprisoned in the last galactic war.

As a Deliverance Counselor and Minister, I had a very troubled soul come to me for Deliverance prayers and counseling who identified as Christian. She had been physically abused by her mother as a child, and now had dreams of her blood being drained away to be stored in vats and canisters. She believed she was under the spell of witchcraft. While, indeed, it was a form of witchcraft and sorcery, she had the classic dream of abductees, which I knew right away was connected to a deeper layer of what was happening to her soul, and that was she was an abductee.

The only reason I do this work is because I was one, and Jesus Christ delivered me from alien abductions. Unfortunately, many do not understand this in DELIVERANCE ministries that what many may perceive as vampirism and witchcraft attacks seem to coincide with those who have been abducted.

The Living Christ is a very real Person and is present to us and through us through the Holy Spirit. He has Supernatural Power to break generational curses, witchcraft, word curses, and stop alien abductions.

Yes, this is why Christians believe that all aliens are demons. But then, who are the Angels? I prove they are not all winged beings, though that is one classification, but are a large group of extraterrestrial humans made in God's image that come from all four corners of the Universe. This is evidenced throughout Scripture.[6] See, Book Three, *Who Are the Angels?* Chapter Fourteen: *The Harvest of Angels*, pp. 289-302.

The Angels play a starring role in End Times Prophecies. The Angels are the Harvesters of the end of this Age. (Matthew 13:39) So, *who* are they harvesting? And why?

As I have discovered through discernment and pronounced in my books, the term *New Age* came out of the Bible. It represents the Age to Come, also known as the Millennial Reign of Christ on Earth, the Thousand Year Reign, the Age of Aquarius, Heaven on Earth.

In my research and experiences, most New Agers came out of some kind of Christian Church, or denomination of Catholicism. What they find amongst New Age circles are friendlier and more accepting people than in the Churches. Those who have read the preceding four books in this Book Series, would know that I have proved it was the evil, demonic, counterfeit, *Religious Spirit* that created the conditions for the apostasy. So, if Christians keep maligning, judging, and rejecting people for being New Agers, more and more people will leave the Church and become New Agers. It is a form of self-sabotage by the Church because of their obvious blind spot, which is the pervasive Religious Spirit that wars against the Holy Spirit.[7]

Most people seek and crave the love, the grace, the supernatural healing power of Christ evidenced through the Presence of the Holy Spirit. But, instead, what people have experienced in the Church are the evils and oppression of the *Religious Spirit*, who comes from the god of this worlds' demonic hierarchy, set up in all religions on Earth. Ironically, that includes the so-called New Age Religion, which is essentially Luciferianism, ruled by the Religious Spirit as well.

So, how can these people be liberated of this obvious spiritual stronghold that creates all kinds of soul bondages? Turn to the Holy Spirit through Christ, He is the only One with the power to deliver from Religious Spirits. With that said, Christians who are really born again with the Holy Spirit should show evidence of the Fruits of the Spirit, which would be understanding, empathy of those who were rejected and disenchanted because of the bad and sinful behavior of so-called Christians.

I visit and do book signings for New Age Groups because I write about them. I have counseled multiple people over the past 35 years, who all seem to have the same things in common. They have had ET/Alien Experiences, they were born and raised in Christian churches and they are planted in New Age circles or were there because they were searching for truth and were hurt by the Church.

Lots of Gray areas are in the New Age! No pun intended, but literal gray and the grays seems to work to deceive most New Age channels. All the esoteric sciences, which then became metaphysical which later became the New Age, are infiltrated by hidden, occult grays. There is a lot of counterfeit fluff inside the New Age, but there are also a lot of people experimenting with the idea of growing beyond Religion. These people are deceived as they take with them from their respective churches a Religious Spirit, are seduced by the promise of enlightenment, and end up with the Luciferian Agenda: an Agenda to implant humans into an enslaved system known as the New World Order.

> "Beware lest anyone cheat you through philosophy and empty deceit, according to the tradition of men, according to the basic principles of the world, and not according to Christ."
>
> (Colossians 2:8)

Discernment: there is a vast difference between the Lord Jesus Christ and Christ Consciousness. Luciferian New Agers embrace Christ Consciousness, but fall short of an intimate relationship with the Living LORD of Life, Truth and Love.

> "Then those who feared the Lord talked often one to another; and the Lord listened and heard it, and a book of remembrance was written before Him of those who reverenced and worshipfully feared the Lord and who thought on His name."
>
> (Malachi 3:16)

CAN FALLEN ANGEL ETS BE SAVED?

They are called *fallen* because they rebelled against the Creator God. They are extraterrestrials, not exactly angelic angels. They were all cursed, some imprisoned, some bound to the Earth to torment humans. What if they are masquerading as good ETs? What if they are influencing the New Agers and Contactee ET Cults who are being misled to believe they are in contact with Pleaidian, Arcturian, Light Workers, but really they're all fallen angels, rebel ETs who created bad karma thousands of years ago and have been stuck in the Earth ever since?

Can these beings be saved? Can they repent? Could they become reconciled with the Creator who in His mercy let them live, but demoted them, even put them under a curse, as He did in Genesis?

What if we are a key to their Redemption? What if, as they watch us, and witness the profound transformations that take place when we repent and turn towards depending on the Lord, that it causes them to feel jealous?

If God can use the Gentiles by blessing them with signs and wonders to make the Jews jealous, then what about the Fallen Angels? And Rebel ETs? As I concluded in Book One of *Who's Who in the Cosmic Zoo*, what if Earth was a type of reality show for extraterrestrials to collect a library of data on human development, human transformation, and redemption? [5] What if they are watching those whose lives are radically transformed through the power of the grace of redemption? To watch a soul that is full of darkness, hatred, violence, and addiction get turned around through the power of Christ is fascinating, to say the least, to the best of us. How much more are extraterrestrials watching those of us who get switched on to the power of God, and start to learn that that the power of love trumps the love of power? What a powerful testimony that is, even for an ET.

ARE WE ALONE AS SOLARIANS?

> "Imagination is everything. It is the preview of life's coming attractions."
> ~Albert Einstein

As I defined and identified in Book One of *Who's Who in the Cosmic Zoo*, all the beings who live within the boundaries of our Solar System are technically called Solarians because the name of our Sun is Sol. Is Earth the only planet with lifeforms in the Sol System? Does our Sun have a twin? Binary star systems are common throughout the Universe, and the most likely systems observed to host planets, so why would we be any different? [6]

This book is about discerning what has been taking place in our skies and space around our Sun. In recent years, people from all over the world are photographing other objects around our Sun that appear to be exoplanets and exomoons which are spherical in shape. However, we are also seeing exceptionally large spacecraft that are distinctly intelligent, some even look like the USS Enterprise from Star Trek, or the Millennium Falcon from Star Wars. What on Earth is going on?

Let us discern, why does our government have a secret Space Program they call Solar Warden? Could it be that they have known our Sun is a Stargate and they are guarding this solar system, and our Earth along with its neighborhood of planets, Moons and a combination of Exoplanets interloping within our solar system bringing solar cores from an incoming solar system, and spaceships?

What we are seeing is a clash of two systems. How about a clash of two kingdoms?

What if the 10 kingdoms of Revelations refer to ten planetary kingdoms that our Earth and its fellow Solarians will square off against in the near future?

The Lord promised to recreate the Heavens and Earth, because He knew it was no longer a perfect model, the orbits having become irregular after the Flood. There have been clashes that often turned into collisions and cataclysms. I believe there are no accidents in the universe, only error caused by sin to the Curses of the Death Star. The Lord has promised to correct all and wipe it from memory by transforming it all into His Kingdom of Heaven governed by the Kingdom of God.

No coincidence that the ancient Egyptians worshipped a red Sun encircled by a serpent. Many red Sun artifacts have wings which depict Nibiru as symbols of a great red dragon.

THE GREAT DEBATE: DID GOD CREATE ALIENS?

The word *discern* originally meant to smell or sniff out. Discerning was a function that had to do with olfactory organs, one's sense of smell. Animals have this extra sensitive sense of smell, some humans have it too, if they are not defiled with smoke and toxins. For example, bank tellers are trained to discern counterfeit bills from real money. They first must learn what the real bill looks and feels like, so they can spot the subtle differences in the counterfeit. And so, it is with ETs, aliens, angels, gods, and messiahs. This is what I am teaching in my book series, *Who's Who in The Cosmic Zoo?* [8]

Discernment of the presence of the twisted serpent, the Leviathan spirit, has evoked this curse from the Lord:

> "Woe to those who call evil good and good evil, who put darkness for light and light for darkness, who put bitter for sweet and sweet for bitter."
>
> (Isaiah 5:20)

I think most Christians share the same problem, we simply do not understand the Bible as we have been given mistranslations, which has led to misunderstandings. We have been left just plain ignorant of what God's Word is actually saying, who wrote it, who put it together, and who edited it. All this is important in understanding

CHAPTER TWO: LIFE ON OTHER PLANETS

Bible stories, concepts, history, prophesies, and whether God created aliens too, as well as humans.

The Bible does refer to extraterrestrial life and aliens. Not all aliens are demons, some serve the King of the Kingdom of Heaven. Evidenced in the following scriptures:

> "Jesus did many other things as well. If every one of them were written down, I suppose that even the whole world would not have room for the books that would be written."
> (John 21:25)

This passage clearly indicates that there is much more to the Bible than included in our modern Canon. As I have stated multiple times, there are approximately 23 books missing from our Biblical canon which were deleted, rejected, giving us an edited version. It is important to keep this in mind when viewing the Bible, and mistakenly thinking that it is the be all and end all of the Final Word of God. God is infinite, not limited. To suggest that the Word of God contained in the modern Bible are His only words is putting God in a box and limiting the scope of His words and knowledge. The Bible is a guidebook which includes the history of the Jews, His chosen people. Some stories recorded in the scriptures are more detailed than others. But many stories are truncated, or deleted entirely, evidenced in the other missing and rejected texts.

Rejected books, not considered part of the New Testament Canon:
1. The Book of Jubilees
2. Epistle of Barnabas
3. Shepherd of Hermas
4. Paul's Epistle to the Laodiceans
5. 1 Clement
6. 2 Clement
7. Preaching of Peter
8. Apocalypse of Peter
9. The Book of Paul and Thekla
10. Gospel of Mary Magdalene
11. Gospel of Thomas
12. Gospel According to the Egyptians
13. Gospel According to the Hebrews

Some of these books contain true accounts of things that occurred, 1 Maccabees, for example. Others contain some good spiritual teaching like the Wisdom of Solomon.

The gospel of Thomas in particular was a forgery written in the 3rd or 4th century A.D., ostensibly written by the apostle Thomas. It was not. The early Christians almost universally rejected the gospel of Thomas as heretical. It contains many false and heretical things that Jesus supposedly said and did. None of it (or at best truly little of it) is true. The gospel of Barnabas was not written by the biblical Barnabas, but by an imposter. The same can be said of the gospel of Philip, the apocalypse of Peter, and many others. All these books, and the many others like them, are pseudepigraphal, essentially meaning "ascribed to a false author." [7]

Nibiru was mentioned in the newspapers in the early 1983. NASA itself found Planet X. In my next chapter, I am going to discuss the infamous interview of 1994, between NASA's Chief Supervising Astronomer Dr. Robert S. Harrington and Zecharia Sitchin, who confirmed that they indeed have located Nibiru. Sitchin was an expert in linguistics and translated cuneiform tablets into English. It was then that the world learned that the 12th planet was called Nibiru. [8]

The supervising astronomer at the naval observatory died a mysterious death about a year later, as have about all of the astronomers who have spoken too much on the exact location of Nibiru.

We talked about the Second Earth growing up in the 60s and 70s, and here we have its manifestation. We live in interesting times. The Truth is Stranger Than Fiction!

What today astronomers call Planet X is in fact not just one planet, but multiple planets within a solar system. The star they orbit is a brown dwarf star called Nemesis, our binary Suns twin. Most star systems are binary, meaning there are two stars locked in orbit around each other. Our Sun, Sol, has planets which orbit as we all well know. Nemesis also has planets which orbit it, and one of them is Nibiru. Nibiru is also known as Wormwood. Nibiru has plasma wings formed by tails of comets, asteroids, fireballs, and meteorites.

GOD CREATED ALIENS

This question is for my mostly Christian brothers and sisters: Why are you so certain that God only created the humans on this Earth? Why create an infinite universe and only put life on one small spot in this universe? How and why would life elsewhere change your faith or would it? My God is capable of anything, so why would other life-forms or life elsewhere be such a big deal for people?

In *COVENANTS* – Book Four, I presented the facts as based on clinical research, which prove overwhelmingly that Christians are and will have the hardest time with Disclosure due to the cover up of ETs from mistranslations in the English Bibles.[7] See, *Who is God?* Book Two,[12] and *Who Are the Angels?* Book Three.[13]

Ironically, atheists accept the prospect of alien life and other dimensions much more easily because their views are based on science. Many atheists are leading Disclosure movements and work in SETI projects. *COVENANTS* – Book Four [9], documents and analyzes these facts about social science and our present culture. Further, I present solutions to not only Christians but to all belief systems, about how Disclosure of ETs and aliens will ultimately shake out, and I show where ETs and Aliens are found in *God's Divine Plan of Salvation*.

The University clinical research proves hands down that mainstream Christianity and all of its denominations are having the hardest time accepting Disclosure of ETs and aliens. The fault lies in their doctrines, which have unfortunately become a watered down, edited version of the ancient Jewish texts, which happen to be rife with multiple references to alien and extraterrestrial and extra-dimensional beings. The scriptures that were left in our current Bible canon do indeed tell the history of alien and extraterrestrial visitors who intervened with human affairs, but unfortunately the discernment of who these visitors have been lost in translations from Hebrew to English. When English Bibles were translated into other languages the cover-up was only perpetuated through mistranslations. I proved this in *Who is God?* Book Two of the *Who's Who in the Cosmic Zoo* series.[10] Cognitive dissonance prevails in those with fixed preconceived notions that will be, and have been, proven wrong.

Someone implanted the modern evangelical movement with the belief that all aliens are demons who come from the Nephilim and Fallen Angels. The Nephilim are only one kind of alien hybrid, fallen angels are another part of the story. The ancient scriptures prove that God created aliens, which does not preclude that aliens created aliens too. My first book is an A-Z Compendium, and goes over 120 types of ETs and Aliens, including Nephilim, and many more.[14]

Life should not have to be politicized; it is a gift of God. Who are we to deny life to one species yet protect life of another? This is the essence of the ongoing Extraterrestrial warfare that has pervaded Earth for Millennia. Life is now an exopolitical issue.

In Book One, my readers will remember that I laid bare the facts and research about the ongoing hybridization program from decades of alien abductions of humans. These hybrids are slated to be a space-faring new breed of humanoids to replace the present human race.[14]

The Fallen Angels are called so because they rebelled against the Creator God. They were somewhat cursed, some were imprisoned, others were bound to the Earth to torment Humans. What if they are masquerading as friendly ETs? The ones influencing the New Agers Contactee ET Cults? What if we are misled to believe our invisible friends are Pleiadian, Arcturian, Light Workers but really, they are all fallen Angels, rebel ETs who created some bad karma thousands of years ago and have been stuck with the Earth since?

Can these beings be saved? Can these beings repent? Could they become reconciled with the Creator who in His mercy let them live, but demoted them, even put them under a curse, as He did Genesis?

What if we are a key to their Redemption? What if, as they watch us, and witness the profound transformations that take place when we repent and turn towards depending on the Lord, it causes them to feel jealous?

If God can use the Gentiles by blessing them with signs and wonders to make the Jews jealous, then what about the Fallen Rebel ETs?

Notes:
1. Biblehub.com Commentary on 1 Corinthians 15:22 and 45
2. Gary Bates, *Did God create life on other planets? Otherwise, why is the universe so big?* https://creation.com/did-god-create-life-on-other-planets, November 2009
3. Ella LeBain, *Who's Who in the Cosmic Zoo?* Book One – Third Edition, Chapter Three: *From Adam's Failure to the Second Adam's Victory*, pp 71-74 – Tate. 2013
4. Isaiah 53:6,10; Matthew 20:28; 1 John 2:2, 4:10
5. Ibid, *Concluding Words*, pp. 415-446.
6. Ella LeBain, *Who Are the Angels?* Book Three, Chapter Fourteen: *The Harvest of Angels*, pp. 289-302. Skypath Books, 2016.
7. Ella LeBain, *COVENANTS* – Book Four, Chapter Two: *How Disclosure of ETs Impacts Religion*, pp. 5-13; Skypath Books, April 2018.
https://www.amazon.com/COVENANTS-Times-Guide-Aliens-Angels/dp/0692988637/
8. Ella LeBain, *Who's Who in the Cosmic Zoo?* Book One – Third Edition, Tate, 2013. https://www.whoswhointhecosmiczoo.com
9. The Lost Book of the Bible, https://www.gotquestions.org/lost-books-Bible.html
10. Zecharia Sitchin, *the 12th Planet*, Book One of the Earth Chronicles, Avon, NY, 1991
11. Ella LeBain, *COVENANTS* – Book Four, Skypath Books, 2018.
12. Ella LeBain, *Who is God?* Book Two, Skypath Books, 2015.
13. Ella LeBain, *Who Are the Angels?* Book Three, Skypath Books, 2016.
14. Ella LeBain, *Who's Who in the Cosmic Zoo?* Book One – Third Edition, https://www.amazon.com/Whos-Who-Cosmic-Zoo-Third/dp/1629942065/, 2013.

CHAPTER THREE

CONTACT WITH ETS FROM PLANET X

Truth is Stranger than Fiction!

There has been a plethora of films released over the past 50 plus years that have centered on soft disclosure, as if to prepare the public for the arrival of aliens and extraterrestrials, is truly, stranger than fiction. *Contact*, and one of the Star Trek movies, *First Contact*, which was released in 1996, both inspired our culture to imagine the day when we, the inhabitants of Earth meet ET.

Billionaire Robert Bigelow, NASA partner, Bigelow Aerospace CEO and Entrepreneur, recently said, "There has been, and is, an existing presence, an ET presence, and I spent millions and millions and millions of dollars, I probably spent more as an individual than anybody else in the United States has ever spent on this subject."

There are many myths and beliefs regarding the events surrounding the passage of Nibiru by Earth, including that the ancient gods of antiquity, those specifically mentioned in the Bible will return to Earth. From my intense study of end times Bible prophecies, I don't see any way of avoiding contact with extraterrestrials as the Bible is clear that angels, who are extraterrestrial messengers, will fly around the Earth preaching the gospel of grace to a dying world, as one last call to repent and accept Christ as Savior.

> "And I saw another angel fly in the midst of Heaven, having the eternal gospel that he might evangelize those that dwell on the Earth and every nation and kindred and tongue and people, saying with a loud voice, Fear God and give glory to him, for the hour of his judgment is come; and worship him that has made the Heaven and the Earth and the sea and the fountains of waters."
>
> (Revelations 14:6-7)

People need to get prepared for the disclosure of the alien presence that has been on Earth for millennia. People must accept the reality of the return of the ancient gods who are scheduled to meet on Earth in divine appointment to resolve, in a final battle, the conflict they have with the Creator God, who is returning for that same purpose with His vast army of celestial warriors. Christ is called the Lord of Hosts because He commands legions of warrior angels, who are prophesied to return to Earth *in the clouds of Heaven*. I thoroughly proved in Book Two and Book Three that these angels are spaceships, starships, in Hebrew Merkabah or *Chariot*.

The beings associated with Nibiru are referred to as Annunaki, translated by Zecharia Sitchin of the Sumerian tablets to mean *those from Heaven to Earth come*. In other words, they were extraterrestrials, ancient astronauts, who came to Earth from their home world, Nibiru, one of the planetary bodies in the Nemesis-Nibiru System. Many have suggested that Nibiru is actually a large mothership which the Annunaki, inhabit, making them Nibiruans. See, Book One of *Who's Who in the Cosmic Zoo?* pp. 110-119; 339. [1] All we really know is that there are multiple celestial objects in the Nemesis-Nibiru system that have entered into our solar system.

This invasion is not limited to the seven planets, the brown dwarf star that it orbits, along with its dozens of Moons, the system both precedes and is followed by a whole lot of space junk, asteroids, meteorites, fireballs, and comets. Remember, Nibiru is only one of its objects. To suggest that Nibiru may be a mothership, a gigantic space city that is on an irregular, rebellious, tilted, and elliptical orbit around our Sun that takes approximately 3,600 lunar years, or 3,568 solar years, is certainly plausible. Given the mythologies and mounting archeological evidence showing up all around the world, I am going to conclude, anything is possible.

My purpose is to help my readers prepare for the inevitable, which is something many fear, and that is a personal encounter with our Kinsman Redeemer, Lord of Hosts and Savior of All Creation, whom we know as Yeshua, Jesus the Christ. Many people would rather encounter aliens, without any question of whether or not they are benevolent or malevolent towards humans, instead of preparing their hearts, minds, and souls for a *close encounter* with Jesus Christ.

As I presented in Book One of *Who's Who in the Cosmic Zoo?*[2] more people believe in space aliens, and the possibility of alien life, than believe in God or in any religion. I have been laying out the discernment and distinction between having a personal relationship with the Living God, and religion, which is comprised of many man-made theologies, interpretations, and rituals.

> "The fruit of the righteous is a tree of life, and he who wins souls is wise."
>
> (Proverbs 11:30)

ENDGAME DISCLOSURE OF THE ALIEN PRESENCE

Preparation for the Endgame involves disclosure of the alien presence here on Earth. This will reveal an age-old space war between various groups of aliens and ETs, along with their evil intentions for the future of humankind. Such revelations will not only cause cognitive dissonance in the minds of those who have invested all their time and energy in proving that the aliens are real, but real shock when they discover that these very aliens, they have waited so long to meet do not have our best interest at heart, but instead want to exploit us.

Anyone who has run up against the conspiracy theories that the Jews really run this world, its banks, its jewelry trade, its fur market, discover that the real force to contend with on this planet are the Archons, spiritual Realities which will require humankind to have a much stronger Presence on our side if we don't want to succumb to the Draconian Reptilian agenda, those Slave Masters of humankind. They have been in power since the last two Fly-Bys of Nibiru.

> "We war not against flesh and blood beings, but against Powers, Principalities, Rulers of the Darkness of this Present World (Archons) and Spiritual Wickedness in the Heavens (cosmos)."
>
> (Ephesians 6:12)

As we have gone over throughout this book series, this is the battle in which every one of us on Earth as human being must fight. The Good News is that there is a way out. It is promised to all those who put their faith in the One who has come and promises to return to set all captives free. He is the true Messiah, the Savior of humankind, and according to scriptures, all creation.

> "I have come to set the captives free."
>
> (Isaiah 61:1; Luke 4:16)

Future glory is promised to all those who put their faith in the Lord's divine plan of salvation through the exalted Son of Heaven, Yeshua HaMoshiach, the Lord Jesus Christ. Unfortunately, due to the infiltration by the Reptilian Overlords into just about every religion on Earth, people have become too jaded to believe that anything to do with Jesus implies a religion, a cult or being *religious*. This could not be further from the truth. The power and authority coming through Christ sets people free from the bondages of religions and their demonic *religious spirits* which war against the Spirit of Christ, the Holy Spirit. We are promised to *ascend* into *glory bodies* when Christ returns, bodies that never age and are immortal. And it is not just for us, it is for *all of creation*, which groans and waits with expectation for its redemption, which will happen at the Second Coming of Christ.

> "For the creation was subjected to futility, not by its own will, but because of the One who subjected it, in hope that the creation itself will be set free from its bondage to decay and brought into the glorious freedom of the children of God. We know that the *whole creation* has been groaning together in the pains of childbirth until the present time."

(Romans 8:20-22)

Back in the early 1990s, when I was doing astrology counsel for a living, I had a client who said something to me I never forgot. "We're all evolving here, but so are the gods, who are evolving along with us, and through us." Then I was turned on to Zecharia Sitchin's book, *the 12th Planet, Book One of the Earth Chronicles*, in 1991, along with *Genesis Revisited*, which initiated a paradigm shift in my perspective and worldview.

One of the first books I ever read when I was 13 years old was *Chariots of the Gods* by Erich Von Daniken. So, I already had some understanding of the Ancient Astronaut theories, but Sitchin took it to another level, with his revelations through his translations of the Sumerian Cuneiform. I had the honor and privilege of meeting Sitchin three times in the 1990s. He was gracious and kind to allow me to interview him when I first met him in 1994 at the *When Cosmic Cultures Meet Conference* in Washington, D.C. where he was a keynote speaker. I include his message in my next chapter.

I was fascinated with everything he had to say, and afterwards, we sat down and talked further. We found common ground, both being Jewish, and both having history in Israel. Sitchin spoke five languages fluently. When people who are not even able to use correct grammar in their own language today criticize his translations, I just shake my head. I am not saying every word he wrote was perfect, but he certainly communicated the main messages from deep antiquity. I find it to be an effort in futility to waste time arguing over nuances. I think the message was clear in the tablets, that ancient extraterrestrials came to Earth, manipulated the human species to serve them to mine the Earth for gold and other precious metals and resources which they needed to preserve their planet and world.

The gold was needed as a heat shield for their planet or spaceships. This is where the idea and belief that Nibiru was actually a mothership, and not an actual planet, is introduced because of the need by the Annunaki to obtain vast amounts of gold to preserve their world. Gold is used in many areas of similar technology today because it is a stable metal. Gold is mostly used in the electronics to serve the purpose of plating contacts in switches, relays, and connectors, and in electrical wiring for applications that need immense amounts of energy. The use of gold in electronics is perhaps its most important industrial use. It is only logical to deduce that the Annunaki needed for it a similar purpose in their own technology, and according to Sitchin, to preserve the atmosphere in their world. We cannot help but extrapolate that it is more than likely a heat shield for the outer atmosphere.

With that said, we, as a species would be better off if we could prepare ourselves for what is to come upon the Earth. We had best discern the differences between what these ancient beings want with us, considering that they used human women to create a race of humans with just enough intelligence to follow orders and understand how to mine, yet not enough intelligence or power to rebel against these extraterrestrials or even escape their enslavement.

This is why, the Creator God sent a Savior to humankind.

Taking all this into account, perhaps my old client was right, that these so-called *gods* aren't actually gods at all, but created beings, who just like us are on a path of evolution and who need to learn to take responsibility for the consequences of their actions and inactions. They will reap what they have sown on their next fly-by. This is why the Bible tells us that there is more than one rapture in the Last Days of this present age. Christians assume that only Christians are going to be raptured, but the scriptures tell of three separate end-time events.

> "Let both grow together until the harvest: and in the time of harvest, I will say to the reapers (harvesters), gather you together first the tares, and bind them in bundles to burn them: but gather the wheat into my barn."
>
> (Matthew 13:30)

Who are the Tares? The Tares are a metaphor for the Wicked. It is a separation of the Wheat and the Tares (Weeds). The point is the angels, who are extraterrestrials sent from the Lord of Heavenly Hosts, are commanded to gather the Harvest, part of which is separating the fruit from the weeds. The angels that serve Heaven's Kingdom are put in charge of gathering the Harvest.

> "The harvest is the end of the age, and the harvesters are angels."
>
> (Matthew 13:39)

See, Book Three, *Who Are the Angels?*[3] The work of the angels is to be the harvesters at the end of this Age.

> "As the weeds are pulled up and burned in the fire, so it will be at the end of the age. The Son of Man will send out his angels, and they will weed out of his kingdom everything that causes sin and all who do evil."
>
> (Mathew 13:40-41)

The angels are extraterrestrials. So, with the above Scriptures, tell me, how can there not be Contact with ETs at the End of the Age? Prepare! As humans, we need to be able to tell the difference between the benevolent ETs from the demonic aliens. This is the main message of *Who's Who in the Cosmic Zoo?*[4] and throughout this series. We, the collective *we*, must work together to exercise our spiritual muscles of discernment. If we allow ourselves, as we have in the past, to be enticed, sucked into a group of ETs who promise us the world, only to find ourselves, as a species, enslaved, and/or replaced and destroyed, then history has taught us nothing.

History repeats itself, because we learn through repetitions, and it repeats until we learn our lessons. Have we? How are we going to handle it when we meet angels who claim to be working for Heaven's Kingdom, impressive in their technologies or special powers? They may look beautiful to us, and we are strangely attracted to them because they are other-worldly but underneath, we have not realized that they have ulterior motives. Perhaps they may want to use us? Not just biologically, which has been the purpose of the centuries-long abductions, but worse, to be used as pawns in their war against the Creator God, who gave us all souls?

Therefore, we must ask ourselves the question when meeting ET, who do they say God is, at least to them? Who do they say Christ is to them? If the message of Salvation is written into the very stars themselves, which I prove through the ancient records is exactly the case, then surely, extraterrestrials, especially ancient Astronauts, and interdimensional travelers, should be able to say who He is?

As someone who has come full circle, brought up as a hybrid child, half Jewish and half Italian Catholic, I have searched my whole life for answers. The answers I found are shared throughout this book series. But I still have questions, and I'm sure you do, too, so this is being written with the hopes that we can get the conversation started, without negative mocking, cognitive dissonance, or persecutions for being different, because that kind of stuff never gets anyone anywhere, and it certainly will keep you in the dark about how to handle disclosure and what's to come on the divine schedule for the Last Days of our present Age.

I began the discussion in Book One of *Who's Who in the Cosmic Zoo?*[5] which started at the place of acceptance of ETs, aliens, gods, and angels, without argument or any ifs, ands, or buts about whether they exist, as we need to move on from that. I began with the acceptance that not only do they exist, they exist in diversity, there are many kinds and many types, and based on my research, and my experience, they are either serving the Heavenly Kingdom or the Kingdom of Darkness, themselves. Good versus Evil, nothing new about that. However, if we are to move forward with the whole ET question, we need to understand Whose side they are on. Not grammatically correct, I know, but this has impact.

This is why I suggested in my chart that there may be a clash between two Kingdoms both over the inside and on the Earth. There is also an alliance which consists of breakaway groups from both sides who seek peace and cooperation. See, *Who's Who in the Cosmic Zoo?* Book One – Third Edition, pp. 90-91, *The Cosmic Drama/Galactic Warfare/Class Between Two Kingdoms: The Human-Reptilian Wars, i.e., "Draco Wars"*, Chapter Four: *How to Tell Who is Who?*[6]

A daunting task! It is certainly not easy for us when most of us are not even aware that there is a celestial battle going on! We see the ongoing battles on Earth and between humans. We know that the Bible tells us that the real battle isn't exactly between us, but between overlords, archons, rulers of the darkness of this present world, and spiritual wickedness in high places, which implies space aliens.

Perhaps rethinking our history with the understanding that others are behind every major war and battle we have had to fight can show us a clear lesson, which we can then apply to our present exopolitical situation. We who live on the surface of the Earth are at the mercy of God. We may have advanced technologies now, more than ever, but is it really enough to protect ourselves against a more powerful, technologically advanced beast?

This is why the Divine Plan of the Kingdom Heaven, the chief topic of discussion and sermon by Jesus is that the Kingdom of God is within us. If we really understood what that means, both individually and collectively, we would be able to withstand anything, because all power and authority over creation lies within the King of kings, who imparts and authorizes His Authority to those who both know Him and trust in Him.

If Jesus said, "Truly, truly, I say to you, whoever believes in Me will also do the works that I do; and greater works than these will he do, because I am going to the Father." (John 14:12). That is like being given a blank check by your father, and told you can spend it anyway you want, except for one condition, that it must be done according to His Will. "And this is the confidence (the assurance, the privilege of boldness) which we have in Him: [we are sure] that if we ask anything (make any request) according to His will (in agreement with His own plan), He listens to and hears us," (1 John 5:14 -AMP).

So, we need to know is, what is God's will. Well, it's written into the very Stars themselves that ALL Creation should be saved and be reconciled with their Creator, through a Kinsmen Redeemer, a Hero, a Messiah, who will be the Good Shepherd, who will lead everyone back, and bring us together to live as One.

Jesus said, "and I have other sheep that are not of this fold. I must bring them also, and they will listen to my voice. So, there will be one flock, one shepherd," (John 10:16).

Notice that He used the word, *have*. This implies they are already His. He has them. When He asserts His Will, when He brings them together with us, we will be One flock, with One Shepherd, which is Him.

This implies that it is God's Will to save extraterrestrials and aliens. As I stated in Book One, many of them are here watching us. They watch us for many reasons, not just because we're their Reality Show, and we're so interesting to watch, but because they are looking for the humans who "get it," the ones who "know" the power and authority of the King of Heaven's Kingdom. The ones who carry the Kingdom of God within them. The ones whose lives are transformed from devastation, weakness, sickness, and rebellion to one of wholeness, strength, healing, and service to the King.

They are learning from us. If we humans, who have been genetically manipulated, implanted, and enslaved for lifetimes by alien masters who harvest our energy, can be saved from behind enemy lines, then maybe so can they. Our survival is a testimony to the amazing extravagant grace of God, which transforms bitterness to sweetness, jealousy to love, rejections to acceptance, fear to joy. These ET watchers, many of whom fell away during the last galactic conflict and were bound inside the inner Earth, or in the various dimensions, can also be saved and brought back into right relationship with the Creator, too.

Never think for a moment that you are alone. If you believe in angels, you know there are both good and evil among them. The evil angels we call demons, some of whom look human, while others are beast-like, reptilian, alien, hybrids. Some are monsters. Some are miserable within their own existence, accursed, remembering a time in long antiquity when they were loved by the Creator; but now are desolate.

Meanwhile, here on Earth, we argue amongst ourselves over who is the most victimized, and who is a racist. As I made my point clear in Book Four, *COVENANTS*,[8] all racism on Earth was initiated by the ETs and aliens who have been managing us for millennia.

"Meet the new boss, just like the old boss…Then I'll get on my knees and pray, we won't get fooled again, don't get fooled again, no no, Yeah! Meet the new boss, Same as the old boss." [9] (Peter Townsend, *Won't Get Fooled Again*, The Who, Album: Who's Next, 1985)

Do you get where I am going with this? It means that we are a species with amnesia and have forgotten our past. Many have been collectively implanted with false beliefs. For example, the Church of Rome deliberately and intentionally attempted to delete the support for reincarnation from the Hebrew scriptures, in order to implant a lie to prop up their power over the gates of heaven. The Roman Church further inserted a rogue scripture in at Hebrews 9:27, when Torah that clearly says that the Lord God takes people down to the grave and the Lord brings them back and gives then life. "The LORD brings death and makes alive; he brings down to the grave and raises up" (1 Samuel 2:6). See, *Who is God?* Book Two, Chapter Twenty-Eight: *What Happens When You Die?* pp. 455-483.[10]

The point is that without the idea of reincarnation, we, as a species, are disconnected from our collective pasts. Individually, my life would not have made any sense had I not been shown this truth at a young age. Unfortunately, I have found myself in the middle of many religious controversies, being born a hybrid, half-breed. But this is the reason I believe I was born, to be the bridge between two worlds, and shed understanding, so people can have peace – peace of mind and peace in their spirits, and more importantly, salvation and reconciliation with the Creator of all life. The Lord of Spirits, as the Books of Enoch calls Him. The Creator of our Souls. As I pointed out in my thesis in Book One [7], the way I chose to present the taxonomy of ETs and aliens was to categorize them between soul-matrixed beings, and soulless beings.

Today, the biggest threat against our existence is not necessarily the ancient aliens who have been here for millennia, but the proliferation of artificial intelligence and all that it can do. It can literally take over human minds, human bodies, and our planet. Do we, as a species, want that to happen? This is why alliances are formed between us and the ETs who also find that technology is a threat to their existence. Point is, we need to find our common ground with the alien presence, as they say in the Middle East, the enemy of my enemy is my friend.

This is why there is an alliance and type of confederation, which is what I called this in 1998, and published in Book One in 2012 [6], because *confederation* is a union of sovereign states, united for purposes of common action often in relation to other states. It is rooted in the Founding Principles establishing how the United States of America would be governed. It is created by treaty. Confederations of states tend to be established in order to deal with critical issues, such as defense, national security, foreign relations, internal trade, or currency, with the general government being required to provide support for all its members. According to Wikipedia, *Confederalism* represents a main form of intergovernmentalism, which is defined as any type of interaction between states that takes place on the basis of their sovereign independence or government.

This is similar to the alliance between the two Kingdoms, and, to add another layer, the alliance between Earth Governments and the Governments of the Inner Earth. In other words, between us humans and the interdimensional humans, extradimensional humans, antediluvian extraterrestrials, and alien races that live inside the Earth. This is the realm of Exopolitics, which is the ongoing discussion of our interactions, our agreements, and how we choose to handle the alien presence.

I am inserting into the discussion the need to lead some of these beings back to the Creator. You see, according to the Bible, the Lord gave the Earth to humankind, but the alien presence, under the leadership of Lucifer-satan, was here already. He lost his kingdom when the civilization of Atlantis was destroyed. Some of it sank and went deep inside the Earth. Interesting, in keeping with my motto, *the Truth is Stranger than Fiction*, the blockbuster film, *Aquaman*, released in 2018, told this story. Anyone who has studied the stories of Atlantis readily connected the dots here, which were more than obvious. What if it were all true?

What if part of the disclosure is that antediluvians, some of whom took refuge or were shipwrecked inside the Earth, are a part of the Alien Presence we have been living with all these millennia? The Bible record, as I have already proved in this Book Series, does indicate there were two major floods. The Flood of Noah is acknowledged worldwide, but before that time occurred was the Luciferian Flood, as indicated in Genesis 1:2. There was water everywhere when the Lord allowed the Earth to be repopulated to start all over again. The Serpent was a bipedal, luminous, attractive, and intelligent humanoid who enticed the Woman in the Garden of Eden. She was likely talking to Lucifer Himself! After the Lord cursed him, the Jewish scriptures, *The Book of Adam and Eve*, says that he was first made mute, the Lord took away his ability to speak, and then He turned the talking Serpent into a snake.

We need to understand that these beings are still with us, accursed, dejected, banished but nevertheless allowed to live and coexist with what became the Evadamic race of humankind. The Serpent race was jealous that the Lord God would give the Earth to humans, so he has pulled all kinds of deceptive tricks and traps on humans to try to fool them into giving up their power to him, so he can have his kingdom back again.

There is a reason why the Lord God did not destroy them, but let them live in accursed bodies, mostly because the Lord God is merciful, and His loving-kindness crowns Him as the Lord of lords. According to both scriptures on Earth and the Heavenly Scroll, His path of Salvation is open to ALL Creation, and that includes these fallen beings.

Now, just because some of them are deeply entrenched in their wicked ways, and we think there is no way they could ever be saved or repent, does not mean that that they should not be given the chance. This is up to the Lord. What matters is that the Lord, in His everlasting loving-kindness, gave them their chance, which in the end gives glory to Him as being fair, just and merciful, which distinguishes Him as the Almighty, and not just another son of Heaven who's abused his powers.

Our confidence in making it to Heaven is found in the words of Jesus, 'I have other sheep that are not of this sheep pen. I must bring them also. They too will listen to my voice, and there shall be one flock and one shepherd,' in John 10:16, which is the foundation for your confidence that you who have given your life to Christ. 'Do not offer any part of yourself to sin as an instrument of wickedness, but rather offer yourselves to God as those who have been brought from death to life; and offer every part of yourself to him as an instrument of righteousness,' Romans 6:13, and 12:1, 'Therefore, I urge you, brothers and sisters, in view of God's mercy, to offer your bodies as a living sacrifice, holy and pleasing to God—this is your true and proper worship.'— will make it to Heaven, will not perish, but have eternal life (John 3:16). Our crowns that are promised to us as faithful servants are dependent on how we fulfill God's will on Earth, as it is in Heaven. Leading the lost back into the Kingdom of Heaven, which the Kingdom of Light, the Kingdom of Love, and the Kingdom of Peace, will matter most, especially to the heart of the Creator, who is the King of the Kingdom.

Let's continue to do the good work we're all called for, the great commission which is to make Heaven more crowded by leading people to the Cosmic Christ, and yes, that includes those stuck in between galactic wars, who lost their kingdom. If they repented, the Lord would forgive them, perhaps that is why He empowered us humans with His Spirit. Not just for other humans, but for the humans that are otherworldly, the ones who have lost their way and are in rebellion against the Creator, the ones who mistakenly think that if they just build a bigger, more powerful spaceship, they may have a chance to win the next battle. While this may sound like an effort in futility to us who are being saved, it's still, nevertheless, their modus operandi, because they know that He is returning, and they believe they have a divine appointment to defeat Him and take back the Earth from us humans.

My prayer is that those of you who have not given your lives to Christ will be drawn to do so, especially if you want to help save ETs. Otherwise, without the Supernatural Power of Jesus Christ, you can end up deceived and enslaved by them.

This is the promise of God for the Last Days, so don't ever underestimate the power of God to keep His Word, and the Power of God sweeping through the human vessels who have humbled themselves to the Lord and have given their lives to serve His Kingdom of Love, Light and Heaven.

The Old Testament Prophecy: "And it shall come to pass afterward, that I will pour out my spirit upon all flesh; and your sons and your daughters shall prophecy, your old men shall dream dreams, your young men shall see visions: even on the male and female servants, I will pour out my Spirit on them in those days." (Joel 2:28-29)

The New Testament Fulfillment: "In the last days, God says, I will pour out My Spirit on all people. Your sons and daughters will prophesy, your young men will see visions, your old men will dream dreams." (Act 2:17)

It is already happening folks! And it is only going to get stronger as the Last Days get darker. The Last Days of the end of this Age are usually wrought with consternation, frustration, and death. It is the end of a way of doing things, and the transition into a new age is painful, like giving birth. Jesus warned us it would be like labor and birth pains (Mathew 24:8).

Here are a few passages from the End Times Prophecies about what we are to expect in the Last Days of the End of this Age.

> "But realize this, that in the last days difficult times will come. For men will be lovers of self, lovers of money, boastful, arrogant, revilers, disobedient to parents, ungrateful, unholy, unloving, irreconcilable, malicious gossips, without self-control, brutal, haters of good, read more."
> (2 Timothy 3:1-5)

> "Many will be purged, purified and refined, but the wicked will act wickedly; and none of the wicked will understand, but those who have insight will understand."
> (Daniel 12:10)

Notes:
1. Ella LeBain, *Who's Who in the Cosmic Zoo?* Book One, pp. 110-119, 339.
2. Ella LeBain, *Who's Who in the Cosmic Zoo?* Book One – Third Edition, https://www.amazon.com/Whos-Who-Cosmic-Zoo-Third/dp/1629942065/, 2013.
3. Ella LeBain, *Who Are the Angels?* Book Three, Skypath Books, 2016.
4. Ella LeBain, *Who's Who in the Cosmic Zoo?* Book One – Third Edition, https://www.amazon.com/Whos-Who-Cosmic-Zoo-Third/dp/1629942065/, 2013.
5. Ella LeBain, *Who's Who in the Cosmic Zoo?* Book One – Third Edition, https://www.amazon.com/Whos-Who-Cosmic-Zoo-Third/dp/1629942065/, 2013.
6. Ella LeBain, *Who's Who in the Cosmic Zoo?* Book One – Third Edition, pp. 90-91, *The Cosmic Drama/Galactic Warfare/Class Between Two Kingdoms: The Human-Reptilian Wars, i.e., "Draco Wars,"* Chapter Four: *How to Tell Who is Who?*
7. Ella LeBain, *Who's Who in the Cosmic Zoo?* Book One – Third Edition, https://www.amazon.com/Whos-Who-Cosmic-Zoo-Third/dp/1629942065/, 2013.
8. Ella LeBain, *COVENANTS* – Book Four, Skypath Books, 2018.
9. Peter Townsend *Won't Get Fooled Again*, The Who, Album: Who's Next, 1985.
10. Ella LeBain, *Who is God?* Book Two, Chapter Twenty-Eight: *What Happens When You Die?* pp. 455-483.

CHAPTER FOUR

CONNECTING THE DOTS TO NIBIRU

"All truth passes through three stages. First it is ridiculed.
Second, it is violently opposed. Third, it is accepted as being self-evident."
~Arthur Schopenhauer (1788-1860)

It seems a lot changed after the 1990 interview between the Supervising Astronomer of the U.S. Naval Observatory, Department of Defense, Dr. Robert S. Harrington, Ph.D. and Archeoastronomer Zecharia Sitchin. Harrington confirmed the existence of the incoming Planet X into our Solar System, thereby confirming the research of Sitchin's translations of the Sumerian Cuneiform Tablets that such a planet Nibiru exists, and, according to Harrington, has life on it. Twenty-eight months after their interview, Harrington mysteriously died at the age of 50 under suspicious circumstances.

Harrington set up an observatory in New Zealand to track Planet X and became convinced in the existence of such a Planet X beyond Pluto. He searched for it and found it through the IRAD probe in 1983. Initially Harrington collaborated with Thomas C. Van Flandern, an American Professional Astronomer who specialized in celestial mechanics and was known for his alternative and fringe views on physics, cosmology, and extraterrestrial life. Both knew and met with Sitchin, who later cited their research as corroborating evidence to his work. I knew and interviewed Sitchin and will share more on my experience later in this chapter.

Harrington became a believer in the existence of a Planet X beyond Pluto and searched for it himself. Apparently, he identified it with the IRAD probe in 1983. When Harrington retired from the Naval Academy, he agreed to give Sitchin an interview. Harrington initially collaborated with Van Flandern; both were pursued by Sitchin. The Harrington-Sitchin interview took place in at the U.S. Naval Observatory Headquarters in Washington, D.C. on August 30, 1990.

Harrington and van Flandern served to collaborate evidence of Sitchin's work. Harrington's work was to study the perturbations of Pluto and its Moon. He observed a perturber beyond Pluto, which he called an Intruder, and that this Planet

CHAPTER 4: CONNECTING THE DOTS TO NIBIRU

X was on an elliptical orbit around our Solar System every 3,600 years. It caused Uranus to be on its side, as its rings are vertical instead of horizontal like those of Saturn. Harrington calculated several parameters of Planet X and its orbit. He started from the perturbations in the orbits of Neptune and Uranus, knowing that Pluto could not be responsible for them.

The observations he used were supplied by the Nautical Almanac Office of the U.S. Naval Observatory and go back as far as 1833 for Uranus and 1846 for Neptune. He also postulated that this perturber caused Neptune's satellites to be knocked out of their orbits, two of which he believes became Pluto and Chiron. During the interview, he compared Sitchin's chart of Nibiru, which the Sumerian tablets show as entering our solar system through the constellation Sagittarius, but rightly pointed out that it would now come through the constellation Libra, due to precession. It appeared at that time they corroborated and completed each other's work. Their interview set a new precedence in not just the search for Planet X but for its discovery.

Dr. Robert Harrington died of esophageal cancer on January 23, 1993, not long after filming the interview and confirming Zecharia Sitchin's life work of properly translating the Sumerian tablets. His wife believes he was murdered for his work on locating Planet X.

Two years after the Sitchin-Harrington interview, a NASA press release publicly embraced Planet X/Nibiru as being real. "Unexplained deviations in the orbits of Uranus and Neptune point to a large outer Solar System body of 4 to 8 Earth masses on a highly tilted orbit, beyond 7 billion miles from the Sun." NASA Official Press Release, 1992.

Also, in 1992, NASA Scientist Ray T. Reynolds confidently went on record proclaiming, "Astronomers are so sure of the 10th planet, they think there is nothing left but to name it." NASA at that time did everything but scream "Nibiru!" from the mountaintops. In this book, I am going to show you how they not only named it, but renamed it, in order to obscure, conceal and marginalize it.

If it were not for Sitchin, none of us would even be referring to Planet X as Nibiru and Marduk. These names are directly from the Sumerian Tablets.

I had the privilege of interviewing Zecharia Sitchin in 1995 while attending a one-of-a-kind conference in Washington, D.C., called *When Cosmic Cultures Meet*, which was supported by the Rockefeller Foundation. It was groundbreaking, we did not have to debate the existence of extraterrestrial and alien life, we discussed where we go from the point of disclosure of it. That is when I began compiling interviews, researching, and writing what has now turned into a five Book Series, *Who's Who in the Cosmic Zoo?* The series begins from the standpoint of acceptance of the alien presence and with identification and discernment of the different types of aliens and extraterrestrials reported and experienced. To me, this was the most logical next step.

Towards the end of the conference, three panels covered *Fear, Hope and Future*, which were the interactive portions between the main speakers and the audience. I was inspired to focus my Book Series on answering the three main questions posed to the panels.

- How to eliminate fear of disclosure while reconciling religious beliefs with the nature, origin, and purpose of extraterrestrials.
- By anticipating the benefits of dialogues between cosmic cultures, how can we ensure that include the best interests of all Earth's inhabitants: plant, animal, human and those who live inside the planet?
- And, in anticipation of the time when cosmic cultures do meet, what long term endeavors should be undertaken and by whom to prepare humanity for future contact? What steps are presently being undertaken? Zecharia Sitchin's answer to the Future panel, was, "to know the future, study the past."

Sitchin was one of the keynote speakers, and he and his wife were so kind to give me an interview to discuss his Nibiru work. This was only a year after the suspicious death of Harrington, so Sitchin was guarded in his responses when asked when Nibiru will pass the Earth. He knew, but NASA told him he could not reveal it. We had a good rapport, because of our shared heritage as Jews, and our time spent in Israel. Sitchin believed that the Bible reflected the Sumerian Tablets. He was a great inspiration to me and my work, for this Book Series, *Who's Who in the Cosmic Zoo?* I devoted Book Two – *Who is God?* to exposing the mistranslations from the original Hebrew to English, and how the truth about who the gods were, whose names are present in the Old Testament, were lost in translation to the English. This is why Christianity does not understand the difference between Yahuah and Lucifer/satan, Enki and Enlil respectively in the Sumerian Tablets.

Sadly, it was not long afterwards that the elite sent out paid trolls to disseminate false propaganda to discredit the Planet X system and debunk truth tellers. These hired shills were planted to smear lies to conflate and destroy our chances of full Disclosure. It's important to understand that all End Times False Flag Events are manufactured by the Elite (Illuminati) to discredit the Truth, and purport a New World Order Agenda, which includes a One World Religion, Global Government that is Draconian in nature and a one world currency, all controlled by digital technology, artificial intelligence in a myriad of forms, including cutting edge holographic technology.

WHEN COSMIC CULTURES MEET

The following are key points Zecharia Sitchin made at the *When Cosmic Cultures Meet Conference* in 1995 taken from my notes and proceedings.[3]

Sitchin was one of the keynote speakers, and we were all eager to listen to him speak. At that point, I had finished reading, *Genesis Revisited*, and *The Twelfth Planet*. Sitchin's talk was titled: *The Past Holds the Key to the Future*. His findings from archeological evidence, especially Sumerian clay tablets, asserted that cosmic cultures have already met once, are on a 3,600-year cycle and are destined to meet again. That is the news. The question: Is this good news or bad news? It appears that it depends upon the returning Anunnaki's' judgment of our moral progress and discipline in controlling human population.

He began with discussing the theme of the conference, *When Cosmic Cultures Meet*, which he said must have been chosen with great forethought, because it doesn't say, IF cosmic cultures meet, thereby implying that a meeting is a certainty whereas only its time is uncertain. While implying such a certainty, there's still ambiguity about time. He made a point about, "it does not say, when cosmic cultures WILL meet, which while acknowledging the possibility (or certainty) projects the event into the boundless future. Rather it chooses 'meet' which can mean the present – a suggestion ripe with possibilities and evaluations that allows the discussion of UFOs and abduction reports, which are indeed the subject of a good many sessions of this Conference."

Sitchin told us,

"Well, having devoted a lifetime to the subject, I can tell you this: The cosmic cultures HAVE ALREADY MET. The records of the PAST are replete with information and evidence on the subject; and **IF WE WANT TO KNOW WHAT THE FUTURE HOLDS – WE MUST STUDY THE PAST.**

"These records of the past have been available to us all along; yet they have been ignored because they had been labeled 'myths.' My own questioning of what had really happened in antiquity began with wondering, as a schoolboy, why the 'Sons of ELOHIM' who had intermarried with the 'daughters of the Adam' in the days before the Deluge (Genesis 6) were called in the Bible NEFILIM, and why we have been told that the term meant Giants' when in fact it meant, 'Those who have descended,' who have come down to Earth from the Heavens. Biblical scholars explain that this enigmatic segment in the Bible echoes 'pagan myths.'

"Years of study, and a review of a century of archaeological discoveries, trace the origin of all those 'myths' – of the Romans and Greeks, the Egyptians and Babylonians, Assyrians and Hittites and all others, to the SUMERIANS. Their civilization had appeared suddenly and as if out of nowhere circa 3800 B.C., in the great plain between the Euphrates and Tigris rivers (today's Iraq). Their civilization produced virtually all the 'firsts' that we deem essential to a high and modern civilization: the wheel and the kiln, intensive agriculture and metallurgy, navigation and commerce; writing, arts, music, dance, laws and courts of law, kingship and royal courts; mathematics, chemistry, medicine, astronomy; high rise buildings, temples; a priesthood, religion.

"Most amazing of all was their knowledge of astronomy. They knew of, described and even depicted on 'star maps' all the planets we know of today (even those discovered by us only in the past century or two); and what is pertinent to our subject today, that 6,000 years-old knowledge included insistence that there is one more planet in our Solar System, which they called Nibiru – a planet whose great elliptical orbit brings it to our vicinity once every 3,600 Earth-years, when it passes between Mars and Jupiter.

"It was from that planet, the Sumerians asserted in all their records, that intelligent beings had come to Earth some 450,000 years ago. It was not a one-time visit, a crash landing, an accident. The visitors kept coming and going every time their planet was in our part of the Heavens, every 3,600 years.

"The Sumerians called them ANUNNAKI, literally meaning, 'Those who from Heaven to Earth came' – the NEFILIM of the Bible. According to numerous Sumerian texts it was the Anunnaki who jumped the gun on evolution and mixing their genes with those of a female hominid, created 'Adam' – Homo sapiens, you and me. Later, the Anunnaki gave mankind civilization and scientific knowledge, 'All that we know as taught to us by the Anunnaki,' the Sumerians stated. Indeed, only this can explain how Sumerian knowledge had attained its heights without our modern instruments. That is why my book *Genesis Revisited* carries the subtitle, 'is modern science catching up with ancient knowledge?'

"An examination of human progress since the Deluge, some 13,000 years ago, reveals that the marked advanced, the transitions from Paleolithic to Mesolithic to Neolithic to the Sumerian civilization, are spaced every 3,600 years. Does this suggest that each time Nibiru neared Earth, each time the comings-and-goings had taken place, and by implication also the next time – we will be given one more dose of knowledge, one more upgrade of civilization?

"Here is where my admonition, that the Past is the Future, must be studied very closely. For, as the Sumerian and Biblical records point out, the Anunnaki judged the product of their genetic engineering sternly. Indeed, on one occasion, they sought the total destruction of the Children of Adam through a global catastrophe, an avalanche of water – the Deluge of Biblical renown.

"Will the Anunnaki come again on Nibiru's next approach, and what will the results of the next encounter be?

"The loss in 1989 of the Soviet spacecraft Phobos-2 at Mars, the circumstances of which were first revealed in my book *Genesis Revisited*, led me to the conclusion that an ancient space base on Mars has been reactivated. What happened in 1989 may well explain the loss of the U.S. spacecraft Mars Observer in 1993. Both incidents bring to mind the tale of the Tower of Babel, when mankind was building a launch tower and the Lord came down and said to unnamed colleagues: We cannot let mankind do that. Such a reactivation of the space base on Mars could well be the explanation for the UFO phenomenon.

"I have been asked what Mankind can do to assure a benevolent outcome of the next encounter with that other 'cosmic culture,' that of the Anunnaki. Indeed, some have wondered, is it at all up to us, in view of the obviously superior technology and perhaps even biology of the Anunnaki?

"In answer, I can only invoke the lessons of the Deluge. According to the Biblical version, the decision to wipe Mankind off the face of the Earth was made because 'the wickedness of Man was great on Earth.' The much earlier Mesopotamian version gave as a reason the excessive proliferation of a tumultuous humanity. These two causes – the evil non-Earth and the population explosion – are the two most conspicuous aspects of the twentieth century. The past teaches that these two phenomena could well be again the criteria by which we will be judged; and in both respects, mankind controls its own direction and destiny.

"The realization that we are not alone, in our own Solar System, and the coming re-encounter with another, more advanced civilization, should be at the top of humanity's agenda. In this respect, a Conference such as this is an important step in the right direction."

Sitchin concluded, "to know the future, study the past." ³

Sitchin was also one of the three speakers on the Future Panel, which took place at the end of the Conference. Day one was the Fear Panel, Day two was the Hope Panel, and Day three, to conclude the conference, was the Future Panel. The following questions were discussed between Sitchin and Jerome Glenn, the Executive Director of the American Council/United Nations University. This organization provides a point of contact between the United Nations University (the UN's think tank) and the United States. Glenn is coauthor of *Star Trek: The Endless Migration*, and author of *Future Mind*. He is also coordinator of the Millennium Project (on Global Future Studies) for the United Nations University's World Institute for Developing Economies. During his talk, he questioned if the United Nations should be the point of contact for communications between Extraterrestrials and Earth.

I remember having multiple discussions with conference goers and Col. Donald Ware, retired Air Force Fighter Pilot, Nuclear Engineer who devotes his retirement to UFO Investigation, who told me that the United Nations was going to go through a major transformation in the future, and when it does, the ETs will have arrived. They will turn the United Nations into the Galactic Federation of Planets, and the United Nations will be the focal point of communications on Earth. So, Glenn was not too far off in his futuristic projections. Col. Ware and I became good friends, as he contributed to my Magazine/Newsletter, "Astro-Logical Liaisons" which later became *Celestial Liaisons* through the late 1990s and early millennium. Col. Ware lectures and wrote on UFOs and Human-Alien Interactions, the long-term government covert educational program on UFOs, telepathic communications and UFO Phenomena, paths to world government as he relates to his own close encounters by witnessing eight craft that were not of this world. He has studied UFOs since he saw seven alien vehicles over Washington, D.C. on July 26, 1952. I learned a lot from the Colonel, a kind gentleman, who understands the connection between the scientific and the spiritual, a rare quality in anyone these days.

Back in the 1990s when I was doing interviews as the Cosmic Reporter, Col. Ware told me there will come a day when the United Nations will be transformed into an outpost for the Galactic Federation of Planets, when the world experiences Disclosure of ETs and Aliens. While it may sound like the stuff of science fiction, as anyone who is familiar with my work knows, *the Truth is Stranger than Fiction!* Yet, December 16, 2017, was the U.S. Day of Disclosure when the Pentagon released a video of a UFO that it was studying. "Disclosure has begun, and there's a much more to be revealed."

Bible Prophecy appears to be unfolding. Zechariah 12:3, says, "On that day, when all the nations of the Earth are gathered against her, I will make Jerusalem an immovable rock for all the nations. All who try to move it will injure themselves."

In other words, the UN will at some point become impoverished, leaving it vulnerable to be taken over by *outside* forces. Interestingly enough, in 2017, President Donald Trump decided to withhold U.S. contributions to the United Nations. What kind of political teeth would an impoverished UN have? Totally open for takeover by outside forces. Could this be a set up to transform the UN in the future? Only time will tell, but in 2017, the UN lost U.S. funding over their attacks to the U.S. Foreign Policy over Jerusalem.

In 2017, The UN condemned the U.S. on recognizing Jerusalem as the capital of Israel. UN Ambassador Nikki Haley told the United Nations, "we will remember this!"

The United Nations General Assembly on December 21, 2017, approved a resolution 128 to 9 calling President Trump's recognition of Jerusalem as the Capital of Israel coupled with his plans to move the U.S. embassy there "null and void" — however, it does nothing to change U.S. plans. On May 14, 2018, the U.S. Embassy officially moved from Tel Aviv to Jerusalem.

The UN vote came after UN Ambassador Haley warned member nations she would be "taking names" of those who vote against the United States. Here are some key points from Haley's speech in response to the UN vote:

> "But we'll be honest with you. When we make generous contributions to the UN, we also have a legitimate expectation that our good will is recognized and respected. When a nation is singled out for attack in this organization, that nation is disrespected. What is more, that nation is asked to pay for the 'privilege' of being disrespected.
>
> "In the case of the United States, we are asked to pay more than anyone else for that dubious privilege. Unlike in some UN member countries, the United States government is answerable to its people. As such, we have an obligation to acknowledge when our political and financial capital is being poorly spent.
>
> "We have an obligation to demand more for our investment. And if our investment fails, we have an obligation to spend our resources in more productive ways. Those are the thoughts that come to mind when we consider the resolution before us today.
>
> "Instead, there is a larger point to make. The United States will remember this day in which it was singled out for attack

in the General Assembly for the very act of exercising our right as a sovereign nation. We will remember it when we are called upon to once again make the world's largest contribution to the United Nations. And we will remember it when so many countries come calling on us, as they so often do, to pay even more and to use our influence for their benefit.

"America will put our embassy in Jerusalem. That is what the American people want us to do, and it is the right thing to do. No vote in the United Nations will make any difference on that.

"This vote," she said, "will change the way Americans look at the UN."

The Israeli ambassador followed Haley at the podium and called out the hypocrisy of the UN. He asked why there was no resolution condemning Arab nations for calling for "days of rage" and violence following President Trump's announcement.

"Those who support today's resolutions are like puppets pulled by the puppet strings of your Palestinian puppet masters," he said calling them "blind to the light" and "unaware of the manipulation surrounding you."

JERUSALEM AND THE RETURN OF THE ELOHIM

Sometimes Exopolitics and American policies on foreign affairs overlap. With the shocking words of the UN in reaction to the U.S. decision to move its Embassy to Jerusalem, we are certainly witnessing a turning point that reveals the Bible prophecy true. This was an important revelation of the UN's true colors in that they are prepared to lead the Nations of the Earth against Jerusalem and its rightful place in world affairs. Our transition into the next Age will be dependent on what culminates over Jerusalem because this is where the Lord returns to end the ancient war against Him.

Why does this eclipse the rest of the news? Because the return of the Elohim depends on the arrival of Nibiru, which then launches the return of Christ, who according to the quintessential End Time Prophecy then sets up His Heavenly Kingdom on Earth. What the Book of Revelation describes as (what I'm calling a Phantasmagorical Mothership) called the New Jerusalem, The Heavenly City, which descends from the Heavens and overlays the old-scorched Jerusalem, opening up into twelve gates i.e., portals, each named after the twelve 'blessed' Tribes of Israel. (See, Revelation 7:5-8) This immense craft lands over Jerusalem, Israel, as the center of the Lord's Throne, and begins His rule over all the Nations of the world. Revelation 21:12 describes these twelve gates of heaven as being made of pearls.

CHAPTER 4: CONNECTING THE DOTS TO NIBIRU

Each individual gate is made of one single enormous pearl, which I am saying will act as dimensional portals.

According to Revelation 21:12, the Heavenly City (Mothership) has 12 Gates, 12 Angels, 12 Names. "The New Jerusalem "had twelve gates, and at the gates twelve angels, and names inscribed, which are the names of the twelve tribes of the sons of Israel." Each gate will represent the death and resurrection of Jesus Christ, as we enter into New Jerusalem. Angels are at these gates. Angels are "ministering spirits, sent forth for service for the sake of those who are to inherit salvation" (Hebrews 1:14). Ultimately, our inheritance is the New Jerusalem, the Angels (Extradimensional Extraterrestrials) who serve as the ministering spirits, who watch over the gates (portals). The New Jerusalem comes out of the Heavens and lands on Earth all the way at the very end of when all End Times Prophesies are fulfilled, i.e., Tribulations, Raptures, Armageddon wars. It's very arrival marks the beginning of the Millennial Reign of Christ on Earth.

However, between now and then, we can conclude that the ultimate transformation of the United Nations may, in fact, go through several changes before the Lord returns and that it may be called at that time, the Galactic Federation of Planets after Disclosure becomes global, which will lead up to the final battle which will be the real Star Wars over the ancient portal of Jerusalem.

Now, back to Zecharia Sitchin's remarks at the *When Cosmic Cultures Meet Conference*, at the final conclusion in the Future Panel. The question by the moderator was: "How do you explain to world leaders as well as average people what happens if, in fact, a contact with ET comes?"

> Sitchin responded, "The question really has two parts. They do not deserve the same answer. One is how to explain this whole business to world leaders and to the average people. The average people – we covered that in a previous question, and that is as best as possible – education, movies, television, whatever means are used these days to convey information and educate the public at large. As far as the world leaders are concerned, that, of course, raises the first issue of who are the world leaders? There could be some debate or disagreement on that, not just between the remaining major powers. I once heard a committee on the subject say that those visitors to Earth who finally decided to show themselves and make official contact, etc., first of all they debate where to land, in Moscow, in Washington, the UN, the Vatican, and they land in Washington and the being comes out and says, 'Take me to your leader' and they take him to Hillary. So, there could be some disagreement who the world leaders are today. Though I see many, many faults in the UN organization that need a good revamping, perhaps this is the only place that we can look to once it is put in

shape. The world leaders already know the truth. They know who is out there, whether contact the way we are discussing it here has been made or not. I am talking these days; I am not talking in the past. The planet is on its way back. Its robotic emissaries are here or on the way back. This is well-known. In my opinion all evidence points to the fact that their waystation or space-station on Mars has been reactivated.

"I would draw the attention of all here, and if you're not familiar with the details, it's in one of my books — the so-called Phobos Incident – in 1989, when two Soviet spacecraft were sent to Mars. One of them was lost on the way, they said, because somebody leaned on the console and pushed the wrong button in Moscow. I have never been at the launch site, but I saw it on television and movies with the hundreds of people who must do something so the next one could do something. So how could one just lean on a button and lose a spacecraft? But that is what was claimed – that Phobos One was lost. But Phobos Two did make it to Mars, orbited Mars for about a month in March of that year, sent back incredible photographs of features on Mars that appeared both in visual light, meaning regular camera film, and infrared. Among them was a shadow of an elliptical object, according to all those who were willing to make comments at the time on the subject, and some of them appear on camera in a video documentary based on Genesis Revisited. The video title is, Are We Alone? They said it is the shadow of an object that should not be there.

"After orbiting Mars for 26 days and sending some of those unbelievable photographs, the spacecraft was redirected to its main objective. Mars has two small Moons or Moonlets, the larger of them is called Phobos. There are many indications that it is an artificial object, not a natural object, hollow inside. One of the first things that Phobos Two, after leaving Mars orbit and examining the Phobos Moonlet, was to fly in tandem with the Moonlet, or this artificial object about 35 meters above it, and bombard it with laser beams. As it neared Phobos Two, some object that looked like a missile, and the Soviets, finally, after two years of not releasing the photograph, released it, hurdling toward the spacecraft, at which point it went out of commission, fell silent, and disappeared. Obviously, it was shut down.

"When we sent in the Mars Observer, I was asked by some of my friends, 'Well, what will it find?' I said – and that was in July – 'first let us see if it gets there.' A month later, it did not get there. Again, something happened as if somebody there says, 'I don't want you to take a look at me.' 'You can fly to Venus, but not Mars.' I think the world leaders are aware of what is going on. I think that some of the changes in the geopolitical situation, the dissolution of the Soviet Empire, the end of the Cold War, has to do with their knowing what is happening. We do not have to explain it to them. Whether they are making the right decisions based on their information or not, this I don't know."

On science, Sitchin answered, "I do not view science as an obstacle to understanding, discover, and preparing humanity for what is undoubtedly about to happen. I think, on the other hand, science is the key and means and the clue to understanding and achieving what the question puts the finger on. The only problem with science is that there is so much establishment, that the dogma can be challenged or changed only at the risk of the innovator. I think that while science is the key to the future, scientists should have more open minds."

As I started off this Book Series, in Book One of *Who's Who in the Cosmic Zoo?*, with my initial dissertation on 'Science is the Mind of God,' proving that when the Age of Reason emerged, all the greatest scientific minds of our history, all trusted God and were all men of faith. In 2019, one of our leading physicists, Michio Kaku, Ph.D., said that "science is evidence of Divine Design." Kaku agrees with Albert Einstein that God is like a mathematician. Considering that physics uses the language of mathematics to reach conclusions, this was a pretty big paradigm shift for someone who started out as an atheist.

Science and religion do not always see eye-to-eye, but Kaku believes otherwise. Most top physicists believe in a God because of how the universe is designed. He says the universe is one of order, beauty, elegance, and simplicity. "Believe me, everything that we call chance today won't make sense anymore. To me, it is clear that we exist in a plan which is governed by rules that were created, shaped by a universal intelligence and not by chance." [4]

Kaku explained the universe did not have to be this way. It could have been ugly and chaotic. In short, the order we see in the universe is evidence of a Creator. This physicist seems to prove the passage from Psalms 90:2, "Before the mountains were brought forth, or ever you had formed the Earth and the world, from everlasting to everlasting you are God."

DISCERNING DISCLOSURE

Disclosure of Extraterrestrial and alien life on Earth will literally jumpstart our planet into the final chapter of Bible prophecy, and we will finally understand the truth of our history, and why it was kept secret from us for millennia.

Sitchin taught me, through his legacy of the knowledge that unravels our past: in order to know the future, we must study the past. We must discern the history and understand *who is who* in the cast of characters of so-called gods, who most people accept now as extraterrestrials. Now, some of them may have been more powerful than others, and yes, as I discerned in Book One of *Who's Who in the Cosmic Zoo?* they also seem to fall within a hierarchy.

Sitchin was asked "What's the most important thing we can do to get from here to there in the future, where we are already working with extraterrestrials, and have since adjusted our institutions to teach a new paradigm?"

He responded, "from an historical or prehistorical vantage point, look at the lessons of the past and for the present and future. On several occasions, as I mentioned, each time the contact was close, they gave us more knowledge, more civilization, more ways to improve ourselves, to heal ourselves. On at least one occasion, which is recorded not only in the Sumerian and Babylonian texts but also in the Bible, the great flood, or deluge, the commander in chief on Earth, Enlil, sought to wipe mankind off the face of the Earth. There are two reasons given for that decision, which he tried to force on his colleagues. In the Bible, the one most familiar to all of us, the reason given is, and I quote, "because the wickedness of men was great on Earth." This is the reason the Bible records the decision to do away with mankind. In the Sumerian texts, which are much more extensive and are the origin of the tale of the deluge, the reason given is that Enlil could not stand the noise of mankind. He complained to his colleagues that mankind increased in numbers too great, filling the Earth, making a tumult, and enough is enough, feeling: 'I've had it.' Now, if you try to assess the prospect of the next close encounter, not just the signals and the preliminaries for it, and what might happen – first of all, the question is: *"Who is now in charge?"*

"On Nibiru, and thus on the decisions affecting us, ... there was a constant divergence of opinion, a conflict when Enlil who was strict, who wanted things to be right and exact, and according to the rules and regulations. His half-brother, and sometimes adversary, Enki, the first one to land on Earth, the Chief scientist, the one who created us through this genetic engineering, considered mankind as his children, as his offspring. He was responsible for mankind, and thus he connived, and the tale is told in great detail in Sumerian text – he connived with his faithful servant called Noah in the Bible to save him and family and, thus, the seed of mankind and other life on Earth and preserve it even during the deluge.

"Now who is now in charge? Enlil or his line of descendants? There was a conflict about the succession, or that of Enki, whose successor on Earth was Marduk, the Babylonian national god. Many adverse occurrences happened after the rise to supremacy of Marduk in Babylon. So, who is in charge and thus what kind of attitude would be taken, I do not know? But if the criterion would be one of the two, or both – one is the increasing wickedness of mankind, and the other is the proliferation, the getting out of hand, the number of humans on Earth, I think we are guilty of both of those deficiencies.

"Certainly, after WWII and the Holocaust, what ensured afterwards, and even in Chechnya and so on, you can compare human wickedness and not compare it, but certainly mankind has illustrated that it's wicked, or capable of wickedness. This probably did not go unnoticed. The other is the proliferation, the increase in population that is really the cause of so much poverty, suffering, and disease in so many parts of the world. Few efforts have been made to contain or control the growth of mankind. Some of those under the auspices of the UN have been thwarted by other powerful entities – in human affairs – religious and others. So, it is quite possible we will find one thing on both counts, and this part – this part is entirely up to us. I mean, none of those who are in Nibiru is forcing mankind to get out of hand in numbers, or to slaughter each other. Whether this is up to governments, or the change is to begin with individuals, each one should decide. If we are to do anything in preparation for the encounter, meeting no doubt with superior beings, let's recall that these are the two areas: human wickedness and human uncontrolled proliferation, and it's up to us to prepare for a more benevolent result, or a less benevolent result of that encounter."

THE EXOPOLITICS OF CONTACT WITH ET

Another question posed and discussed at the *When Cosmic Cultures Meet Conference*, during the Future panel, Sitchin was asked about moving forward with, "Who are we and how do we deal with all this (Disclosure)? Do you think our ability to communicate with and travel to cosmic cultures in the future depends upon a change of consciousness? Do we need other new values to make contact and do aliens have values that we do not understand? Many of us in various walks of life on this planet base our lives on service to others; yet there are many on the same planet who are self-focused, and fear based. How do you convey dualities that are integral to the nature of our planet? The underlying point in the question is, is there a new set of values, is there a new set of ideas that need to attend this transition or this kind of epic event in time?"

Sitchin's response was: "As a matter of fact, when I arrived and spoke briefly to Scott Jones (the President and Founder of the Human Potential Foundation), I said the subject, *When Cosmic Cultures Meet*, implies one party to the meeting is us, people, humanity, Earthlings. Now two questions; one – who are we going to meet, and here I want to comment that for the first time I disagree with you John (Petersen), because you use the term *aliens*, and this term *aliens* immediately implies something different, something to be afraid of, it doesn't look like us, you know – watch ET or other movies, ... it is this, it is that, it has other emotions, etc.

"The text, again I keep returning to my sources, which are the sources for the Bible and for the so-called myths and knowledge of the ancient peoples, that knowledge, or depiction of what happened, is summed up in the Biblical verse where Elohim, which is translated God, with a capital G as a singular, but literally in Hebrew it's a plural term meaning the Deities. Elohim said to unnamed colleagues who must have been there according to the Sumerians, even their names are given, said, 'Let us make the Adam in our image and after our likeness.' In our image means physically to look like them, and after our likeness means internally, emotionally, intelligently, etc. Therefore, when people ask me frequently 'How do they look? Do they look like us?' I say No, we look like them. So, they are not aliens. In some respects, they are not just our creators; we are their offspring. Especially in the story of the deluge where it was proceeded by the intermarriage between some of their young ones; the Bible calls them Nephilim, those who have come down, with the daughters of men and they had children. So, mankind, homo sapiens, are their offspring, in more ways than one.

From the initial mixing of the genes to the intermarriage, through what mostly the Egyptians call the demigods, who were offspring of some of the Annunaki that were mostly females, but not always. Some of their females also had human mates. So, they are not exactly 'aliens' but humans.

"Now when we then discuss *When Cosmic Cultures Meet*, whom are we discussing at this conference, and we're all from the same side. Now to whom are we talking? I think I am the only one who is capable of being their spokesman. Not because I have met, but because I have read the records of, those who have met them and listened to them and wrote down the tales, the records of those ancient events. If you compare even biologically, humans, homo sapiens, with earlier hominids, with our apelike cousins, you find that first the ability to speak makes us different due to the structure of the larynx, etc. I am not talking about the brain, which, of course, is different. The ability to speak, the ability to have a language. According to the Sumerians, they are the ones who taught us music. So, they like music. They wanted the people to play for them, to dance for them. They are the ones who taught mankind art, plastic art, painting, other works of art. So, when the question arises, 'Whom are we going to meet?' we are going to meet ourselves. But at a much earlier and more advanced stage."

I was immensely impressed and influenced by Sitchin, who inspired me to put my *Who's Who in the Cosmic Zoo* Book Series together. I am convinced that those who oppose him, based on linguistics, especially with respect to the above Creation scripture about the Elohim in Genesis 1:26, don't understand the specificity of the Hebrew language, which is based on singular, plural, feminine and masculine participles. On a Hebrew linguistic level, there is no doubt that Genesis 1:26 is written in its plurality, not just the word, *Elohim,* which is plural in and of itself, but each of the words are using pluralities, let 'us' make Adam, in 'our' image, according to 'our' likeness, in the first part of Genesis 1:26, all written in plurality, specifying a group of gods speaking, who are creating people called 'Adam,' which become humankind. But wait, there is more, in the second part of the verse, the words transition into the plurality of the Adam, mankind, when they immediately refer to Adam as 'they.' "And let 'them' rule, over the fish of the seas, and over the birds of the air, over the livestock and over all the Earth, and over every creeping thing that creeps on the Earth."

They essentially created a new race of humans to steward the planet and "lord" over the animals and wildlife. It was not just one man, or one woman, which is why in Book One, I refer to them as the Evadamic Race, which would be humankind that Genesis 1:26 created on Earth.

I understand the cognitive dissonance this core truth causes in Funda-mentalists, who have been reading the Bible all their lives in English and totally miss out on all the nuances, details that can only be discerned in the Bible's original language, Hebrew. This is what caused me to write, *Who is God?* Book Two of *Who's Who in the Cosmic Zoo?* series, in which I attempt to discern through linguistics as well as spirit, who all the major gods are, and why they are different from each other. In 540pp, I outline *Who* the cast of characters of gods are in the Bible. Those who read the English Bible, and other languages that are translated from English, are not being told the true story that was recorded in the Bible. I prove through various Bible stories that it was a group of gods, and satan was among them, the adversary who tested humankind. I also prove and distinguish the God of gods who rules over all of them. It is quite clear if you follow the rules of the Hebrew language.

I absolutely disagree with Mike Heiser on this issue, which just proves that having a Ph.D. in Hebrew from the University of Wisconsin does not buy you understanding of how Hebrew linguistics work from a Jewish perspective. So, for the record, I agree with Sitchin on the correct translation of Elohim is plural for gods. Sitchin is not wrong! Yes, there are a few instances where the word *Elohim* is used to mean God, like the word *sheep* can be used for a single sheep or a group of sheep, but the majority of the times the word *Elohim* shows up in the Jewish Bible, it is referring to a group of gods in its plurality. Genesis 1:26 is one of them, as the entire sentence is written in its plurality, so nothing is taken out of context there.

I distinguished in Book Two, *Who is God?* that the LORD, whose name is Yahuah, a name mentioned 7,000 times throughout the Old Testament, is completely different than the Elohim, and I proved He has the power to punish the Elohim. This leads us to the logical conclusion that the Elohim is not "the God or Lord", but a group of gods, or deities, and the Book of Job lists them as a type of council of gods that meet in the Courts of Heaven and petition before the God of gods. This discerns a hierarchy of gods, and reveals there is one more powerful, who is also called *The Almighty*, who is above all the Elohim.

Hebrew is a poetic language, and often throughout the Torah, truncations of the names of the LORD are given for poetic reasons, as well as a reminder to use the title Adonai, which means LORD, instead of actually saying His full name out loud. They feared misuse of His Name, to the point of superstition. So, if you have not had a chance to read *Who is God?* I recommend you refer to it to get all these important points.[5]

Sitchin was a brilliant yet humble man. I had the privilege of interviewing him in 1994. He was definitely one of my inspirations for my writings, research, and books. If it were not for his translations of the Sumerian Cuneiform stone tablets, we probably would not be calling Nibiru by its Sumerian name – Nibiru – although it does go by many other names from around the world. It was described and depicted by the ancients as the red/gold planet with wings.

CHAPTER 4: CONNECTING THE DOTS TO NIBIRU

Because Sitchin was a Jewish scholar who did not believe that Jesus (Yeshua) was the Messiah, he was blinded to some historical points. There is one major point I disagree with Sitchin on, and that is that the Anunnaki created humankind. This is false! I began my argument against this falsehood in my Concluding Chapter[6] of Book One of *Who's Who in the Cosmic Zoo?* which I completed in greater detail in *COVENANTS* - Book Four. I prove that humans are pawns in the middle of a genetic war between various ET types. The Anunnaki genetically manipulated and downgraded the original human blueprint as created by the Elohim, the group or council of gods who serve the Lord in the Courts of Heaven.

We were created in their image and likeness according to the Genesis account, which is written about a plurality making a plurality. This suggests that the original humans were created with 12 strands of DNA, just like the Elohim, but at the Fall of Mankind a curse was pronounced, and this group of Anunnaki were allowed to carry out Mankind's fall based on the spiritual legal ground of that curse by genetically downgrading humans to be used as slaves. They disabled 10 strands just to give humans enough intelligence to follow orders and perform tasks. There were several human prototypes, which is why we have Australopithecus, Homo Habilis, Homo Erectus, Homo Neanderthalis, and finally homo sapiens. There were multiple genetic experiments. This is why today's geneticists call the 10 disabled strands junk DNA. They see it, but do not understand why it is there.

Jesus promised to restore humankind back into our original Glory Bodies, according to the original human blueprint, in the image and likeness of the Elohim gods. Today's humans still fall short of that image and likeness, genetically speaking. Jesus is the Second Adam; He came to restore what the first Adam lost.

I also find it coincidental that Nibiru sounds like Hebrew, and the planet Nabu that Senator/Queen Amidala represented in Star Wars is a derivative of it as well. Later in this book, I am going to connect many dots to Star Wars Jediism and Ancient Judaism, whether it be coincidental or by design, you can decide. See Chapter: *The Real Star Wars*.

It is written in the Jewish Torah and Zohar that when a red star appears with seven planets, it heralds the arrival of the Messiah. Jesus talked about the signs in the Heavens in Matthew 24:30.

The Trappist-1 system discovered in 2017 and disclosed by NASA as having seven Earth-sized planets is no coincidence and matches the Zohar text along with the Sumerian Cuneiform Tablets.

> "Immediately after the tribulation of those days the Sun will be darkened, and the Moon will not give its light; the stars will fall from Heaven, and the powers of the Heavens will be shaken. Then the sign of the Son of Man will appear in Heaven, and then all the tribes of the Earth will mourn, and they will see the Son of Man coming on the clouds of Heaven with power and great glory." (Matthew 24:29-30)

These are not regular, normal stars and weather clouds! This is Heavenly language. As I proved in great detail in *Who is God?* Book Two, the Clouds of Heaven[5] are, in fact, starships that glow with the Glory Cloud of the Lord, as well as using Cloud Cloaking as a cover.

Planet Nine is believed to exist somewhere in the void of space between the Kuiper Belt and the Oort Cloud, billions of miles from Earth. Professor Michael Brown, a professor of astronomy at the California Institute of Technology (Caltech), originally proposed the Planet Nine theory in 2016. With the astronomer colleague Konstantin Batygin, they have since dedicated their lives to pinpointing the mystery planets' location. Brown is now looking to prove the planet exists by focusing the Subaru Telescope Mauna Kea Observatory on Hawaii at the night skies.[7]

Notes:
1. Skywatch Media News, YouTube Channel, *Nibiru-Planet X System & It's impact on our Solar System*.
2. R. S. Harrington, *The Location of Planet X*, The Astronomical Journal, Volume 96, Number 4, U.S. Naval Observatory, Washington, DC, October 1988.
3. *When Cosmic Cultures Meet*, An International Forum presented by the Human Potential Foundation, held in Washington, D.C., May 27-29, 1995, *The Proceedings*, Zecharia Sitchin, p. 163-166, *The Past Holds the Key to the Future*, The Human Potential Foundation, 1996.
4. Dr. Michael Layne, *Renowned Physicist Says, 'There Is A God.'*, https://godtv.com/renowned-physicist-says-there-is-a-god/, February 28, 2019.
5. Ella LeBain, *Who is God?* Book Two of *Who's Who in the Cosmic Zoo? A Guide to ETs, Aliens, Gods and Angels*, Skypath Books 2015.
6. Ella LeBain, *Who's Who in the Cosmic Zoo?* Book One, Trafford, 2012.
7. Sebastian Kettley, Planet Nine SHOCK: Telescopes in Hawaii are hunting down mystery planet larger than Earth, https://www.express.co.uk/news/science/1109962/Planet-nine-proof-subaru-telescope-hawaii-planet-9-discovery-planet-x-michael-brown/amp, April 8, 2019.

CHAPTER FIVE

THE NEW ASTRONOMY

> "The LORD has established his Throne in the Heavens,
> and His Kingdom rules over all."
> Psalms 103:19 ESV

This Bible Prophecy in both Old and New Testaments states that the Lord, the Maker of Heaven and Earth, is going to create a new heaven and a new Earth (Isaiah 65:17; Revelation 21:1). Our solar system regularly collides with a neighboring solar system which intersect and interlope every 3,600 -20,000 years. It would be logical to conclude that this is why the Lord intends to recreate the heavens. There are always cataclysms and Earth changes that occur when these two systems pass each other, and often some of their objects collide.

The new heavens will be the subject of our new astronomy which we are seeing now, with the presence of seven exoplanets along with our Suns twin, Nemesis. This brown dwarf has been called the *Death Star* by the ancients because in perihelion with our Sun, it not only drains the solar energy to itself, but brings its exoplanets too close for comfort creating cataclysm, climate change, and Earth changes, as our historical records prove In addition, Immanuel Velikovsky, author of *Worlds in Collision*, postulated that the passing of these exoplanets through our solar system caused the destruction of a planet, the remains of which can be found in the asteroid belt.

Every so often, this comet-like planet flies through our solar system and literally wreaks havoc by upsetting the orbit of every planet, including our own. It has turned one planet upside down, sent two spinning retrograde, and can even knock planets off their axis, or put them on a collision course with other objects.

This is why the Lord of the Cosmos has spoken several times through His prophets that He intends to recreate the heavens and the Earth, because the old model just does not work anymore. However, this book is focused on the *spiritual discernment* of the Heavens, and the fact that the Lord often uses the movement of the stars, Sun, Moon, and planets to bring about His Divine Will. The End Times of this present age is certainly no exception.

I go into great detail to compare the Four Living Creatures of Ezekiel to that of Revelation, and I analyzed from original Hebrew that these living creatures are not angels but alien beings that serve the thrones. The thrones in Ezekiel are spaceships. The thrones in Revelation are in Heaven. They both serve the Lord. They are not demons but alien creatures.2 See, Book Three – *Who Are The Angels?*

In Revelation 4:6–8, four living beings are seen in John's vision. These appear as a lion, an ox, a man, and an eagle, similar to Ezekiel but in a different order. They have six wings, whereas Ezekiel's four living creatures are described as only having four wings.

The Earth plane is for testing. We are being tempted, tried, and tested here. This is the ground upon which we must overcome. God does not tempt. Satan is the tempter. He is the god of this world according to scripture. Satan does not live in Heaven, therefore there will not be any tempting or testing there as it is all in the presence of the Lord.

James 1:13, "When tempted, no one should say, God is tempting me. For God cannot be tempted by evil, nor does he tempt anyone."

I go over this very issue packed with corroborating scripture in Book Two, Who Is God? which discerns who is the Lord God versus who is Satan, and how Bible translators confuse what is revealed in original Hebrew. 3

The angels that were tempted chose to follow Satan in his rebellion. God did not tempt the angels. He created us and the angels with free will. One third of heaven's angels fell into temptation, while two thirds remained faithful.

Firstly, let us break down the word, *Heavens*. In Hebrew, the same word for Heaven and skies is *shamayim*. This is relevant because many verses actually relate to beings that come out of the skies, most important example would be the return of Jesus Christ who is coming out of the Heavens (the skies) with the clouds of Heaven which I prove in great detail in Book Two, *Who Is God?*, are starships. 3

In Hebrew, HaShamayim, literally translates to The Heavens which would be the Hebrew name for the title of this book. In Genesis 1:1, "In the beginning the Elohim created the Heavens and the Earth." So, what are the Heavens really made of? Rashi, the great medieval Jewish scholar, suggested HaShamayim could be *fire in waters*, because the word for water is *mayim*, which is inside *hamayim*, meaning *Heavens*. This implies that there is some element of water within the Heavens. Hebrew scholars have taught that the word for sky can be broken into two words, *sham* meaning *there* and *mayim*, meaning, *there is water*. Multiple Bible commentaries deal with the relationship between these two words and two worlds. Meaning that there is water in the sky, and you could say, sky in the water. The reason for the plurality in both words is that the sea and the sky are two halves of the same component making them whole, if we look at the horizon, the midpoint between the sea and the sky. In addition to the fact that both the sky and water are made up of many molecules.

Skypath Books, Publisher of *Who's Who in the Cosmic Zoo?* proclaims: *The Sky is the Limit!* as its motto. In Hebrew this is *HaShamayim Hem Hagvul*. The word *gvul* means *border* and is found throughout the Bible. In English, the word *sky* is singular; in Hebrew, however, the word *shamayim* is plural. This is but another example of how Hebrew language is a language of physics. Physics includes mathematics, which encompasses precision, discernment, and accuracy. So, to move forward in our basic understanding of the Heavens, we must think of it as a plurality. This means there are many aspects and dimensions of the Heavens. Please keep that in mind, as we delve deeper into the multiverse, which includes, according to the Books of Enoch, ten heavens, or ten dimensions, which I will prove to you in Book Six, *Heaven's Gospel*, through connecting Enoch with modern day physics.

Perhaps there is something else to it? What if *hashamayim* could be a *sea of names*, since the word for *name* is *shem*, while the plurality of names is *shemot*, it is root *shem* is found within the word *shamayim*. It's also interesting to note that the first time the word *HaShamayim* appears in the Bible, it coincides with the first mention of the first name for God, which is *Elohim*, both in their Hebrew plurality, in Genesis 1:1. In English Bibles, you will most likely see, "in the beginning, God created the Heavens and the Earth." The only word in English that is in its plurality is the word, 'Heavens.' However, as I proved extensively, in *Who is God?* Book Two2, and in Book One, that the Hebrew word, *Elohim*, is in its plurality, literally translating to: gods. The singular would be El, which means god, and is often at the end of the names of angels, meaning *of God*.

The Hebrew words of Genesis 1:1 is pronounced, *"Beresheet bara Elohim et HaShamayim v'et HaAretz,"* which should be translated as: "In the beginning, the gods created the Heavens and the Earth." Here *Elohim* is considered to be a name of God, despite it being a title, and not an actual name. The significance is deep.

Secondly, understand that the word *angels* mean *messengers*, also a plurality in Hebrew, *Malachim*, so the history of the rebellion of the *Elohim* gods and the fallen angels *Malachim Nephilim*, all makes logical sense. They were actual extraterrestrials that rebelled against the Almighty Creator God. They followed one of heaven's fallen sons, who is Lucifer the Satan, who according to Isaiah 14 challenged the Almighty for power over this universe and the Earth itself.

God created angels and humans with free will. This is clear throughout ALL the ancient scriptures, including the great rejected Jewish texts like the Books of Enoch, Jasper, Jubilees, Giants, War Scrolls, all corroborating this fact.

As I have said throughout this series, testing and tempting does not come from the Lord God who is the Creator and Savior made of love and light. Testing and tempting come from the satans, also a plurality. In Book Two, *Who is God?* I explain this in detail from original Hebrew translations that prove that Abraham was NOT tempted by the Lord Yahuah, but by one of the Elohim, the council of gods, Satan being one among them. [3]

One of the reasons I was led to put this series together was to clarify discernment between who is God the Creator and who is Satan the god of this world and the great counterfeiter and deceiver.

This lies in first understanding of *who is who* in the Bible's cast of characters and how spiritual legal ground works in the Courts of Heaven (an actual place, not the skies) and the spiritual realms. This is understood by focusing on the laws of the Creator God written in His Word.

The Bible scriptures contain the Word of God, but they also contain the words of Satan, evil kings, evil queens and fallen men. I make this distinction because Christian fundamentals think the Bible is God's Word in totality. It contains God's Word, and historical records of others' words which simply are not God's Word.

> "Study to shew thyself approved unto God, a workman that needs not to be ashamed, rightly dividing the word of truth."
> (2 Timothy 2:15)

My contribution to Bible believers is the correct translations of the Hebrew Scriptures that were lost in English translations, specifically discerning and exposing the identities of *Who* the cast of characters are in the Bible and why they get to do what they do, based on the history and/or myths of fallen beings from the Heavens, the cosmos.

The Jewish texts are clear that there are evil stars and death stars and that the heavens are under a curse. This is why Jesus Christ promised in both Old and New Testaments to recreate the Heavens and the Earth in the Age to come.

> "See, I will create new Heavens and a new Earth. The former things will not be remembered, nor will they come to mind."
> (Isaiah 65:17)

> "Then I saw "a new Heaven and a new Earth," for the first Heaven and the first Earth had passed away, and there was no longer any sea."
> (Revelation 21:1)

Because we are in the End Times, these last days are also the Disclosure Era when knowledge is increased. "But you, Daniel, shut up the words and seal the book, until the time of the end. Many shall run to and fro, and knowledge shall increase." (Daniel 12:4)

The Bible prophesies have been 100% accurate thus far. The history of the Bible is also accurate. The difference in doctrines, not so much.

The editing of the Old and New Testaments —deleting twenty-three books quoted in the modern Bible Canon – is problematic to say the least. However, the biggest problems are the gross mistranslations from Hebrew to English, which have created huge misunderstandings due to mistranslations and mistransliterations. I exposed some of the biggest ones in my 2nd book *Who is God?* [3]

The Starfleet of Star Trek is at the point of becoming a reality. Again, what are they hiding from us?! The fact that aliens are real? Are they an enemy?[4] The Truth is Stranger than Fiction!

THE BIRTH OF ASTROLOGY ON EARTH

In ancient times, astrology and astronomy were considered one language and one study. This fact annoys people, but it is the truth, the study of astrology is what gave birth to what we call astronomy today. The word *astrology* means the language (logos) of the stars (astra). The word *astronomy* means the naming or identification of stars.

The Book of Enoch recorded that the fallen angel Baraqiel, Barâqîjâl, Baraqel (Aramaic: ברקאל, Greek: Βαραχιήλ) was the 9th Watcher of the 20 leaders of the 200 fallen angels that are mentioned in an ancient work called the Book of Enoch. The name means *Lightning of God*, which is fitting since it has been said that Baraqiel taught men astrology during the days of Jared or Yered. The fact that this ancient manuscript records that one of the 200 *Bene HaElohim* or fallen angels or rebellious sons of the gods, came down to Earth to teach men astrology is indicative of the fact that astrology is a cosmic language that the Creator uses even now. Why did they teach men this language if it did not first exist already in the Heavens?

Of course, the original meaning was lost after it was plagiarized over and over again, from the Babylonians, then the Chaldeans, the Egyptians to the Greeks and then the Romans, to the pop astrology entertainment horoscopes online today. Chapter 8 of the Book of Enoch assigns certain teachings to specific fallen angels.

> "And Azâzêl taught men to make swords, and knives, and shields, and breastplates, and made known to them the metals and the art of working them, and bracelets, and ornaments, and the use of antimony, and the beautifying of the eyelids, and all kinds of costly stones, and all coloring tinctures. And there arose much godlessness, and they committed fornication, and they were led astray, and became corrupt in all their ways. Semjâzâ taught enchantments, and root-cuttings, Armârôs the resolving of enchantments, Barâqîjâl, (taught) Astrology, Kôkabêl the constellations, Ezêqêêl the knowledge of the clouds, and Sariêl the course of the Moon. And as men perished, they cried, and their cry went up to Heaven . . ."
>
> (Enoch Chapter 8:1-2)

For centuries, traditional or classic astrology consisted of the basic seven planets. Then in 1781, William Herschel discovered Uranus, followed by the discovery of Neptune in 1846, Pluto in 1930, and then there were 10 planets incorporated into astrological lore, catching up to what the ancient Sumerians etched in stone over 5,000 years ago. When humankind became aware of all 10 planets again, it not only made us conscious of our celestial neighborhood, but we began to understand the messages left to us by the ancients.

Over the years, many research studies collected tons of data on how the movement of these known planets affected human events, human relationships, and general social trends. The average accuracy rate of using traditional astrology in forecasting was at best 75%. Why?

A few reasons, one which has to do with the fact that there is an incoming celestial system of planets, Suns and Moons that periodically intersects the boundaries of our solar system. This system renders traditional astrology incomplete.

Secondly, the hand of God, coupled with the element of Free Will in humans, act as wild cards when it comes to forecasting.

However, we now know that seven planets and four Moons intersect our solar system, that regularly eclipse our Sun and Earth. I am calling this the Nemesis-Nibiru System. I also want to emphasize, that there are seven exoplanets attached to this system, and only one of them is considered rogue and has an unusual orbital pattern from the rest, which is what Archeoastronomer Zecharia Sitchin referred to as Nibiru. [3]

When these days overtake us, people generally will respond to these electromagnetic energies by becoming more agitated than usual, impatient, ill-tempered, rude, and offensive to others. Some are challenged by anxiety and depression. And then there is the elephant in the room – Earth changes, climate changes, bizarre and extreme weather, Earthquakes, and volcanic eruptions.

This is the reason I no longer put any credence into astrology, is because of this 'new astronomy,' the presence of our binary twin, Nemesis and its 7 planets and moons that cyclically intersect within our solar system space. For example, the radiation and plasma exchanges between it and our Sun, is causing perturbations in our solar system, especially as it causes solar flares and coronal mass ejections to wave over the earth, raising what we call, the Schumann Resonance, which is earth's natural heartbeat to become elevated, that it's literally affecting all life forms on Earth. The fact that in the past 2 years, there have been several cracks in Earth's magnetosphere, it's causing the Schuman Resonance to become elevated on a regular basis, which trumps any astrological transit that we might have paid attention to in the past.

No Truth is hidden to those with an open mind. Remember, it is perfectly ok to have an open mind – your brains will not fall out. Your mind, however, will expand and your soul will grow closer to its Creator.

In order to take in new information, there is a psychological change called cognitive dissonance. The concept of cognitive dissonance was first introduced by psychologist Leon Festig (1919-1989). In the late 1950s, he, along with other researchers, proved that when people are confronted with challenging, new information, most people seek to preserve their current understanding of the world by rejecting, explaining away, or avoiding the new information or by convincing themselves that no conflict really exists. This occurs when there is a mental conflict

CHAPTER 5: THE NEW ASTRONOMY

that occurs when beliefs or assumptions are contradicted by new information. Cognitive dissonance is, nonetheless, considered an explanation for attitude changes.

In this book, you can expect to see new information on the actual meanings and words in the scriptures that were both misconstrued, deliberately obscured and lost in translations. These new meanings may create cognitive dissonance. [4]

The secret of genius is to carry the spirit of the child into old age,
which means never losing your enthusiasm.
~ Aldous Huxley

Notes:
1. Immanuel Velikovsky, *Worlds in Collision*, Macmillan Publishers, April 3, 1950.
2. Ella LeBain, *Who Are The Angels?* – Book Three of *Who's Who in the Cosmic Zoo? A Guide to ETs, Aliens, Gods or Angels*, Skypath Books, 2016. https://www.whoswhointhecosmiczoo.com
3. Zecharia Sitchin, *The 12th Planet (Book One of the Earth Chronicles)* Bear & Company, NM, January 1, 1977
4. Ella LeBain, *Who is God?* Book Two, Skypath Books, p.7, 2015. https://www.amazon.com/dp/0692911529

CHAPTER SIX

ALL THINGS NIBIRU

Jesus said, "there would be signs in the Heavens prior to His return."
You are seeing them now.

"There will be signs in the Sun and Moon and stars, and on the Earth dismay among the nations bewildered by the roaring of the sea and the surging of the waves. Men will faint from fear and anxiety over what is coming upon the Earth, for the powers of the Heavens will be shaken."
Luke 21:25-26

"The third angel blew his trumpet, and a great star fell from Heaven, blazing like a torch, and it fell on a third of the rivers and on the springs of water. The name of the star is Wormwood. A third of the waters became wormwood, and many people died from the water, because it had been made bitter."
Revelations 8:10-11

SOMETHING IS AFFECTING THE ENTIRE SOLAR SYSTEM
The sign in the Heavens is the Nemesis-Nibiru system intersecting with our solar system. Wormwood comes out of the Nemesis-Nibiru system. Everyone assumes that we should have only one Sun, but why are people all over the world sharing the photos on social media of two Suns in the skies? Many are capturing a red planet. What they are seeing is a large planet that orbits the Brown Dwarf Star known as Nemesis, our Suns binary twin. Nemesis cycles around our Sun, which is why for a cycle we do not see it, because it is on the other side of the Sun. Nemesis has seven planets, one a Dwarf Planet called Sedna. NASA has known about this for years, and even imaged it back in 2007-08. It reveals that the Nemesis system is intersecting and consequently interacting with our system. Sedna has a huge orbit, which intersects within our solar system, from 2007 till now has been literally inside of it.

CHAPTER 6: ALL THINGS NIBIRU

WHAT'S IN A NAME?

Names are used for identification. However, if you want to cover something up from the past, as I have proved extensively in *Who is God?* Book Two, about the Names of God, then changing the name is done to obscure its history and deny its previous existence.[1] This has been done with the planet Nibiru. NASA denies its existence publicly, yet privately they have been tracking and researching it for over 50 years. How do I know this? Well, I had the privilege to interview Zecharia Sitchin in 1994. He told me things about his interactions with NASA, and it's no secret that he was also interviewed by the late Astronomer, Dr. Robert Harrington, who worked for the United States Naval Observatory (USNO), who confirmed the existence and location of what they started calling Planet X or the Tenth Planet.

NASA then confirmed his research, "Unexplained deviations in the orbits of Uranus and Neptune point to a large outer Solar System body of four to eight Earth masses on a highly tilted orbit, beyond 7 billion miles from the Sun." Since 1981, astronomers have researched Planet X beyond Pluto, and confirm something else is out there.

In 2018, there were multiple disclosures about Planet X, or Planet Nine, confirming its existence in and beyond the Kuiper Belt. In 2017, the big Astronomical Disclosure of the Year was the Disclosure of seven Earth-sized planets orbiting around a Dwarf Star that was only 39 light years from Earth, they named it Trappist-1. TRAPPIST-1 is an ultra-cool Dwarf Star, and its 7 planetary system was initially discovered by the Belgian optic robotic telescope with the same name in May 2016 but was officially announced by NASA on February 22, 2017.

The name TRAPPIST-1 comes from the abbreviations from The **TR**ansiting **P**lanets and **P**lanetes**I**mals **S**mall **T**elescope–South (TRAPPIST) which is the name of a Belgian optic robotic telescope that specializes in scanning the sky in search of comets and exoplanets. So, who is to say it wasn't the 7-planet system written about by the ancient Sumerians, only because it was rediscovered by this team of astronomers at the University of Liège in Belgium, it was named after the telescope that located it? The actual Belgium telescope was located at the La Silla Observatory in the Atacama Desert, Chile. The telescope went online in 2010, and after six years of operation, it confirmed the discovery of a red dwarf star with planets in the habitable zone which came to be known as TRAPPIST-1, in May 2016.

Then not too long after, on August 24, 2016, they disclosed planets with water on Proxima Centauri b, which is only 4.24 light years from Earth, found in the constellation Centaurus, making it the closest exoplanet to our solar system, which also coincidentally orbits a Red Dwarf star system.

More recently, the most obvious example of *The Truth being Stranger than Fiction*, that coincidentally aligned with the announcement of Proxima b, the newly discovered and potentially Earth-like planet orbiting Proxima Centauri, our solar systems nearest neighboring star, was first described in a 2013 Science Fiction Novel titled, *Proxima*, by Stephen Baxter. This book detailed human exploration and colonization of a remarkable similar world that he described exactly as the

91

astronomers who discovered Proxima b. Astronomers determined Proxima b was a little more than the planet's estimated mass—perhaps just a third greater than Earth's—and its orbital distance from its star, which is slightly more than seven million kilometers.

Additionally, they suspected it might be tidally locked, meaning the world eternally turns the same face to its star, leaving its trailing hemisphere forever shrouded in night. As small as they are, these observations coincidentally closely matched the description that first appeared in the 2013 Sci-Fi novel *Proxima*. Was Stephen Baxter's imagination prophetic? Or did he know something about the past?[2]

My point is, what if these red dwarf stars were the ones the ancients wrote about in their stone tablets, but due to their language, called it something else? So, if NASA can use plausible deniability by saying Nibiru does not exist, but they know it really does, they just renamed it under another system, like Trappist-1 or Proxima-b. Therefore, it would behoove us who study patterns, cycles and history, to connect these dots, and not readily assume, just because NASA denies Nibiru, doesn't mean it doesn't actually exist, regardless of what name they attach to it.

Nibiru means *the Planet of the Crossing* to the Sumerians and Ancient Akkadians. The word *Nibiru*, is Akkadian. The word *Nibiru* comes from the ancient Akkadian language, meaning crossing, such as in fording a river. Because the ancient Israelites also came out of the same cradle of the Mesopotamian Civilization, the word *Nibiru* is the same in Hebrew. *Nibiru* also implies *transition point*. It was with this meaning that *Nibiru* came to be used in Babylonian astronomy to refer to an equinox. It can also have other meanings, such as a reference to the constellation Libra, which was the location of the autumnal equinox in the first millennium BC. *Nibiru* can also refer to locations in the sky in conjunction with certain stars or planets. *Nibiru* is associated with the Babylonian god Marduk, which in turn is identified with the planet Jupiter. The Planet Nibiru would have a very intense gravity.[3]

The multi-lingual Archeoastronomer, Zecharia Sitchin is well known for his translation of the Sumerian cuneiform stone tablets. He published in *The Earth Chronicles* that the Sumerian Tablets dated back at least 6,000 years literally written in stone which state that the ancient Sumerians knew of a planet beyond Pluto. They counted it as the 12th Planet, because in their system of reckoning, the Sun and Moon were counted along with every other celestial object in the sky. The 12th Planet's Sumerian name was NIBIRU.[4] It meant the Planet of Crossing, because of the way it crossed through the ecliptic and into our solar system in an elongated elliptical orbit.[4]

The name *Nibiru* would have remained obscure in the modern world if not for Zecharia Sitchin (1920–2010), a Jewish Russian-born American author, who grew up in Israel. In his 1976 book, *The 12th Planet* that was followed by six more volumes in his *Earth Chronicles* series, Sitchin claimed that thousands of years ago extraterrestrials from the planet *Nibiru* established the Sumerian civilization. His hypothesized planet *Nibiru* has a very elongated orbit with a period of 3,600 years.

Nibiru spends most of the time far from the Sun, well beyond the orbit of Neptune, but it enters the inner solar system once each orbital period. Collisions with some of Nibiru's satellites and other catastrophes on earlier passages through the inner solar system created havoc on the Earth, Moon, set up the asteroid belt, and increased comets. Sitchin claimed that the history he revealed came from writings from Sumerian and other ancient Mesopotamian civilizations; however, archaeologists and others who study the ancient Near East universally dismiss this. It is easy to connect the dots to elements and hypothesis of Immanuel Velikovsky's catastrophism, *Worlds in Collision* and Erich von Däniken's alien astronaut thesis, *Chariots of the God*, in Sitchin's work. [4]

The Sumerian Tablets show a solar system with a Sun, Moon and ten planets. In Sumerian *Nibiru* means Planet of Crossing, which has a double meaning that can be interpreted in two ways; 1. because it crosses the ecliptic plane as it orbits around our Sun and passes our Earth, 2. It has four Moons, that orbit around it, and looks like a cross in the sky. I have pictures of Nibiru with its four surrounding Moons, taken in 2017 from a telescope, that appears to be in the shape of a cross, with a large red object with a tail, that seems to have a hot golden core/hole. The tail is made of comets and iron oxide; hence it is bright red color. The ancients depicted Nibiru in their stone carvings as a planet with red wings. The wings are made of iron-oxide plasma and debris. It has been called *Destroyer* by the ancient cultures who lived to tell the tale and wrote about its passing Earth. [5]

The Native Americans also had oral traditions of what I am calling the Nemesis-Nibiru system according to the Hopi Prophesies, which describes two celestial bodies, the Blue and Red Kachinas. Kachina is their word for deified ancestral spirit, which can also refer to a star or planet. They also call the bright blue star Sirius the Blue Kachina. The Hopi Prophesies specifically relates to a planetary body passing our Sun which they call the Blue Kachina, that precedes the Red Kachina or Fiery Red Planet. Similarly, the ancient Chinese called it the Red Dragon because of its fiery tail. The ancient Greeks called it Hercolubus, also known as the Destroyer.

So, if the ancients were trying to warn us of this cosmic maniac, perhaps it would behoove us, as a species, to listen and take it seriously, instead of trying to debunk and deny it? Even a broken clock is right twice a day.

NIBIRU IN THE ZOHAR AND JEWISH BIBLE

The Zohar speaks of an ancient red-hot planet surrounded by seven revolving stars that will pass the Earth in the end of days that will appear as *Fire in the Sky*, and then the Moshiach (Messiah) will be revealed.

The Zohar recorded that during the Floods of Noah, God used two planets from the Pleiades to collide with one another and cause the floods on Earth. The Jewish Scriptures reveal that the Lord uses planetary bodies as He rules over all of them to judge the inhabitants of Earth. He has done it in the past, and according to the Bible, He is scheduled to do it again.

The Lord is going to use these exoplanets to wake up sleeping humanity and to turn people back to Him. The fact that the powers that be have been going to great lengths and spending tons of money to keep this information from the general public should wake people up as to why.

But we have arrived at a time when it is becoming harder to hide these exoplanets. The abundance of chemtrails in the sky, along with other types of technology such as Project Blue Beam, created to beam clouds onto the blue sky, to cover up and obscure these exoplanets, are nothing but flawed technologies. Just as your computer systems and cell phones do not perform 100% of the time, neither do these technologies. Otherwise, why are so many people still able to view these exoplanets and their brown dwarf star? People have reported seeing these exoplanets and the second Sun all over the globe.

There is cognitive dissonance, that state of disbelief when new information or knowledge is disclosed, that keeps us from facing the truth. Then there are outright lies, mind control and yes, magic spells being used to keep people in the dark, oppressed by spirits of unbelief and doubt. Meanwhile, you do not need to be a rocket scientist to figure out that something very unusual is happening in our skies and with our Sun.

The Sun controls climate changes, but the passing and presence of these exoplanets hastens pole shifts which are happening right now, like birth pangs. These contractions are occurring in the Earths inner core, causing volcanic eruptions that are happening more frequently lately, as well as Earthquakes and climate changes.

However, the Lord is patient, and wants all to come to Teshuva (Repentance) to Him so that everyone can be saved. The Lord is merciful, and His loving-kindness is everlasting, but those who rebel and reject Him will perish, while those who turn to Him in repentance Him will be saved.

The recurring message of the Prophets throughout the Scriptures is that the righteous will inherit the Earth, but the wicked will perish, be removed from the Earth and thrown into the fires as the Tares. (Matthew 13:30)

The Earth will be transformed. Those whose souls are safely tucked away under the Shadow of the Almighty will be saved and will inherit the New Earth. That is the Gospel as written into the very Stars of Heaven as well as the Scriptures on Earth. The Truth Is Stranger Than Fiction!

NIBIRU – WHAT'S IN A NAME?

NASA has been calling Nibiru – Planet X. However, as I have laid bare the fact that the name Planet X has been used multiple times for multiple orbs, even for Pluto, in order to renumber the planets within our own solar system. Planet X has not become Planet 9. There has been a lot of back and forth in order to obscure the facts about the past. It is certainly no coincidence that NASA announced in 2017 that there are seven Earth-sized planets around a dwarf star close to Earth. This is the Nemesis/Nibiru system which has seven planets and is intertwined with our solar system. [6]

There is a lot of confusion amongst researchers over the difference between Nibiru and her giant sister planets. All the name changes are no help either. Planet X, Planet Nine, Blue Kachina, Red Kachina, Red Dragon, Hercolubus, Vulcan, and how many more? Then in 2016, NASA disclosed their discovery of 7 Earth sized planets orbiting a brown dwarf star, they named it the Trappist 1 system. Many of us have connected the dots to the Nemesis system which is our Sun's Twin Brown Dwarf that has 7 planets, in its orbit, one of which is Nibiru, which is approaching our Sun. Many of us believe announcing the discovery of Trappist 1 was NASA's way of diverting attention from what is actually taking place here. Nemesis is a brown dwarf star that is also known as our Sun's twin, but it is dark. It has been observed to have a total of twelve objects.

One of the first scientists to model the world's climate cycles in depth was Eric Wolf, a researcher here at University of Colorado, Boulder. They are now claiming that the conditions for life do indeed exist on the Trappist-1 System. The ancients have recorded this on the Sumerian cuneiform tablets, the Bible scriptures, extra-biblical scriptures, and recorded from Native American history and prophesies. I for one do not as well as being believe in coincidences. It is all intelligent design. [6]

I hold the position, and prove in my books, that Nibiru was always depicted by the ancients with wings and is red/gold. There are several researchers who are calling the Blue Planet Nibiru which is not accurate according to the Sumerian, Biblical, Greek, and Chinese accounts, which insist it was red. The Chinese called it the Red Dragon because of its red fiery tail. The Greeks called it Hercolubus which was described as a giant red planet that caused Earth changes.

In my 3rd Book, *Who Are the Angels?* Chapter 16: *The Second Coming and Nibiru*, I proved that it was the Blue Planet's eclipse of our Sun that caused three hours of darkness during the Crucifixion of Jesus Christ, according to the biblical account. [7] The Blue Planet is now known as Planet Nine by astronomers. It is believed to be a part of our own system extending out beyond the Kuiper Belt, orbiting around our Sun yet passing between the Earth and Sun. This would explain the 3-hour eclipse which brought darkness upon the land during the Crucifixion of Christ, as the book of Matthew simply says the Sun failed without indicating the cause. "Now from the sixth hour darkness fell upon all the land until the ninth hour" (27:45).

Mark and Luke write, "darkness fell over the *whole land*" (Luke 23:44; Mark 15:33). It was a sudden and great darkness. Unexpectedly the whole scene at Calvary shut down and a death silence covered the land, yet the miraculous darkness appeared to have vanished as quickly three hours later just before Christ died. While scholars insist this wasn't an eclipse of our Sun by the Moon, as no lunar eclipse could last three hours. A planet that is believed to be 5-6 times the size of Earth, that orbits between Earth and the Sun roughly every 2,000 years, could very well have been the celestial object that eclipsed the Sun for those three hours.

When researching the Nemesis-Nibiru System, or the Planet X system, we must always be mindful that it is not just one planet, but a system. Each planet orbits at a different rate and speed. I believe this causes researchers to get confused as to why there are different events that happen every 360 years, every 2,000 years, and the big cataclysmic events, every 3,600 years.

Nibiru comes around our Sun every 3600 years, initiating shifts and cataclysmic changes. Immanuel Velikovsky, best pals with Albert Einstein, author of *Worlds in Collision*, tracked these cataclysms to a 3600-year cycle.[8] His work from the 1950s was done purely through the mathematics of physics, much of which has since been confirmed through the NASA technology and Hubble photos as accurate.

The fact that this binary system intersects and collides with ours is the reason the Lord promised to recreate the Heavens and Earth in Isaiah 65:17; Revelations 21:1. "For behold, I create new Heavens and a new Earth; And the former things will not be remembered or come to mind."

> "The words, "Once more," signify the removal of what can be shaken--that is, created things--so that the unshakable may remain."
>
> (Hebrews 12:27)

The Zohar [the foundational work in the literature of the Jewish Torah & mystical thought] states explicitly that the Messianic process will be accompanied by "several stars appearing." The Zohar goes into great depth, describing how many stars, and which colors they will be.

The Zohar coincides with the Torah verse, Numbers 24:17: "I see him, but not now; I behold him, but not nigh; there shall step forth a star out of Yaakov, and a scepter shall rise out of Yisrael, and shall smite through the corners of Moab, and break down all the sons of Seth."

NIBIRU: RED OR BLUE?

There is a lot of confusion on the internet on whether Nibiru is blue or red. My position is rooted in what the ancients recorded, and is consistent that the Winged Planet, the Messianic Star, was Red to the ancients. Now, the Hopis had recorded it as well that when the Blue Kachina shows up, the Red Kachina will follow. Kachinas was a word for living star.

The confusion may lie with the fact that Nibiru was believed to be blue up until a certain point in history, after which it was referred to as red. It was associated with Sirius or the Dog Star which some say was where Nibiru emerged after being ejected from the Sirius Cluster. I believe the Blue and Red Kachina are part of the same system, and both orbit around the dark star Nemesis, our Sun's brown dwarf twin. It has been called the Red Dragon by the Chinese, because of its ruddy color. The Red Star in the Zohar is self-descriptive and depicted as a red disc with wings by the ancient Sumerians. I think the Blue Kachina is what NASA has renamed Planet Nine, a gigantic blue planet. [7]

In April of 2018, I had a dream. I was looking up to the sky and saw three Moons side by side against a bright blue sky in broad daylight! Woke up, talked to God about it and then went back to sleep. When I woke up again for good, I realized we were on the cusp of the full Moon lunar eclipse, which could relate to a triple whammy with a comet passing, however, I think I was seeing Moons connected with one of our neighboring planets. Perhaps a future event. They were whitish/gray from my perspective.

Then exactly one year later, April 6 of 2019, I woke up out of a dream that was clear as day. I saw three Suns rising over a mountain range far on the horizon as I was traveling on a long flat desert-like road. It wasn't where I live, as the Sun sets against our mountain range while the Sun rises on the eastern plains. It looked like a desert with the mountains on the horizon, the sky was blue, the three Suns were bright white/yellow. I gasped in my dream and woke myself up.

It was a similar dream to what I dreamt the prior year. I was outside, looking up and saw the three Moons in a row against a blue sky. My first thought upon awakening was the Nemesis-Nibiru System has arrived within our system. And it is getting closer. These are the Signs in the Heavens predicted by Jesus Christ that will appear before His return.

Luke 21:25, "There will be signs in the Sun, Moon and stars. On the Earth, nations will be in anguish and perplexity at the roaring and tossing of the sea."

Then, I during a prayer session, I heard an audible voice, which I identified as the Lord's voice, tell me, 'I am returning with Nibiru'. I've heard His audible voice a few times in my life, the first time, when I met Him in the desert of Sde Boker in the Negev desert of Israel, when He said, "I am the Messiah, follow Me." So, I recognized His voice, mainly His vibration within my being, that made me discern it was Him, in no uncertain times, because He knew I was searching for answers. Well, this book is the outcome and conclusions that I have come to, in my 40+ year search. See my Chapter on *The Messianic Star* for more scriptural connections to the Lord and Nibiru.

The anomalies happening to the Sun now in Sunrise and Sunsets around the world are being caused by Nemesis and her planets, which are close to transiting our Sun. There is a huge planet called the blue planet with stripes which can be seen as blue or green depending on the angle, which may also be the Blue Kachina of Native American prophecy. The prophesy states that when the Blue Kachina appears, the end will come after the Red Kachina (Nibiru) appears. This is the main cause of Earth changes, historically and presently.

We are in the middle of said prophesy. All is well and in Divine Order. The Lion of Judah is coming soon. When we see Nibiru, I do believe that is the prophesied sign in the Heavens that Christ warned about before His Second Coming. It was the Blue Kachina, Nibiru's Red Kachina sister. The Blue Kachina or Planet Nine was the planet that eclipsed during His crucifixion as recorded in Mark 15:33. Our solar and lunar eclipses only last for minutes and are generally over with from start to finish in an hour. The Blue Kachina is one of the signs in the Heavens Jesus told us to watch for. It precedes Nibiru.

The jury is out on exact timing. The consensus amongst most astronomers tracking Nibiru, and yes, it can be seen through a telescope, is that it will be visible to the naked eye this decade. Then it will slowly transit towards its first passing of Earth on its orbit around the Sun. Some say it may take five months or seven years to complete its orbit around the Sun and then pass Earth for its second deadly pass. Well, for starters, seeing is believing. There are many doubters out there who will finally see what others have been seeing through telescopes.

Time will tell whose calculations are correct, and whose are off, but know this, it is coming, and it is the harbinger of the end of this age and the Second of Coming of the Lord, who will return to judge the Earth, and then rule over all the nations from the New Jerusalem.

Revelation 6 is the forerunner rider on the white horse, who is Elijah to return. As chaos increases on the Earth, so will the Lord's messengers. "The harvest is the end of the age, and the harvesters are angels." (Matthew 13:39)

I do not know how long it will take before it passes Earth. Some say 2029, others 2092. We will see. Important thing is to be ready to meet our Maker no matter what happens, death or Rapture.

> "And the great dragon was cast out, that old serpent, called the Devil, and Satan, which deceives the whole world: he was cast out into the Earth, and his angels were cast out with him."
>
> (Revelation 12:9)

God is a time traveler. Scripture says that a day is like a thousand years to the Lord and a thousand years is like a day. So how long in our linear timeline are a few minutes on the prophetic clock?

Lay people began to see Nibiru in the skies after September 23, 2017. It was photographed through telescopes somewhere around Neptune, coming in at a southwest angle moving through the Constellations of Virgo and Libra.

Recently there are a plethora of photos taken through telescopes by various astronomers of the incoming Nibiru - Nemesis system. It has become the only section on Google Sky that is blocked out. On my Facebook page, I showed what it looked like before they blacked it out and after as well through infrared. Just go to the Album: Signs in the Heavens, which is publicly posted on *Who's Who in the Cosmic Zoo Facebook Fan Page*. Its tails are made of iron oxide, full of comets, meteorites, and fireballs. Revelation 12 is the Prophecy of Nibiru and is written in astronomical language. The Virgin mentioned is not Mother Mary but is the constellation of Virgo. On September 23, 2017, Jupiter and Nibiru passed through Virgo. Astronomers said sometime after that transit is when people will begin to be able to see the red winged planet, the planet of Crossing, with our naked eyes, just as we can view Jupiter in the night sky without binoculars. However, the powers that be have made that difficult by a number of different space technologies, including daily Chemtrails, being used these days to obscure and obfuscate the Nemesis system.

> "And I shall give signs in the Heavens, miracles in the Earth: blood, fire and clouds of smoke."
> (Acts 2:19)

Nibiru will pass the Earth not once but twice. September 23, 2017 became the turning point and astronomical transit of Nibiru, the Red Dragon, will transit with Jupiter in the sign of Virgo, which represents Israel. As predicted in Revelation 12 the Virgin will give birth to the man child who is threatened by a dragon. The Dragon is Nibiru, the Virgin is the Constellation of Virgo.

See my rather lengthy chapter on Nibiru and all of its many historical names, through astronomical viewpoints connecting the dots to Bible prophesies and historical scriptures of previous Nibiru transits, in *Who Are the Angels?* Chapter Sixteen: *The Second Coming and Nibiru* pp.338-363.[7]

In *Who Are the Angels?* Book Three, I postulated that it could take a "Shemitah" which is a seven-year cycle from the time it passes the Earth on its way around the Sun until it passes Earth for a second time. I found it highly coincidental that almost all the astronomers, including NASA's unnamed whistle-blowers are predicting it to take 24-36 hours to pass the Earth, whereas I pulled multiple scriptures throughout the End Times Bible Prophesies that refer to it as a 'day', 'the Terrible Day' of the Lord.[7]

NIBIRU AND THE INTERNET

There are so many amateur astronomers posting photos and videos on their YouTube Channels now who have been documenting and tracking Nibiru, its sister exoplanets and exomoons. It appears since its September 23, 2017 transit, that tracking Nibiru has garnered more attention and a growing audience on the internet, which created a paradigm shift with many people on social media. I started my Facebook account in 2009, and any time I mentioned Nibiru or Planet X, I was laughed at, mocked and challenged up until 2017, when the passage occurred which many saw on all kinds of sky watching programs, telescopes, and photos. People started watching weather cams all over the world and posting them where you can see the authenticity of the sky prior to Sunrises and during Sunsets. Then more groups started focusing only on sharing photos of the skies. Granted, the objects people were seeing were not all the comet planet Nibiru, but its sister exoplanets and exomoons along with its brown dwarf star, Nemesis.

The consciousness grew and opened up acceptance in the minds of many, as more and more photos and videos were circulated, "seeing" turned into "believing."

The naysayers started to take a back seat, and those who thought every object was a lens flare also diminished, as people were seeing planets and exomoons that totally challenged what we were all taught in our respective science classes.

The trend surged on YouTube Channels where you can find daily postings of the exoplanets, twin Suns, and possible sightings of Planet X, Nibiru. Some hot webcams were from Alaska and Cancun's *Webcamsdemexico* that videographed Sunsets revealing multiple light.[9]

And then the silliness ensued, as people claimed that each time we went through a solar or lunar eclipse it was a Nibiru eclipse and viral false claims would circulate over the internet that the Rapture was going to happen, or the Tribulation would begin in Israel. Yet, to date, none of these Blood Moons brought the Rapture nor the Tribulation. That certainly doesn't mean it's not going to happen, but many are jumping the shark, so to speak, in trying to fit every Blood Moon Lunar Eclipse into Bible Prophecy, just as much as those who think every object they see around the Sun is Nibiru. Yes, the word "Nibiru" has become a generic term for second Suns, comets, exoplanets and exomoons photographed around the Sun. But the truth is, not all of these exo-objects are Nibiru.

In fact, as of 2019, Nibiru was not even around the Sun, because it is still entering our system coming in at an elongated elliptical orbit believed to be between Jupiter and Neptune. When we see two Suns, we see Nemesis, the brown dwarf star that intersects with our solar system every 2,000 years. We are also seeing other objects that are part of its system, exoplanets and exomoons. Nibiru is one of its planets, but it is the oddball one, because it has an odd orbit from all the rest. It swings around our Sun as well as its own star, Nemesis, approximately every 3,650 years give or take a decade.

People ask why they cannot see the exoplanets in their skies. While geographical location and timing are key, due to the Sun's angles, which can either illuminate exoplanets or obscure them, there's also been a concentrated effort by the powers that be to cover up the exoplanets, because one of them, the red rotating cratered one, eclipses our Sun daily. So, they put up a Sun simulator to obfuscate through a lensing system. (See my section on the Solar Simulator and Project Blue Beam for more details.) In addition to a prism solar simulator lensing system to obscure and obfuscate the eclipses of the exoplanets, there is also a worldwide effort to chemtrail the skies. That creates cloud cover to confuse us Earth humans as to what we are exactly seeing when we look up at the sky. This leads to many debates on the internet over photos of our skies, because the lensing system started to create petal effects over and around the Sun, which are actually flares along the lens arrays that are intentionally being projected onto the sky from the prism Solar Simulator.

With that said, it all depends on your location on the planet. Those closest to the South Pole and the equator seem to get better views, due to the Earth's angle in relation to the Sun, as well as those located around the Arctic Circle. Personally, I spent about a year watching weather cams situated at multiple locations around Alaska, and posted those photos, screenshots of the videos to both my Facebook

Albums, *Sign in the Heavens*, and *Twin Suns*. Anyone who follows me personally, or follows my book page, *Who's Who in the Cosmic Zoo?* on Facebook, has access to hundreds of these photos posted over the years. So please feel free to visit to see for yourself. To date, Facebook has not censored them. They are all up for public viewing.

What we are witnessing are approaching planetary and celestial objects (brown dwarf stars, gas giants, planets, Moons, and comets) that are part of another solar system that is intersecting orbits within ours.

Most solar systems with planets are binary systems. Our Sun has always had a twin, which has been called an evil twin, known as Nemesis, and within it exists the Destroyer, a comet planet that has an elongated orbit that interlopes into our space approximately every 3,650 years. It has been known to sometimes collide with our planets and Moons.

A great resource book to understand our own systems history is *World's in Collision* by Immanuel Velikovsky, who was a colleague of Albert Einstein. Velikovsky said that the asteroid belt was formed by a planet breaking up, its pieces were locked in a similar orbit, i.e., an asteroid. [8]

Every passing of Nibiru is different, it all depends on how close it comes to us. Nibiru is a comet planet, with a wing-like extrusion and a tail made of plasma. One thing to keep in mind when viewing objects around the Sun is that Nibiru's objects are gravitationally drawn to the Sun, not Earth. However, due to Nibiru-Wormwood orbit, they will enter through our system and pass Earth not once but twice as they orbit around our Sun.

I believe we are seeing parts of the Nemesis-Nibiru system. According to the theories of Sumerian cosmology, Nibiru was the twelfth planet in our solar system family of ten planets, the Sun, and the Moon. Its catastrophic collision with Tiamat, a planet that was between Mars and Jupiter, would have formed the planet Earth, the asteroid belt, and the Moon.[12] Could these ten planets be the 10 Kingdoms mentioned in the End Times Prophecies in Daniel Chapter 7 and then repeated in Revelations?

INSIDER'S CLAIMS

For all intents and purposes, I am going to refer to a whistleblower, an insider working for one of the alphabet soup agencies and had access to NASA's data on the status of this incoming planet, Nibiru. He is an unnamed voice on YouTube interviews, someone who cares about the public being informed, while remaining anonymous, but clearly an insider. Researcher have been saying the poles are shifting 42 miles a year, but insiders say, the Earth's poles are shifting 42 miles per day! Insider says Comet Elenin and Planet X are part of the same system. Extinction level event occurs when Nibiru is here. He is predicting a three-mile wave from the coastlines all around the world as it passes Earth, while our Sun is pulling it in through gravity. When it is behind the Sun, it is going to orbit around and go back out as the Sun pulls it in like a magnet. It does not collide with the Sun or the Earth but disrupts everything. Hence its ancient name, the Destroyer.

Here is what we know for sure: Nibiru has its own system of planets and Moons and a dwarf star, with a debris field of hot iron. This is enough to destroy a third of the Earth.

Nibiru's mass is five times the mass of Earth. When 70 pounds of iron rock come down upon the Earth like hail everything will be destroyed. Earths tectonic plates are stretched like fabric by the gravitational and magnetic pull. Sea levels rise as high as the crustal layer shifts. Who could survive such an event?

The Great Lakes end up in the Gulf of Mexico according to Al Bielek's accounts in which he described his time-travel journey to the future. Andy Basiago, who participated in the 1980 CIA Time Travel Project Pegasus, saw Washington DC in the future under a hundred feet of water. This is what prompted the powers that be to duplicate all the federal buildings in Denver, Colorado, which is slated to become the makeshift capital of the United States of America, if and when Washington DC goes under water.

The Insider saw a blazing hot red ball of fire coming through our solar neighborhood. Asteroids five hundred miles across. Meteorites in the tail of this intruder. Insider says meteorites extend millions of miles in the tail. Says Nibiru will come 14-20 million of miles to Earth. It will not hit the Earth, but it is passing will flip Earth upside down. Read Isaiah 24 when the Lord turns Earth upside down. (See my chapter on Pole Shift.)

There will be crustal shifts, but the core remains undisturbed. Objects come from 20 million miles away travelling at three kilometers a second will lock like a magnet onto the Earth and the Earths plates and crust will move. This will complete the Pole Shift that has already begun. (See my Chapter herein, on Pole Shift.)

The Year 2018 was unprecedented in the frequency of fireballs and Earthquakes, which are signs of the approaching Nemesis-Nibiru system that will cause the poles to flip. The last fly-by of Nibiru occurred during the Exodus from Egypt, 3600 years ago, that precipitated the ten plagues. Each passing marks some kind of global catastrophe. Locusts came because everything was dying around them. They lost their usual source of food, so they end up being a plague to everything around them.

> Behold, The Lord makes the Earth empty and makes it waste, *distorts its surface.*
> (Isaiah 24:1)

> The Earth is *violently broken*; the Earth is split open. The Earth is shaken exceedingly. The Earth shall *reel to and fro like a drunkard* and shall *totter* like a hut. [Earth's 23.5-degree tilt]
> (Isaiah 24:19-20)

Planet X and the Nibiru system is a scientific explanation for Isaiah's End Time Prophesy. These objects are part of an approaching celestial system that will greatly affect the Earth during the Great Tribulation as stated in the End Times Bible Prophecies.

DISINFORMATION AGENTS

Nancy Lieder, founder of the website zetatalk.com, claims to have received messages from extraterrestrials from the Zeta (ζ) Reticuli system, a visual double star 39 light years away and consisting of two stars like the Sun. Yes, that is the same Zeta Reticuli Gray Aliens who showed their star chart to Betty and Barny Hill, back in their famous and dramatic 1961 Alien Abduction. (See, Book One Chapter on the Grays) [10]

In 1995, Lieder announced that the Zetas, as she calls these extraterrestrials, warned her that in 2003 a large planet would pass remarkably close to the Earth with devastating consequences that would destroy civilization. Of course, that date came and went without incident. Then Lieder had the chutzpa to tell her followers that this date was "a white lie. . . to fool the establishment." She says that this Planet X, as she calls it, is still coming, and has since made multiple predictions over the years that have not come true. [11]

While Lieder never called her Planet X Nibiru, she announced on her website in 1996 that "Planet X and the 12th planet are one and the same." Given the widespread knowledge of Sitchin's thesis, and the title of his first book, *The 12th Planet*, Leider clearly intended to equate her Planet X with Sitchin's Nibiru. However, Sitchin strongly disagreed with this equation. For one thing, Sitchin dated Nibiru's last pass through our solar system was 556 BC; with an orbital period of 3,600 years, which means Nibiru would not be due to pass through again until around 3000 AD. Lieder connected her Planet X and Sitchin's Nibiru, and many began to equate the two, as today many people use the terms Nibiru and Planet X interchangeably. [11]

Samuel Hoffman, who is known on Facebook and YouTube as the Montana Sky Watcher, was allegedly taken by government sources well over 35 years ago because of a high school paper he wrote about the incoming Nibiru system and its planets. They did a number on him, destroyed his life, and in my opinion, he was implanted. He runs a Facebook Group that produces unhealthy cult mentalities. His posts circulate because he does hand drawings and color maps of the Nibiru system with its sister exoplanets. However, and here is the kicker, he claims to have received his information from a channeled source which contradicts everything the ancients recorded. As I proved through multiple credible historical sources, the ancients referred to Nibiru, Hercolubus, the Red Dragon, as Red. Why would they call it the Red Dragon if it were not red? Planet Nine or the Blue Kachina is Blue. The Hopi specified the difference between the Blue and Red Kachinas.

The name Nibiru was first translated in Zecharia Sitchin's work. We would not be using that name if it were not for his ground-breaking research.[12] When the U.S.

government took over it changed the names. Hoffman was held up for 30 years by them. They are the ones putting out disinformation with the aliens. The ancients and the Bible are not wrong!

The research I presented in my 3rd book, *Who Are the Angels?*, in my Chapter: on *The Second Coming and Nibiru*, I included all the names used by ancients for this celestial body that is described as the Destroyer.[7] Each of the major ancient empires and their cultures recorded this planet, and gave specific names to it, all describing Nibiru as red. The information Hoffman circulates that Nibiru is the big Blue Planet, which professional astronomers are calling Planet Nine, that is four times the size of Earth. This piece of disinformation goes against the history, the astronomy, not to mention the actual photos taken through multiple telescopes, which prove that Nibiru, the comet planet with wings, actually has a golden core that is surrounded by red iron oxide, making it look red, not blue. When I had the honor to interview Zecharia Sitchin in 1994, he spoke about Nibiru being a red comet planet.

It is extremely important to exercise discernment, not just spiritually, but also in the knowledge available to us through ancient history, astronomy, and the actual facts. Changing names due to so called channeling from aliens is disinformation, which is put out to confuse everyone.

Dr. Claudia Albers, Planet X Researcher and Astrophysicist, calls it the Planet X System, which is made up of comets and stellar cores of dead Suns which are increasingly attracted to our own planet. The system is drawing energy from the Earth in the same way as it does from our dying Sun. She describes how larger stellar cores will be coming toward our atmosphere now that our Sun is no longer shining and cannot maintain them in orbit around itself as it did in the past. She believes that the Nibiru system is the Planet X system of stellar cores, dust, and debris and in fact, does not believe that the planet Nibiru exists as it has been previously described. She estimates that our Sun was already dark in 2011 and provides evidence in one of her videos to support that claim.[13]

Dr. Albers claims that our Sun is already dead and is no longer giving light. She lost her position in a South African Witwatersrand University, teaching physics when she began investigating Planet X. She writes articles on Planet X, solar cores, brown dwarf stars and our Sun's present situation. She reads them daily on YouTube. Even a broken clock is right twice a day. She does give interesting reports and analysis, but she links everything on Earth including plagues of locusts to Planet X. When I asked her how insect infestations are connected to Planet X, she replies that it is because of its proximity that they are here. She claims the locusts come from the Planet X system. When I questioned her as how they got here, did they fly through space or did they arrive on a spaceship, she missed the tone and replied that they come through space.

The problem many people have with Dr. Albers is that not a single astrophysicist backs or agrees with her. That does not make her wrong about her research on solar cores. She stands pretty much on her own, except for a co-author of her books,

Scott Cione, and ex-con who provides her with photos of the exoplanets and twin Suns to analyze daily on her YouTube Channel.[14] When I questioned her statement that the Sun is no longer emitting light, then why do we still have daylight, despite the Sun Simulator, I got crickets.

I am not saying she is right or wrong, but the evidence is simply not there – yet. I do agree with her assessment that our Sun is dimming and eventually will die. But she seems to believe it is already dead. I do not. I think this is a key point, because typically when stars die, they implode, and when that happens, it will be the long awaited and prophesied solar flash that has been predicted by the ancients. See my Chapter: *Nemesis & the Death of Our Sun*, section: *The Solar Sneeze*.

Notes:
1. Ella LeBain, *Who is God?* Book Two, Chapter: *Names of God, What's in a Name?*, Skypath Books, 2015.
2. Lee Billings, *The Book That Predicted Proxima b*, https://www.scientificamerican.com/article/the-book-that-predicted-proxima-b-excerpt/, September 8, 2016.
3. Zecharia Sitchin's Book *The Lost Book of Enki: Memoirs and Prophecies of an Extraterrestrial God*, Bear & Company; First Edition (November 15, 2001)
4. Zecharia Sitchin, *The 12th Planet, Book One of the Earth Chronicles*, Avon Books, NY, 1976
5. Dr. Danny R. Faulkner, https://answersingenesis.org/astronomy/solar-system/Nibiru/, March 24, 2017
6. Sarah Lewin, *New Model: Nearby Exoplanet TRAPPIST-1e May Be Just Right for Life*, https://www.space.com/36349-trappist-1e-just-right-for-life.html?utm_source=sp-newsletter&utm_medium=email&utm_campaign=20170405-sdc, April 2017.
7. Ella LeBain, *Who Are the Angels?* – Book Three, Chapter 16: *on The Second Coming and Nibiru*, pp. 338-359, Skypath Books, 2016.
8. *World's in Collision* by Immanuel Velikovsky, Doubleday; Second Edition, 1951.
9. http://webcamsdemexico.com/webcam-punta-cancun-poniente
10. Ella LeBain, *Who's Who in the Cosmic Zoo?* Book One – Third Edition, pp. 248-289, Tate, 2013.
11. https://answersingenesis.org/astronomy/solar-system/Nibiru/
12. Zecharia Sitchin, *Book One of The Earth Chronicles, The 12th Planet*, Avon Books, 1976.
13. https://www.drclaudiaalbers.com/
14. https://www.youtube.com/user/claalb1

CHAPTER SEVEN

NIBIRU: MYTH OR SCIENCE?

*"There shall be signs in the Sun, and in the Moon,
and in the stars; and upon the Earth distress of nations,
with perplexity; the sea and the waves roaring."*
Luke 21:25

In the 1950s, there was an absolute storm as astronomers were all watching for Nibiru or Planet X. It was written about in astronomy magazines without censorship. It has been located and seen, the Nemesis-Nibiru System with all its seven planets that orbit around a brown dwarf star known as Nemesis. Brown Dwarves need to be seen in infra-red light. Wormwood is only visible with infra-red equipment except during ecliptic plane crossing events. Through a telescope, Nibiru, one of its biggest planets, is seen as a bright orange-red ball surrounded by red iron oxide, spewing its dust into its atmosphere, that forms a tail, like a comet. This is why the esteemed Chilean Astronomer, Carlos Munoz Ferrada, referred to it as a Comet-Planet, because it has the size of a planet but speed and elliptical orbit of a comet.

Is there life on Nibiru? Who were the Annunaki? See, Book One of *Who's Who in the Cosmic Zoo?* [1]

Wormwood orbits in a tilted, elliptical direction opposite that of our planet. What could possibly go wrong with this scenario? Ferrada was the first to speak about this Comet-Planet and identified it as a red planet with a dark star. In 1940, Ferrada predicted accurately that the powers that be would attempt to cover-up Planet X when it comes barreling towards the Earth.

A Brown Dwarf is a natural phenomenon with a mass 13 to 18 times that of Jupiter at 3 to 5 times that its size. Generally speaking, any object that orbits close to the Sun tends to be pulled into the Suns gravitational field, which causes it to accelerate. As it speeds up, it gets sucked into the Suns orbit, and speeding objects tend to catch on fire, as is the case with comets and meteorites. They burn.

The Physicist John DiNardo said Planet X broke through our heliosphere in 2003 based on measurements of electrical discharges on the Sun. It would take years for it to move through the solar system, which is understandable, and this will delay the object in passing us, because of its elliptical orbit. When these planets enter our system, they typically do not strike the Sun, they orbit around it due to centrifugal force. They tend to loop around celestial objects, change direction, making a U turn.

NASA'S ACKNOWLEDGMENT OF NIBIRU
An official NASA Press Release in 1992 stated the following:

> "Unexplained deviations in the orbits of Uranus and Neptune point to a large outer Solar System body of 4 to 8 Earth masses on a highly tilted orbit, beyond 7 billion miles from the Sun." (NASA Press Release, 1992)

> "Astronomers are so sure of the 10th planet they think there is nothing left but to name it." (Ray T. Reynolds, NASA Press Release, 1992)

HAARP: Secret Weapon Used for Weather Modification Electromagnetic Warfare

> "It isn't just conspiracy theorists who are concerned about HAARP. The European Union called the project a global concern and passed a resolution calling for more risks. Despite those concerns, officials at HAARP insist the project is nothing more than a research facility." (from TV Documentary on HAARP by Canadian Broadcasting Corporation – CBC)

GROWING EVIDENCE OF INSTABILITIES IN OUR SOLAR SYSTEM
- Recent solar activity highest in 8000 years;
- Sun's magnetic field has decreased in size by 25%;
- 300% increase in galactic dust entering our Solar System;
- Mercury's magnetosphere experiencing significant increases;
- Venus exhibiting a 2500% increase in its "green glow;"
- Mars showing a rapid appearance of clouds and ozone;
- Mars observations reveal up to 50% erosion of its ice features within a 12-month period;
- Jupiter plasma torus increasing; its Moon Io exhibiting the same changes;
- A 200% increase in the density of Io's plasma torus;
- Jupiter's Disappearance of White Ovals since 1997 – recent increase in storms;
- Io's ionosphere is ~1000% higher;
- Jupiter's Moon Europa much brighter than scientists expected;
- Jupiter's Moon Ganymede is 200% brighter;
- Saturn's plasma torus is ~1000% denser;

- Aurorae first seen in Saturn's polar regions in recent years;
- Uranus was featureless in 1996 – exhibiting huge storms since 1999;
- Uranus in 2004 was also markedly brighter than in 1999;
- Neptune is 40% brighter in the near-infrared range based on observations from 1996 – 2002;
- Pluto observations reveal a 300% increase in atmospheric pressure.[2]

NEMESIS THE DEATH STAR

NASA scientists have been searching for an invisible Death Star that circles the Sun, which throws potentially catastrophic comets at the Earth conducted a study: *Search for Death Star Throws out Deadly Comets.*

In the March 11, 2010 Space.com article, *Sun's Nemesis Pelted Earth with Comets, Study Suggests*, NASA's Wide-field Infrared Survey Explorer, or WISE, was able to uncover many failed stars, or brown dwarfs, in infrared light. They confirmed that our Sun's Nemesis pelted Earth with comets in the past. Nicknamed "Nemesis" or "The Death Star," this undetected object could be a red or brown dwarf star, or an even darker presence several times the mass of Jupiter.

Originally, Nemesis was suggested as a way to explain a cycle of mass extinctions on Earth. The paleontologists David Raup and Jack Sepkoski claim that, over the last 250 million years, life on Earth has faced extinction in a 26-million-year cycle. Astronomers proposed comet impacts as a possible cause for these catastrophes.

Our solar system is surrounded by a vast collection of icy bodies called the Oort Cloud. If our Sun were part of a binary system in which two gravitationally bound stars orbit a common center of mass, this interaction could disturb the Oort Cloud on a periodic basis, sending comets whizzing towards us.

OORT CLOUD: THE FOOTPRINT OF NEMESIS

A recently discovered dwarf planet, named Sedna, has an extra-long and usual elliptical orbit around the Sun. Sedna is one of the most distant objects yet observed, with an orbit ranging between 76 and 975 AU (where 1 AU is the distance between the Earth and the Sun). Sedna's orbit is estimated to last between 10.5 to 12 thousand years. Sedna's discoverer, Mike Brown of Caltech, noted in a Discover magazine article that Sedna's location doesn't make sense. "Sedna shouldn't be there," said Brown. "There's no way to put Sedna where it is. It never comes close enough to be affected by the Sun, but it never goes far enough away from the Sun to be affected by other stars."[?] Perhaps a massive unseen object is responsible for Sedna's mystifying orbit, its gravitational influence keeping Sedna fixed in that far-distant portion of space?

John Matese, Emeritus Professor of Physics at the University of Louisiana at Lafayette, suspects Nemesis exists for another reason. The comets in the inner solar system seem to mostly come from the same region of the Oort Cloud, and Matese thinks the gravitational influence of a solar companion is disrupting that part of the cloud, scattering comets in its wake. His calculations suggest Nemesis is between 3 to 5 times the mass of Jupiter, rather than the 13 Jupiter masses or greater that some scientists think is a necessary quality of a brown dwarf. Even at this smaller mass, however, many astronomers would still classify it as a low mass star rather than a planet, since the circumstances of birth for stars and planets differ.

The Oort Cloud is thought to extend about 1 light year from the Sun. Matese estimates Nemesis is 25,000 AU away (or about one-third of a light year). The next-closest known star to the Sun is Proxima Centauri, located 4.2 light years away.

Binary star systems are common in the galaxy. It is estimated that one-third of the stars in the Milky Way are either binary or part of a multiple-star system. Red dwarfs are also common — in fact, astronomers say they are the most common type of star in the galaxy. Brown dwarfs are also thought to be common, but there are only a few hundred known at this time because they are so difficult to see. Red and brown dwarfs are smaller and cooler than our Sun, and do not shine brightly. If red dwarfs can be compared to the red embers of a dying fire, then brown dwarfs would be the smoldering ash. Because they are so dim, it is plausible that the Sun could have a secret companion even though we've searched the sky for many years with a variety of instruments.

NASA's newest telescope, the Wide-field Infrared Survey Explorer (WISE), may be able to answer the question about Nemesis once and for all.[3]

WHO WAS CARLOS MUNOZ FERRADA?

Carlos de la Fuentes Marcos, scientist at the University of Central Missouri and co-author of the study said, "This excess of objects with unexpected orbital parameters makes us believe that some forces are altering the distribution of the orbital elements of the European Telecommunications Network Operations (ETNO) and we consider that the most probable explanation is that other unknown planets exist beyond Neptune and Pluto."

In 1940, this Chilean astronomer, one of the most brilliant and respected astronomers in the world, correctly predicted that the Elite would attempt to cover up Planet X as it comes towards the Earth. Ferrada referred to Nibiru as a *Comet Planet* because it has the size of a planet, but speed and elliptical orbit as a planet. It has the great mass as a planet but an elliptical orbit as a comet. It also has a tail. Its large gravitational field brings a new star. These things have not happened in 13,000 years, but 13,000 years ago saw the destruction of Atlantis. Such resonance and effects produce reflections and weaknesses will cause a catastrophic change that may include a sizable shift in geophysical and human structure.

Ferrada said, "There are so many things, which is unfortunate, mankind is not ready. The change comes, the destruction comes, and above all will affect humanity in its existence in its production and in its own subsistence."

Nemesis is for the most part invisible due to its surrounding red iron oxide dust clouds.

Carlos Munoz Ferrada, born in Chile, January 1909 - died October 17, 2001, was a self-taught sailor, astronomer, and geophysicist. He was known for his ability to predict Earthquakes within a margin of error two hours, along with other surprising events such as the arrival of Hercolubus or Planet X in the end times. He was greatly respected as an astronomer and will be remembered as the one who has been accurate in his predictions. Hercolubus or Planet X, which he calculated will eventually pass 14 million of kilometers away from Earth accelerating the great geophysical changes, is already in progress.

In 1999 he warned in an interview that Spain and the Mediterranean form a great catastrophic triangle, along with Sumatra in the Asian Pacific, and the territory around Chile. These zones were demarcated at the beginning of his calculations, but we know that each day new zones of high seismicity are added until all Earth will be shaken by the global cataclysm. The Chilean astronomer specified the mass, velocity, orbital time, trajectory, and the terrible consequences of Hercolubus on our solar system.[4]

The interview was published on June 28th, 1999 on Channel 4 on the Television of Puerto Rico and has subsequently been broadcast in Venezuela as a part of an Extraordinary Documentary by producer and director, Joseph Landon, which can be found on YouTube. [4]

Chile recognized Ferrada for his mathematical accuracy in predicting the Earthquake that devastated in the south-central region of Chile in 1939. Ferrada even set the date for the Earthquake, which was published in the Journal of Conception. He was within a margin of error of two hours. The disaster claimed the lives of 60,000 people. He also announced the Chilean Earthquake of 1960, which was the most powerful Earthquake ever recorded, a 9.5 magnitude on the Richter scale. 1985 there was an 8.0 mag, yet the Chilean Earthquake of 2010 came close, as some recorded it was a 9.0, but was later downgraded to an 8.8.

Ferrada was a genius in geophysics, who discovered numerous stars, galaxies and comets, and accurately forecasted Earthquakes, climate changes, volcanic eruptions, and changes in the orbit of Halley's Comet.

Here are some scientific characteristics of Hercolubus according to the calculations of Carlos Munoz Ferrada:

> "Hercolubus is a RED PLANET that travels an elliptical orbit of a comet yet has the great mass of a planet; in other words, a planet with a tail, hence he coined the phrase *comet-planet*.

> "The star is charged with cosmic energy, as to say rarefied radiation which according to scholars will alter human health and behavior; it will provoke incurable epidemics, intense irritability and cause wars.

"Hercolubus does not comply with the conventionally established celestial laws of physics. It travels between our Sun and the Black Sun Nemesis which is found 32 trillion kilometers away. It will pass as close to 14 million kilometers from Earth.

"Hercolubus is approximately SIX TIMES BIGGER THAN JUPITER, thus its approach of 14 million kilometers from our world will create a disastrous attraction with the incandescent liquid minerals inside the Earth, precipitating tremendous internal pressures and therefore volcanoes and Earthquakes.

"It will end up penetrating our Solar System and be visible to the naked eye as well as photographed. "Its arrival will cause human and geophysical changes, bringing much change and destruction." [5]

Ferrada did not say that the great final catastrophe would occur on August 11th, 1999, but that from that date onwards, geophysical and climatic changes would intensify, especially Earthquakes in the oceans of a volcanic origin with seaquakes, underwater Earthquakes, resulting in tsunamis. He said that Hercolubus could be photographed from August 1999 onwards and that its appearance would seem bigger than our Moon. Even though the mainstream media have suppressed reports and photographs, these days, you can see for yourselves a plethora of photos and videos of two Suns all over YouTube.

There are internet and Facebook groups which have grown exponentially since I started on Facebook, in 2009, and photos of the twin Suns were first mocked and rejected, now they are all over the newsfeeds on a daily basis. There is daily if not hourly YouTube videos documenting the skies, from all over the world, as well as various weather cams that focus on the sky from Sunrise to Sunset in remote locations in Alaska and Mexico. It is certainly no secret anymore. It is there, it is around our Sun, and its accompanying sister planets and their exomoons are seen as well.

Just go to my Facebook page and peruse my Albums: *The Heavens, Signs in the Heavens, and All Things Nibiru* and you can see for yourself. [6]

Hercolubus glows in the infrared range, which makes it difficult to see with the naked eye. Only in special conditions of light refraction through the clouds, and at certain times of day when the Sun is at a certain angle, can get it illuminated at locations closest to the poles. It may be seen through special filters.

We are approaching a time when Hercolubus will be permanently visible as a second Sun and in common conditions in broad daylight. This is why it was rediscovered in the 1980s by IRAS, an orbiting infrared telescope NASA.

The reports were published in North American newspapers and magazines and there are videos which offer testimonies of the affirmations of Dr. Robert Harrington, such as the interview that was done with him by the writer Zecharia Sitchin in 1990. See, my Chapter on *Connecting the Dots to Nibiru,* for details.

Muñoz Ferrada earned the respect and admiration of the Chileans for his ability to anticipate climatic and geophysical disasters, but he was often in prison for "disturbing the peace and causing anxiety." According to testimonies of the inhabitants of Valparaiso he was doggedly pursued by the authorities who practically did not allow him to appear in public. The posthumous report dedicated by The Journal of Today of El Salvador, reads as follows:

> "In order to explain his observations and predictions, Muñoz Ferrada theorized Geodynamics, which is based on the attractions of the heavenly bodies, large explosions in the Sun, and the cycles of the geophysical disturbances. With this method he predicted Earthquakes, volcanic eruptions, climate changes, discovered new planets and comets while deciphering their trajectories and other features.
>
> "According to Muñoz Ferrada the Earthquakes are caused by the gravitational influence of a few stars on others, rather that the accidental movement of the tectonic plates.
> "On June 28th, 1999, on Channel 4 on the Television of Puerto Rico, after 50 years of investigation and scientific documentation, the great Chilean Astronomer, at 90 years of age, revealed with absolute precision the three velocities that Planet X has in its orbit. Its velocity as it passes close to the dead Sun or the black star, another in the middle of its elliptical orbit and a third as it passes close to our Sun, to a thousandth of the speed of light, and 14 million kilometers from Earth."[5]

Muñoz Ferrada was a scientific genius ahead of his time, misunderstood by many and, according to those who knew him, a gnostic in his doctrine and practical life. The daughter of Ferrada, Mrs. Marina Muñoz spoke of her father: "He was a man way ahead for the century in which he lived. There was so much to learn from him, but his knowledge was not always fairly valued."

THE ORDER OF OUR SOLAR SYSTEM

I am reprinting a 1960 Newspaper Article on the Discovery of the 10th Planet, so you can see not only the transparency back then, but also the numbered order of the planets, along with their awareness and discovery of a 10th Planet, aka Planet X in 1960:

"Russ Report Finding 10th Planet [7]
New York, February 12, 1960 (AP) Reprinted in The Milwaukee Sentinel – February 13, 1960:

"A Soviet newspaper says Russian Astronomers have discovered a new planet wheeling far out around the Sun. This would mean our Sun has 10 – instead of nine – sons or daughters.

American Astronomers polled Friday by the Associated Press say the Soviets may well be right.

They all expressed a desire for more details so they could look for themselves in the international check-and-recheck spirit of science.

The nine known planets are Mercury, Venus, Earth, Mars, Jupiter, Saturn, Uranus, Neptune, and Pluto.

Pluto is the outermost planet, wheeling in an orbit more than 3-1/2 billion miles from the Sun. It was discovered by U.S. Scientists in 1930. Its existence was guessed from perturbations or influence of its gravity pull upon other planets. The reasoning was that something had to be out there. By careful hunting, Pluto's existence was found.

Now the provincial Soviet newspaper Kazakhstan Pravda says Astronomers at the Kazakhstan Astrophysical Institute have confirmed the existence of a small planet even farther out than Pluto.

The newspaper says this planet's existence was first detected in 1957 by accident while photographing stars in the constellation Capricorn, and that subsequent studies have confirmed that it is a planet. No name has been given to it. Nothing further could be learned from inquiries made in Moscow.

Dr. Gustav Bakos of the Smithsonian Institute in Boston in 1960 said the irregularities in Pluto's orbit have suggested that there might be another planet out there. So, the Russians may have found one.

If the Soviets tell exactly where the new planet is, or is supposed to be, American Astronomers could soon check it, said Fr. Paul Herget of Cincinnati Observatory.

"It can be verified, and it would be folly to announce it, if it weren't so," he remarked." [7]

IDENTIFYING THE PLANET X SYSTEM

Perhaps this explains why we do not see easily see the system behind or in front of us?

Nemesis is the Brown Dwarf Star, which is 56 times the size of Earth, which is 171,000 Earth mass. The Nemesis-Nibiru system makes a repeated appearance into our system approximately every 3,600 years, however, information received from a group of Gray Zetas, claims it is exactly 3,567 years.

Nibiru is the largest planet in the Nemesis Solar System. Nibiru is 6 times the size of Earth. It has seven Moons.

Helion is the second largest planet in the Nemesis system, a gaseous planet that is 3-1/2 times the size of Earth. It is pink or red and the one we see the most in weather cams around the world. It has craters and spins quickly, and not only is its orbit intersecting with our Sun, but it has also been known to eclipse it.

Harrington is a Moon of Helion, named after the late U.S. Naval Astronomer Robert S. Harrington. It is comparable to our Moon.

Ferrada is a Moon of Nemesis, named after famed Astronomer Carlos Munoz Ferrada. It is the same size as our Moon.

Arboda is a purple planet and the smallest of the Nemesis planets. Arboda is 2-1/2 times the size of Earth. Arboda orbits around Nemesis fast and has no Moon. It shows up in photos and weather cam videos around the world.

Nibiru moves through the skies like nothing we have ever seen before. Its orbit is unusual even in light of all the others in the Nemesis system.

Ecliptic The ecliptic is the plane of Earth's orbit around the Sun. From the perspective of an observer on Earth, the Sun's movement around the celestial sphere over the course of a year traces out a path along the ecliptic against the background of stars. The ecliptic is an important reference plane and is the basis of the ecliptic coordinate system. The point of perihelion is when Nemesis comes closest to our Sun. This is why we do not see it behind the Sun each year.

Pluto orbits our Sun from around the Kuiper Belt every 248 years. It takes eight hours for Sunlight to reach the dwarf planet. Pluto's orbit is inclined. When Neptune was first sighted and identified in telescopes, they saw there was another object in orbit influencing its orbit and that of Uranus, which turned out to be Pluto. It orbits around the Sun in an inclined and elliptical orbit that appears to be tilted from the rest of the planets.

Pluto was downgraded to dwarf planet status with its Moon Chiron, which were both used to explain the wobbles and perturbation of Uranus and Neptune's orbits.

CHAPTER 7: NIBIRU: MYTH OR SCIENCE?

Errors were made in these calculations. Bottom line, there are other larger and dwarf planets that visit our system on a cycle that could be said to be part of ours. What the ancient Sumerians recorded were cataclysms among the planets, one such collusion causing the asteroid zone between Mars and Jupiter. Another knocked Uranus into a retrograde spin on its axis, causing its rings to become vertical, instead of horizontal like Saturn's rings.

Conspiracy to cover it all up goes higher than NASA, which are the Illuminati who control the information because they own the media. They do not want the public to know, thereby causing a panic. They don't think panic makes any difference now as people are searching for information on Planet X's arrival, and watching videos, weather cam, and photos of the exoplanets in our skies daily, on what has turned into a new competitive genre on YouTube.

Being friends with multiple Planet X researchers, it did not take me long to figure out how highly competitive these YouTubers have become, and how this genre has grown exponentially in recent years. I think this is mainly due to the fact that people are catching on, seeing exoplanets and exomoons in our skies, and noticing that there has been a huge change in our Sun.

Disturbances Throughout the Solar System:

1. Recent Solar Activity Highest in 8000 Years
2. Sun's magnetic field has decreased in size by 25%
3. 300% increase in galactic dust entering solar system
4. Mercury magnetosphere experiencing significant increases
5. Venus – 2500% Increase in Green Glow
6. Mars – Rapid Appearance of Clouds, Ozone
7. Mars – Up to 50% Erosion of Ice Features in one year alone
8. Jupiter Plasma Torus increasing – Moon Io observing same changes
9. 200% Increase in Density of Io's Plasma Torus
10. Jupiter's Disappearance of White Ovals since 1997 – recent increase in storms
11. Io's Ionosphere 1000% Higher
12. Europa "Much Brighter Than Expected"
13. Ganymede 200% brighter
14. Saturn's Plasma Torus 1000% Denser
15. Aurora First Seen in Saturn's polar regions in recent years
16. Uranus featureless in 1996, now exhibiting huge storms since 1999
17. Uranus 2004 also markedly brighter than in 1999
18. Neptune 40% Brighter, Near Infrared Range – 1996 – 2002
19. "Global Warming on Pluto" – 300% increase in atmospheric pressure. [8]

CATASTROPHIC UPHEAVAL

The Nemesis-Nibiru system is just that, a solar system. In the past, astronomers thought it was one planet, hence the search for Planet X. But as I've already documented in astronomer's historical search for the ever-elusive Planet X, it has turned into the discovery of not only multiple planets and dwarf planets but has also led to the discovery and understanding that the fly-by that was documented in the Kolbrin Bible and other related ancient scriptures is not just one object, but many.

Let's start with comets. Astronomers recently discovered that a comet has broken off from the orbit of Nibiru and is headed straight to Earth. The Book of Revelation mentions a blazing star named Wormwood that falls to Earth and turns one third of its waters bitter, resulting in loss of life. Likewise, a blazing red star is mentioned in the Zohar, a collection of books written by rabbis during the Inquisition that expound on the Torah and offer added insight into the mystical and Kabbalistic teachings of the Torah, the Jewish Bible. Moses wrote about it in what is known as the Five Books of Moses, the Torah.

During his lifetime, Sitchin (1920-2010) claimed Nibiru was, in fact, heading towards our planet. 8 But the powers that be refused to tell the public. Nevertheless, they sent up various deep space telescopes, and probes to locate it. NASA's Neowise Mission, a Wide-field Infrared Survey Explorer (WISE), is a NASA infrared-wavelength astronomical space telescope active from December 2009 to February 2011. In September 2013, the spacecraft was reactivated, renamed NEOWISE and assigned the new mission, to assist NASA's efforts to identify and characterize the population of near-Earth object (NEO).

NEOWISE has analyzed over 100 asteroids and is tracking large distant comets. It has spied on two comets. We are seeing an increase in comets, asteroids, meteorites, and fireballs that are coming into our system from the incoming Nemesis-Nibiru System.

It is my opinion, NASA has found the system, and they believe Nibiru will *not* hit the Earth. However, other objects, such as comets, asteroids, meteorites, and fireballs in its tail, will.

There was a time on Google Sky in 2007 when the only portion of the sky was blacked out, and mysteriously covered up with a black square. That section of the sky is where Nibiru was located. In keeping with the media blackout, Planet X was X'd out of the Google Sky grid map. Then in 2015, the blacked-out section was removed and replaced with a section of the sky in which Planet X can clearly be seen at coordinates: 5h 42m 21.0s 22 deg 36' 45.7. Google Sky mysteriously reinserted back into the missing grid, that fully depicts the flaming winged Nibiru, revealing a red object with a golden core and what appears as golden wings, which are actual plasma tails of debris of iron oxide.

The fact that Google Sky chose to black out that particular section of the sky which appears to contain Planet X that is strongly suspected to be the planet Nibiru remains a mystery, except to researchers, like myself, who know that it's because of

CHAPTER 7: NIBIRU: MYTH OR SCIENCE?

the cover up. This proves that Google Sky was in fact trying to hide something for eight years. So, it is now visible from both the actual sky and Google Sky, indicating more evidence that the tide is turning toward the truth about Planet X.

This also proves my point that something has shifted in the narrative. There was a time when astronomers mysteriously disappeared and/or were found dead for their research on Nibiru. Now, a plethora of YouTubers present all kinds of evidence on a daily basis that Nibiru is real, and it is orbiting within our vicinity, along with a trail of other objects and debris.

In mid-April of 2012, the European Space Agency invited astronomers, physicists, nuclear engineers, mathematicians and even soldiers in the Strategic Space Defense from around the world to a conference in Rome to discuss the topic of an asteroid striking the Earth, with the proposed hypothesis "not if, but when." This dovetails into the Nibiru cataclysm scenario, minus the forbidden planet, of course.

This Planetary Defense conference strategized on how to collectively handle Near-Earth Objects (NEOs) plunging toward us using space weapons technology and has turned into an annual event. 12,700 asteroids have been identified as NEOs with orbits coming within 121 million miles from our main Sun. [10]

Sitchin said that Nibiru orbits around our Sun and passes our Earth every 3,600 years.[9] He was basing this on lunar years. According to the solar calendar, it is actually 3,548 years, which is close enough, as far as I am concerned, considering that our calendars have been manipulated and changed multiple times. When Jesus told us in Matthew 24:36, "that no man knows the day or hour," He knew that mere humans could not possibly figure out when He would return, because the timeline had been manipulated several times. One of the reasons I've spent years writing this book is because during one of my prayers to the Lord while asking Him when He would be returning, I heard Him answer me, in a very clear voice, "I am returning with Nibiru." That got me researching and writing, this book being the result. See my Chapter on the *Messianic Star* and the return of Christ – it is all connected.

The dawn of day can begin an hour before actual Sunrise. That means the dark night sky begins to light up slowly until the light obliterates the darkness and obscures the stars. Then the Sun rises. Similarly, we can call the dawning of the Age of Aquarius to have an orb of influence of 50 years, give or take, like we would give our dawn an hour. Processional Ages typically last 2,160 years, which means that the ending of one age and the dawn of the next age can easily eclipse each other by 50-100 years, coming and going. Likewise, we can extrapolate that same dawn to the return of Nibiru, which coincidentally, the discrepancy between 3,600 lunar years and 3,548 solar years, is only a difference of 52 years, similar to the orb of influence dawn astronomers and astrologers give to Processional Ages.

The Hebrews, Sumerians, Arabs, and Mayans all used lunar calendars. Sitchin's *Lost Book of Enki* developed the lunar calendar into twelve lunar cycles, divided into two seasons - summer and winter. [11] The 10 Plagues of Exodus mark the last fly-by when Nibiru passed the Earth.

Likewise, the day we see Nibiru come out from behind our Sun will be known as the Terrible Day of the Lord, referenced by Old Testament Prophets Zephaniah, Isaiah, and Jeremiah. NASA insiders who claim to have seen Nibiru through the Space Telescope Hubble, said they estimate that after Nibiru orbits around and behind our Sun, it will be energized, taking just 24 -36 hours for Nibiru to pass the Earth. Again, I am asserting that they are estimating it comes close to the Bible Prophecies around the Terrible Day of the Lord. That is going to be a long day, and may well turn the Earth upside down, which they estimate will take the magnetic poles approximately 20 minutes to completely flip. Isaiah 24 predicts that the Lord will turn the Earth upside down. It is referring to this event which will complete and finalize our next pole shift.

> "Behold, the LORD makes the Earth empty, and makes it waste, and turns it upside down, and scatters abroad the inhabitants thereof. And it shall be, as with the people, so with the priest; as with the servant, so with his master; as with the maid, so with her mistress; as with the buyer, so with the seller; as with the lender, so with the borrower; as with the taker of usury, so with the giver of usury to him. The land shall be utterly emptied, and utterly spoiled: for the LORD has spoken this word. The Earth mourns and fades away, the world languishes and fades away, the haughty people of the Earth do languish. The Earth also is defiled under the inhabitants thereof; because they have transgressed the laws, changed the ordinance, broken the everlasting covenant. Therefore, has the curse devoured the Earth, and they that dwell therein are desolate: therefore, the inhabitants of the Earth are burned, and few men left."
>
> (Isaiah 24:1-6)

During the Great Deluge, also known as the Floods of Noah, Nibiru was on our side of the Sun. Nibiru came flying by Earth during the catastrophic flooding that destroyed the ancient civilizations of Atlantis and Lemuria, which according to the Bible, is known as Lucifer's Floods. See, *Who is God?* Book Two.[12] However, according to researchers, this upcoming fly-by will be on the same side of the Sun as Earth, whereas during the Exodus Plagues it was not on the same side of the Sun as Earth.[13]

Marshall Masters asserted that what the Mayans were actually looking for when they ended their Long Count Calendar on December 21, 2012, was the next passage of Nibiru. That date sky-rocketed into the realm of conspiracy theories, false flags, and a whole lot of doomsday authors.

NIBIRU AND ANTISEMITISM

When studying historical trends, I also observe that each time these fly-bys occur, there is some kind of catastrophe not related to Earth changes, but a change in the cosmic order of our own solar system. Immanuel Velikovsky postulated in his books, *Worlds in Collisions* (1950), *Ages in Chaos* (1948), and *Earth in Upheaval* (1952), that the presence of these rogue planets coming into our solar system are responsible for the formation of our asteroid zone and the changes in the order of our planets, Venus, Mars, Saturn, Uranus, and Pluto. [14, 15, 16] He believed Venus was ejected from Jupiter as a comet or comet-like object around 15000 BC. He was violently opposed for his viewpoints. But what if he were right? What if the fly-by of Nemesis-Nibiru caused the disruption in planets and Moons to be knocked out of their respective orbits?

For years, forecasters were solely focused on Earth changes, understandably so, as we see things from our perspectives, and we are on Earth. However, when I connect the information coming from various researchers, astronomers, insiders and whistleblowers with the historical accounts and the theories of Velikovsky and Sitchin, I think to myself, maybe these old Jewish scholars had a point. I also find it interesting, that physicist Albert Einstein, and Immanuel Velikovsky were friends, pen pals, and debated these theories back and forth through letters and Velikovsky's final works, *Stargazers & Gravediggers: Memoirs to Worlds in Collision*, published posthumously by his wife, Elisheva Velikovsky, in 1983, revealed that Albert Einstein was reading *Worlds in Collision*, before he passed away.[17]

I was born in 1961 and grew up reading *Chariots of the Gods* in 1974,[18] which influenced me greatly, opening my mind to the possibilities, and to history. Living and going to school in Israel in 1976-1979; and 1983 was no doubt an education that taught me a completely different worldview than that in the United States of America, or Europe, which was seeing the world through the aftermath of the Jewish Holocaust, the very reason why Israel was established as a homeland for the Jews in the first place. I was naturally led to learn from these Jewish geniuses, Zecharia Sitchin and Immanuel Velikovsky. Both were violently opposed by the scientific communities. With that said, I cannot help but intuit that a great deal of the opposition and criticism that was directed towards them, both during their lifetime and posthumously, was due to antisemitism.

Antisemitism is not just hatred of Jews, it is rooted in jealousy of Jewish works, Jewish scholars, and Jews who pave the way through science, psychology, language, and theology. So much of the criticism is motivated by jealousy, not to mention the frustration with not understanding the concepts the savants spent their lifetimes communicating.

Today there are those who are still motivated by jealousy of Jewish scholars, to seek to make a name for themselves by choosing to debunk them, which essentially is the ancient spirit of Replacement Theology, that which the Church of Rome began with Constantine's Creed. They are so-called Christian scholars who have exalted themselves as Hebrew experts, who reject Jewish works in order to extol their own. Mike Heiser, who holds a PhD in Hebrew Studies from the University of Wisconsin, a Gentile Christian, has spent an awful lot of time and space trying to debunk *Ancient Aliens*, and has attacked Sitchin's scholarship, due to his lack of understanding of Jewish History.

Too many Hebrew Roots Cult members have not a clue about Jewish Roots, the history of when Jews became Jews, and why they have been attacked by Gentiles and satan from the beginning of their seeding on Earth. Following Hebrew Roots does not make you Jewish, as most of these cult members are antisemitic and lack knowledge of Jewish history, which is Jewish roots, not necessarily limited to the language of Hebrew, although that is inclusive. They are quick to retort, that you do not have to be Jewish to study Hebrew. This is true but learning a language does not equate to understanding a group of people and their culture which has historically been rooted in antisemitism, persecution, and suppression. Heiser spent hours making a video attempting to debunk Sitchin, in order to make a name for himself, titled *Sitchin Was Wrong*. I am not even going to quote or cite it here, because I refuse to circulate its lies and antisemitism.

I'm not suggesting that Sitchin was 100% right about everything, but he was a credible Jewish scholar who was proficient in 5 languages and was the first Archeoastronomer who attempted to translate and interpret the Ancient Sumerian Cuneiform Tablets, something Heiser has not done. Hence his blatant jealousy. Furthermore, Sitchin was right about a lot of things, which as time marches forward, is being proved through the confirmation of astronomers validating the Nibiru system, despite renaming it Trappist-1.

As a researcher, lifelong investigator, and a retired paralegal of 25 years, I have learned to incorporate a standard in all my research, and that is to eat the meat and throw away the bones. Meaning, extract what is relevant and worthy, and discard the rest. No one is perfect, and all humans are flawed, but if they present a thesis, and a theorem (proven theory), and others continue to reject and debunk them, it is not a matter of science application, it is a matter of cognitive dissonance, in this case, blatant antisemitism and disrespect for the scholarship of Jewish authors and researchers because they beat them to it, or wrote the books that they wished they had written, or, they have been brainwashed to believe a lie and the only way they can handle the truth is to debunk it, which is a toxic combination of jealousy, pride that causes cognitive dissonance.

What Mike Heiser has done is create a false narrative on the internet through the Christian Fringe and Hebrew Roots Cult, that triggers them, like some kind of mind control programming, that whenever the name *Sitchin* is mentioned, ignorant people of his actual works jump in to debunk and post, *Sitchin was wrong*, along with all kinds of mean-spirited, inaccurate and inappropriate responses. This reveals their ignorance, brainwashing and Jew-hatred. Most, if not all, of these posters have not even read Sitchin's exhaustive research, yet they blindly follow Heiser, because let's face it, it's cool to hate Jews, right? Antisemitism is racism.

CHAPTER 7: NIBIRU: MYTH OR SCIENCE?

I have no doubt that when Heiser becomes aware I've criticized his defamation and slander of Sitchin, he will attack me too because I too am Jewish, only I'm a Jew for Jesus, a Messianic Jew, which a lot of Christians can't embrace because I reject their Christian cults, religious spirits and Replacement Theologies that are based on the doctrines of antisemitism. I am willing to take that risk because he will only confirm his blindness of Jews and Jewish history, regardless of his knowledge of the Hebrew language. Suffice to say, his criticism of Sitchin was petty which says more about Heiser than Sitchin's imperfections.

Picking apart Sitchin's exhaustive 12 books of research on Ancient Sumerian Cuneiforms, by criticizing a word here and another word there, which in my opinion and discernment was motivated by jealousy and cognitive dissonance, is exactly what Modern Day Pharisees of Christian Hebrew Roots do (Replacement Theology), which is what got the Ancient Pharisees rebuked by Jesus for their jealous religious spirits by scolding their petty blindness, who said, "you strain a gnat, yet swallow a camel?" (Matthew 23:24) It reminds me of the song by Louis Armstrong and Ella Fitzgerald, *You say Tomato, and I say Tomato*. By the way, I'm married to a Brit, and we often joke that America and England are two countries separated by a common language – English! Yet we speak differently, we have different ways of saying the same thing. In laymen's terms, those who focus on the small things, will never see the bigger picture. My approach to research, is, *eat the meat and throw away the bones!* Or, if you're a vegetarian, *drink the juice and spit out the pits*.

If it is any help to know that others have proved, via numerous YouTube videos, in rebuttal to Heiser's intention to prove Sitchin wrong, actually proved him right! Ah, the irony of it all! So perhaps, we can put it down to confusion? Or just not understanding the *true history* of the Hebrews?

Nobody knows it all. Like I say all the time in my panels and talks, "I know what I know, and what I don't know, I don't know, you know?" As researchers that is what we do, we learn, we investigate, we gather knowledge to connect dots so we can achieve understanding. Granted, Sitchin's work is heady, the ancient languages are like Greek to most of us, well, actually ancient cuneiform, and many are confused. What's important is to get the gist of it.

I had a conversation while working alongside the late great Apollo 15 Astronaut Al Worden in the summer of 2019 during the SPACEFEST Convention, about Zecharia Sitchin's work. Worden was influenced by his work on the Annunaki to the point that whenever the media posed the question to him if he believes in aliens, his spiel was, "of course I do, I see one every morning when I look in the mirror!" Worden believes he is the alien. He shared with me, that, "even if Sitchin is half right," he felt that was huge. Coming from the mind of a true scientist, space explorer and a kind human being.

Unfortunately, Astronaut Al Worden died on March 18, 2020. Worden is best known as America's first astronaut to perform a Deep Space EVA (Extra Vehicular Activity) on the return from the Moon aboard Apollo 15. He died with his secrets of what he saw on and around the Moon. Not even I could get him to talk about it.

His energy at SPACEFEST was effervescent. He was spunky, cheerful, and approachable. My first impression when he greeted me first sitting at my table, was that he was a super nice guy. We worked together for two days, as my Author Table was 6 feet away from his. I was literally the elephant in the living room, talking up my books about Disclosure of the Alien Presence, which he obviously overheard and lived with my banners for the weekend. I felt him watching me, and while he knew that I knew what he knew, I knew he wasn't able to say anything. Nevertheless, he was respectful, courteous, and a perfect gentleman to me.

Due to the Non-DISCLOSURE agreement by which all NASA astronauts are bound, he had to remain silent about it. When we got chatting, it was like it was a given, I didn't bother him with questions about it, and he didn't mention what I was doing either, but we connected and bonded through discussing 'who' created us, and just how much of the ancient Sumerian material was true. He even reiterated to me his running spiel that he used whenever the media questioned him if he saw aliens or their UFOs in Space, which I didn't even have to ask him, but after talking about the Creator and the Anunnaki aliens, he wittingly retorted with, "Whenever I look in the mirror, I realize, I'm the Alien!" Al had great wit and humor and he was a brilliant rocket scientist. At SPACEFEST, he was in high demand, and he led the auction. His energy for an 88-year-old was inspirational and awesome. I was sad to learn of his passing through our shared Facebook Group. He was a great friend to our Space Hipster community.

Until the Messiah returns to set everyone straight on the manipulations of history, it is an effort in futility to argue about stuff that is tens, if not hundreds of thousands of years old. Suffice it to say, a lot has been lost in translation. The Chief Naval Astronomer, Robert Harrington confirmed Sitchin's work on Nibiru, so perhaps Sitchin was not so wrong after all.

Furthermore, I knew Sitchin. He granted me an informal sit-down interview with him and his wife at the *When Cosmic Cultures Meet Conference*, in Washington, D.C. in 1995. They were kind and hospitable to me, we shared a heritage, and bonded over Israel. I knew little about his work, but I was intrigued and wanted to learn. So, after our interview, I was so impressed with the kindness, integrity, and humility of this brilliant man, that I submitted to his mentoring through his International Bible Studies and attended his lectures and classes on a regular basis as often as I could, especially after I moved to Colorado, where he returned often to give lectures and classes. When you get to know someone, it helps a great deal in understanding their written works and scholarship. I think Heiser missed out on these encounters when Sitchin was alive, which differentiates someone being authentic, who spent his entire life creating a body of language, that very few can match.

Anyone can learn a language, but in order to understand the roots of antisemitism, you have to know and understand Jewish history, and why Jews have been targeted, persecuted, and maligned by Gentiles since their extraterrestrial DNA was introduced on Earth. With that said, there are many Gentile Christians who support and respect Jews, and their works, along with their generous support of Israel, the Jewish State.

I am here to connect the dots and to offer knowledge that leads to understanding. I wish people would stop arguing about this and come to the common knowledge that we would not be talking about Nibiru if it were not for Zecharia Sitchin. Again, I'm not saying he was 100% correct on every word he wrote, but he certainly influenced researchers, like myself and many others, to follow his path of bringing ancient knowledge forward, where it can hopefully be applied. The same needs to be said for Immanuel Velikovsky, whose works now, are looked upon as 'prophetic.' As during the advent of the formation of NASA, and then the deep space Hubble Telescope, has proved many of Velikovsky's theories correct.

To those who prefer to pick apart their theories instead of learning how to extrapolate their insights, truths, and knowledge to be applied today, consider this: even a broken clock is right twice a day. And with the amount of Calendar changes, manipulations to time, and different calendrical systems, we really need to face the music here: we are dealing with a broken clock. A type of cosmic disorder, trying to connect the dots to our past and more importantly, our history with this ancient binary Nemesis-Nibiru System. I think both Sitchin and Velikovsky were certainly on to something.

According to Sitchin's interpretation of the Sumerian tablets, to the Annunaki, who inhabit the Nibiru system, the 3,600-year cycle, was considered to be "One Shar" to the Annunaki. This implies time travel, similar to what the Bible implies describing time to the Lord Yahuah, "a thousand years is like a day to the Lord, and a day is like a thousand years." (See, Psalm 90:4; 2 Peter 3:1)

Notes:
1. Ella LeBain, *Who's Who in the Cosmic Zoo? A Spiritual Guide to ETs, Aliens, Gods & Angels*, Book One, Chapter Five: *Annunaki (Nibiruans)*, pp. 110, Third Edition, 2013.
2. Mountain Wolf Blog: *Evidence for Nibiru or "Planet X"*, http://mtnwolf63.wordpress.com
3. Leslie Mullen, Sun's Nemesis Pelted Earth with Comets, Study Suggests, https://www.space.com/8028-sun-nemesis-pelted-earth-comets-study-suggests.html, March 11, 2010
4. *Chilean astronomer CARLOS MUÑOZ FERRADA Predicts Hercolubus aka Planet X*, Excerpts Reprinted from the JOURNAL OF TODAY, of El Salvador, https://www.rawgist.com/chilean-astronomer-carlos-munoz-ferrada-predicts-hercobulus-aka-planet-x/, May 19, 2015.
5. *Carlos Ferrada Predicts Planet X*, Aug 20, 2013 https://www.youtube.com/watch?v=N9f-Bhub0Lg
6. The *Who's Who in the Cosmic Zoo* Facebook Page, https://www.Facebook.com/Whoswhointhecosmiczoo
7. *Russ Report Finding 10th Planet* (AP) February 12, 1960, New York, NY, Reprinted in *The Milwaukee Sentinel* – February 13, 1960.
8. *Solar System Warming–and its Implications*, http://prof77.wordpress.com/2010/08/07/why-all-planets-in-our-solar-system-are-warming-and-its-implications, August 7, 2010.
9. Zecharia Sitchin, *The 12th Planet, Book One of the Earth Chronicles – Astonishing Documentary Evidence of Earth's Celestial Ancestors*, Avon Books, New York, 1976.
10. Planetary Defense Conference Exercise – 2019 – NASA, https://cneos.jpl.nasa.gov/pd/cs/pdc19/

11. Zecharia Sitchin, *The Lost Book of Enki – Memoirs of an Extraterrestrial God*, Bear & Company, Vermont, 2002.
12. Ella LeBain, *Who is God?* Book Two, Skypath Books, 2015
13. Marshall Masters, *Being in it For the Species*, CreateSpace, 2014
14. Immanuel Velikovsky, *World's in Collision*, Doubleday; Second Edition, 1951.
15. Immanuel Velikovsky, *Ages in Chaos*, (1948), Doubleday Reprint Edition, February 1, 1952.
16. Immanuel Velikovsky, *Earth in Upheaval*, (1952), Dell; Mass Paperback Edition, 1972.
17. Elisheva Velikovsky, *Stargazers & Gravediggers: Memoirs to Worlds in Collision*, 1983.
18. Erich Von Daniken, *Chariots of the Gods*, Bantam Book, Putnam's Edition, February 1970.

CHAPTER EIGHT

OUR SOLAR SYSTEM IS PERTURBED

"From a far-away land they came, from the end-point of Heaven does the Lord and his weapons of wrath come to destroy the whole Earth. Therefore, will I agitate the Heavens and Earth shall be shaken out of its place. When the Lord of Hosts shall be crossing, the day of his burning wrath."
Isaiah 13:1

"Out of destruction comes life."
Bhagavat Gita

In Sumerian, the name *Nibiru* means *Planet of Crossing*. The Sumerians and Egyptians talked about how the planet Nibiru had an elliptical orbit, as opposed to a normal horizontal orbit. The planet took around 750,000 years to come between Mars and Jupiter, and when it did, it created devastation on all the planets during its passing.

There are eight planets and four Moons in this adjacent solar system that surrounds a brown dwarf star, now intersecting within our solar system. This system is draining our Sun. It is also attracting solar cores that are draining our Sun as well, acting like little vampires. This is one of the reasons contributing to the cooling effect on the Earth. There are going to be some more events happening with our Sun that will continue to change Earth including all the other planets in our solar system.

Nibiru is the only object with a rebellious elongated orbit, that orbits around our solar system approximately every 3600 years. It has entered into our solar system. I believe our Sun get eclipsed regularly by one of these objects, which explains the darker dimmer skies around Sunset, and in the middle of the day, which is seen on weather cams in Alaska and Mexico.

Burak Eldem of Turkish origin was influenced by Zecharia Sitchin's work on decoding the past. Eldem published a book called *2012: Rendezvous with Marduk*

(Inkilap (2003) which became a best-seller in Turkey. It is the first of a trilogy of books called *The Hidden History*. [1]

Like Sitchin, Eldem refers to another distant planet in our system, which they both call Marduk and/or Nibiru, presumed to have an orbit of 3,661 solar years. Eldem suggests that the orbit of 3,661 years was written as three wedges based on the sexagesimal (based on 60) ancient mathematics used in Mesopotamia, through reckoning by sixtieths. Sexagesimal became the ancient Babylonian decimal system relating to the number 60, using a fraction system based on sixtieths (i.e., with a denominator equal to a power of 60), as in the divisions of the degree and hour. Sexagesimal (base 60) is a numeral system with 60 as its base. It originated with the ancient Sumerians in the 3rd millennium BC, was passed down to the ancient Babylonians, and is still used — in a modified form — for measuring time, angles, and geographic coordinates.

The first wedge is 60×60 (3600), the second is 60×1 (60), and the third is 1, thus 3600+60+1 which equals 3661. Eldem argues that this was misunderstood by the Babylonian Jews which is why they created the enigmatic 666. Hence the number of the Beast was the number of solar years of this imagined planet's orbit. Conclusion: Nibiru is the Beast.

Eldem argues that civilization was born around 3100 BC and that the Mayan Fifth World Age and Hindu Kaliyuga simultaneously starts around 3100 BC. I do not know anything about the Kaliyuga, but the fifth age of the Maya is nothing but a transfer of Aztec beliefs upon the Maya who had no such ages. He goes on to suggest that "the eve of the Egyptian Kingdom with King Narmer (or Menes) also coincides with this year" (never mind that there existed a long process of politicization before this king). Further, "Harappan civilization of Indus, is believed to have started around that time. Ancient Sumerians had captured the Obaid cities (beginning with Uruk) and established their city-states exactly around 3100 BC. The ancient La Venta civilization of Meso-America is believed to have its beginning around these years." This is revealing, because he believes that La Venta civilization is something preceding the Olmec, despite the premise that the Olmec claimed they built La Venta. This is likened to the Egyptians claiming they built the pyramids, which clearly were built by giants and other worldly beings.

He goes on, and the 3100 BC date includes Caral in Peru and Stonehenge. Eldem concludes that this "was the time of relief" after the Great Flood; which allowed people to build sophisticated cities…the author suggests a date around 3150 BC for the cataclysmic events which eventually became the main inspiration for the popular myth known as "The Deluge."

Eldem focuses on a history of catastrophes, the most important series of catastrophes he discusses appeared around 1649 BC. He suggests that "the Olmecs appeared as the cultural heir of the ancient La Venta civilization who vanished around 1650 BC.".

It is the "tenth" planet that causes these catastrophes, while incorporating the traditional model with Pluto as the ninth planet. Its last passage was 1649 BC and the very reason why the Maya stated 12.21.2012 as their date for the 'end of the Fifth Sun, ("Fifth World Age").

He goes on and says that "the correlation between Marduk's orbital period (3661 years) and World Ages, also helps us to decode the mystery of Tzolkin, the 260 days sacred calendar. One "World Age" was 13 Baktuns; and one Baktun was 20 Katuns. So, Tzolkin was formularized as 13 "Uinals" (20 days "months"). This was a miniature model of a "World Age" in fact! And else? Let's change the "year" unit in a "World Age" and use "Tzolkin" (260 days period) instead of a solar year: 5125.36 x 260 = 1,332,593.6 days, 1,332,593.6 / 365.24 = 3648.54 years…This value is very near to 3661 years orbital period of Nibiru." See my Chapter later in this book on, *Calendar Shenanigans*.

THE NUMBER OF THE BEAST

The research of Burak Eldem cannot be ignored; he claims the orbit of Nibiru should be interpreted as a Sumerian sexagesimal number. Thus, we should read the orbital timeframe of 3600 years according to a "60 based" system. We then get; $60°$ (1) + 60^1 (60) + 60^2 (3600) which adds up to 3661 years! Three glyphs in cuneiform writing resemble "the number of the beast" (666) that describes the period of Nibiru's revolution.

There is really no need to distinguish between "long years" (365 days) and "short years" (360 days) in the aspect of the 3661-year period; it is a solar year any way you see it. And the 3661 number is a measure of Earth years, based on our planets orbit around the sun. However, the length of the year has many other implications to the daily life; no wonder there are various solar calendars in stone erected all over the globe.2

Eldem seems to reiterate in agreement what Immanuel Velikovsky laid claim to in his theses on *World's in Collision*, about the change in the orbit of the planet Venus. This makes sense, because both Venus and Mars are similar to Earth, and at one time had surface life, despite the fact that researchers today believe Mars still does. Anyway, Eldem said, "during one of its orbital passages in 5310 BC, Marduk caused Venus to change its orbit and wander in our Solar System for almost 2,200 years. Prior to this confrontation, Venus was probably at a point between Mars and Jupiter, now called as "the asteroid belt." The event was a "stunning" one for the inhabitants of the Earth; and the first impact's result was severe: The Black Sea was flooded around 5300 BC…The following 2,000 years was a real chaos in Heavens. Early Neolithic settlements like Catalhoyuk, Hacilar and Jericho, became shelters for frightened human beings.

Zecharia Sitchin and Burak Eldem also suggested this planet, which looks a lot like CR105, because of the similar orbit; however, the crossing path of CR105 is very different than Nibiru or Planet X. 80% of Nibiru's orbit lies much further from the Sun, whereas CR105's orbit is 60% away from the Sun, and 40% near the rest of the planets. According to historians, the Ancient extinct now, Planet Tiamat, that existed between the orbits of Mars and Jupiter around 510 million years ago, was a victim of Planet X, when Tiamat collided with one of the Moons of Nibiru. The collision caused it to split into two halves, one half broke into many pieces which became the asteroid belt as well as the Moons of Mars, while the other half became our home, planet Earth.

CR105 is labeled as (148209) 2000 CR105 discovered by Marc Buie, February 6, 2000. CR105 is a trans-Neptunian object and the 10th-most-distant-known object in the solar system as of 2015. Considered a detached object, it orbits the Sun in a highly eccentric orbit every 3,305-3,439 years at an average distance of 222 astronomical units.

This strange planet is known to be 20 times bigger than Jupiter, with a burning Moon that acts like Nibiru's personal Sun. Since Nibiru goes much, much further away from our Sun, this theory makes sense and stands out. The Annunaki, who are supposedly the citizens of Nibiru, came to planet Earth around 25,000 years ago and gave a lot of knowledge and detail to the developing humanoids, hence our present mythologies and core beliefs of the world's major religions. The Mayans even predicted the existence of Nibiru, or according to them, a certain dark energy in shape of a planet which would be coming near Earth in the distant future. Every time this planet came around, entire civilizations from planet Earth were wiped out.

THE OORT CLOUD OF NIBIRU

Swedish researcher Joachim Nilsson claims there is a difference between the magnetic effect and the gravitational effect of Nibiru, the magnetic effect is what causes the Pole Shift. When considering the effects of gravitation, we must also address the solar wind of charged protons that both Nibiru and the Sun emanate. In the balance point between the outward directed solar wind and the gravitational force directed inwards, there is a field of dust, stones and asteroids referred to as the "Oort cloud."

The Sun has this hazardous field, but the same characteristics can surely be applied to Nibiru. Nilsson estimates that the Oort cloud of Nibiru precedes the arrival of the star itself with 63 years. He concludes that the entire tribulation of its passage lasts close to 130 years, as it is the time span of Nibiru Oort cloud from one edge to the other.

As the Oort cloud of the star Nibiru reaches Earth it will trigger an enormous earthquake. During that time, Earth will also be hit by a number of rocks, ranging from dust particles up to, in the worst case, large asteroids. This great earthquake will have severe consequences to the cities of Earth, there will be major damages to infrastructures. The following Pole Shift, on the contrary, will completely re-write the map of planet Earth; there will be quakes, volcanoes, tsunamis and possibly a severe deluge.

CHAPTER 8: OUR SOLAR SYSTEM IS PERTURBED

TIMELINES AND TIMEFRAMES

Is it possible to predict when Nibiru will return if there have been so many calendar shenanigans? Let's go back to the Bible, which most historians and bible scholars can agree there was a devastating earthquake in the years of King Uzziah of Judah. The earthquake was so severe that it profoundly impressed the prophets such as; Amos 1:2, Zechariah 14:5 and Isaiah 24:19. The year of the great earthquake 750 BC, was a precursor that our Sun's Binary Brown Dwarf, Nemesis-Nibiru once again was about to leave the northern skies and move in below the ecliptic. Shortly after that time, the calendars all over the globe was changed to a 365-day year, from previously 360 days. Today, in order to catch up with the actual 360-day solar year, we have leap year, which adds back a full day every four years, to make up for the quarter day it now loses every year. Oy vey! More *Calendar Shenanigans*.

Nilsson believes according to his calculations that the passage after the 750 BC quake occurred around the year 687 BC. We are told from Chinese records, that "stars fell like rain" in this time, maybe an indication of a major cosmic event. In the year 702 BC, 15 years earlier, we are told in 2 Kings 20:11, that Hezekiah received a sign from God: "And Isaiah the prophet cried unto the LORD: and he brought the shadow ten degrees backward, by which it had gone down in the dial of Ahaz." In that time there was a disturbance of the Earth axis and rotation, even if it wasn't as severe as in the days of Joshua. Nibiru magnetic influence halted and reversed the Earth rotation for a brief time. Joshua 10:12 states: "Sun, stand thou still upon Gibeon; and thou, Moon, in the valley of Ajalon." Most researchers, including Nilsson agree that this episode was clear evidence that a Pole Shift occurred despite the fact that its exact year remains elusive and in dispute due to in my opinion, all the *Calendar Shenanigans*.

EARTHQUAKES AND NIBIRU

The grand earthquake caused by the Oort Cloud passing, is believed to come first, then the Pole Shift follows, later followed by the actual passage of the star. Shortly after the 750 BC earthquake, the Babylonian calendar was changed from 360 to a 365-day year, marking a period that represented the beginning of the 7th Baktun of the Mayas. Other sources suggest the city of Rome was founded around this time and that it was the beginning of the Olympiads in Greece of 2x4 years.

Nilsson points out that there is a time period between the initial earthquake and the following Pole Shift which he puts around 702 BC when Earth experienced a minor Pole Shift marked by the rotational and magnetic properties of Earth and Nibiru still being in harmony with each other. That at the passage of 687 BC, several thousands of King Sennacherib's massive army died, not from "natural causes" but rather by an intervention from God, a celestial thunderbolt.[2]

Sennacherib is one of the most famous Assyrian kings with Anunnaki roots, for the role he played in his campaign in the Levant's ancient Israel. After Sennacherib inherited the throne in 705 BC, Marduk-apla-iddina retook Babylon and allied with the Elamites. In 701 a rebellion, backed by Egypt, instigated by Merodach-Baladan (2 Kings 20:12–18; Isaiah 39:1–7), rebellion broke out in the Levant. Sennacherib reacted firmly, by taking the rebel cities, except for Jerusalem, which, though besieged, was spared on payment of a heavy indemnity (2 Kings 18:13–19:36; Isa. 36:1–37:37).

The return of Nibiru, implies karmic retribution for past passages and rebellions on Ancient Israel, which is why Bible Prophecies all point to the final End Times Star Wars over the ancient portal in Jerusalem. As Nibiru approaches from below the ecliptic, we can expect similar earthquakes that happened in 702 BC. [2]

According to Burak Eldem, planet Earth had an extremely severe period around this time. It was the beginning of several years of tribulation and upheavals, undoubtedly connected to Moses and the Exodus. There was a strong earthquake, a "pillar of fire" in the sky, it rained "blood," an impenetrable darkness appeared, the seas roared, and huge waves engulfed the shorelines.

Nilsson points out, three major events are connected to the approach of Nibiru. We have a great earthquake and an accompanying calendar change. The rotational stoppage and occasional Pole Shift follow. Finally, the actual Passover (Transit) of Nibiru, when the heavenly bodies scorch each other with tremendous thunderbolt discharges. The time difference between the Earthquake and the Pole Shift is approximately 48 years. From the Pole Shift to the Passover of Nibiru, we count 15 years. If we take the entire approach into account, from earthquake to Passover, we get 48 + 15 years, resulting in a total period of 63 years. This timeframe gives clues as to when the "Long Day of Joshua" might have taken place. We simply add 48 years to 1642 BC, ending up at 1594 BC. [2]

NIBIRU'S RETURN

It takes 2769 years for the dark star Nemesis-Nibiru to move along an elliptical orbit below the ecliptic, then it rises above the ecliptic, following a smaller orbit for a period of 892 years. Then after descending below the ecliptic again, the star crosses its own path. Maybe this is why Nibiru was called by the Ancient Sumerians: "Planet of the Crossing."

Nilsson posited, that Jesus Christ might be seen as the human embodiment of the star Nibiru and of Anu, the God of the planet Anu, from where the word Anunnaki is derived. He questioned, if the hazardous crossing of Nibiru's path filled with large asteroids, stones, and gravel happened during the crucifixion? [2]

As I wrote in Book Three, *Who Are the Angels?* in my Chapter on Nibiru and the Second Coming of Christ, that the crucifixion of Christ seemed to be one part of this the Nemesis-Nibiru system crossing Earth's path with our sun, as the Bible records that there was 3 hours of darkness during midday that followed the crucifixion. This is no ordinary eclipse of our Moon over the Sun, which was impossible, as scripture records it was a full moon phase, which follows the Jewish Calendar of Passover.

Our Sun is already being influenced by dust and debris associated with the Nibiru complex. Strong indications point to our Sun seems to be reacting to a cloud of "red iron oxide" that is emanating from the solar wind of Nibiru. As a result, the sun gets less active, hence the last decade of no sunspots, which causes climate change as our weather on our planet has been immediately affected. Was the deployed Sun Simulator compensating for this factor? At the same time, infrared waves from Nibiru makes the climate warmer. This is what's behind the warming not just on Earth but all the planets in our solar system. Earth's glaciers have begun to melt, with unprecedented monster storms. Climate changes cause rain to increase in some areas while causing drought in other places, the balance of nature is severely disturbed.

What is to be expected during the upcoming passage? Nibiru has most likely performed its customary tilting at the Aphelion point and moved back towards the solar system. It's clear that Nibiru is approaching us again, as our magnetic pole fields are shifting and are heading for reversal.

Can humans accurately calculate the arrival of Nibiru? There have been a lot of miscalculations based on misunderstandings of how the system works, which is very different than tracking an incoming comet or asteroid, that seem to follow an ordered orbital path. Nibiru is the cosmic maniac, it is the red headed stepchild of our galaxy, and it's this reason why its dark star Nemesis is given the moniker as our Sun's evil twin.

Matthew 24:36 states; "of that day and hour no one knows, neither the angels of heaven, nor the Son, but the Father alone."

According to the calculations of Nilsson, the dreaded Pole Shift is NOT expected to happen until sometime around 2068. But the times leading up to that year can be troublesome enough. While there is no need to spiral into despair, anyone with a flexible mind will manage and not only survive but can still have a good life. His advice is, don't focus too much on stocking up food piles, as you will only attract thieves. Instead, stock up on resourcefulness; and by all means, get your life right with God through Christ, then you will overcome the situation.[2]

When observing and studying these celestial events both past, present and future, one must assume there are "blind forces at play." However, there are many people who equate Nibiru with the celestial manifestation of Christ and I am not the only one. Let's all be humble and abide in faith, the Lord stands by our side. "I am the Alpha and the Omega who is, and who was, and who is to come, the Almighty." (Revelation 1:8)

HISTORY OF PLANET X

James McCaney, an expert on Planet Nibiru and Mayan history, explained that around 10,000 years ago major devastation occurred which destroyed many civilizations on our planet. He also explained how ruined cities in South America vanished, not because of war or plague, but major physical destruction on Earth. He also went on to say that before Nibiru passed us by 10,000 years ago, the North Pole was somewhere in the state of Wisconsin, while the South Pole was somewhere in the Pacific Ocean. If he is right and this event did occur because of Planet X or Planet Nibiru, then we should not worry about it for the next 740,000 years or so, right? Wrong. Remember, even if Nibiru crossed its path from between Jupiter and Mars, it is now surging upwards to make its longest route around the Sun. This is due to its elliptical orbit which goes in a roundabout tilted entry as it comes remarkably close to the Sun on one end, while 80% of orbit is furthest away from the Sun. Therefore, the Earthquakes happening in Japan, Chile and other places could be due to the magnetic pull from Nibiru that is increasing as it nears our plane. The pull from Nibiru will increase the gravitational forces of each planet in a rubber band effect.

The last Fly-By falls during the Floods of Noah:

> "...For as in the days that were before the flood they were eating and drinking, marrying, and giving in marriage, until the day the Noah entered the ark."
>
> "And knew not until the flood came, and took them all away; so, shall also the coming of the Son of man be."
>
> (Matthew 24: 38-39-KJV)

The whole solar system is being affected by the passing of the Nemesis-Nibiru system, which is not just limited to magnetic shifts and climate changes, Earthquakes, volcanic eruptions, floods, tsunamis, tornadic activity, and monster storms. All the planets in our solar neighborhood are feeling and showing signs of changes in their atmospheres, in the way of an increase of comet activity which is a sign that Nibiru is here. When it will pass the Earth, is everybody's guess. Those who know, are not telling.

Dr. Mike Lockwood from Rutherford Appleton National Laboratories in California, who has been investigating the Sun, reports:

- Since 1901 the overall magnetic field of the Sun has become stronger by 230%. This means changes to all the planets and objects that orbit around the Sun.
- Our Moon is growing an atmosphere. There is now a 6,000-kilometer-deep layer of Natrium that was not there before.
- Mercury shows unexpected polar ice, discovered along with a surprisingly strong intrinsic magnetic field.

CHAPTER 8: OUR SOLAR SYSTEM IS PERTURBED

- Venus: There is a 2500% increase in auroral brightness, and substantive global atmospheric changes in less than 40 years.
- Mars: Global Warming, huge storms, disappearance of polar ice caps.
- Jupiter: Over 200% increases in brightness of surrounding plasma clouds. Huge belts in the giant planets' atmosphere have changed color all the while space rocks have been hurtling into the gas giant.
- Saturn: Major decrease in equatorial jet stream velocities in only 30 years, accompanied by a surprising surge of X-rays from equator.
- Uranus: Big changes in brightness, increased global cloud activity, formerly a very calm atmosphere.
- Neptune: 40% increase in atmospheric brightness.
- Pluto has been scapegoated by its apparent demotion as a planet in order to confound those who are researching Planet X, (the 12th Planet or the tenth Planet or Planet Nine) by taking Pluto out of the planetary count. If it were not for the anomalies in Pluto's unusual elliptical tilted orbit from the rest, scientists would not have discovered Planet X/Planet Nine which affects its orbit.[2]

Scientists at the Japanese Kobe University believe another planet – up to two-thirds the size of the Earth – is orbiting in the far reaches of our solar system. "The possibility is high that a yet unknown, planet-class celestial body, measuring 30% to 70% of the Earth's mass, exists in the outer edges of the Solar System. If research is conducted on a wide scale, the planet is likely to be discovered in less than 10 years," Kobe University said in a statement. The research by Kobe University comes two years after school textbooks had to be rewritten after Pluto got demoted from the list of planets by the IAU (International Astronomical Union).

Pluto was discovered by the American Astronomer, Clyde Tombaugh, in 1930 in the so-called the Kuiper belt, (a chain of icy debris in the outer reaches of the solar system). In 2006, the IAU (International Astronomical Union) ruled that the celestial body was merely a dwarf planet. Astronomers concluded that Pluto's oblong orbit overlapped with that of Neptune. They excluded it from being a planet because of its unusual elliptical solar orbit, and defined the Solar System as consisting solely of the classical set of Mercury, Venus, Earth, Mars, Jupiter, Saturn, Uranus, and Neptune.

Planet X would have an elliptical solar orbit as well, the Kobe University team said, estimating its radius was 15 to 26 billion kilometers. The team noted that more than 1,100 celestial bodies have been found in the outer reaches of the solar system since the mid-1990s. "In coming up with an explanation for the celestial bodies, we thought it would be most natural to assume the existence of a yet unknown planet," the university stated. [2]

PLANET X OR NIBIRU?

The phrase, *Planet X* was first coined back in 1781 by Perceval Lowell. It was believed to be a huge planet that was outside the Kuiper Belt that perturbs the orbit of our known outer planets from Saturn, however, it turned out to be the planet Uranus.

On March 3, 1781, William Herschel discovered Uranus, which was the first planet discovered through a telescope. This discovery launched the search for Planet X. Then in 1846, French Astronomer Alex Bouvard first observed the perturbations of Uranus to the existence of Planet X, but it turned out to be the planet Neptune. Unfortunately, Bouvard died before the exact coordinates of Neptune could be confirmed by French Mathematician Urbain Leverrier and British Astronomer John Couch Adams. Using Bouvard's observation, on September 23, 1846, Urbain Le Verrier discovered Neptune. Neptune was the first planet discovered by mathematical precision. Calculations on the planets Uranus and Neptune indicated the presence of another planetary body, believed to be Planet X which was well beyond Neptune's orbit.3

The search for Planet X ceased in Europe but continued in America. In 1894, the Lowell Observatory was established in Flagstaff, Arizona by Perceval Lowell, who was an independent businessman, Astronomer, and mathematician. Lowell originally founded the observatory to study the planet Mars, but he kept observing perturbations with Neptune. This led Lowell to discover the planet Pluto in 1929.

Nevertheless, the search for Planet X continued on the fringe of Astronomy. It was not until 1978, when it was determined that Pluto lacked the mass to be Neptune's Perturber, the search for Planet X picked up more steam. In 2006, the IAU demoted Pluto to a Dwarf Planet. Pluto was not alone. They discovered five satellites orbiting around the dwarf planet, the largest known as its Moon, Charon. The smaller four were later discovered through the Hubble Space Telescope in 2005, 2011 and 2012, and officially named, Nix, Hydra, Kerberos, and Styx in 2013 by the IAU.

Then the search for Planet X continued, because of the perturbations observed in Pluto's orbits, which led to the discovery of Trans-Neptunian Objects (TNOs). There were so many TNOs being found to rival Pluto in size. In the quest to locate Planet X, aka Nibiru, the following exoplanets were discovered in our Solar System since 2012. They are classified as Dwarf Planets, also called, Trans-Neptunian Objects (TNOs). They all tend towards elliptical and slightly tilted orbits in our Solar System. Pluto (Moons: Charon, Styx, Nix, Hydra, Kerberos); Sedna; Eris (Moon: Dystonia); 2007 Or 10, Makemake, Haumea (Moons: Namaka, Ki'iaka); Quaoar (Moon Weywot); Orcus (Moon: Vanth).

Astronomers detected that there was something perturbing Pluto's orbit, and found these TNOs within orbits of our Solar System. All these outer Solar System objects share a trait that continues to prove there is another large unknown planet perturbing its orbits. The search for Planet X has been going on for over a century.

In 1950, Immanuel Velikovsky published his groundbreaking book, *Worlds in Collision*. He was good friends with physicist Albert Einstein. They wrote letters back and forth comparing each other's math and hypotheses, known as the Einstein-

Velikovsky Correspondence. It has been said and believed, that when Einstein died, he was reading Velikovsky's book, *Worlds in Collision*. The publication of Velikovsky's book acted as a bombshell that rocked both the literary and scientific realms, and these controversial reverberations continue to this day. [4]

Velikovsky suggested a catastrophic history of our solar system, which at once resolved innumerable, historical, astronomical, and geological enigmas, and challenged multiple scientific dogmas of his time. The academic community was intensely polarized. Several denounced his book as rubbish, without even reading it because he was Jewish, while other scientists and scholars praised Velikovsky's method and revolutionary conclusions. Velikovsky wrote a book before he died, titled *Stargazers and Gravediggers: Memoirs to Worlds in Collision*. It was later published in 1983 by his surviving widow, Elisheva Velikovsky. In it, he tells of the unprecedented treatment he endured at the hands of his critics, some motivated by cognitive dissonance and disbelief, while others clearly rejected his theories due to antisemitism. Velikovsky was viciously attacked by mainstream science for his theories, many have been proved by the advent of NASA's space technologies and deep space telescopes. He used ancient accounts to question the "Darwinism of Evolution" from evidence of previous cataclysms, which opened interest in Planet X System by both mainstream science and the media. [5]

On March 3, 1972, NASA launched Pioneer 10 Spacecraft to explore outer planets. According to insiders, it was really a secret mission to find Planet X. The mission carried with it the golden disc which produced and organized by the late astronomer Carl Sagan with the intention and forethought if Pioneer 10 encountered extraterrestrials. Pioneer 10 communications are no longer possible, so there is no way to know if extraterrestrials found Carl Sagan's golden disc. Only time will tell, as disclosure of extraterrestrials and aliens ensue, and more humans begin to have contact.

Following Pioneer 10, Archeoastronomer and linguistics author Zecharia Sitchin published his translations of the Ancient Sumerian Cuneiform Tablets in 1976. The Sumerian texts revealed a race of extraterrestrials he called Annunaki living on a planet called, "Nibiru." Somehow, he concluded they left a record, written in stone, that they bioengineered early humans on Earth so as to turn them into a slave race. They needed humans with enough intelligence to follow orders, be capable of the tasks at hand, and mine the Earth for gold for them. The texts revealed they needed it to protect their home, Nibiru. [6]

Here is where we're all split as researchers as to what he meant, what may have been misinterpreted to mean a planet, but may have been a gigantic spaceship city that they maintained, and the gold was needed to protect their heatshield from space radiation. Here is why that is plausible; we know now that there are Moons, satellites that are artificial. Our Moon has been determined to be an artificial hollowed-out sphere. They tested it, and it rang like a bell.5 Just because the Apollo lunar modules were built for the sole purpose of landing two men on the surface of the Moon, their resourcefulness didn't end when NASA scrubbed the Apollo missions, after aliens intimidated them and essentially sent the message no to return. But that is a story for another chapter.

NASA used the retired spacecraft for science. They sent unmanned modules for controlled crashes into the Moon. These crashes caused Moonquakes, and scientists measured the vibrations moving through the Moon and found it rings like a bell. [7]

The fact that the Moon rang like a bell implied it was hollow. Many believe it is an artificial structure that serves as a type of mothership, an alien base where the Archons can monitor Earth and humans. In this book series, we have gone over all the mythologies that originated with the Moon and Moon cults. Including the beings on the Moon, which I called *Mounians* in Book One, could be a collaboration of Annunaki, Reptiloids and Grays. [8]

We can extrapolate from this truth that other solar systems may also have artificial satellites, planetoids that are artificial spheres, which we can just call motherships. Therefore, it is possible that Nibiru was such a type of object, but not exactly Moon-like, but glowing hot gold and red.

The other belief is that it is a red-hot, blazing star with tails of red iron oxide that has a hot golden core. We are seeing photos of it now through telescopes and it has been exposed on Google Sky.

Another possibility for my readers to consider is that it can be all of the above. It can be both. What some Planet X researchers fail to realize, is that we are dealing with a system, not a single planet or object. When you are dealing with a mini solar system that is not so mini after all, then you are dealing with multiple objects within it. Just because it has a central Sun, maybe a brown dwarf star, which is smaller than our Sun, it is not exactly small, and neither are its planets, Moons, and objects.

For example, one of the objects we are seeing is a Moon with a hole in it, a chalky white grey color that looks remarkably similar to the Death Star in *Star Wars*. I show these photos in my presentations and talks, that the Truth is Stranger than Fiction! There are artificial constructs orbiting around our universe. They are around Earth, within our neighboring binary twin stars system. So, the Nemesis-Nibiru System is a group of objects, not all artificial. The blazing red-hot star, that is gradually orbiting in our direction, may or may not be Nibiru, but belongs to Nibiru's system.

This may be why the Book of Revelation calls it Wormwood, which Revelation 8:11 says, means bitterness. The Jewish texts calls it the Destroyer. So, we can connect the dot that destruction is bitter.

Sitchin said based on the lunar calendar that the Sumerians used, that their planet Nibiru returns around our Sun and passes our Earth, not once but twice, every 3,600 years, which causes cataclysms. Those who translate this into the solar calendar, know it comes to 3,568 years.

When I met with Sitchin, he was very tight-lipped when asked the question, "when will Nibiru return again? Did you figure out what year based on your research?" He always evaded the question. He admitted he did not know. But when

CHAPTER 8: OUR SOLAR SYSTEM IS PERTURBED

I sat with him and questioned him myself, his energy was telling me something different. He knew, but he was not allowed to tell me.

I respected that, I thought he was a truly kind, sweet old Jewish scholar, whom I thoroughly enjoyed learning from on three separate occasions. He was gracious enough to sit down with me in an informal interview for my books, along with his lovely wife Rina, both of whom I enjoyed chitchatting with, because of our shared backgrounds in Israel. He was a fascinating speaker, well-versed in the ancient literature, and spoke five languages. He was subject to intense scrutiny and vicious attacks from many who were Antisemitic, jealous and threatened by the knowledge he was bringing forth. Fundamental Christians were top of that list of Antisemitic attackers, who secretly wish they could be like Sitchin, who was able to decipher the ancient tablets, and they were not. Most antisemitism is rooted in jealousy of Jews. This clever Jew figured out something that no one could, and they hated him for it. This, in my opinion was the 'spirit' behind most of his attacks.

During the 1990s, there were worldwide *Sitchin Study Groups*, which I had participated in and after meeting Zecharia, kept in touch through his mailings, which were rather lengthy but informative newsletters most of which I still have. While I'm not as knowledgeable as Sitchin in the ancient Sumerian or Akkadian languages, nor are those who have spent quite a bit of time attacking him on some of the claims that emerged from his research, mainly because his conclusions clearly threaten the status quo. I will say however, that as time goes on, we are seeing that much of what he brought forth is turning out to be true. For example, no one would be using the names Nibiru or Annunaki if it were not for Sitchin's research.

After 30 years of research, in 1976, Sitchin finally published his first and most important book, *The 12th Planet*, which articulated his theory about the Annunaki and the Nephilim in the Jewish Bible. He went on to write multiple sequels which became known as the *Earth Chronicles*. He was proponent of the Ancient Astronaut Theory also known as the Paleocontact Theory. His style was different from that of Erich von Däniken and other proponents of the Ancient Astronaut or Paleocontact theory. Unlike them, Sitchin had a complete theory; he was not just saying that aliens from somewhere in the cosmos visited Earth in ancient times. He was revealing who the Annunaki or Nephilim were. He gave details about where they come from, how often they come to Earth and what they did when they were here. He described their reason for setting up a terrestrial colony and a way-up station on Mars.[1]

Sitchin claimed that the Annunaki genetically manipulated the creation of mankind, the "adamic race," out of a primitive ape man species that had developed naturally on this planet. Those of you who have read my Conclusion in Book One of *Who's Who in the Cosmic Zoo?* know that this is the point where I differ with the order of events here.[9]

I agree with Sitchin about the Annunaki genetically manipulating humankind, which I renamed in my book series as the "Evadamic race," which Sitchin called the "adamic race" because Eve was a big part of the story, and well, I'm against sexism, and believe that it's clear in Genesis that Elohim created male and female as equals

137

in their sight, which I prove, through the Torah and the Words of God is the truth in Book Two, *Who is God?* Chapter Twenty-Nine: *Who Created Sexism?* [10]

My conclusions are that there was a much more advanced race of humans living on Earth that predated the Annunaki, known as the Antediluvian Ancient Civilizations of Atlantis and Lemuria, that was lost in the first great deluge, that Genesis 1:2 starts off in its aftermath. That the story of Creation in Genesis is but a synopsis of what really happened in the Garden of Eden, which a much more detailed and dramatic account of the drama that ensued between Adam, Eve, Lucifer, Yahuah and Yahuah's Angels, can be found in the rejected Jewish Apocryphal texts, *The First and Second Books of Adam and Eve*. [11]

Taking the entire story together, we can conclude from the Genesis account, that all became put under a curse. The curse gave way to the spiritual legal ground for Lucifer, who was the old Lord of the Earth during his reign and fall of the kingdom of Atlantis. After losing the Earth to Adam, and then after Adam failed in the experiment, those events brought a curse on all of them, including the Earth. See, Genesis 3:14-19. In Book One, I connected the dot from the first Adam to the Second Adam which is Christ, who was sent to redeem us from the Curse and that includes the Curse on the Earth!

I do agree with Sitchin on the genetic manipulation, but I believe it was not an upgrade, but a downgrade. That the first Evadamics were much more advanced, and that when the Annunaki came they used the already existing women on the Earth to implant their hybrid and genetically modified Earth's first test tube babies, as Sitchin called them. The women used as incubators were not ape women, but more advanced. The curse is called the Fall of Mankind, but this fall, was more of a genetical downgrade, which the Annunaki performed in order to create a slave race. This required taking the already existing 12 strands of DNA that the Elohim created male and female with, and disabling 10 strands, leaving the new human with two strands, in which they can basically function, follow orders yet be demoted from having any of the superpowers that came with the other 10 strands of DNA.

This is what the Second Adam has come to restore. The Promises of Jesus Christ are that when He returns, we are made into immortal, eternal beings, like the angels. The angels are also the Elohim. Angels in the Hebrew means extraterrestrial messenger because they are ordered and commanded from the Lord of Heaven's Armies, who are extraterrestrials.

Sitchin taught us when and why the Biblical floods took place, and there were two floods, the first the sinking of Atlantis, which is where Genesis starts in its aftermath, and the second was the floods of Noah. He described the relationship between the so-called gods and the first human kingdoms, he gave us history and knowledge, and to many of us who have studied his works and listened to him teach, he gave us a deeper understanding of who we are, why we're here, and who's coming back and why.

I was inspired to learn under Zecharia and, while he was not perfect, as none of us are, he was right on track on most stuff. This is what the work of a researcher is supposed to be about, to inspire more research, to answer some questions, yet create more questions, and so on. Suffice it to say, Sitchin was more loved than he was hated in the end, and his legacy is the movement towards the truth of our past, history and, more importantly, of what's to come, that was started by him. So, while there are those who want to pick his work apart, because of their cognitive dissonance or antisemitism, or just plain stubborn ignorance, even a broken clock is right twice a day. Sitchin was not wrong!

While this book is not about analyzing every detail in Sitchin's theory or pointing out its flaws, you can refer to critics for that, but to connect a major dot or two, in which Sitchin's work is foundational, and that is the search and discovery of Planet X aka Nibiru.

Sitchin's theory can be divided in two categories: the astronomical and the mythological aspects. The astronomical is probably the more controversial one. It was Sitchin's argument that the Annunaki did not just come from some unknown planetary body in space, but from a precise celestial body within our own Solar System. He called *The 12th Planet*. To us, this would be the 10th planet. The reason he called it the 12th Planet, was because the Sumerians incorporated the Sun and Moon in counting Celestial Objects, which also conforms to the classical tradition of 12 zodiacal constellations, 12 months, etc. So, suffice it to say, the 12th Planet, is the 10th Planet, but as I point out in my expose` here on the deliberate agenda to cover-up Planet X, they demoted Pluto, and are now calling our extra planet, Planet Nine. This is also to confuse and confound researchers, as well, in my opinion, to invalidate Sitchin's research.

Those who go to such elaborate lengths and hoaxes in order to conceal the truth from us, are also antisemitic in my opinion, because it is always the Jewish scholars whom these types steal from, then try to debunk. Meanwhile, scholars such as Sitchin and Velikovsky inadvertently started a whole new movement and way of looking at our past.

Sitchin used the name Marduk for the 12th planet in his first book, after the chief god in the Babylonian cosmogony, but in later books he referred to it by the Sumerian name Nibiru, in which Sitchin described its extremely elongated elliptical orbit that takes 3,600 years to make one full orbit around our Sun. Not long after Sitchin met with Dr. Robert S. Harrington, the Chief Astronomer from the U.S. Naval Observatory, who in his search for Planet X, ended up confirming Sitchin's research.

I met Sitchin about a year after Dr. Harrington's mysterious death, so whenever I postured my questions to him, in more ways than one, on Nibiru's return, his refusal to answer, told me more than any answer I could expect. That was because he was threatened to keep it a secret. His wife too, was uncomfortably silent, and on an intuitive level, I picked up fear from both of them, that their lives and their lifestyles had been threatened. It is my opinion that he had an idea, maybe not an exact date, but a year, and that after Harrington's death, the powers that be were going to take over from there. This is why there has been lots of ups and downs in our collective search for Planet X since then.

The fact that another planet exists that orbits in the outer limits of our Solar System is not that farfetched, but his assertion that a planet that far away from the Sun could possibly inhabit life is too controversial for most mainstream scientists to accept. With that said, that does not make him wrong.

The mythological aspect of Sitchin's books is fascinating, which contributed to his popularity. Collectively between all the volumes of *The Earth Chronicles*, you come to understand the main psycho-spiritual traits of the Annunaki gods, who are really just extraterrestrials, with a god complex. The Sumerian Sun God Utu, the feuding brothers Enki, Enlil and the goddess Inanna (better known by her Babylonian name Ishtar) along with all their plots, counterplots and dysfunctional family was played out as a Cosmic Drama or even a Clash of the Titans. It is captivating stuff for most of us, who enjoy mythology and who wonder about the origins of life and how that fits into the Bible.

After he published *The 12th Planet* in 1976, he followed by publishing six other volumes in the Earth Chronicles book series: *The Stairway to Heaven* 1980, *The Wars of Gods and Men* 1985, *The Lost Realms* 1990, *When Time Began* 1993, *The Cosmic Code* 1998, and *The End of Days: Armageddon and Prophecies of the Return* 2007. He then went on to publish seven companion volumes to *Genesis Revisited, Divine Encounters, The Lost Book of Enki, The Earth Chronicles Expeditions, Journeys to the Mythical Past, The Earth Chronicles Handbook* and his last book published in 2010, *There Were Giants Upon the Earth: Gods, Demigods, and Human Ancestry*. There are dozens of translations of Sitchin's books in all the major languages, English, Russian, Hebrew, Spanish, German, French, Japanese, Italian, etc. In addition, there was a very well-produced European documentary based on *Genesis Revisited* titled *Are We Alone in the Universe?* which is available on Amazon Prime Video and DVD formats.[12]

Of course, he's not presenting all the answers, but neither did Erich Von Daniken, who admits his first international best-selling book, *Chariots of the Gods*, had more questions than answers at the time, which he says over 50 years later, now has those answers. [13] The point of researchers who author books, is to present a body of knowledge, share their own research and conclusions, and inspire others to take it from there. Those who curse the tree at its roots, do not get any fruit. However, those who appreciate and nurture the roots of the tree, and are discerning of how things evolve, may prune a branch or two, here and there, and end up with an abundant harvest. That is how responsible researchers need to approach this body of knowledge, which is too vast to be ignored.

Remember, the 1992, NASA Press Release, "Unexplained Deviations in the Orbits of Neptune and Uranus, point to a large outer Solar System body of 4-8 Earth masses, on a highly tilted orbit, beyond 7 billion miles from the Sun."

In 1997, a groundbreaking movie and disclosure film, *Contact*, was released, the film adaptation of Carl Sagan's 1985 book of the same name. The film was a science fiction drama directed by Robert Zemeckis. Sagan and his wife Ann Druyan wrote the story outline for the film. Carl Sagan, however, died in 1996, before the film was released, which was dedicated to him. While *Contact* was based on a science fiction novel, the film did incorporate real situations and scientists. [14] They added science and religious analogies as a metaphor of philosophical and intellectual interest in the search for the truth of both human and alien contact.

Contact often suggests that cultural conflicts between religion and science would be brought to the fore by the apparent contact with aliens that occurs in the film. A point of discussion is the existence of God, with several different positions being portrayed. A description of an emotionally intense experience by Palmer Joss, which he describes as seeing God, is met by Arroway's suggestion that "some part of [him] needed to have it"—that it was a significant personal experience but indicative of nothing greater. Joss compares his certainty that God exists to Arroway's certainty that she loved her deceased father, despite her being unable to prove it. [15]

At the end of the film, Arroway is put into a position that she had traditionally viewed with skepticism and contempt: that of believing something with complete certainty, despite being unable to prove it in the face of not only widespread incredulity and skepticism (which she admits that as a scientist she would normally share) but also evidence apparently to the contrary.[16]

Zemeckis stated that he intended the message of the film to be that science and religion can coexist rather than being opposing camps, [17] as shown by the coupling of scientist Arroway with the religious Joss, as well as his acceptance that the "journey" indeed took place. This, and scattered references throughout the film, posit that science and religion are not nominally incompatible: one interviewer, after asking Arroway whether the construction of the machine—despite not knowing what will happen when it is activated—is too dangerous, suggests that it is being built on the "faith" that the alien designers, as Arroway puts it, "know what they're doing." [18]

Notes:
1. Burak Eldem, *2012: Rendezvous with Manduk*, Book One of the Trilogy: *The Hidden History* (Inkilap (2003), Turkey.
2. Joakim RS Nilsson, Nibiru, Stockholm, Sweden, used with permission, https://astromantra.se/english/nibiru2.html
3. World News Australia, *Scientists discover Solar System's 'Planet X'*, February 28, 2008 http://news.sbs.com.au/worldnewsaustralia/scientists_discover_solar_system39s_39planet_x39_541620
4. Immanuel Velikovsky, *Worlds in Collision*, 1950.
5. Immanuel Velikovsky, *Stargazers and Gravediggers: Memoirs to Worlds in Collision*, William Morrow & Co; 1st edition (March 1, 1983)
6. NASA, Jet Propulsion Laboratory, California Institute of Technology, *When The Moon Rang Like a Bell*, Podcast, Episode 2: *Music of the Sphere*, https://www.jpl.nasa.gov/podcast/content, November 5, 2018.
7. Amy Shira Teitel, *Does the Moon Sound Like a Bell?*, May 27, 2016 https://www.popsci.com/does-Moon-sound-like-bell

8. Ella LeBain, *Who's Who in the Cosmic Zoo?* Book One, Trafford, 2012.
9. Ibid, Conclusion of *Who's Who in the Cosmic Zoo?* Book One, 2012
10. Ella LeBain, *Who is God?* Book Two, Chapter Twenty-Nine: *Who Created Sexism?*, Skypath Books, 2015.
11. Rutherford H. Platt, *The First and Second Books of Adam and Eve*, Kessinger Publishing, LLC (September 10, 2010)
12. Zecharia Sitchin, *Are We Alone in the Universe?* Amazon Prime Video, 1978
13. Erich Von Daniken, *Chariots of the Gods*, Bantam Books; Unabridged edition (1972)
14. Carl Sagan, Ann Druyan, *Contact*, Robert Zemekis, Director, 1997 Film.
15. Norman Kagan (2003). "Contact". The Cinema of Robert Zemeckis. Lanham, Maryland: Taylor Trade Publishing. pp. 159–181. ISBN 0-87833-293-6.
16. Robert Zemeckis, Steve Starkey, DVD audio commentary, 1997, Warner Home Video
17. Norman Kagan (2003). "Contact". The Cinema of Robert Zemeckis. Lanham, Maryland: Taylor Trade Publishing. pp. 159–181. ISBN 0-87833-293-6.
18. Contact (1997 American film)

CHAPTER NINE

REPEAT AFTER ME, PLUTO IS A PLANET

*"Just so you know, Pluto is a Planet.
You can write that the NASA Administrator declared
Pluto a planet once again!
I'm sticking by that, it's the way I learnt it, and I'm committed to it."
~NASA Administrator Jim Brisendine,
during a Tour of the Aerospace Engineering Sciences Building
at the University of Colorado Boulder, August 24, 2019*

I studied astronomy at NYC's Hayden Planetarium in 1986 only to have Neil deGrasse Tyson downgrade Pluto from its planetary status in 2006. That is when I thought they are messing with us. A lot of us got mad at that. Now, there is talk of Pluto being restored to its planetary status, there is literally a movement to do so, which includes NASA Director Jim Brisendine. The Great Pluto Cover-Up was really a conspiracy to cover up Nibiru. Zechariah Sitchin's 1976 groundbreaking book, The 12th Planet? included Pluto as a planet from the Sumerian tablets approximately 6,000 old. If the ancients considered Pluto a planet, then who are we to disagree? It is nuts, like saying the Earth is flat. Just another distraction from the Truth. [1]

Then the US Naval Observatory proved there was indeed another planetary body affecting the orbit of Pluto, so they named it the 10th Planet. For those who are now thoroughly confused, the discrepancy is because Sitchin's work accurately defined how the Ancient Sumerians viewed the Solar System, they counted the Sun and Moon as planetary bodies, hence the #12. The US Naval Observatory does not count the Sun and Moon, so they called Planet X the 10th Planet.

Then in 2006, they demoted Pluto from planetary status, so then they started calling Planet X, Planet 9. Ok, so if you are confused, that is exactly where they intended you to be, because they do not want us knowing or discerning Nibiru. Pluto has always been our 9th Planet, and that is why this 9th chapter of this book, is all about our 9th Planet Pluto.

This is another reason why NASA debunked Nibiru, which they renamed as Trappist-1. So, they can say, that silly Jew Sitchin was nuts, there is no such thing as Nibiru. Does not matter what you call it, the point is, it is there and heading our way. It will swing by Earth, not once but twice as it orbits around its own brown dwarf star Nemesis and intersects around our Sun.

FACTS ABOUT PLUTO

- Pluto is named after the Roman god of the underworld.
- This was proposed by Venetia Burney an eleven-year-old schoolgirl from Oxford, England.
- Pluto was reclassified from a planet to a dwarf planet in 2006.
- This is when the IAU formalized the definition of a planet as "A planet is a celestial body that (a) is in orbit around the Sun, (b) has sufficient mass for its self-gravity to overcome rigid body forces so that it assumes a hydrostatic equilibrium (nearly round) shape, and (c) has cleared the neighborhood around its orbit."
- Pluto was discovered on February 18th, 1930 by the Lowell Observatory.
- For the 76 years between Pluto being discovered and the time it was reclassified as a dwarf planet it completed under a third of its orbit around the Sun.
- Pluto has five known Moons.
- The Moons are Charon (discovered in 1978,), Hydra and Nix (both discovered in 2005), Kerberos originally P4 (discovered 2011) and Styx originally P5 (discovered 2012) official designations S/2011 (134340) 1 and S/2012 (134340) 1.
- Pluto is the largest dwarf planet.
- At one point it was thought this could be Eris. Currently the most accurate measurements give Eris an average diameter of 2,326km with a margin of error of 12km, while Pluto's diameter is 2,372km with a 2km margin of error.
- Pluto is one third water.
- This is in the form of water ice which is more than 3 times as much water as in all the Earth's oceans, the remaining two thirds are rock. Pluto's surface is covered with ices, and has several mountain ranges, light and dark regions, and a scattering of craters.
- Pluto is smaller than a number of Moons.
- These are Ganymede, Titan, Callisto, Io, Europa, Triton, and the Earth's Moon. Pluto has 66% of the diameter of the Earth's Moon and 18% of its mass. While it is now confirmed that Pluto is the largest dwarf planet for around 10 years it was thought that this was Eris.
- Pluto has an eccentric and inclined orbit.

- This takes it between 4.4 and 7.3 billion km from the Sun meaning Pluto is periodically closer to the Sun than Neptune.
- Pluto has been visited by one spacecraft.
- The New Horizons spacecraft, which was launched in 2006, flew by Pluto on the 14th of July 2015 and took a series of images and other measurements. New Horizons is now on its way to the Kuiper Belt to explore even more distant objects.
- Pluto's location was predicted by Percival Lowell in 1915.
- The prediction came from deviations he initially observed in 1905 in the orbits of Uranus and Neptune.
- Pluto sometimes has an atmosphere.

When Pluto elliptical orbit takes it closer to the Sun, its surface ice thaws and forms a thin atmosphere primarily of nitrogen which slowly escapes the planet. It also has a methane haze that overs about 161 kilometers above the surface. The methane is dissociated by Sunlight into hydrocarbons that fall to the surface and coat the ice with a dark covering. When Pluto travels away from the Sun the atmosphere then freezes back to its solid state.[2]

Just because Pluto is not a giant but a dwarf planet, does not mean it is not a planet! In the early 2000s, Tyson, who directs the Hayden Planetarium at the American Museum of Natural History in New York City, endorsed a new exhibit describing Pluto as an object more akin to icy bodies in the outer solar system, together known as Kuiper Belt objects. [3]

It was sheer madness what deGrasse Tyson started, now scientists must fight to restore Pluto as a planet. [4]

There is more than a dozen known dwarf planets. Only two of these bodies, Ceres, and Pluto, have been observed in enough detail to demonstrate that they fit the IAU's definition. The IAU accepted Eris as a dwarf planet because it is more massive than Pluto.[5]

Then, in 2006, everything changed and astronomers from around the world declared that Pluto did not meet the criteria for being called a planet. The issue at hand was Pluto's mass, which just was not high enough to give it what astronomers claimed was necessary for all true planets: a clean orbital path around its host star. Pluto had everything except for this clear neighborhood requirement, since debris from the nearby Kuiper belt spilled over into Pluto's own orbit and the much larger Neptune occasionally tugged on Pluto.

After that messy list of criteria was cemented, planetary scientist Philip Metzger of the University of Central Florida in Orlando says astronomers should seriously rethink their decision to snub Pluto.

"The IAU definition would say that the fundamental object of planetary science, the planet, is supposed to be a defined on the basis of a concept that nobody uses in their research, and it would leave out the second-most complex, interesting planet in our solar system," said Metzger.

Metzger's argument isn't that Pluto meets the stated requirements for being considered a planet — everyone agrees that Pluto doesn't fit the description set forth by the International Astronomical Union — but rather that the list of criteria is just plain broken, and astronomers are simply not following it.

"We now have a list of well over 100 recent examples of planetary scientists using the word planet in a way that violates the IAU definition, but they are doing it because it's functionally useful. It's a sloppy definition," he says.

His stance is that the one sticking point for those who wanted to strip Pluto of its planetary status — the "clear" orbit requirement — is not useful in determining status at all. Instead, Metzger says, the real defining feature of a planet should be whether or not it is massive enough, and creates enough gravitational force, that it becomes spherical.

"It turns out this is an important milestone in the evolution of a planetary body, because apparently when it happens, it initiates active geology in the body," he explains.

Whether or not the IAU will take this new argument into account is anyone's guess. Pluto is presently classified as a "dwarf planet," but Metzger's reasoning regarding the seemingly arbitrary definition of a planet seems pretty solid that restores Pluto as the ninth planet in the solar system.[6]

THE PLUTO COVER-UP

We are at a critical turning point in the Planet X saga, ever since the ALMA Observatory, the world's largest Earth-based observatory, sighted Planet X December of 2015, that busted the ruling elite's lie that there's-no-Planet X, forcing them to react with a damage control publicity campaign. Resulting in the Planet 9 – Planet Pluto bait and switch.

The world's awakening astronomy/astrophysics community no longer trusts the U.S. government/NASA propaganda machine, especially now, with the increasing buzz among Astronomers about a dwarf star lurking deep in our outer Solar System.

The California Institute of Technology chose Dr. Michael Brown to regain the science community's lost trust. The ruling elite have long been grooming Dr. Brown, lime lighting and popularizing him as their admired "Pluto Killer," having him lead the revolt which excommunicated Pluto from planetary status.

PLUTO DEMOTED AS A BAIT AND SWITCH

Yes, Pluto is a smaller than all the other planets. However, Pluto is still a planet no matter how small. It is a dwarf planet with its own dedicated Moon, Chiron. Pluto's demotion is all about the Planet X exopolitical psychology that emerged from the backdrop of the demotion of Pluto's publicity. This exopolitical stunt publicity stunt was designed to change the astronomy community's attitude about the quantity of outlying objects. Which was motivated to confuse Planet X researchers, and discredit Zecharia Sitchin's Research, *The Twelfth Planet*.[7]

CHAPTER 9: REPEAT AFTER ME, PLUTO IS A PLANET

This stunt was nothing but an effort to cover up the fact that the Nemesis-Nibiru System, which is just that, was intersecting within our system, which it does, like clockwork in its respective cycles and orbits in its dance with our Sun. The Elite, the Powers that Be, influenced their science advisers to orchestrate this elaborate exopolitical hoax, to convince Planet X Researchers who know that Planet X is gravitationally magnetizing into our space comets, asteroids, meteorites, fireballs, solar cores, planetoids and Moons and a whole lot of space junk in its orbits.

So, by demoting Pluto into that class of newly found outlying objects, what they call the Trans-Neptunian planets, which are Moons, asteroids, they have successfully conveyed the impression to the astronomical community that this horde of previously unknown outlying objects has always been there, so astronomers shouldn't start getting suspicious about the recent influx of comets and other objects surging in from the outer bounds of the solar system. In other words, what they are saying by changing the celestial order of the planets, is, this has always existed, so move along now; there is nothing to see out there that has not always been out there.

Dr. Brown's-CalTech-Planet 9 report (notice the contrived significance of that name "Planet 9") is intended to stealthily replace the new Planet X through deception upon the mindset of the world astronomical community. Their report signified a major turning point in the incoming Planet X saga because now the diabolical ruling elite are preparing to shelter themselves from the coming Planet X-consequent megastorms, megaquakes, mega eruptions, while keeping its coordinates, and all the data they've been collecting on it for years, through their own deep space telescopes.

Why would they keep this to themselves? Because the Elite, the New World Order Globalists, who are hell bent on ruling the world, feel that the world's population is overgrown, and after the cataclysm, when or if they emerge from their luxurious underground bunker communities, the entire population will be greatly reduced to a half billion. How do we know this? It is literally written in stone, in what is known as the Georgia Guidestones, erected in 1980 at this address: 1031 Guide Stones Rd, Elberton, GA 30635. Look it up and check it out for yourself.

They are allowing about 7 billion people to die, unprepared for what is about to come upon the Earth. These ruling elites must deceive the populations and, to do so, they must deceive, for just a little longer, the world astronomical community. They have already murdered the Chief Astronomer of the United States Naval Observatory, Dr. Robert S. Harrington, after he had taken time-lapse infrared photographs of Planet X, and after he had announced his intention to reveal the finished photographs to the world.[8]

John Dinardo reported in his 2016 article, *The Planet 9 bait-and-switch involving NASA, Caltech and Planet X / Nibiru*,[8] that he had a photocopy of the then-uncensored Washington Post report states that astronomers were then concerned that this massive outlying object was naturally being pulled in to menace our inner solar system by our Sun's great gravitational force field. And now, last month's sighting of Planet X by the greatest radio telescope on planet Earth, the ALMA Observatory, has forced these diabolical ruling elite to fall back to a new please-don't-worry-about-Planet X propaganda platform. This is quite ominous because they know what we will soon find out: that more and more astronomers will begin to sight this Sun-pulled dwarf star because, quite simply, gravity works, and gravity cannot be stopped from pulling in this dwarf star on its inexorable loop-around rendezvous with our Sun.

When the populations find this out, they will lose their fear and their obedience to their diabolical elite rulers, and they will start spending on survival preparations, which will topple the already toppling economic house of cards. Why, the people might be so emboldened as to not pay their taxes. And that is why the elite (which includes the World Bank/International Monetary Fund, to whom taxes really funnel) are so feverishly desperate to scrap the old there's-no-Planet X propaganda platform and to adopt the new private CalTech-Planet 9 yes-there-is-a-large-object-out-there-but-it-ain't-incoming propaganda platform.[8]

The Heavens declare the glory of God (Psalm 91) The cosmos was created by Divine Design. Even the Big Bang was sparked by a Creator who initiated it.

Just to prove that Planet X is not a New Age phenomenon, is the esteemed science reporter, Thomas O'Toole published his article, *Possibly as Large as Jupiter, Mystery Heavenly Body Discovered*, in The Washington Post on December 30, 1983.[9]

The purpose of me reprinting this this article written in 1983, is that a great deal of information about Planet X developed because it could be observed better because of its closeness and development in technology, as well as to assert the point that astronomers knew of Planet X's existence for quite some time. Planet X is not fallacy.

> "A heavenly body possibly as large as the giant planet Jupiter and possibly so close to Earth that it would be part of this solar system has been found in the direction of the constellation Orion by an orbiting telescope aboard the U.S. infrared astronomical satellite.
>
> So mysterious is the object that astronomers do not know if it is a planet, a giant comet, a nearby "protostar" that never got hot enough to become a star, a distant galaxy so young that it is still in the process of forming its first stars or a galaxy so shrouded in dust that none of the light cast by its stars ever gets through.
>
> "All I can tell you is that we don't know what it is," Dr. Gerry Neugebauer, IRAS chief scientist for California's Jet Propulsion Laboratory and director of the Palomar Observatory for the California Institute of Technology, said in an interview.

CHAPTER 9: REPEAT AFTER ME, PLUTO IS A PLANET

The most fascinating explanation of this mystery body, which is so cold it casts no light and has never been seen by optical telescopes on Earth or in space, is that it is a giant gaseous planet as large as Jupiter and as close to Earth as 50 trillion miles. While that may seem like a great distance in Earthbound terms, it is a stone's throw in cosmological terms, so close in fact that it would be the nearest heavenly body to Earth beyond the outermost planet Pluto.

"If it is really that close, it would be a part of our solar system," said Dr. James Houck of Cornell University's Center for Radio Physics and Space Research and a member of the IRAS science team. "If it is that close, I don't know how the world's planetary scientists would even begin to classify it."

The mystery body was seen twice by the infrared satellite as it scanned the northern sky from last January to November, when the satellite ran out of the super cold helium that allowed its telescope to see the coldest bodies in the heavens. The second observation took place six months after the first and suggested the mystery body had not moved from its spot in the sky near the western edge of the constellation Orion in that time.

"This suggests it's not a comet because a comet would not be as large as the one, we've observed and a comet would probably have moved," Houck said. "A planet may have moved if it were as close as 50 trillion miles, but it could still be a more distant planet and not have moved in six months' time."

Whatever it is, Houck said, the mystery body is so cold its temperature is no more than 40 degrees above "absolute" zero, which is 456 degrees Fahrenheit below zero. The telescope aboard IRAS is cooled so low and is so sensitive it can "see" objects in the heavens that are only 20 degrees above absolute zero.

When IRAS scientists first saw the mystery body and calculated that it could be as close as 50 trillion miles, there was some speculation that it might be moving toward Earth. "It's not incoming mail," Cal Tech's Neugebauer said. "I want to douse that idea with as much cold water as I can."

> Then, what is it? What if it is as large as Jupiter and so close to the Sun it would be part of the solar system? Conceivably, it could be the 10th planet astronomers have searched for in vain.
>
> It also might be a Jupiter-like star that started out to become a star eon ago but never got hot enough like the Sun to become a star. While they cannot disprove that notion, Neugebauer and Houck are so bedeviled by it that they do not want to accept it. Neugebauer and Houck "hope" the mystery body is a distant galaxy either so young that its stars have not begun to shine or so surrounded by dust that its starlight cannot penetrate the shroud.
>
> "I believe it's one of these dark, young galaxies that we have never been able to observe before," Neugebauer said. "If it is, then it is a major step forward in our understanding of the size of the universe, how the universe formed and how it continues to form as time goes on."
>
> The next step in pinpointing what the mystery body is, Neugebauer said, is to search for it with the world's largest optical telescopes. Already, the 100-inch diameter telescope at Cerro del Tololo in Chile has begun its search and the 200-inch telescope at Palomar Mountain in California has earmarked several nights next year to look for it. If the body is close enough and emits even a hint of light, the Palomar telescope should find it since the infrared satellite has pinpointed its position." [9]

The mystery about the object is that astronomers do not know and can't agree if it is a planet, a giant comet, a nearby protostar that never got hot enough to become a star, a distant galaxy so young that it is still in the process of forming its first stars or a galaxy so shrouded in dust that none of the light cast by its stars ever gets through, or now, they're presuming Planet 9 (not Pluto) is actually a Primordial Black Hole.

Certainly, if the late Dr. Harrington, the chief officer of the naval observatory, was correct, then the minute it became clear that Planet X was moving toward Earth, the flip-flop began.

More recent discoveries from the Chilean observatories of Sedna have confirmed the existence of a massive perturber (to use the old language) or a shepherd (to use newer lingo).

This article from The Smithsonian's Air & Space publication makes it clear that the Caltech Astronomers Mike Brown and Konstantin Batygin used a March 2014 issue of Nature to guide them to their discovery of Planet 9:

"When the two Astronomers published the announcement, in the journal Nature in March 2014, they noted something odd about 2012VP, Sedna, and 10 other super-distant objects. They were all clustered in an unexpected way; their orbits all crossed the ecliptic plane—the conceptual flat disk around which the eight planets orbit the Sun—awfully close to the spot where they came closest to the Sun. So odd was the uniformity that the pair suspected something was causing it; they wrote, '[A]n unknown massive perturbing body may be shepherding these objects into these similar orbital configurations.'"

"Here is the bait-and-switch: Instead of proclaiming this object "Planet X / Nibiru," Brown and Batygin dub the gravitational mass "Planet 9." From Sugarman's language, it is clear that the mainstream scientific and journalistic community is in lockstep in removing from the public consciousness an object "possibly as large as Jupiter" and replacing it with an object "10 times the size of the Earth":

"In January 2016, Brown and Batygin published a paper outlining the evidence for a hypothetical Planet 9. They described a world with a mass about 10 times that of Earth and two to four times its diameter. They defined its likely orbital path, postulating that it would orbit the Sun every 10,000 to 20,000 years. (By comparison, it takes Pluto just 248 years.) They proposed that, like the Solar System's other giants, it was made up of a rocky, icy core enveloped by gas. They believe it is located in the direction of the Orion constellation, and is currently near its farthest point from the Sun—as many as 1,200 astronomical units away (Earth's distance from the Sun, 93 million miles, is one astronomical unit, or AU)."

Here is the distinction: Jupiter is much more massive than just 10 times more than the Earth; and, if Planet X is "possible as large as Jupiter," then please read these quotations:

"Jupiter's diameter is 11.2 times larger than Earth. In other words, you could put 11.2 Earths side-by-side to match the diameter of Jupiter."

"And Jupiter's volume is even bigger. It would take 1321.3 Earths to fill up the volume of Jupiter. In terms of surface area, Jupiter is 121.9 times bigger than the Earth. That's how many Earths could be flattened out to cover the surface of Jupiter."

"Jupiter has 317.8 times the mass of the Earth."

According to this 2017 Space.com article, Nemesis Star Theory: The Sun's 'Death Star' Companion:[10]

"In 1984, Richard Muller of the University of California Berkley suggested that a red dwarf star 1.5 light-years away could be the cause of the mass extinctions. Later theories have suggested that Nemesis could be a brown or white dwarf, or a low-mass star only a few times as massive as Jupiter. All would cast dim light, making them difficult to spot."

Here is the new coalescence between Muller's suggestion the discovery of Sedna:

"The dwarf planet Sedna lends further credence in the eyes of some to the existence of a companion star for the Sun. With an orbit of up to 12,000 years, the planet presents a puzzle to many. Scientists have suggested that a massive object such as a dim star could be responsible for keeping Sedna so far from the Sun."

The Space.com writers suggest that Planet 9 or Nemesis could be responsible for shepherding Sedna:

"Another theory is there is a huge ice giant, nicknamed 'Planet Nine,' that is at the edge of our Solar System. Researchers Konstantin Batygin and Mike Brown (both from the California Institute of Technology) suggested in 2016 that such a body might be stirring up smaller icy bodies in the Kuiper Belt. A search for Planet Nine is ongoing. Notably, Brown was part of the research team that found Sedna and several other icy bodies in the Kuiper Belt, and he was one of the lead advocates for reclassifying Pluto (once considered a planet) to a dwarf planet in 2006." [10]

Either way you slice it, something really big—Red Dragon, Nemesis, Nibiru—is coming. It is bringing the Tribulation. It is causing an uptick in slinging comets, asteroids, and meteors our way—and its trailing Moons and planetoids.

Although the Washington Post reported NASA's discovery of "Planet X" in 1983, most people cannot imagine why the brown dwarf star took so much time to be gravitationally pulled in by our Sun. To understand why, we need to realize how vast our solar system is, and that this celestial body actually decelerates after it overshoots the Sun at the underside of our solar system's ecliptic plane, as it makes a huge U-turn through the outer edge, up and over our Sun.

The late Dr. Harrington discovered that Planet X was coming in way below the southern skies of the South Pole, so in its natural elliptical trajectory, obeying the laws of Newtonian mechanics, it would overshoot the Sun below the ecliptic, and would begin to decelerate as the Sun would then begin to pull back on it during its overshoot. The resulting vast looping U-turn could, depending on its mass and velocity during its overshoot, would take it beyond Pluto before it could decelerate to a speed component of its velocity vector equal to zero miles per second, as it would gradually be reversing its direction, beginning to loop up into the Ecliptic plane, then above the Ecliptic, on a trajectory over the paths of all the planets, including Earth, and then over the north pole of the Sun.[11]

Notice the great many record-breaking floods, hot and cold temperature extremes, droughts, windstorms. Well, these are all caused by the repetitious, uniformly periodic, cometary return of this dwarf star (every thirty-six and a half centuries) which ancient witnesses called Nibiru and which N.A.S.A. actually called "Planet X" in at least one of its internal documents, several decades ago.[12]

Researchers at California Institute of Technology have found new evidence of the existence of a ninth planet in our solar system that they described as a super-Earth. The latest findings come three years after the same team proposed for the first time the idea of Planet Nine. The same group that demoted Pluto, causing the planet count to go from 10 to 9. Suffice it to say, it was a bait and switch to prepare for Planet Nine, discredit Sitchin's research and confuse and confound the Planet X Researchers by switching the order of our solar systems planetary line up. It was deliberately done to cover up what I am calling the Nemesis-Nibiru system intersecting within our solar system.[12]

"At five Earth masses, Planet Nine is likely to be very reminiscent of a typical extrasolar super-Earth," Konstantin Batygin, an assistant professor of Planetary Sciences, said. He added the unknown planet may serve as the solar system's missing link of planet formation. "Planet Nine is going to be the closest thing we will find to a window into the properties of a typical planet of our galaxy," Batygin said.[13]

Speculating and theorizing about the existence of yet undiscovered planets in our solar system has been bounced around for centuries. Prior to each new discovery of another outer planet has come detection of anomalies in the erratic, inexplicable motions of the outermost known planet. For instance, before Neptune's existence was determined, for decades Astronomers had been theorizing that Uranus' (discovered in 1781) irregular movement may have been caused by the presence of yet another undiscovered planet. Indeed, that was the case in 1846 when Neptune was first sighted and identified.

The now dethroned ninth planet Pluto discovered in 1930 (relegated in 2006 to minor dwarf planet status) and Pluto's later found Moon Chiron were then used to explain the observed "wobbles" in Uranus and Neptune's respective orbits. Thus, errors in calculating precise positions of known planets hold an enduring pattern of later confirmation of cause determined by each newly discovered planet. Hence, for over a century, scientists have debated that yet more major planets and dwarf planets belonging to our Solar System are still out there in space waiting to be found and existing anomalies to be explained.

SCIENTIFIC EVIDENCE

Rather than dig into ancient mythology or biblical prophecy as so much of the current speculation about Nibiru already covers, this presentation will limit its focus to reviewing the most tangible, credible pieces of scientific evidence supported by Astronomers and astrophysicists who in recent years have risked destroying their careers, reputations and their very lives promoting their controversial findings and conclusions. This year more than ever the movement's growing authenticity of the Planet X story is, in fact, gathering momentum, at least garnering lukewarm support from some rather high-profile notables within the scientific community. This article will chronicle this growing body of empirical evidence validating not only Planet X's existence but its eminent approach towards Earth.

This highly controversial Planet X system entering our solar system with potentially catastrophic implications was first featured in a Washington Post article way back in 1983. NASA's infrared astronomical telescope found the mysterious planet "possibly as large as Jupiter" near the constellation of Orion. As distant a time as 32 years ago, it was the closest Heavenly body to the Earth beyond Pluto.

Researcher-activist John Moore has long asserted that a number of his inside ex-military and government intelligence sources have independently confirmed that a top-secret meeting took place in 1979 in a New Orleans briefing room where highest ranking U.S. Naval flag officers were first informed of the coming inevitable Planet X disaster that would occur within their lifetime.

In October 2003, a highly significant and telling Department of Defense sponsored paper was published called "An Abrupt Climate Change Scenario and Its Implications for United States National Security." This paper notes that oceanic saline levels were dropping, causing "thermohaline circulation collapse." A serious disruption of the Atlantic conveyor belt that pushes warm saltwater near the surface circulating northward from the south while deeper cooler water flows southward would have a devastating impact on global climate. The paper concludes with the following sobering predictions: wars over energy, food, and water resources, increasing draught over wider land mass, and violent climate change effects producing more frequent, higher magnitude natural catastrophes of all kinds. A dozen years later all of these most disturbing developments are clearly manifesting.

Of course, the current standard politically correct scientific dogma is that global warming aka climate change emanating from CO_2 greenhouse gasses is the obvious culprit. This mass deception is designed to conveniently obscure the multiple dire effects caused by many decades of the globalists' geoengineering, HAARP, weather modification and weather wars, not to mention the hidden, "nonexistent" Planet X system hurtling towards us deeper inside our Solar System that's likely causing tremendous changes at the surface of all our Solar System planets of course including our own Earth's atmosphere and global surface.

In March 2010, the mainstream UK newspaper The Telegraph ran a story entitled "Search on for Death Star that Throws out Deadly Comets." The sub-headline announced that NASA is looking for the twin Sun that is the brown dwarf star Nemesis slowly circling our Sun and periodically "catapulting" deadly comets to the Earth. So as tight-lipped NASA has been over the years, little tidbits reflecting that it's still keeping an ever-present eye out for the cataclysmic Planet X system still manages to seep through to the public every now and then from so called "reputable" news sources. The article goes on to state that the star that's "five times the size of Jupiter" is the prime suspect that wiped out the dinosaur 65 million years ago.

The then latest NASA trophy piece surveying the Heavens, the Wide-Field Infrared Survey Explorer (WISE), was capable of detecting the faint heat emanating from the dwarf star 25 times the distance between Earth and the Sun, or one-third a light year away. Launched in January 2009 WISE discovered at least 1000 dwarf stars located within 25 light years from the Sun until October 2010 when its coolant was scheduled to run out. Twice as far out as Nemesis is the Oort Cloud which is the sphere of icy bodies surrounding our Solar System.

With the Planet X system entering our Solar System, its gravitational force is believed to launch trajectories of comets and asteroids that are large-sized space debris objects of icy rocks and dust that bombard our Solar System planets including the Earth. Professor John Matese of the University of Louisiana-Lafayette believes that this Nemesis that includes planet Nibiru as part of the Planet X system is mainly responsible for supplying such a concentrated volley of comets from the Oort Cloud into our inner Solar System.

In May 2012 livescience.com and examiner.com both featured headline stories heralding the probable discovery of Planet X. Astronomer Rodney Gomes of Rio's National Observatory of Brazil had just presented new evidence of the mysterious planet at a meeting of the American Astronomical Society. Gomes demonstrated that the icy objects outside Neptune's orbit in the Oort Cloud displaying irregular movement that cannot be explained by known mathematical laws of astrophysics but can be by the gravitational influence of so-called Planet X. Additionally, the dwarf planet Sedna's peculiar orbit can also only be accounted for by the presence of a large unknown planet. Gomes findings were very well received by his peers, believing that he got his math right. But of course, not wanting to endorse Planet X completely, they dutifully towed the scientific/government line of status quo by tempering their enthusiasm with a call for more research and posing alternative theories that may not involve another large planet.[14]

Then in November 2014, iflscience.com reported evidence that two new undiscovered planets may be lurking in our Solar System. The possibility of a planet lurking in the outer reaches of the solar system has gained new ground, based on the orbits of recently discovered objects. There is a new twist to the latest evidence, however, with suggestions of not one but two large planets at mind-bending distances from the Sun. They acknowledged, that the quest for a Planet X beyond Neptune has been going on for more than a century. Recently, two dwarf planets Sedna and 2102 VP113 have been identified with orbits extending to distances hundreds of times further from the Sun than our own. [15]

A January 2015 article written by Mike Hall from Space.com also addressed the possibility of two more planets beyond Neptune and Pluto moving in the outer limits of our Solar System subtly influencing the orbits of the dwarf planets. Carlos de la Fuente of Madrid's the University of Complutense stated:

> "This excess of objects with unexpected orbital parameters makes us believe that some invisible forces are altering the distribution of the orbital elements of the ETNOs [extreme trans-Neptunian objects], and we consider that the most probable explanation is that other unknown planets exist beyond Neptune and Pluto." [16]

Still another team of Astronomers in March 2014, Chadwick Trujillo, and Scott Sheppard, announced the discovery of 2012 VP113, another ETNO joining Sedna as two known members of the "inner Oort Cloud," the comet-filled sphere that lies just beyond the Kuiper Belt and Pluto. Trujillo and Sheppard claim these two objects' orbits comply with the gravitational presence of a big "perturber," up to even 10 times the mass of the Earth.

Major unheard-of changes are currently appearing on the surface and atmosphere of the Sun and all the planets. Astronomer Mike Lockwood from California's Rutherford Appleton National Laboratories found that since 1901 the Sun's overall magnetic field has increased by 230%. Solar flares and storms have also increased. The Earth's Moon now has a 6000-kilometer-deep atmosphere of Natrium that it never had before. Mercury, the planet closest to the Sun, now has grown polar ice. Venus' brightness has skyrocketed by 2500% and experienced major changes in its atmosphere in the last four decades. Meanwhile, further away from the Sun than the Earth, Mars has melted polar icecaps and its magnitude of storms has strengthened exponentially.

Jupiter has had the brightness of its plasma clouds increase 200% accompanied by a pattern of intermittent thickening then weakening. The largest planet in our Solar System's enormous belts have been changing color with radiation levels erratically waning and then flaring up again. An increase in asteroid and comet activity crashing into Jupiter likely from Nibiru's entrance into our Solar System has also been observed. During the last three decades the jet stream on Saturn's equator has slowed in velocity while X-rays at the equator have spiked. Uranus has become brighter with a huge flare-up of storm activity when its surface before was always placid. Neptune has also increased in brightness by 40%.

The atmospheric pressure on Pluto has jumped up by 300% despite it moving further from the Sun. The glowing plasma on the edge of our Solar System has increased 1000%. The profound changes being measured in space are unprecedented. An increase in energy emission is changing the fundamental structure of all matter throughout our Solar System. Taken together, these massive, across the board changes most likely are all effects from Planet X moving through our Solar System.

Of course, changes on the Earth have been pronounced as well, most notably with electromagnetic changes, yearly incremental polar axis changes, and extreme weather changes. For instance, North Pole, Alaska recently saw its late spring temperatures soar to mid to high 80's F. From 1963 to 1993 the frequency of natural disasters globally has leaped up by 410%. And in more recent years the rate of disasters has soared even higher.

With an increasing number of volcanic eruptions along the Pacific Ring of Fire and Earthquakes felt globally particularly along the San Andreas and the New Madrid fault lines, coupled with the Nepal 7.8 Earthquake a couple months ago, major volcanic and Earthquake activity in 2015 is rapidly on the rise. As early as November 2013 10 different long-dormant volcanoes went active, first suggesting Planet X's presence. The rate of volcanic eruptions from 1875 to 1975 has increased by 500%, but even greater now.

For the last 2000 years the Earth's magnetic field has been decreasing gradually. But in the last 500 years that decrease has become much more dramatic. Effects from a diminishing magnetic field would double the amount of radiation causing an epidemic of skin cancer deaths, speed up climate change and magnify extreme weather. Solar winds would eliminate ions that allow the Earth to retain water and air. All these horrific changes are happening right now. What next to no scientist who wants to stay alive is revealing is that this torrid rate of ecological degradation and life killing Earth surface change are more than likely caused by Planet X racing ever closer toward us.[17]

Another rather peculiar development to the Planet X story has been Google Sky's strange actions. Several years ago, no doubt in literally keeping with the media blackout, Google did black out the grid in the sky to ensure that Planet X was X'd out of its sky map. Then about a week ago Google Sky mysteriously reinserted back in the missing grid fully depicting the flaming winged Nibiru. So, it is visible from both the actual sky, and the Google Sky now… more evidence that the tide is turning toward the truth about Planet X.

Back in mid-April, the European Space Agency invited Astronomers, physicists, nuclear engineers, mathematicians and even soldiers in space defense from around the world to a conference in a Rome suburb to discuss the topic of an asteroid striking the Earth, with the proposed hypothesis not if, but when. Hmm, sounds familiar to the Nibiru cataclysm scenario, minus the forbidden planet of course.

This Planetary Defense conference strategizing how to collectively handle Near-Earth Objects (NEOs) plunging toward us using space weapons technology has been an annual event the last six years. 12,700 asteroids have been identified as NEOs with orbits coming within 121 million miles from our main Sun.[18]

Notes:
1. Zechariah Sitchin, *The 12th Planet*, Avon Books, NYC, NY, 1976.
2. Pluto Facts, https://space-facts.com/pluto/
3. Elizabeth Howell, *Neil deGrasse Tyson Rejects Pluto Planethood Proposal on 'Colbert'*, https://www.space.com/36126-pluto-planethood-slammed-by-neil-degrasse-tyson.html, March 20, 2017.
4. Tim Prudente, The Baltimore Sun Writer, *Scientist leads effort to restore Pluto's planetary stature*, https://www.seattletimes.com/nation-world/scientist-leads-effort-to-restore-plutos-planetary-stature/, March 17, 2017.
5. Elizabeth Howell, *What Is a Planet?*, https://www.space.com/25986-planet-definition.html, April 7, 2018.
6. Mike Wehner, *Pluto might become a planet again because astronomers can't make up their mind*, https://bgr.com/2018/09/07/pluto-planet-argument-definition/, September 7, 2018.
7. Zechariah Sitchin, *The 12th Planet*, Avon Books, NYC, NY, 1976
8. John Dinardo, *The Planet 9 bait-and-switch involving NASA, Caltech and Planet X / Nibiru*, https://planetxnews.com/2016/01/27/the-planet-9-bait-and-switch-involving-nasa-caltech-and-planet-x-Nibiru/
9. Thomas O'Toole, *Possibly as Large as Jupiter, Mystery Heavenly Body Discovered*, The Washington Post, https://www.washingtonpost.com/archive/politics/1983/12/30/possibly-as-large-as-jupiter/1075b265-120a-4d40-9493-a8c523b76927/, December 30, 1983.
10. *Nemesis Star Theory: The Sun's 'Death Star' Companion*, https://www.space.com/22538-nemesis-star.html, July 21, 2017.
11. Christian deBlanc, *How the Media separated Nibiru from Planet X from Nemesis, Planet X*, Writers, August 10, 2017, https://planetxnews.com/2017/08/10/media-separated-Nibiru-planet-x-nemesis/
12. John DiNardo, Planet X, Washington Post flip-flops on Planet X / Nibiru, January 8, 2016, https://planetxnews.com/2016/01/08/washington-post-flip-flops-on-planet-x-Nibiru/
13. Darwin Malicdem, *Planet Nine: Scientists Find 'Solid' Evidence Of Super-Earth Planet Beyond Neptune*, Mar 8, 2019, https://www.medicaldaily.com/planet-nine-scientists-find-solid-evidence-super-Earth-planet-beyond-neptune-430846
14. The Planet X / Nibiru system and its potential impacts on our Solar System. July 12, 2016 ·Planet X·, https://planetxnews.com/2016/07/12/planet-x-Nibiru-system-potential-impacts-solar-system-2/
15. Stephen Luntz, *Astronomers Find Evidence Of Two Undiscovered Planets In Our Solar System*, https://www.iflscience.com/space/signs-planet-x-and-maybe-y/, November 16, 2014.
16. Mike Wall, *Mysterious Planet X May Really Lurk Undiscovered in Our Solar System*, https://www.space.com/28284-planet-x-worlds-beyond-pluto.html, January 16, 2015.
17. Joachim Hagopian, Global Research, *The Nibiru Planet X System and Its Potential Impacts on Our Solar System, A Scientific Case for Nibiru/Planet X System* https://www.globalresearch.ca/the-niburu-planet-x-system-and-its-potential-impacts-on-our-solar-system/5459788, September 6, 2015.
18. Amber William, Nasa Warning – Clear Signs That Planet X is Affecting Earth in a Bad Way, https://www.mydailyinformer.com/nasa-warning-clear-signs-that-planet-x-is-affecting-Earth-in-a-bad-way/, June 13, 2018.

CHAPTER TEN

TRACKING AND CLOAKING

"Knowledge makes a man unfit to be a slave."
~Frederick Douglass

"There will be signs in the Sun and Moon and stars,
and on the Earth dismay among the nations,
bewildered by the roaring of the sea and the surging of the waves.
Men will faint from fear and anxiety over what is coming upon the Earth,
for the powers of the heavens will be shaken."
Luke 21:25-26, Berean Study Bible

Now to make matters a bit more complicated and convoluted, I have seen a plethora of evidence in studying weather cams from all over the world, which are online from Sunrise to Sunset, that there's technology in place that somehow the Nemesis-Nibiru System is being hidden by NASA. See, Chapter on the Solar Simulator.

In 1984, California Berkeley Astronomer Richard Muller suggested that a red dwarf star named Nemesis was responsible for the periodic mass extinctions that had affected Earth. This would be the red dragon (red from iron oxide dust) that is being scrubbed from Google Sky. I managed to find these coordinates (09:47:27, 13:16:27) if you go to Google Sky look for featured observatories and then IRAS infrared sky, do not zoom too much or you will not see it. Right Ascension (RA): 5h 53m 27s, Declination (D): -6 10" 58". For a period of time, it was the only object blacked out on Googly Sky.

THE VATICAN'S L.U.C.I.F.E.R SPACE TELESCOPE
Meanwhile, the Vatican, who owns and operates two of the most sophisticated and largest telescopes on Earth oddly named, LUCIFER. Why? LUCIFER is a convex lens binocular telescope that views in the infra-red spectrum and is allegedly exposing countless spacecraft, exoplanets, and planet anomalies. Nemesis, the brown dwarf, and the black star are visible only in the infrared range on the light spectrum. There are not many telescopes with that capability. The Vatican leading

this technology and science, has always caused many to wonder why and since when does the Vatican have such an interest in the Heavens?

L.U.C.I.F.E.R. located at the Mount Graham Observatory in Arizona is an acronym for "Large Binocular Telescope Near-infrared Utility with Camera and Integral Field Unit for Extragalactic Research". Surely, they know this is not a coincidence. Speculation infers that the Vatican has these high-powered telescopes to monitor biblical warnings such as "Wormwood," aka Nibiru or Planet X:

> "And the third angel sounded, and there fell a great star from Heaven, burning as it were a lamp, and it fell upon the third part of the rivers, and upon the fountains of waters; And the name of the star is called Wormwood: and the third part of the waters became wormwood; and many men died of the waters, because they were made bitter."
> (Revelation 8:10-11)

The Vatican has been monitoring extraterrestrial activity, incoming spacecraft and has been tracking Nibiru. They are looking for the aliens, the gods of old, the fallen ones, to return to Earth. As I laid out in *Who is God?* Book Two, by connecting the dots to ancient Babylon, Mesopotamia to the Roman Empire which became the Church of Rome, today's Vatican.[1] Their ancient gods of old were these ancient aliens that wore miters. The Vatican worships these ancient aliens, just look at the pope's hat, the miter was fashioned after ancient Babylon. It is the same exact hat as the clay tablets with the same symbol on them, that was used to cover up these beings with large, elongated skulls. This stuff is wild.

The Vatican are waiting on their Lord who is Enlil and the fallen angels. See, *Who is God?* on subchapter, *Who is Enlil?* Enlil became Baal Ilyah who today is known as Allah.[1] The Jesuits await their alien saviors, who are nothing more than satan's army. This will be one of the greatest deceptions ever for the comatose sheep.

The Pope's miter hat also symbolizes the fish of Pisces. Not only is Jesus returning at the end of the Piscean Age, but so are His enemies to compete with Him over the harvest of the Earth, which would be us humans. The Vatican knows we are approaching the end of the age and entering into the Processional Age of Aquarius, so they are prepared to meet their gods and baptize them into their church, while claiming that they created us humans, when it was truly Yahuah, Yeshua, and the Holy Spirit that created us, but many are already deceived by this deception, because they can't discern who is who in the Cosmic Zoo. Hence, the purpose of this book series, which is an attempt to establish the fact that, one, not all the gods are the same, nor are they all on the same side, and two, that there's a hierarchy to both Kingdoms, which clash with each other over humankind, and thirdly, humankind has been genetically manipulated and downgraded from the original human model, that was created in the image and likeness of the Elohim, who serve the God of gods, who is Yahuah. See, *Who is God?* Book Two.[1]

The Ancient Church of Rome already made contact with extraterrestrials hundreds of years ago, which explains why they banned dozens if not hundreds of Jewish Scrolls detailing Ancient Alien accounts, along with why their religious belief in "God," reveals that their own religious texts points to an extraterrestrial and not the ultimate Source or Creator. This is evidenced by why they use priests, a tradition carried over from ancient Babylonian religions, and that today's Catholic Church does not promote a direct line to God through Jesus Christ, but instead teaches praying to saints, archangels, Mother Mary and going to priests. The Bible clearly says that we are to call no one on Earth Father, except our Heavenly Father, who listens to all our prayers through Christ. The Catholic Church continues to keep crucifixes which are idols of crucified Christs on their altars, which is an act of Voodoo. In Voodoo, whenever you nail anything down, it's symbolic of thwarting its power.

The Vatican believes Lucifer is God, and regularly invokes his name in masses and in deed. The Vatican is the stronghold of the Ancient Church of Rome established by the Roman Empire, who oppressed Jews, Israel, and all true believers in who Yeshua, aka Jesus Christ was.

The Vatican came out and said that the existence of Aliens would NOT disprove the existence of God. They have one of the strongest telescopes on planet Earth right now, ironically called L.U.C.I.F.E.R. It has been postulated that they made this announcement only because they know something that they are not openly disclosing. In addition, Erich Von Daniken, the founding father of the Ancient Astronaut theory, claims he is deeply religious, he believes in God and has shared on panels that he still prays every night. His works include revealing how the Bible is a record of angels who are really extraterrestrials, descending to Earth in some type of spacecraft. Von Daniken has said, "My God doesn't need a vehicle" which got him thinking and researching, proving the Ancient Astronaut theory, that the so called "gods" our ancestors worshipped, were in fact, Extraterrestrials.

> "Jesus said, "I have other sheep that are not of this sheep pen. I must bring them also. They too will listen to my voice, and there shall be one flock and one shepherd."
> (John 10:16)

There are other humans God created that do not live on Earth. Jesus consistently referred to his human followers as sheep multiple times. In John 10:16, he is referring to other humans that do not belong to Earth. It is His task as Messiah and as the Cosmic Christ to bring all of us together under the Kingship of His Heavenly Kingdom. I have proven this scripture – along with multiple others throughout this Book Series – that the Jewish Bible absolutely indicates God created both good and evil aliens. Christians only have half the story, and their present mantra that "all aliens are demons," is only half true. All demons are aliens, but not all aliens are Demons. Scripture says God Almighty created aliens, to serve Him.

There are alien creatures located in His Throne Room on His Starships. See, *Who is God?* Book Two, Chapter: Two – *Ancient Technology and Biblical Astronauts*, p. 21-46.2

Yes, God has His own fleet of Spaceships! When you read the Bible in the original Hebrew, it is plain as day. See the Book of Ezekiel, Daniel, Zechariah, Isaiah, that all correlate to Revelation. Unfortunately, most of this knowledge has been covered up with mistranslations in English Bibles, yet it is described in detail in the Hebrew Scriptures.

WHO IS MYSTERY BABYLON?

Let us first lay some biblical groundwork, as I am clearly not referring to Babylon, Long Island, New York. Geographically speaking, Ancient Babylon was located were today's region of Iraq is. Think back to the Iraqi war of 2003, when the U.S. Marines had a mission, to storm the Baghdad Museum, and retrieve the Ancient Stargate that Muhammar Khadafi gave to the then now deceased Iraqi leader, Sadaam Hussein as a gift. That Stargate was retrieved, and brought back to NORAD, in Colorado Springs. The U.S. government perceived this ancient artifact as a 'weapon of mass destruction' once it figured it its potential.

I find it no coincidence that Colorado Springs has an actual Stargate in its *America the Beautiful Park* located in downtown Colorado Springs, which is actually used as Fountain. Nevertheless, a replica is hiding in plain sight, along with two obelisks that face it. Truth is Stranger than Fiction!

Bible Prophecy follows these symbols:
1. A woman/virgin as a Church.
2. A beast as a ruler or ruler of that nation.
3. A Harlot/whore as a false or backslidden church.

Bible Prophecy points to 6 clues as to who Mystery Babylon is:

Clue number 1:
> "And there came one of the seven angels which had the seven vials, and talked with me, saying unto me, come hither; I will show until there the judgement of the great "WHORE" that sits upon many waters." . . . "Mystery Babylon the great, the mother of all Harlots."
> (Revelation 17:1 and 5)

The mother of all Harlots is spiritual name that refers and symbolized the false, rebellious, counterfeit, and backslidden church. Always in Scripture, the Church is referred to a woman, who is either the Bride of Christ, who has been made pure and is ready to meet her Bridegroom (who is the Lord), or she is the Whore, the Prostitute, the Harlot, who symbolized the rebellious, counterfeit, false Church, who

CHAPTER 10: TRACKING AND CLOAKING

is not serving the Lord, but are instead controlled by the god of this world, who is Lucifer/Satan.

Clue Number 2:
> "And he said unto me, the waters which thou saw, where the whore sits, are peoples, and multitudes, and nation's, and tongues."
> (Revelation 17:15)

So, if the Harlot is a church, then that church rules over multitudes, nations, and tongues. Which Church has the largest congregation of 1 billion people? It is 3rd world largest owner of land, roughly 177 million acres. With its priceless art, gold, investments around the world. It is one of the wealthiest institutions in the world.

Clue Number 3:
> "And here is the mind which hath wisdom. The seven heads are "MOUNTAINS," (hills) on which the woman sits."
> (Revelation 17:9)

There is only one famous city that sits on seven hills and that is Rome. The seven hills of Rome are:
1. Aventine Hill
2. Caelian Hill
3. Capitoline Hill
4. Esquiline Hill
5. Palatine Hill
6. Quirinal Hill
7. Viminal Hill

Clue Number 4:
> "And the woman (CHURCH) was arrayed in purple and scarlet (RED) color, and decked with gold and precious stones and pearls, having a golden cup in cup in her hand full if abominations and fornications."
> (Revelation 17:4)

There are two governing bodies in the Catholic Church, and it is their official policy that they wear red and purple to distinguish themselves from each other. [1.] The College of Cardinals wear Red, the Scarlet Sector. [2.] The College of Bishops wear Purple, which include the Catholic news world and the Arch Dioceses of Chicago. We all know of the abominations, pedophilia, and fornications of Catholic Priests.

Clue Number 5:
> "And I saw the women drunken with the blood of the saints, and with the BLOOD of the MARTERS of Jesus: and when I saw her, I wondered with great admiration."
> (Revelation 17:6)

In the history of the Catholic Church, it has slaughtered millions of Christians for not believing in their perceived false doctrine. Walter Montano in his book *Behind the Purple Curtain* stated that there were 50 million people/Christians killed in its 700-year reign of terror. Definitely has the blood of the saints on her hands. [3]

Clue Number 6:
> "And this woman (CHURCH) which thou saw, is the great city (ROME) which reigns over kings of the Earth."
> (Revelation 17:18)

The Vatican, which is a sovereign city and nation unto itself, dominates Rome, and Italy, has its own police force, its own bank and is surrounded by a thick stone wall. The Vatican believes it has the right to guide any and all of its members in their political thinking. If the Pope chose to issue an order to particular heads of states who are Roman Catholic, these leaders are bound to obey the Pope. Many nations have leaders, politicians and extremely wealthy people who are Roman Catholic. Imagine the Pope issuing an obeisance to the global governing authority. Between 500-1500 AB the Pope removed and replaced kings in Europe at will many times. The harlot ruled over kings.

Revelation's End Times Prophecies gives us clues as to who MYSTERY BABYLON is. Contrary to popular mistaken belief, it is not the United States of America, it isn't New York, and it isn't Mecca. Mystery Babylon is Rome and the Vatican. The Pope will be the false prophet and leader of the One World Religion. As I've proved in COVENANTS – Book Four, Pope Francis is already working hand in hand with the spirit of Antichrist as he has set up conferences and task forces to establish Chrislam in his bid to rule over a One World Religion, and thinks he is the leader of the One World government. [4]

Too many New-Agers get sidetracked with the false belief that focusing on the negative only attracts more negative. Problems must be exposed to light and acknowledged, no denied, in order to be solved and overcome. Suffering, violence and injustice towards humans, animals and the planet will not be eradicated until we are willing to face reality of it, and *who* and *what* is behind it. [5]

PLUTO IN CAPRICORN

For years, Pluto was considered Planet X. See my chapter herein, "Repeat After Me, Pluto is a Planet." But we know today that Pluto is no longer Planet X. Its energy field, known as the Destroyer or the great transformer, has been known to cause revolutions in its revolution around our Sun, which takes approximately 268 years to complete. The last time Pluto was in Capricorn was in the 1700s during the last American and French Revolutions, which let's face it, changed the world. Pluto in Capricorn cycles are known to transform economies, governments, and religions.

Pluto entered Capricorn in 2008 and will remain there until 2023 according to the Tropical Zodiac. If you look around the world, you will see revolutions all around the world, creating all kinds of governmental upheavals, through protests, riots, and outright civil wars.

In 2008, when Pluto entered Capricorn, we saw the collapse of hundreds of banks, through the mortgage crisis, these crises happen on schedule!

In recent news, we saw a lightning bolt directly hit the Vatican. Does anyone else see the symbolism here? Even more recent, the new Pope released two doves (of peace), who were both attacked by a crow and a seagull. The symbolism would suggest that there will be no peace found within the Church of Rome. The gig is up for the Powers That Were!

You will find master astrologers at the highest levels of government and secret societies because these are the people who know the cycles of astronomy and what each particular energy will bring. For example, American President Ronald Reagan would not sign any document before consulting his master astrologer.[6]

I have been persecuted by Christians all my life for even studying astrology because they are implanted with blindness and misunderstanding held in place by religious spirits. Those who are world leaders and high-ranking secret society members know all about this incoming Nemesis-Nibiru system interloping without space, they know about the precession of the equinoxes, Pluto in Capricorn, and the Age of Aquarius.

A changing of the guard is coming. The world as we know is breaking down, so that it can be rebuilt with competing ideologies. One will fail, the other will live long and prosper for a millennium, and then the universe is scheduled to be re-ordered. The Antichrist group will continue to poison our water, air, and food supply, because many are slaves and feel like they cannot speak up, but we need to speak our truths. In other words, when we remain silent, we are acquiescing to the decisions they make for us, which includes GMO's, fluoridated water, chemtrails, vaccinations, and who knows what else.

TIME FOR A RESET?

Through the discovery of OOPARTS, out of place artifacts, we know that humankind has been here for hundreds of thousands of years, which seems to contradict the Bible's timeline of approximately 6,000 years. The reason people mistakenly think the Earth is 6,000 years old is because that is when the Bible story began, which clearly picks up after the First Great Deluge – the sinking of Atlantis, Lucifer's Flood. This was long before the last Deluge of Noah's Flood. God in His mercy deliberately left a lot of prehistory blank to us, drawing a curtain over the ancient past. Depending on the closeness of Nibiru, on its next passing, it could cause an extinction level event where a reset of humanity is not only possible but prophesied, indicating its part of a grander plan of predestiny. [6]

Nibiru, the comet-red planet with a golden core and a large tail of iron oxide and debris, which can make it look like wings, is being cloaked. Once it reaches the orbit of Jupiter and Mars, it will cause the solar winds to become concentrated enough for the tails dust to start to light up. The cloaking cannot hide the light behind it. Right now, at the publication of this manuscript in 2020, it has been tracked to be behind the Sun, coming in through the southwest, entering our Solar System between Neptune and Jupiter, now yet Jupiter and Mars.

The larger the mass, the slower the orbit, as it must negotiate the gravitational field of our Sun, and our own planet. When it begins to light up as it approaches closer between Earth and the Sun, it will no longer serve any purpose to hide it, as it will cause the Earth to become perturbed and be literally moved to its core, causing a pole shift. In which case, there will be coastal displacement, flooding in the form of tidal waves, Earthquakes, and volcanic eruptions, that few will survive. This is known as the Terrible Day of the Lord in the books of Isaiah and Zechariah.

So, why invest billions into hiding and cloaking this system? The powers that be, working through various agencies around the world, feel it may be the only way to silence millions of amateur astronomers from collecting photographic evidence of Nibiru. Despite their efforts, YouTube percolates daily with all kinds of photos and videos taken all of the world of Exoplanets, Exomoons from the Nemesis-Nibiru System intersecting within our Solar System, that are seen around our Sun, and eclipsing our Sun, some daily.

This is one of the main reasons, a sophisticated lensing system, along with Craft, Chemtrails, and holographic projections are being used to obfuscate the daily motion of a multiple objects from the Nemesis System, one of which has been orbiting between the Sun and Earth, causing eclipses of our Sun that last for a few minutes a day. I have witnessed it with my own eyes, multiple times in person and recorded on Weather Cams. The Alaskan skycam's video footage has been revealing several planets being hidden by the Chemtrails and the Sun simulator's sophisticated lensing operations. The cams are pointed in the southwest direction to track Planet X (Nibiru).

I tell you, here, *the Truth is most definitely Stranger than Fiction*. There has also been a great deal of activity around the Sun which has been recorded on SOHO. The

Solar and Heliospheric Observatory (SOHO) is a spacecraft built by a European industrial consortium led by Matra Marconi Space that was launched on a Lockheed Martin Atlas II AS launch vehicle on December 2, 1995 for the purpose to observe and study the Sun. Since then, SOHO has discovered over 3,000 comets. SOHO is an international project consisting of joint cooperation between the European Space Agency and NASA. Both agencies have public websites that publishes video recordings from SOHO of the Sun. This allows scientists to study Sunspots and solar flares, however, many of us have been seeing much more than solar flares and coronal mass ejections coming from the Sun.

In addition to SOHO, there is NASA's mission of STEREO (Solar TErrestrial RElations Observatory) which consists of two identical spacecraft which were placed in two different orbits around the Sun. STEREO Ahead was placed slightly closer to the Sun than Earth and STEREO Behind slightly further away than Earth. This causes them to respectively pull farther ahead of and fall gradually behind Earth's orbit, thereby exposing the far side of the Sun. Both spacecraft are equipped with telescopes that are focused on solar activities.

Through STEREO's photos and videos which are published publicly on https://www.nasa.gov/mission_pages/stereo/main/index.html, and https://www.spaceweatherlive.com/en/solar-activity/solar-images/stereo.html, we're witnessing exoplanets, Exomoons and spacecraft, and lots of spacecraft.[7, 8] Some days, there's none, other days, there's what appears to be massive ships that are shaped like the USS Enterprise or the Stone carved Faravahar in Persepolis. Faravahar is one of the best-known symbols of Zoroastrianism, which is essentially a winged object with a god, known as the Ahura Mazda who appears to be riding inside the winged object etched in stone in. This symbol represents a winged guardian or fravashi, who is also believed to be an angelic being of the Zoroastrian religion. This so-called god or angelic being, many of us believe was an ancient astronaut. The objects showing up around the Sun, coincidentally appear to be shaped just like Faravahar.

Besides solar eclipses from exoplanets, the cover up of UFOs, which at times appear to not be alone, as photos surfaced in 2018 of what appeared to be an armada of spacecraft around the Sun. Most stars are gateways into other universes, and our Sun is no different. Nevertheless, these objects are being cloaked, obfuscated, and covered up for the most part by the Elite who are running these space agencies. Just lookup cloaking on YouTube. The U.S. research is 30-60 years ahead of what is published. During the President Regan's Administration, he initiated the SDI, "Strategic Defense Initiative" also called the "Star Wars Defense" system, that was implemented with lasers in space.

Likewise, there's tons of videos on YouTube proving laser and holographic technology. Many believe it is possible for lasers to project a fake star pattern projected over the area in question. This is what we're seeing almost daily in weather cams, when the skies are clear of clouds, different shapes that appear and disappear, in order to cover up eclipses of exoplanets with our Sun, or other objects they are hiding from the general public. We consistently see large pink triangles, sphere shaped objects with stripes, that look like planets, but are actually projections, and star shaped objects as well. We see all the effects of an approaching planet, but due to the ongoing obfuscations and efforts to conceal it, many are dangerously unaware, and in denial that these objects are closer than most think. Also, why does the Bible imply a short period for the Earth being warned? In modern time with telescopes, this should not be possible without interference.

WHO WAS MALACHI MARTIN?

Malachi Martin (1921 – 1999) was an Irish Catholic priest and writer on the Catholic Doctrine. Originally ordained as a Jesuit priest, he became Professor of Paleography at the Vatican's Pontifical Biblical Institute. From 1958 he served as secretary to Cardinal Bea during preparations for the Second Vatican Council. Disillusioned by reforms, he asked to be released from certain of his Jesuit vows in 1964 and moved to New York City. His 17 novels and non-fiction books were frequently critical of the Catholic Church, which he believed had failed to act on the third prophecy revealed by the Virgin Mary at Fátima.

> "Because the mentality, the attitude, amongst those who, at the higher levels, the highest levels, of Vatican administration and Vatican geopolitics know that, now, knowledge of what's going on in space and what's approaching us, could be of great importance in the next 5, 10 years." [9] (Malachi Martin, 1996)

In the 1990s, Father Malachi Martin was interviewed by Mr. Bernard Janzen of Triumph Communications. In this interview Father Malachi Martin says "There was this consecration, this enthronement of satan within the Vatican, of Lucifer by the way. It is an historical fact. It was done one day by a certain group of people representing Luciferians all over the world, especially American Luciferians. It was done. Therefore, in a certain sense, Lucifer has power. He doesn't own yet, but I'm sure he hopes to own some Pope as his man." [9]

Martin's work often dealt with satanism, demonic possession, and exorcism, liberation theology, the Second Vatican Council, the Tridentine liturgy, Catholic dogma, modernism, the financial history of the Church, the New World Order, and the geopolitical importance of the Pope. Martin continued to offer Mass privately each day in the Tridentine Mass form, and vigorously exercised his priestly ministry all the way up until his death. He was strongly supported by some traditional Catholic sources and severely criticized by highly liberal sources. He was a periodic guest on Art Bell's radio program, Coast to Coast AM, between 1995 and 1998.

In the final years before his death, Martin was received in a private audience by Pope John Paul II. Afterwards, he started working on a book with the working title Primacy: How the Institutional Roman Catholic Church became a Creature of the New World Order. Martin said, "Observant Catholics, traditionalist Catholics, will become hunted like doves."

His discernment and admonition were prophetic, in that the Church of Rome was corrupted and became sheep in wolves clothing, as today's Pope is well known by many Christians and Catholics, as the False Prophet. Malachi Martin was tuned into the warnings of God as written into the scriptures:

> "Take heed to yourselves, and to the whole flock, wherein the Holy Ghost hath placed you bishops, to rule the church of God, which He hath purchased with His own blood. I know that, after my departure, ravening wolves will enter in among you, not sparing the flock. And of your own selves shall arise men speaking perverse things, to draw away disciples after them."
> (Acts 20:28-30)

> "When these things begin to take place, stand up and lift up your heads, because your redemption is drawing near."
> (Luke 21:28)

Nibiru was claimed by Marduk, and Marduk is also what and 'who' we refer to as Lucifer. They know our origins come from the beings of this planet. They know we were genetically manipulated from these beings.

WHAT'S IN A NAME CHANGE?
According to the Enuma Elish, dubbed the Epic of Creation, we have the following history of who is who in the Nibiru Zoo:
- Marduk = Nibiru=Lucifer
- Marduk is offspring of Enki. Mother was Isis, Father was Enki (Osiris)
- Marduk has been at war with Enlil, Allah, and Yahweh. Marduk claims to be Most High. Anu was his grandfather. Anu father of Enki and Enlil.
- LUCIFER telescope is looking for Marduk or Nibiru.
- Nibiru rested on top of the Sun for 9,000 years. [10]

According Gerald Clark's research found on his website and corresponding YouTube, [8] Nibiru's size was re-examined from data points to the biblical number 666 mentioned in the Book of Revelation Chapter 13:18. Clark uses Sitchin's research along with his own, to measure the diameter of Nibiru using the measured linear analysis data shows that Planet X average diameter of the three results is Nibiru is 66.00 times bigger than the diameter for Earth! He claims it is a coincidence that the number is a sexagesimal base 60 result! 600 is 10x60, the Anunnaki link between the sexagesimal and Anu's rank of 60 and the base 10 system we use on Earth. See, Zecharia Sitchin, *When Time Began*, for detailed analysis using the mathematical relationship between the Earth and Nibiru correlation methods, relating the sexagesimal to the base ten number system we now use.[11]

Combining 600+66 gives us the wicked number of the beast, 666 from Revelation 13:18. Marduk was also known as a great dragon given, he was born of Enki and his half-sister Ninmah. The dragon was a sign of kingship in the Sumerian culture.

ANUNNAKI NAMES ASSOCIATED WITH PLANETS:

The Sumerian Tablets counted our Solar System's Planets from the outermost to the Sun.

1. Pluto's name in the Enuma Elish was Gaga;
2. Uranus name was Anu in both the Enuma Elish and the Anunnaki Name in the Sumerian Tablet;
3. Neptune was called Nudimud in the Enuma Elish and Ea/Enki by the Anunnaki;
4. Saturn was called Anshar by Enuma Elish;
5. Jupiter was called Kishar by Enuma Elish and Enlil by the Anunnaki;
6. Mars was known as Lahmi by Enuma Elish and Ninhurta or Nirgal by Anunnaki;
7. Earth was called, Ki by Enuma Elish and Urash by the Anunnaki;
8. Venus was called, Lahamu by Enuma Elish, and Ninmah/Damkin by Anunnaki;
9. Mercury was called Mummu by Enuma Elish and Ningishzidda by Anunnaki;
10. Moon was called Qingu in Enuma Elish;
11. Nibiru was called Marduk in Enuma Elish and Bel/Moloch by Anunnaki. [10]

Then after that, the Grecian and Roman Empires renamed the planets after their gods and goddesses.

The Sumerian Tablet 1 tells a story of a group of beings called the Igigi, who were slaves/servants of the Anunnaki. Tablet 1 tells of a minor revolt of the Igigi that was led by their leader named, "Allah" who led a group of rebels to surround the Egal Fortress that was led by Enlil. Enlil's influence ruled over the Dynasty of Ur. Allah and Wife Nirgal had a son, 'Nanu' who became King of the Dynasty of Ur. Nirgal was associated with Mars, therefore researchers believe that life on Earth originated on Mars. Artifacts found on Mars are identical to those found on Earth, i.e., three pyramids that are 19 degrees apart, and a sphinx. The only difference is that the Sphinx on Mars is facing upwards towards the Heavens, whereas the Sphinx on Earth is facing East. The late author David Flynn proved through the history and artifacts in his book, Cydonia: The Secret Chronicles of Mars, that Sidon on Earth was connected to Cydonia on Mars.[10]

The Igigi mysteriously disappeared, and the offspring of Enki (Marduk) were in opposition to Enlil over the Avatars, humankind that they were genetically manipulating and using as slaves to replace the Igigi. Inanna was Isis in Egypt. Enki was Osiris. The destruction of the City of Ur became the genesis of the Jewish Race and Judaism.

Marduk, Enki's son, became a megalomaniacal Lord of Babylon. Marduk and Enlil were arch enemies. They were at war until Marduk beat Enlil in 2,000 B.C., when Alexander the Great took over Babylon.

This ancient war between these two ET giants continues today, as the followers of Allah who actually are in bondage to Enlil, are at war with the followers of Yahuah, who used to be called Enki. Marduk and Enlil are responsible for influencing humankind into constant divisions, conflicts, and wars. It is the core issue that is the foundational cause of generational sins and generational curses between Jews and Arabs. This is why Mohammed was chosen to have an alien encounter with an extraterrestrial fallen angel called Gavril, who was sent by Enlil aka Allah, who influenced Mohammed to start a new religious cult to compete with Christianity and Judaism, who were both under Yahuah aka Enki.

In 586 BC, Marduk's forces conquered Jerusalem, taking Jews captive into Babylonian captivity. God had decided to make the Hebrews his chosen people. He wanted them to be set apart so that they could be used to tell the people of the Earth about who he is and how he wants them to live. This is when Judaism really was established as a 'set apart' religion. This is when Jewish-Hebrew traditions and rituals took root, which served to keep the Jewish people together during their 50 years of Babylonian captivity until being led and given the Promised land. God had given the Hebrew people the fertile land of Canaan to the Hebrew people after he drove out the other nations that occupied the territory. The reason he took the land away from these people was due to their wickedness, deviant sexual practices, and idolatry. Once the Hebrew people settled into Canaan, God had to constantly warn them about getting involved in the pagan practices that went on in other nations.

Marduk read the Enuma Elish every year in Babylon, which the captive Hebrews were forced to listen to. This is influenced the writing of the Jewish scriptures, which is why Genesis is essentially a "synopsis" of the Epic of Creation from the Enuma Elish. The Hebrew scribes actually wrote a much more detailed account of the celestial and supernatural drama that took place in the Garden of Eden, that is known today as the Apocryphal writings, The First and Second Book of Adam and Eve, which tells in much greater detail, what went on between Lucifer/satan, the Lord Yahuah, Yahuah's Angels (extraterrestrial messengers) and their first models of their Evadamic Race, we know as Adam and Eve. [13]

Marduk built a gateway of the gods in Babylon. There were twelve gates, and he overthrew Enlil over their space facilities to communicate with their home planet, Nibiru. Marduk was the god in that region, and was called, "Baal", which in Hebrew means Lord. See, War of Gods and Men, Zecharia Sitchin. Marduk and Yahuah were at war. [14] This is when during a conflict, they confused humans by scrambling the languages, to block communication with the gods via the Tower of Babel which is believed by Ancient Astronauts Theorists to be an Ancient Space Port to the Heavens.

Jesus said, "The Kingdom of God is within you." This implies intelligent design in the genome, which allows the spirit of God to live inside humans and use humans as vessels to be set apart for the eventual Kingdom of Heaven coming to land on Earth. See, Revelation 22.

Intelligent design is evidenced in our genetic codes. The evidence of the Creation is in the Creation (Romans 1), every 'hu'- man is God breathed souls housing human bodies. HU means the breath of Spirit/Soul, chi, prana, human energy, which is the breath of the Creator. This is why the sound HU is in the middle of God's name.

Ya-'HU'-ah, is the Creator, the Almighty, the one who gives life and the one who takes it away, and the one who resurrects from the grave, and gives life back again.

> "The LORD (Yahuah) gives both death and life; He brings
> some down to the grave but raises others up."
> (1 Samuel 2:6)

Marduk is at war with Yahuah. He won battles over Babylon, Sidon on Earth, and Cydonia on Mars. Marduk infiltrated Judaism as well, he separated the real Jews from those who were claiming to be Jews, who lived in Khazaria, along Crimea and Bulgaria. They in scripted Judaism under Babylonian rule, this is known as the Masoretic text, which essentially was written by Babylonian Jews who wanted to protect the Sacred Name of Yahuah from the Pagans and those stealing their identity. So, they took the Torah, originally written in ancient Hebrew/Aramaic, which did not have any vowels, and rewrote the Bible with vowels throughout. It deliberately took the vowels from the words Lord and God, which are Adonai and Elohim, and transposed those vowels over the Sacred Name of God, the Tetragrammaton. This altered the pronunciation of the sacred name so as to confuse the Gentiles from invoking it. They essentially created a false name, Yehovah or Jehovah. The oral tradition about the correct pronunciation and invocation of the sacred name of God was passed on orally from generation to generation of the tribes. In fact, one of the lost tribes of Israel, who migrated to North America, the Cherokee Nation, use the name Yahuah as the name of Great Spirit, who they believe is the Creator.13 See, Book Two in my book series from 2015, *Who is God?*

Marduk is still at war with both Enlil and Yahuah over control of the human experiment on Earth. Marduk who became Lucifer, is going to be appear as the god and messiah of this world under Anunnaki control. All countries were conquered by Anunnaki gods. The entire India saga evidenced in the Vedic literature, Bhagavatam, were all Anunnaki gods. Brahma, Vishnu, Shiva, Nephilim gods and goddesses who were depicted as chimeras, with animal mixed with human parts, blue skin, and who had technological tools that could destroy flying cities and cities on Earth with nuclear energy, were Anunnaki.

Marduk's plan was to divide and conquer by separating humankind by language. This confused and alienated us from each other. The same Machiavellian strategies are continued today, to divide and conquer, through religions, through politics, through race and through culture. This tactic prevents humans from achieving unity and one-mindedness which would make us more powerful. The evil demonic Religious Spirit attacks every religion, including those who identify as atheists and agnostics, in order to divide people within groups. People who even belong to the same religion do not agree with one another, because of the implant of the religious spirit.

In the end, Marduk who is Lucifer will manifest as the Counterfeit Messiah, the final Antichrist, who will attempt to pull off a One World Religion and establish a One World Government. This will position him for the real endgame confrontation with the Real Messiah, Yahushua, the Lord Jesus Christ, who will end the battle with a breath and word when He too is scheduled to return to Earth, with legions of His celestial extraterrestrial angels on the clouds of Heaven. This is an armada of highly advanced starships, also known as Merkabahs. [15]

Notes:
1. Ella LeBain, *Who is God?* Book Two, Chapter Twenty-Three: *Babylon History: Where it All Began*, pp.367-386, Skypath Books, 2015.
2. Ella LeBain, *Who is God?* Book Two, Chapter Two: *Ancient Technology & Biblical Astronauts*, pp. 21-46, Skypath Books, 2015.
3. Walter Manuel Montano, *Behind the Purple Curtain* Hardcover, Literary Licensing, LLC, July 27, 2013.
4. Ella LeBain, *COVENANTS* – Book Four, Skypath Books, 2016.
5. Mark Passio, Lecture, *Free Your Mind 2 Conference*, April 27, 2013, New Age bullshit and the suppression of the sacred masculine. http://youtu.be/Q51l_E8Tlp0, July 2, 2013.
6. Origins: 11 Unreal Places and their Artifacts Explained, http://hidden-truth.net/2017/07/origins-11-unreal-place-artifacts-explained/, July 2017.
7. Solar Terrestrial Relations Observatory, STEREO, https://www.nasa.gov/mission_pages/stereo/main/index.html
8. Solar Terrestrial Relations Observatory (STEREO), https://www.spaceweatherlive.com/en/solar-activity/solar-images/stereo.html
9. Lucifer in the Vatican – Father Malachi Martin, https://thewildvoice.org/lucifer-vatican-malachi-martin/
10. Gerald Clark, YouTube, Abrahamic Religions, 2017, https://youtu.be/-ygSDL7qdvA?list=UUFUH_0JRBPG9g5k5M3AgJ4g, http://www.geraldclark77.com/uploads/4/6/0/7/46076627/7th_planet_broadcast_outline_episode_003.pdf
11. Zecharia Sitchin, *When Time Began – The First New Age*, Book Five of the Earth Chronicles, Avon Books, NY, 1993.
12. David Flynn, *Cydonia: The Secret Chronicles of Mars*, (used with permission) 2002. http://www.mt.net/~watcher/stones.html
13. The Apocryphal First and Second Books of Adam and Eve

14. Zecharia Sitchin, *War of Gods & Men: Evidence of Extraterrestrial Warlords who destroyed Ancient Civilization*, Book Three of the Earth Chronicles, Avon Books, NY,1985.
15. Ella LeBain, *Who is God?* Book Two, Skypath Books, 2015.

CHAPTER ELEVEN

THE MESSIANIC STAR

Why Wish Upon a Star when you can pray to the One who made the Stars?

"When I look at your Heavens, the work of your fingers, the Moon and the stars, which you have set in place, what is man that you are mindful of him, and the son of man that you care for him? Yet you have made him a little lower than the Heavenly beings (angels) and crowned him with glory and honor."
Psalm 8:3-5

The last appearance of Nibiru through the heavens over Earth was told in the Sumerian accounts of a celestial battle between Nibiru and the watery monster Tiamat, the Earth's predecessor, who identifies Nibiru as a fiery hulk:

"In front of him he set the lightning,
with a blazing flame he filled his body;
He then made a net to enfold Tiamat therein…
A fearsome halo his head was turbaned,
He was wrapped with awesome terror as with a cloak."[1]

This description of Nibiru translated by Zecharia Sitchin, and the Sumerian texts correlate to the Book of Job. There are parallels between the Sumerian text about Nibiru, and of the Lord's battle with the sea-monster Leviathan.

"Its snortings throw out flashes of light; its eyes are like rays of dawn. Burning lamps go forth from its mouth, and sparks of fire leap out. Smoke goes forth from its nostrils as from a boiling pot or cauldron. Its breath sets coals ablaze, and flames go forth from its mouth…"
(Job 41: 18-21)

In the Mesopotamian viewpoint, the Lord Almighty and Victorious defeated the monster, the watery Tiamat, Earths forerunner. This is not just in the Sumerian texts but also in the Bible. The description of Nibiru as a flaming star and its brown dwarf Nemesis appears in Bible scriptures on account of its passage evidenced in Job above.

Nibiru, however, is an extraordinary planet. It is cloaked and surrounded by nets, halos, lightning, and flames. The imagery described its appearance prior to the impact of one of its Moons with Tiamat, which ruled out these references as being explosions. This fiery entity is not exactly bright enough to blind us under normal viewing conditions. Unlike our Sun with its bright corona, Nemesis-Nibiru is a brown dwarf. This brown dwarf has seven planetary bodies in its orbit, poetically referred to as *winds* in the Sumerian texts. The word in Sumerian literally means: "Those that are by the side:"

> "He sent forth the winds which he had created,
> the seven of them; to trouble Tiamat within
> they rose up behind him."[1]

Sitchin described the cataclysmic destructive effect of Tiamat due to Nibiru's close passage. The watery planet Tiamat must have been larger than the Earth, as the texts state Earth was formed from Tiamat splitting in half from the power of Nibiru's passing. Therefore, Nibiru must be an even larger celestial body than Tiamat to cause that kind of destruction described here:

> "As the two planets and their hosts of satellites came close enough for Nibiru to 'scan the inside of Tiamat' and 'perceive the scheme of Kingu,' (soon to become the Moon) Nibiru attacked Tiamat with his "net" to "enfold her," shooting at the old planet immense bolts of electricity ("divine lightnings"). Tiamat "was filled with brilliance" – slowing down, heating up, "becoming distended." Wide gaps opened in the crust, perhaps emitting steam and volcanic matter. Into one widening fissure Nibiru thrust one of its main satellites, the one called "Evil Wind." It tore Tiamat's "belly, cut through her insides, splitting her heart."[1]

The Sumerians and Egyptians left behind records reporting that planet Nibiru had an unusual elliptical orbit compared to normal horizontal orbiting planets. They say it took Nibiru around 750,000 years to even come between Mars and Jupiter, yet when it did, it created cataclysmic destruction on all the planets it passed.

The planet Tiamat was between Mars and Jupiter around 510 million years ago and became a victim of Nibiru passing. It is believed, according to Sitchin's translations, that Tiamat collided with one of Nibiru's Moons. They crashed into each other in orbit, broke into half, one half shattered into what became the asteroid belt, with Phobos, which became a Moon locked into Mars, while the other half became our home, planet Earth. This is why the Vedic literature *Bhagavat Gita* recorded, "Out of destruction comes life." [2]

From this account alone, it is clear that Nibiru is a cosmic maniac. Its orbit is eccentric, elongated, and different from all the rest. It comes through our system like a bull in a china shop. According to some historians, not all passings of Nibiru are exactly alike. It all has to do with whether it is perihelion or aphelion to Earth or

Sun. It always goes around the Sun and swings around Earth. Because its orbit is tilted away from the ecliptic plane, on a completely different path than all the other neighboring planets, it can cause great cataclysm, as its mass alone creates an huge aura of energy, that disrupts everything in its path. I honestly believe the fact that this thing causes such destruction and upset is the reason the Lord promises in the ancient Bible Prophesy to recreate the Heavens and the Earth.

When it comes to Nibiru, size matters! To be able to harness that kind of destructive-cataclysmic power without ever touching Tiamat, we can assume it must be the size of gas giant proportions. However, when in proximity to other planets – and especially our Sun – gravity kicks in, and there is an inevitable plasma exchange that explains the descriptions of the lightning strikes that point to a mini-Sun or dark star. This giant comet-like planet is not a terrestrial-type world. In fact, Sitchin said that life was on its Moons.[3]

I agree with Andy Lloyd on this analysis[4]: To imply that a body the size of the Moon would destroy the Earth during a close passage, without even touching it, is a ridiculous notion. Therefore, Nibiru must be a much larger planetary body than Tiamat, which in turn was proportionately larger than the Earth. In my opinion, the reason why Sitchin thought Nibiru was a terrestrial-sized planet that supported life was most likely an error in judgment. Nibiru's mass and fiery, comet-like nature, certainly does not seem conducive to supporting life, in fact, just the opposite as its gravitational attraction has proved destructive. It is the Exomoons and Nibiru's sister planets where alien life is.

The Sumerian texts imply that Nibiru is a Titan, a fiery planet with many star-like attributes. It also has seven planets and Moons, that were visible to observers on Earth. I agree with Lloyd's pointing out that the Exomoons of Nibiru would therefore have to be far brighter than the Galilean Moons of Jupiter, which can only be seen through binoculars. This implies one or more of the following things: They are bigger than Io, Europa, Calisto and Ganymede; they are lit by both our Sun and the fiery light of Nibiru; and/or they are seen when Nibiru is closer to us than Jupiter.[5] Seeing Nibiru and her seven planets and Moons has to be an eye-popping dramatic event to witness from Earth. It is unlikely its appearance in our skies could have been missed, especially because the ancient religions expected its return. Back then, it was not covered up by government astronomers, propagandists or chemtrails as it is today. Our ancestors were far more aware of the signs in the heavens than most of us are today. Back in ancient times, astrology and astronomy were considered one science, one language, unlike today. Therefore, Nibiru's expectation was interwoven into their belief systems and understanding of the Cosmos.

Nibiru is a brilliant red star whose halo during perihelion would be swept back from the Sun giving the appearance of fiery wings and/or a fiery tail. It would subsequently display a line, or Saturn-like ellipsoid disc, of its orbiting Moons, creating an extraordinary visual phenomenon in our skies. At the time of this publication, it already is being photographed around the world before and during Sunsets and Sunrises, which often illuminate it.

Its dark star, Nemesis, is certainly playing a role with respect to how our Sun is behaving, which due to its magnetism, is causing the Sun to be drained from its solar cores. That may be connected to the lack of Sunspots recently in our Sun's cycle which precipitates climate change, not just on Earth, but on all the planets in our Solar System. Climate Change is real, but it is NOT being caused by Humans. Humans cause pollution, Climate Change is governed by the Sun. And if our Sun is being drained by the intersecting transits of the Nemesis-Nibiru System, which is causing stress to our Sun, then Climate Changes are not limited to Earth, but to the entire Solar System, which Humans have no control of.

When you see its plasma wings, one can easily understand why the ancient mythologies depicted it as a celestial fiery bird and dragon. The Phoenix of Egyptian mythology, that gets reborn periodically in the sky, or the celestial dragon that makes up so much of Chinese and Arthurian mythology, may have more to do with Nibiru's passing than actual flying birds and dragons. This may have something to do with what Sir Laurence Gardner called the *Messianic Dragon* mythology revealed in his *Bloodline of the Holy Grail Series*.

THE PHOENIX

The Phoenix, also known as the bennu bird, was a mythical creature that symbolized rebirth to the ancient Egyptians. It was strongly associated with the benben stone of Heliopolis, which was likely to have been a massive conical-shaped meteorite worshipped at the temple there. The original benben was the primeval hill which, in Egyptian cosmology, first emerged from the waters of chaos to become Atum, the God of Creation.[5] This indicates its astronomical identity, especially as its period of return is measured in either hundreds, or even thousands, of years. Archeoastronomer E.C. Krupp linked the Phoenix with the Sun and Sirius:

> "The Egyptians attached particular importance to a heron-like bird they called the bennu, and the Greeks identified the bennu with the phoenix. According to Herodotus, the red and gold bennu was reported to return from Arabia to Heliopolis after five hundred years' absence, or, more curiously, according to others, after 1,461 years. In some versions of the legend the bennu dies at Heliopolis, and from either its nest or its own burned remains a new bennu arises to start the cycle of life anew." [5]

The length of time of the Sothic Cycle is 1,461 years. This is the time it takes for our Solar Calendar to catch-up with Sirius again. Sirius has an annual cycle of 365 days compared with our Sun which is 365.25 days, which is why we have leap year every four years to essentially play catch up with that extra quarter day. This is what led Krupp to identify the Phoenix with Sirius. He claimed that the Phoenix also clearly has a solar dimension:

> "The bennu must have been associated with the Sun, for the bird, in Egyptian myth, was said to be the soul of Ra... Contradictory aspects of the myth of the bennu and the very idea of the benben have made it difficult to understand the relationship between these very important symbols and the Egyptian solar religion... The bennu was the guide of the gods in the Duat and came from the heart of Osiris." [5]

The Duat was the word Egyptians used for the underworld or netherworld, the world of the Dead. Researcher Andy Lloyd rightly discerns that the Egyptian's confusion between important stars and the Sun will be a running theme throughout our investigation.[4] Amongst Egyptologists the solution lies in the appearance of a previously unidentified celestial body, that has a mixed identity of being a star, a planet, a comet planet and a dark Sun all in one. It was represented by the Phoenix, or bennu, because of its rebirth in the Heavens after its extended absence.

THE MESSIANIC DRAGON

The symbolism of this Messianic Dragon finds an astonishing consequence within the Dark Star Theory by looking at the constellations of Aquila and Serpens, respectively the Eagle and Serpent. When you combine the two together, you have a Flying Serpent, otherwise known as a Flying Dragon. This reminds me of the Central American Quetzalcoatl, Kukulkan gods who were known as Feathered Serpents.

Horus is a messianic figure and is identified as a hawk. However, his archenemy, Set, is symbolized with evil serpents. The Messianic Dragon suggests a religious duality based on the astronomical pathways in the constellations. This house of the Heavens was called the Domain of the Dragon, not just for poetic reasons, but because of the origin and territory claimed by the Draconian Serpent Race. (See Pole Star: Alpha Draconis.) The connection between these two constellations merge into the form of the dragon and designated the location of the brown dwarf when it came through the Oort Cloud, at the outermost region of our solar system, well beyond Pluto.

This is the region also called the Kuiper Belt that Astronomers are actively studying perturbations coming from Pluto and the Trans Neptunian Objects, which led to the discovery of Planet Nine, which is actually Planet Ten, because they demoted Pluto from being a planet. However, Pluto is a dwarf planet, and a dwarf planet is still a planet. I am going to get into this controversy and why the IAU (International Astronomical Union) deliberately renumbered our Solar System planets.

The Kuiper Belt, also called the Kuiper Cloud, can be seen beyond Saturn's orbit. The Oort cloud is a mass of trillions of comets and dust that circle the Sun on our outermost orbit around our Sun. While the Kuiper belt is disk-shaped, the Oort cloud is spherical shaped. When Nibiru enters through the Oort Cloud, it behaves like a bull in a china shop, as it dislodges comets out of their regular orbits around the Oort Cloud and Kuiper Belt, by inevitably dragging some inside our Solar System, which has already begun in the last few decades, as we're seeing a marked uptick in comets, fireballs, meteorites and asteroids passing by Earth in their magnetic attraction towards our Sun.

The reason Nibiru has been called the Messianic Dragon, or the Red Dragon by the Chinese and the Destroyer by the ancients, is approximately every 3,500-3,700 years, it emerges from its Celestial Domain and its passageway through our orbits of planets within our Solar System inevitably ends up having catastrophic consequences for humankind. We need to remember when we study ancient mythology, that myths are just allegorical, archetypal, and poetic, they are based on real historical events that happened in our ancient past. Reports and stories were passed down through the generations, and this is how mythologies were formed that nevertheless held core truths and clues to our past.

Let us look at the last perihelion passing to help us to extrapolate Nibiru's journey from the past into the future.

PERIHELION OF MESSIANIC STAR

Sitchin initially proposed that Nibiru appeared in 3,760 BC, marking the beginning of the Nippurian calendar and the Jewish count of years.[6] In later books he shifted his timeframe, but I tend to think that his initial idea was the correct one. He argues convincingly that this, and previous extrapolated passages of Nibiru, marked transition periods in humankind's development, essentially indicating input by the Annunaki.

The passage of about 11,000 BC, which Sitchin points to as the date of the flood, fits well with Bauval's theories[7] regarding the Archeoastronomical dating of the pyramid field at Giza (10,450 BC), which marks the post-Diluvial reconstruction of the prehistoric world's lost civilization.

According to Zetatalk,[8] a site that listens to Gray aliens allegedly from Zeta Reticuli, Nibiru follows a 3,597-year cycle. Ok, so that means Sitchin was close by three years, in his estimation of a 3,600-cycle based on the Nibiru years, or Shars, according to Sumerian texts. So, what happened in Earth's history, 2,178 years ago give, or take three to five years either side? Most researchers agree that the last passing was marked by the Biblical Exodus as recorded in the Old Testament Book of Exodus.

Rabbi Yuval Ovadia is an Israeli filmmaker on YouTube. The key point of his films is to urge Jews to return to Israel as soon as possible to be in Israel as the apocalypse nears. He claims that the rise in natural disasters like Earthquakes, tsunamis, volcanoes, and hurricanes suggests Nibiru is on its way. Ovadia says, "All the natural catastrophes we are seeing now are disturbances caused by Nibiru coming closer, like flotsam being pushed ahead of a ship. Nibiru will tear down all the craziness we have built up, all to prepare us for the great era that follows." He asserts that the only safe spot on Earth, when Nibiru passes, will be Israel. He claims that Nibiru is described by the medieval scholar Maimonides in the prophecy of Balaam in the Bible. The Zohar, the basis of Jewish esoteric learning, describes the Star of Jacob as a Nibiru-like world, he suggests.[9]

THE DARKEST DAY IN HISTORY

The Darkest Day in History happened approximately 30-33 A.D. (give or take a few). There was evidence of Nibiru's Sister Planets Eclipsing Our Sun -- Cosmic Event During the Crucifixion known as the *Darkest Day in History* in Matthew 27:45.

The miraculous darkness that surrounded the cross during the Crucifixion of Christ not only symbolized the depth of His suffering for bearing the Sins of the World but was evidence of a scheduled "cosmic event" that took place, which has perplexed Bible scholars for centuries. There is really only one scientific explanation for what the Bible record states was three hours of darkness between noon and 3:00 p.m.: Some kind of eclipse.

Extensive Darkness

How extensive was this darkness? How much of the Earth did it cover? Did it cover the whole of the daylight half of the Earth or just the land of Judea? We are not told in this passage. But it was extensive and compelling enough to grab the attention of the people of Israel for three hours.[11]

"All the land" can refer to "all the Earth" or to "the whole country" as opposed to just the land of Israel. I think Matthew would have us think this was an enormous event that cast darkness over the entire creation. It was a cosmic event. It was an Eclipse of one of the Nemesis-Nibiru Giant Planets that passed between our Sun and Earth.

The cycle of this planet is connected to the Second Coming of Jesus Christ. It marked His Crucifixion, and it will serve as the "Signs in the Heavens" that Jesus Himself predicted would be seen in the skies preceding His Return to Earth.

During my near death experience which caused a spiritual train wreck that changed my life, my testimony will be shared in more detail in an upcoming book, *CinderElla's Shadow*, however, suffice it to say, during one of my many episodes of having the veil torn from my eyes of what the spiritual realm was really about – demoniac, nephilim spirits and aliens – I heard the Lord's voice loud and clear point to the times when the antichrist will rule the Earth for his appointed time during the last Shemitah of this precessional Age, which I heard will be during the year 2045. I also heard the voice of the Messiah, who had just saved my life during an NDE

(Near Death Experience) from a heart attack, stating that He will be returning with Nibiru. Those are the exact words I heard, "I'll be returning with Nibiru."

This communique caused me to compile this Book Series, and conclude with this offering, *The HEAVENS – Book Five of Who's Who in the Cosmic Zoo?* So, it's time to connect the dots to the ancient Annunaki that are reported to have come from the Nibiru system, and who the gods and God of gods, King of kings and Lord of lords is, and why did Yeshua tell me, very clearly, He was returning with Nibiru?

I believe, as laid out in my thesis in Who is God? that the God of Israel, who is Yahuah, is likened to the compassionate Sumerian Extraterrestrial Anunnaki gods, Anu and Enki. Whether they're one and the same god remains to be seen; but considering the fact that the Enuma Elish was read weekly while the Jews were held in captivity in Babylon for 70 years, obviously influenced the scribes who penned the Five Books of Moses, it's likely that the Yahuah, is the God of gods amongst the Anunnaki. [12]

During the Birth of Christ, we have a passage of the Star of Bethlehem, classically thought to have appeared at the turn of the first millennium, which I have proved in *Who is God?* Book Two to be a Spaceship[12], is not a conjunction of Jupiter and Venus. Stars typically do not move across the desert over days. Fixed Stars can take up to fifty years to move just one degree. The Star of Bethlehem was no ordinary star, it was an intelligent moving star, that was a cloud by day and a fire by night, which was none other than a starship.

Author and researcher Andy Lloyd who wrote *The Dark Star*, posited that a description of a very bright red star in Canis Major, seen during the time of Christ, is entirely verifiable.[4] He said there's a Roman text that mentions this astronomical anomaly, and the veracity of that text is credible. Astronomers tried to explain this anomaly for centuries yet have never once linked it with the Messianic Star! I agree with Lloyd that the upsurge in anticipation of the emergence of the Messiah at that time was triggered by the appearance of this legendary star.

The appearance of the Messiah had been expected for a couple of centuries before the time of Christ, as explained by Robert Bauval and Graham Hancock:

> "…in circa 330BC, when the vernal point was beginning its precessional drift into the 'Age of Pisces'…the conquests of Alexander the Great (356-323BC), and the resulting merger of the Eastern and Western worlds, triggered great expectations of a messianic 'Return' in the East. First at Alexandria, then across the Levant, a great agitation began, as if triggered by some prophetic 'device,' which culminated in the great messianic events of Christianity."[13]

The last time Nibiru passed Earth was during the time of the Exodus in 1476 B.C.14

Sitchin said that the 3,600-year cycle served only as an approximation. He taught that the Sumerian mythology and the return of The 12th Planet cycled around our Sun approximately every 3,600 (Earth) years.15 When questioned, he always stressed that he arrived at this approximation based on a mathematical number using the Nibiru year, which he said, was called a shar in Sumerian, and because Nibiru's actual orbit fluctuates, as Halley's comet does on a smaller scale, it was an approximation. The fluctuation could be up to 100 years, making the approximation of Nibiru's cycle intersecting throughout our Solar System between 3,500-3,700, which would be a give or take of about a hundred years on either side of the 3,600 approximation. This is why, in my opinion, a lot of researchers get tripped up on this number of 3,600. They take it as literal, but do not consider, all the calendar fluctuations that this system has inevitably caused in the aftermath of its passing. If the fluctuation is anything up to 200 years, then the Messiah's return in the East could have happened any time between about 350 B.C. and 50 A.D.

The mythologies of Mesopotamia and Egypt, which integrated the belief about a Winged Disc into their religions, probably anticipated the coming of the Messianic Star during the time of Alexander the Great, as Bauval and Hancock claim. The anxiety coupled with this anticipation must have continued for well over 300 years, as many generations of people hoped for the return in vain. As it is not just Nibiru they are waiting for, what they really expected, was the Messiah Himself, who is expected to return alongside this Messianic Star. Hence the reason it has been called the Messianic Star. The fact that people may have misread the signs or miscalculated is in my opinion due mostly, in part, to the fact that time was manipulated multiple times along our timeline, evidenced in the multiple calendrical systems that allege to tell time on Earth.

Nibiru was associated with ancient civilizations that were set up and ruled by these so-called gods, who were extraterrestrials from the Nibiru System. A promised return of a King of the Jews was likely a reference to the promise of a return of Kingship on Earth when Nibiru next appeared. During the times when Yeshua/Jesus walked the Earth, He spoke more about the Kingdom of Heaven and the Kingdom of God, more than any other topic, and referred to it as His Kingdom is not from this world (John 18:36).

All the kings and subsequent royal bloodlines followed the will and laws of these absent gods, who left their bloodline on Earth to rule their kingdoms for them. This is why I have been saying throughout this book series in my essays and discussion chapters on *Rapture, Ascension or Abduction?* that corroborates with Bible Scriptures, specifically saying that the End of Times, Last Days Harvest will be done through the agency of angels. This is that Prophecy spoken by Jesus in Parables, for those who have ears to hear and understand the language of sowing and reaping:

> "The field is the world, and the good seed represents the sons of the kingdom. The weeds are the sons of the evil one, and the enemy who sows them is the devil. The harvest is the end of the age, and the harvesters are angels." Just as the weeds are sorted out and burned in the fire, so it will be at the end of the world. The Son of Man will send out his angels, and they will weed out of his kingdom everything that causes sin and all who do evil. They will throw them into the blazing furnace, where there will be weeping and gnashing of teeth. Then the righteous will shine like the Sun in their Father's Kingdom. Anyone with ears to hear should listen and understand!"
>
> (Matthew 13:38-43)

Because of this Prophecy, I have concluded that the End Times Raptures are in three parts. First the weeds are taken, which may be perceived to those left behind as a mass abduction, second the dead in Christ rise first, and lastly, those who belong to the Lord and are still standing in the end, get taken up to meet the Lord in the air, which is the space above the Earth. Here is the Prophecy on the final two raptures or ascensions. It is noticeably clear by the language, and for those who want to study, I recommend going to Biblehub.com so you can see all the different English versions and translations from Greek and Hebrew.

The following passages come from the Berean Study Bible. Notice the emphasis on those who are alive and remain, and those who have died or fallen asleep.

> "By the word of the Lord, we declare to you that we who are alive and remain until the coming of the Lord will *by no means precede* those who have fallen asleep. For the Lord Himself will descend from Heaven with a loud command, with the voice of an archangel, and with the trumpet of God, and the dead in Christ will be the first to rise. *After that*, we who are alive and remain will be caught up together with them in the clouds to meet the Lord in the air. And so, we will always be with the Lord."
>
> (1 Thessalonians 4:15-17)

THE ABSENCE OF THE GODS

When we piece all the historical pieces together of previous passings of Nibiru that coincided with visitations by the extraterrestrial that came from Nibiru, we see not only their influence and control over humankind as they set up kingdoms on Earth, we also can see how those who became dependent on their visitations, went through deep disappointment and abandonment issues when they were absent, almost to the point of giving up on their promised return, turning to atheism, and a life that is lost without hope or faith. As a matter of fact, a lot of us have reincarnated now, to heal these deep-rooted wounds that stem from our abandonment from God or the gods.

In my humble opinion, this is one of the reasons Yeshua/Jesus incarnated and lived the humble life He did, fulfilling the prophecies of Isaiah 61:1; Luke 4:16, which the role of the Messiah, to set the captive free. What Christ gives us, besides the blessed hope of His return, is the victory over captivity by the gods of this present world, who have given up hope, that the Lord will return, not just for us, but to put things right over them as well.

Therefore, the Scriptures are clear about the seeds of satan at enmity with the seeds of God. There will be two harvests, one will be the weed and the tares, which are the genetic bloodlines of Set, and the Enlil, satan's seeds that were inserted into humankind long ago during the genetic wars between the Anunnaki half-brothers. Secondly, the seeds of Yahuah/Yeshua who may be connected to Enki, the compassionate half-brother who wanted to save humankind. There's evidence in the Sumerian tale that both genetically manipulated, altered, and intervened in the re-creation of humans on Earth, after both of its deluges.

I am going to get into this in more detail as I unpack these issues of harvest, rapture, ascension, and abduction in the following chapters. During the times of Christ, most of the world was ruled by the Romans, who had no understanding of the Middle Eastern traditions of the Winged Planet. Therefore, the promised return of the ancient gods or god to rule the world would amount to rebellion against the armies of Rome. Therefore, the emergent King of the Jews would capture the attention and imagination of the oppressed Jews under Roman rule, and the downtrodden Egyptians of that time, who understood the mythology of the gods and the Winged Planet. If the eminent Messianic Star returned as promised, then it would also give them hope that the gods would return to end the cruelty and oppression of Roman rule, that would establish a divine kingship of days gone by. Another reason they killed Him; He threatened their kingdom with an extraterrestrial Kingdom not of this world. (John 18:36)

However, there is yet another End Times Prophecy that we must integrate into this mix that is supposed to happen under the Messianic Star, and that is the collaboration of the Messianic League. I published this piece in *COVENANTS –* Book Four, p. 510, Chapter Twenty-Nine: *Messianic Prophecies and the Lost Tribes of Israel,*[16] and feel it's relevant to insert into this chapter as well, in order to connect the dots for my readers.

THE MESSIANIC LEAGUE

In the book of Isaiah chapter 19 in the Amplified Bible it states that in the future, Egypt, Israel, and Assyria (Syria), will become the Messianic League. Imagine that! Here is the entire passage as it appears in the Amplified Bible. Emphasis and commentary are mine:

> "Listen carefully, **the Lord is riding on a swift cloud** and
> is about to come to Egypt.
> The idols of Egypt will tremble at His presence,
> And the heart of the Egyptians will melt within them.
> "So, I will provoke Egyptians against Egyptians;
> And they will fight, each one against his brother and each one against his neighbor,
> City against city, kingdom against kingdom.

The swift cloud is the Lord's spaceships. I established this in Book Two: Who Is God? that when scripture refers to the Lord coming in a cloud, or riding on a cloud, it is referring to His spacecraft. In this case, a 'swift cloud,' which means it will move very quickly, zip in and out in the blink of an eye, just as most UFOs do.

> "Then the spirit of the Egyptians will become exhausted within them and emptied out; And I will confuse their strategy, So that they will consult the idols and the spirits of the dead, And mediums and soothsayers.
>
> "And I will hand over the Egyptians to a hard and cruel master, And a mighty king will rule over them," declares the Lord God of hosts.
>
> The waters from the sea will dry up,
> And the river will be parched and dry.
>
> The canals will become foul-smelling,
> The streams of Egypt will thin out and dry up,
> The reeds and the rushes will rot away.
>
> The meadows by the Nile, by the edge of the Nile,
> And all the sown fields of the Nile
> Will become dry, be blown away, and be no more.
>
> The fishermen will lament (cry out in grief),
> And all those who cast a hook into the Nile will mourn,
> And those who spread nets upon the waters will languish.
>
> Moreover, those who make linen from combed flax
> And those who weave white cloth will be ashamed.

[Those who are] the pillars and foundations of Egypt will be crushed; And all those who work for wages will be grieved in soul.

The princes of Zoan are complete fools.
The counsel of the Pharaoh's wisest advisors has become stupid. How can you say to Pharaoh,
"I am a son of the wise, a son of ancient kings?"

Where then are your wise men?
Please let them tell you,
And let them understand what the Lord of hosts
Has purposed against Egypt [if they can].

The princes of Zoan have acted like fools,
The princes of Memphis are deluded [and entertain false hope]. Those who are the cornerstone of her tribes
Have led Egypt astray.

The Lord has mixed a spirit of distortion within her.
Her leaders have caused Egypt to stagger in all that she does,
As a drunken man staggers in his vomit.

There will be no work for Egypt
Which head or tail, [high] palm branch or [low] bulrush, may do.

In that day, the Egyptians will become like [helpless] women, and they will tremble and be frightened because of the waving of the hand of the Lord of hosts, which He is going to wave over them. The land of Judah [Assyria's ally] will become a terror to the Egyptians; everyone to whom Judah is mentioned will be in dread of it, because of the purpose of the Lord of hosts which He is planning against Egypt.

In that day, five cities in the land of Egypt will speak the language of [the Hebrews of] Canaan and swear allegiance to the Lord of hosts. One [of them] will be called the City of Destruction.

In that day, there will be an altar to the Lord in the midst of the land of Egypt, and a memorial stone to the Lord near its border. It will become a sign and a witness to the Lord of hosts in the land of Egypt; for they will cry to the Lord because of oppressors, and He will send them a Savior, a

[Great] Defender, and He will rescue them. And so, the Lord will make Himself known to Egypt, and the Egyptians will know [heed, honor, and cherish] the Lord in that day. They will even worship with sacrifices [of animals] and offerings [of produce]; they will make a vow to the Lord and fulfill it. The Lord will strike Egypt, striking but healing it; so, they will return to the Lord, and He will respond to them and heal them.

In that day, there will be a highway from Egypt to Assyria, and the Assyrians will come into Egypt and the Egyptians into Assyria; and the Egyptians will worship and serve [the Lord] with the Assyrians.

In that day, Israel will be the third party with Egypt and with Assyria [in a Messianic league], a blessing in the midst of the Earth, whom the Lord of hosts has blessed, saying, "Blessed is Egypt My people, and Assyria the work of My hands, and Israel My heritage."[17]

In order for Israel, Egypt, and Assyria to be aligned in a Messianic League, they all must come together through common unity through the Messiah. This sounds like the Millennial Reign of Christ, when He returns to Earth to set up His Kingdom in the New Jerusalem that will arrive through the ancient portal in Jerusalem, and overlay the old Jerusalem, which according to scriptures, will be destroyed by the Gentiles. He comes to reign as King of kings, Lord of lords of all the nations. To bring these ancient enemies together is nothing short of a miracle of God and the specific work of the Messiah.

The word *Revelation* in Greek means "unveiling" or "disclosure." The purpose of the Prophetic Book of Revelation unveils Jesus Christ and discloses *who* his Celestial Army will be. The prophesies of Revelation expose to us the public appearance of His coming *with the Clouds* in great *Glory*. The prophesies point to a time when Jerusalem will be trampled over by the Gentiles that will lead up to, as well as include, the last days' oppression and Great Tribulation (see Revelation 11:1-2). It is only when Yeshua/Jesus Himself appears, that the Gentiles will be shattered and broken like pottery with His rod of iron (see Revelation 19:11-16; Psalm 2:9). It will only be when His feet again stand on the Mount of Olives, that ancient space portal, which is the exact location where He ascended into Heaven in a Cloud (see Acts 1:9-11), when the warring nations that are gathering up against Jerusalem will be defeated once and for all.[18]

Then the city will no longer be trampled underfoot by the Gentiles (see Zechariah 12:2-3; 14:2-7; Joel 3:12-17). We are on our way to that glorious moment. The times of the Gentiles are coming to an end.

CHAPTER 11: THE MESSIANIC STAR

> "There will be great distress in the land and wrath against this people. They will fall by the sword and will be taken as prisoners to all the nations. Jerusalem will be trampled on by the Gentiles until the times of the Gentiles are fulfilled."
> (Luke 21:23-24)

Notes:
1. Zecharia Sitchin *Genesis Revisited* p30, 324-328, 34 Avon Books, NY, 1990.
2. Kapiel Raaj, *Planet Nibiru Facts & Secrets Revealed: Can you feel its Presence?* http://www.krschannel.com/Nibiru-.htm
3. Stephen Wagner, *Is Nibiru Approaching?* May 24, 2019 http://paranormal.about.com/library/weekly/aa021102a.htm
4. Andy Lloyd, *Dark Star*, 2005, *Nibiru the Messianic Star*, http://www.darkstar1.co.uk/ds6.htm
5. E.C. Krupp *In Search of Ancient Astronomies* pp215-219, Penguin, NY, 1984.
6. Zecharia Sitchin *The Lost Realms* p 268, Avon Books, NY, 1990.
7. Laurence Gardner, *Genesis of the Grail Kings*, Bantam, NY, 1999.
8. Nancy Lieder, Zetatalk, http://www.zetatalk.com
9. Rabbi Yuval Ovadia, Planet X-Nibiru, February 8, 2018, https://www.youtube.com/watch?v=xS2FknRh3t8
10. Ella LeBain, *Who Are the Angels?* Book Three, Chapter Sixteen: *The Second Coming and Nibiru*, p. 338-347, Skypath Books, 2016.
11. http://www.abideinchrist.com/messages/mat2745.html
12. Ella LeBain, *Who is God?* Book Two, Skypath Books, 2015.
13. Robert Bauval and Adrian Gilbert *The Orion Mystery*, p.202 Mandarin, 1994.
14. Brad Aaronson, *When Was the Exodus?* Jerusalem Institute of Ancient History, (JIAH), http://ou.org
15. Zecharia Sitchin, *The 12th Planet,* The First Book of the Earth Chronicles, Mass Market Paperback, 1976
16. Ella LeBain, *COVENANTS* – Book Four, Chapter Twenty-Nine: *Messianic Prophecies and the Lost Tribes of Israel,* p. 510, Skypath Books, 2018.
17. The Amplified Bible, Zondervan (February 1, 2001)
18. Rev Willem J.J. Glashouwer, *Jerusalem, the UN and the times of the Gentiles*, http://www.whyisrael.org/2017/01/12/jerusalem-the-un-and-the-times-of-the-gentiles/, January 2017.

CHAPTER TWELVE

NIBIRU IN BIBLE PROPHECY

Who? What? Where? Why?

Revelation 12 reveals prophetic constellations in the Heavens. The Heavens show a Red Dragon which is Nibiru, and the woman clothed with the Sun is the virgin in the house of Virgo. The crown of 12 stars are those above Virgo, in the constellation of royal regal Leo, around the fixed star Regulus. The Moon at her feet, means that it is low on the ecliptic, and she is pregnant, in pain, because the Red Dragon will pass through her house (body). The Red Dragon is also known as Marduk, son of Enki, both considered Great Dragons.

It was the forces of Enlil/Allah who were behind the editing of the Bible Canon at Nicaea. Parts of the Old Testament were taken from the Sumerian Tablets.

A CLASH OF TWO KINGDOMS
Combination of Exoplanets, Solar Cores from an incoming Solar System, and spaceships.

Let's Discern, why does our government have a secret Space Program they call Solar Warden? Could it be they have known our Sun is a Stargate and they are guarding this Solar System, Earth and its neighborhood of planets and Moons?

What we are seeing is a clash of two systems. How about more specifically a clash of two ancient kingdoms?

What if the 10 kingdoms of Revelations refer to 10 planetary kingdoms that our Earth and its fellow SOLARIANS, will square off to in the near future?

The Lord promised to re-create the Heavens and Earth, because he knew it was not a perfect model as the orbits are irregular. There are clashes, that often turn into collisions and cataclysms. Maybe by design or maybe by accident? I was led to believe there are no accidents in the universe. Nevertheless, there's error, and the cause and effect of those errors or curses of the Death Star, are enough for the End Time Prophecy to end with the Lord recreating the Heavens and the Earth that will be transformed into His Kingdom of Heaven governed by the Kingdom of God.

One thing I have observed as a repetitive pattern in history of all kingdoms and governments past and present, is that all kingdoms are Hierarchical. I established they are all run as a Hierarchy in Book One of *Who's Who in the Cosmic Zoo?* in my

Chart discerning the Cosmic Order of the Kingdom of Darkness (the Draconian Kingdom) versus the Kingdom of the Human Vine (the Kingdom of Heaven). The Kingdom of Heaven is governed by the Kingdom of God, which is ruled by the King of the Heavens and Earth, who is known on Earth as the King of kings, Lord of lords and God of gods. Likewise, the Draconian Kingdom is structured like a military compartmentalized hierarchy.

In the book, *Shouting at the Wolf*, by Anderson Reed, she reveals that the entire spiritual realm is run as a hierarchy, all the way down to family unit, the city, the state, the nation, corporations, churches. The leader sets the spirit for the group. If the leader is corrupt, so will the group experience disunity, division, and corruption as a result of it. [1]

THE THREE DAYS OF GROSS DARKNESS

In order to understand the future, we need to understand and learn about the past. Three days of darkness occurred during the Ten Plagues of Exodus, which are memorialized annually during Passover Seders. The Ninth Plague was Darkness. For Darkness to have covered the Earth for three days, an eclipse of a huge plane had to occur to have covered the Sun for three days. This proves what many researchers accept as a Fly-By of the Nemesis-Nibiru system, whose orbit intersects within our Solar System and comes close to the Earth, every 3,600 years.

> "Then the LORD said to Moses, "Stretch out your hand toward Heaven, so that darkness will spread over the land of Egypt—a palpable darkness." So, Moses stretched out his hand toward Heaven, and total darkness covered all the land of Egypt for three days. No one could see each other, and for three days no one left his place. Yet all the Israelites had light in their dwellings.…"
>
> (Exodus 10:21-23)

Multiple prophecies in ancient scriptures predict three days and three nights of darkness that come upon the face of the Earth at the end of this age. Before I present them to you, let me preface by saying nobody exactly knows the dates when this will occur. Many false rumors, which are completely false, circulate annually over the internet about the three days of darkness being predicted by NASA. Now, that is not to say that NASA isn't aware of this future event, because it is written into the scriptures, and NASA is tracking Nibiru, Planet X, Planet Nine and its entire system. As scientists, they know it is only a matter of time when one of the bigger gas giants orbits between our Earth and the Sun and creates a three-day eclipse plunging Earth into three days of darkness. But, thus far, all the predictions on timing have proven to be false. That is not to say, that it is not going to happen.

To be fair, those studying and tracking the Nemesis-Nibiru System are breaking new ground scientifically, as this is something that our generation has never seen before. So, for this to get into scientific models and be a sure thing is something of great controversy amongst Astrophysicists who are working for NASA and other Space Agencies tracking this. However, with that said, they know something big is coming, and they're taking all kinds of actions in secret to protect themselves and the chosen elite who they are trying to save from the coming cataclysm that will cause the Earth to change in dramatic ways, resulting in great loss of life on multiple levels.

This is where alien technology comes into play, which will be part of these upcoming events. But where is this in the scriptures? The Days of Darkness were prophesied by the Old Testament Prophet Isaiah.

> "Darkness shall cover the Earth and *gross darkness* the people, but the Lord shall arise upon you and His glory (His True Character) shall be seen upon you."
> (Isaiah 60:2)

The gross darkness that covers the whole Earth and its people is spiritual darkness for following the god of this world, who is Marduk-Lucifer-satan, and rebelling against the Creator Lord, the One who created all of our Souls. The gross darkness is also a gross misunderstanding of God's Glory thereby misunderstanding who the Creator God really is. The true picture of His character has been tarnished by the Dark Spells of Mistranslations, Religious Spirits, Superstitions, Traditions, Spiritual Competition by the god of this world, TV, and Hollywood depictions.

There are three ways to get true knowledge of who the Creator Lord God is, from His Words contained in the Bible Scriptures, (including the Great Rejected Jewish Texts) and connecting to His through His Holy Spirit, and through the Son, Yeshua, aka Jesus the Christ.

Many other more contemporary prophets and seers spoke of the three days of darkness, like St. Faustina, Padre Pio, the seer of Garabandal Spain, who all seem to have similar visions and messages, that everything on Earth will cease to work – electronics, cellphones, airplanes, cars, televisions, a complete and utter worldwide power outage. It will be a blackout time as one of the last calls for Grace in the Last Days to reconcile with the Creator Lord, as an opportunity to repent and accept the salvation of Christ.

A more recent messenger, who claims to have been given a word of knowledge from the Holy Spirit, Linda Courtney (who can be found on YouTube) said that the Lord told her during the three days of gross darkness, the scientists will confuse people and themselves, as the Lord promises to maintain an average temperature of 55 degrees Fahrenheit all around planet Earth, to protect it from entering into a mini-ice age from being blocked by the Sun. She said to stay in your homes with your loved ones, use pure white candles, unscented, so as not to attract evil spirits which she said would going to be falling in the thousands from the skies to the Earth devouring life on Earth. But those who belong to the Lord will be protected.

This event will happen as another sign in the Heavens that precedes the rapture, the terrible day of the Lord, and the Second of Coming of Christ.

This will be a time of great testing. Today many preppers are preparing for events like this, and even worse, beyond three days, but for months and years in their belief that they could survive the coming intense cataclysmic events that are scheduled to rage upon the Earth. Many of which are already occurring at the time of the writing and publication of this manuscript. There are Earthquakes, tsunamis, volcanic eruptions, massive floods, and monster storms happening all over the planet that have already wiped-out millions of humans and animal life. We are in the End Times, and tribulation is upon us.

Three days of darkness in the Bible

Three days of darkness in 1893?

> "...and then behold, there was darkness upon the face of the land.
>
> And it came to pass that there was thick darkness upon all the face of the land, insomuch that the inhabitants thereof who had not fallen could feel the vapor of darkness;
>
> And there could be no light because of the darkness, neither candles, neither torches; neither could there be fire kindled with their fine and exceedingly dry wood, so that there could not be any light at all;
>
> And there was not any light seen; neither fire, nor glimmer, neither the Sun, nor the Moon, nor the stars, for so great were the mists of darkness which were upon the face of the land.
>
> And it came to pass that it did last for the space of three days that there was no light seen; and there was great mourning and howling and weeping among all the people continually; yea, great were the groanings of the people, because of the darkness and the great destruction which had come upon them."
>
> (3 Nephi 8:21-23)

Metaphysically and metaphorically speaking, the three days of darkness that is to occur in the last days, are parallel to the three days from Christ's Crucifixion and His three days inside the belly of the Earth, preaching to those who disobeyed Him before the floods of Noah. The Bible tells us that many were saved and followed him to the Resurrection, that the graves were opened in Jerusalem, and the dead walked when Jesus was risen from the dead and walked the Earth for another 40 days before He ascended into the Clouds of Heaven, where we are told by the Angels who showed up at His Ascension, that the way He ascended is the way He will return, in the Clouds of Heaven, which are Divine Starships. See *Who is God? Book One, The Motherships of the LORD*. [2]

I found the following scriptures interesting, despite them coming out of the Book of Mormon-Helaman. Even a broken clock is right twice a day.

> "....while the thunder and the lightning lasted, and the tempest, that these things should be, and that darkness should cover the face of the whole Earth for the space of three days."
>
> (Helaman 14:20, 27)

> "But behold, as I said unto you concerning another sign, a sign of his death, behold, in that day that he shall suffer death the Sun shall be darkened and refuse to give his blight unto you; and also the Moon and the stars; and there shall be no light upon the face of this land, even from the time that he shall suffer death, for the space of three days to the time that he shall rise again from the dead.
> And he said unto me that while the thunder and the lightning lasted, and the tempest, that these things should be, and that darkness should cover the face of the whole Earth for the space of three days."
>
> (Helaman 14:20, 27)

> "For Christ also suffered once for sins, the righteous for the unrighteous, that he might bring us to God, being put to death in the flesh but made alive in the spirit, in which he went and proclaimed to the spirits in prison, because they formerly did not obey, when God's patience waited in the days of Noah, while the ark was being prepared, in which a few, that is, eight persons, were brought safely through water. Baptism, which corresponds to this, now saves you, not as a removal of dirt from the body but as an appeal to God for a good conscience, through the resurrection of Jesus Christ, who has gone into Heaven and is at the right hand of God, with angels, authorities, and powers having been subjected to him."
>
> (1Peter 3:18-22)

I am mentioning Three Days of Darkness here because of history repeating itself based purely on the cycles of Nibiru passing our Earth, as was the case during the Egyptian Plagues. However, I am not linking to the YouTubes sites saying that here, because I am suspicious of the origins of those campaigns selling special candles, which are based on Catholic superstitions and witchcraft.

Needless to say, there are multiple reasons that can cause our power grid to go out, especially during solar storms, severe coronal mass ejections and possible asteroids knocking out satellites, as well as cyber-attacks. But we must think of who the enemies are out there and what lengths they would go to. Enemy nations that wanted to sink another nation, could use elecro-magnetic pulse that would freak out another nation by flipping the switch to take down their power grid! We have to at least include the possibility that prophecies can be abused by becoming self-fulfilled prophecies by enemy forces to purport an evil agenda, that will serve as a false flag, and not caused by the real reason the original prophecies were made in the first place, which is when and only when the real Nibiru passes the Earth. Unfortunately, and I'm not naming name, but there are forces on this planet that would counterfeit these prophecies to get people to it is the divine wrath of God instead of bad actors messing with technologies that come from another country.

Jesus tells us to be "wise as serpents and innocent as doves" (Matthew 10:16). I think that applies here to interpreting visions which may or may not be divine in origin.

> "The light shines in the darkness, and the darkness did not overcome it."
> (John 1:5)

JESUS AND NIBIRU

> "There are countless Suns and countless earth's all rotating around their Suns in exactly the same way as the seven planets of our system. We see only the Suns because they are the largest bodies and are luminous, but their planets remain invisible to us because they are smaller and non-luminous. The countless worlds in the universe are no worse and no less inhabited than our earth. For it is utterly unreasonable to suppose that those teeming worlds which are as magnificent as our own, perhaps more so, and which enjoy the fructifying rays of a Sun just as we do, should be uninhabited and should no bear similar or even more perfect inhabitants than our earth."
> (Giordano Bruno, Astronomer 1548-1600)

Giordano Bruno is known for his cosmological theories, which extended the Copernican model. He proposed that the stars were distant Suns surrounded by their own planets, and he raised the possibility that these planets might foster life of their own, a philosophical position known as cosmic pluralism. He also insisted that the universe is infinite and could have no "center". Cosmic pluralism, the plurality of worlds, or simply pluralism, describes the philosophical belief in numerous "worlds" (planets, dwarf planets or natural satellites) in addition to Earth (possibly an infinite number), which may harbor extraterrestrial life.[3]

It's amazing, that four centuries later, there are still people living in the United State of America, who actually believe that the Earth is flat, and that because the

Bible says in Genesis 1, that God created the Sun, Moon and Stars, that they think there is only one Sun. I actually got unfriended by some Christian woman from Alabama, who didn't know that our Sun is a Star, after she quoted scripture to me, that there simply cannot be 2 Suns in our sky, and when I pointed out that the Sun is a star, and it says God created the Sun and the Stars, she blocked me. And we wonder why Giordano was met with a burning? The same ignorant, fearful, antisemitic, judgmental spirit exists in today's Christianity.

The Nemesis-Nibiru system is a seven-planet solar system, that just so happens to intersect within our solar system every 3600 years or so. And it is no coincidence that it was mentioned in the Zohar as a harbinger of the coming of the Messiah King.

> "At that time Melech HaMoshiach (The Messiah-King) will awaken and go out from Gan Eden (Garden of Eden), from the place called Kan Tzippor, and will be revealed in the Galil (Galilee).... A star will arise from the East side, flaming with all colors, and seven other stars will go around this star and make a war with it on all sides three times a day for 70 days, and all the people of the world will see." ~ The Zohar

> "It's going to be a test in Emunah (faith in God). Happy is the person who is going to be in the time of Moshiach. Whoever holds onto Emunah is going to see unbelievable things." ~ The Zohar

> "The task is to remain in Emunah during this time. We're going to see things happening that we don't have any way of understanding. It's going to be things that are beyond belief. It's going to make the Exodus from Egypt look like nothing. Every day it's going to be worse and we're going to forget what it was like the day before. Whoever has Emunah at this time will merit to see the end." ~ Rabbi Shimon bar Yochai, 2nd century sage and author of the Zohar

> "I see him, but not now; I behold him, but not nigh; there shall step forth a star out of Yaakov (Jacob) and a scepter shall rise out of Yisrael (Israel)." (Numbers 24:17)

Israel beware! Those expecting a human Messiah to rise, are being "set up" to receive the antichrist (Revelation 13) as their Messiah. Yes, our Messiah IS coming, but He will come as He left (Acts 1) after His first coming over 2000 years ago, and will bring an end to the antichrist, and all who have received his mark [the mark of the beast]. Jesus our Messiah came to be the sacrifice of atonement [once for all]; to fully fulfil the law, and to establish a new and better covenant; to destroy the law of sin and death. But because He didn't come in the way the Jews of that time expected, or for the purpose they expected, He was rejected. But that rejection was the very means used to bring about that wonderful blood sacrifice, which caused all sin [past, present, and future] to be forgiven, when we just believe, and receive Jesus [the Spirit of Ha Shem in a normal human body] as Messiah and Savior.

What if Marduk, who according to the Sumerian tablets is known as the Lord of Nibiru, is the counterfeit Messiah, that Israel falls for before the real Messiah shows? The planet Mars is named after Marduk. In ancient times, there was a connection between the kingdoms on Mars and Earth that is still evident in the Cydonia region of both Mars and Sidon on Earth.

IS GOD'S THRONE ON NIBIRU?

I can never forget what I heard the Lord tell me when I began researching Nibiru. He told me to study it because it is the sign of His Return to Earth. He said, "I am returning with Nibiru." That statement I heard telepathically, almost audibly, it was that loud in my head, initiated years of research which have culminated with this book series, and this book in particular. Remember the New Jerusalem described in Revelation 21 says that the Lord's throne is situated in its center, and that when it lands on Earth, (a Giant Mothership) opens up all twelve of its gates named after all His Disciples, the Lord who sits in His Throne, will rule over all the nations of the Earth with an iron rod. See, Revelation Chapter 21:9-27.

His Throne will be approaching the Earth from the south pole Revelation 6:12-17. 'Nibiru' coming in from the South Pole. Sun of Righteousness. Remember this is an object with 7 smaller objects surrounding it. There is a black hole with it, which is why lately in the past six months of 2020, Space.com has been running articles that Planet Nine is actually a Black Hole. The Black Hole is behind the red dragon. Something huge and massive as that will have a giant atmosphere in space. Yes, a black hole. You don't hear that a lot, but now even astronomers are saying that a black hole has entered the solar system. The bible tells it this way: and the darkness hideth thee. And darkness is beneath Your feet. (Psalm 139:12)

On August 6, 2020, Space.com published an article, *Thousands of Earthlike 'blanets' might circle the Milky Way's central black hole.* They claim that there are multiple supermassive black holes that dot and pepper our universe which are monstrous gravity wells that bind galaxies together and wreath themselves in whirling cocoons of dust that emit bright X-ray beams. Sometimes, bright columns of matter burst up from their poles, forming jets visible across space. And now some scientists suspect these gravitational monsters might host blanets — tens of thousands of them.[4]

Nope, that's not a typo: Scientists suggest calling these black hole planets by the name "blanets." Such blanets would form from the clouds of whirling dust that circle black holes. And they wouldn't be too different from planets that orbit normal stars. Some would be hard and rocky, like Earth, though likely as much as 10 times larger. Some would be gas giants, like our solar system's Neptune. They would almost certainly be invisible to us, hidden in the disk of matter that birthed them and dwarfed by their supermassive parents. But in a pair of papers published in The Astrophysical Journal in November 2019 as well as in a Japanese Paper, *Formation of "Blanets" from Dust Grains around the Supermassive, Black Holes in Galaxies* on in July 2020, respectively, both teams of researchers laid out the case that these black hole planets must exist.[4]

The fact that the Bible already refers to the darkness that is inherent in Black Hole, is no coincidence, because they were created by the Creator.

> "He bowed the heavens also and came down: and darkness was under his feet."
>
> (Psalm 18:9)

> "He bowed the heavens also and came down; and darkness was under his feet."
>
> (2 Samuel 22:10)

For the past five years, astrophysicists were theorizing that what they're now calling 'Planet Nine' which is actually Planet X, Nibiru, may actually be a black hole.[5]

A growing number of scientists have blamed the weird orbits of distant solar system objects on the gravitational effects of an as-yet-undiscovered "Planet Nine" that lies in the icy realm far beyond Neptune. Physicists are now wondering, what if that supposed planet is actually a small black hole?

Previous studies have suggested Planet Nine, which some astronomers refer to as "Planet X," has a mass between five and 15 times that of Earth and lies between 45 billion and 150 billion kilometers from the sun. At such a distance, an object would receive very little light from the sun, making it hard to see with telescopes.

To detect objects of that mass, whether planets or black holes, astronomers can look for weird blobs of light formed when light "bends" around the object's gravitational field on its journey through the galaxy (simulated image above). Those anomalies would come and go as objects move in front of a distant star and continue in their orbit.

But if the object is a planet-mass black hole, the physicists say, it would likely be surrounded by a halo of dark matter that could stretch up to 1 billion kilometers on every side. And interactions between dark matter particles in that halo—especially collisions between dark matter and dark antimatter—could release a flash of gamma rays that would betray the object's presence.[5]

And in case you're wondering what the Book of Enoch had to say about it? He refers to a solar system, by albeit different names, he names its planets and moons. Could Enoch have been referring to what we're calling the Nibiru/Planet X System?

I am reprinting Chapters 77 and 78 for your convenience: [6]

The Book of Enoch: Chapter 77:

"And the first quarter is called the east, because it is the first: and the second, the south, because the Most High will descend there, yea, there in quite a special sense will He who is blessed forever descend. And the west quarter is named the diminished because there all the luminaries of the heaven wane and go down. And the fourth quarter, named the north, is divided into three parts: the first of them is for the dwelling of men: and the second contains seas of water, and the abysses and forests and rivers, and darkness and clouds; and the third part contains the garden of righteousness. I saw seven high mountains, higher than all the mountains which are on the earth: and thence comes forth hoar-frost, and days, seasons, and years pass away. I saw seven rivers on the earth larger than all the rivers: one of them coming from the west pours its waters into the Great Sea.

And these two come from the north to the sea and pour their waters into the Erythraean Sea in the east. And the remaining four come forth on the side of the north to their own sea, two of them to the Erythraean Sea, and two into the Great Sea and discharge themselves there [and some say: into the desert]. Seven great islands I saw in the sea and in the mainland: two in the mainland and five in the Great Sea."

(Chapter 77:1-8)[6]

The Erythraean Sea was a former maritime designation that always included the Gulf of Aden and at times other seas between Arabia Felix and the Horn of Africa. Originally part of ancient Greek geography, it was used throughout Europe until the 18-19th century.

Enoch names another solar system:

"<u>And the names of the Sun are the following: the first Orjares, and the second Tomas. And the Moon has four names: the first name is Asonja, the second Ebla, the third Benase, and the fourth Erae.</u> These are the two great luminaries: their circumference is like the circumference of the heaven, and the size of the circumference of both is alike. In the circumference of the Sun there are seven portions of light which are added to it more than to the Moon, and in definite measures it is transferred till the seventh portion of the Sun is exhausted.

And they set and enter *the portals of the west,* and make their revolution by the north, and come forth through the eastern portals on the face of the heaven. And when the Moon rises one-fourteenth part appears in the heaven: [the light becomes full in her]: on the fourteenth day she accomplishes her light (full Moon). And fifteen parts of light are transferred to her till the fifteenth day (when) her light is accomplished, according to the sign of the year, and she becomes fifteen parts, and the Moon grows by (the addition of) fourteenth parts. And in her waning (the Moon) decreases on the first day to fourteen parts of her light, on the second to thirteen parts of light, on the third to twelve, on the fourth to eleven, on the fifth to ten, on the sixth to nine, on the seventh to eight, on the eighth to seven, on the ninth to six, on the tenth to five, on the eleventh to four, on the twelfth to three, on the thirteenth to two, on the fourteenth to the half of a seventh, and all her remaining light disappears wholly on the fifteenth.

And in certain months the month has twenty-nine days and once twenty-eight. And Uriel showed me another law: when light is transferred to the Moon, and on which side it is transferred to her by the Sun. During all the period during which the Moon is growing in her light, she is transferring it to herself when opposite to the Sun during fourteen days [her light is accomplished in the heaven, and when she is illumined throughout, her light is accomplished full in the heaven. And on the first day she is called the new Moon, for on that day the light rises upon her. She becomes full Moon exactly on the day when the Sun sets in the west, and from the east she rises at night, and the Moon shines the whole night through till the Sun rises over against her and the Moon is seen over against the Sun.

On the side whence the light of the Moon comes forth, there again she wanes till all the light vanishes and all the days of the month are at an end, and her circumference is empty, void of light. And three months she makes of thirty days, and at her time she makes three months of twenty- nine days each, in which she accomplishes her waning in the first period of time, and in the first portal for one hundred and seventy-seven days. And in the time of her going out she appears for three months (of) thirty days each, and for three months she appears (of) twenty-nine each. At night she appears like a man for twenty days each time, and by day she appears like the heaven, and there is nothing else in her save her light."

(Enoch Chapter 78:1-13)[6]

"See, I will create new heavens and a new earth. The former things will not be remembered, nor will they come to mind."

(Isaiah 65:17)

> "Then I saw "a new heaven and a new earth," for the first heaven and the first earth had passed away, and there was no longer any sea."
>
> (Revelations 21:1)
>
> "And the new city has no need of Sun or Moon, for the glory of God illuminates the city, and the Lamb is its light."
>
> (Revelations 21:23)

There's no coincidence that the ancients called Nibiru the planet of crossing. No coincidence that the Torah/Zohar says that when a red star appears with seven planets, will mark the coming of the Messiah. Jesus said, there will be signs in the heavens that herald His return.

The hope we have is in the place we are promised through our faithfulness in the afterlife and ultimately the Age to Come. Ultimately, for those who believe on the Lord and Creator of all things, our citizenship is in Heaven's Kingdom. The Age to Come, known as the Golden Age, will be when the Kingdom of Heaven comes to Earth.

NIBIRU & THE RETURN OF CHRIST

After the Rapture of the Redeemed, satan will be barred from heaven, no longer able to approach God and accuse the Brethren, in which he does day and night. At that time satan will unleash his wickedness, on the earth like never before. There will be people who will give their lives to Jesus Christ, during this time.

> "And they overcame him by the blood of the Lamb, and by the word of their testimony; and they loved not their lives unto the death. Therefore rejoice, ye heavens, and ye that dwell in them. Woe to the inhabiters of the earth and of the sea! for the devil is come down unto you, having great wrath, because he knows that he has but a short time."
>
> (Revelation 12:11-12)

A woman was praying and asking Jesus when the Rapture was, and she said He told her, "when fire rains down from Heaven, I will rescue you."

The Holy Spirit revealed that the reason the scripture says, "But about that day or hour no one knows, not even the angels in heaven, nor the Son, but only the Father." (Mathew 24:36) that, the reason no one is given a date is because the Father doesn't want Satan to know or figure out. I suppose that goes for the rest of us too. That means that those who try to attach dates to eclipses, blood moons and shooting stars, need to stop serving Satan's kingdom by creating false expectations, false hopes and disappointments to the point of where its caused Apostasy amongst Christians. Just stop it!

When I asked Jesus when the Rapture would be, He gave me a vision and I heard His voice say, "I'm coming back with Nibiru." I received no specific date or time. I never forgot that which spawned my intense research and my writings about it. The connection between Jesus and Nibiru is not something to be ignored. See, Book Three, *Who are the Angels?* contains a rather lengthy Chapter, titled, "Ascension or Rapture?" As I previously wrote in Who Are the Angels? that the three hours of darkness that occurred during His crucifixion, in the middle of the day, was not our Moon, but one of the planets in the Nibiru system. So, for Him to say, He's returning with the coming of Nibiru, makes one wonder, if Nibiru is a Planet or a Mothership? [7]

Needless to say, that history is rife with examples of how this neighboring solar system has intersected with ours marking earth changes, and visitations by extraterrestrials in the past. I think the Lord uses these passings to punish the wicked on the Earth. This is why all the Old Testament Prophets pointed to the 'terrible day of the Lord.'

THE RETURN

> He said, "BEHOLD, HE IS COMING WITH THE CLOUDS, and every eye will see Him, even those who pierced Him; and all the tribes of the earth will mourn over Him. So, it is to be. Amen."
> (Revelation 1:7)

> Daniel mentioned them, saying, "And behold, with the clouds of heaven One like a Son of Man was coming, And He came up to the Ancient of Days And was presented before Him."
> (Daniel 7:13)

Jesus saying it on two different occasions. When He spoke to His disciples of the day of His Return, He said that His sign will appear in the sky,

> "- - and then all the tribes of the earth will mourn, and they will see the SON OF MAN COMING ON THE CLOUDS OF THE SKY with power and great glory."
> (Matthew 24:30)

After He had been arrested Jesus was brought before the high priest to be charged and the high priest asked Him if He was the Christ, the Son of God. Jesus responded,

> "You have said it yourself; nevertheless, I tell you, hereafter you will see THE SON OF MAN SITTING AT THE RIGHT HAND OF POWER and COMING ON THE CLOUDS OF HEAVEN."
>
> (Matthew 26:64)

> "BEHOLD, HE IS COMING WITH THE CLOUDS, and every eye will see Him, even those who pierced Him; and all the tribes of the earth will mourn over Him. So, it is to be. Amen."
>
> (Revelation 1:7)

> "Even the youth will faint and grow weary, and the young will fall exhausted; but those who wait for the LORD shall renew their strength, they shall mount up with wings as eagles, they shall run and not be weary, they shall walk and not faint."
>
> (Isaiah 40:30-31)

As if to make sure that we heard the message, Mark recorded Jesus making those same statements about those clouds in his gospel (Mark 13:26: 14:62).

Why does the scripture include that statement about a cloud receiving Jesus when He ascended? If it was just a normal cloud, He would have ascended "into a cloud". Scripture says that as His disciples were looking on, "a cloud received Him out of their sight." (Acts 1:9).

Would a cloud receiving Him indicates Intelligence in that the cloud, with technology to that had the ability to receive and accept Him?

Two angels spoke to the people who had just witnessed Jesus being taken bodily back into heaven. They said, "Men of Galilee, why do you stand here looking into the sky? This same Jesus, who has been taken from you into heaven, will come back in the same way you have seen Him go into heaven" (Acts 1:10-11).

Zechariah 14:4 says that when Jesus returns his "feet will stand on the Mount of Olives, east of Jerusalem." Matthew 24:30 says "all the peoples of the earth will mourn when they see the Son of Man coming on the clouds of heaven, with power and great glory." Revelation 1:7 says "Look, He is coming with the clouds, and every eye will see Him, even those who pierced Him; and all peoples on earth will mourn because of Him. So, shall it be! Amen."

So.....Jesus left the earth from the Mount of Olives; in His glorified, physical body; as people watched; and disappeared from sight in a cloud. And Jesus will return to the earth to the Mount of Olives; in His glorified, physical body; as people watch; and will come with the clouds. And I say, "So shall it be! Amen."

Paul tells us about how Jesus will be returning. He did not mention Jesus returning in clouds or with the clouds of heaven, but he did say that He would be coming "with all His saints" (1 Thessalonians 3:13).

In the next chapter Paul says that when Christ returns, "God will bring with Him those who have fallen asleep in Jesus." (1 Thessalonians 4:14) If the departed faithful will be returning with Him, are they with Him in heaven now?

He says the Lord will descend from above and the dead in Christ will rise first,

> "Then we who are alive and remain will be caught up together with them in the clouds to meet the Lord in the air, and so we shall always be with the Lord. Therefore comfort one another with these words."
>
> (1 Thessalonians 4:17-18)

Love is all that matters. It is the only thing you take with you when you die. Love for People and all life forms and your love for God. So, let us motivate ourselves to love.

JEWISH PROPHECY: DOES THE NOVA IN 2022 HERALD THE MESSIAH?

The idea that the will of God was written in the heavens was common knowledge in the ancient world. The second part of this book, *Heaven's Gospel -Book Six*, will prove that in great detail. Comets — and nova — were particularly portentous and harbingers of great change to the ancients who kept track of time and marked harbinger type events.

The Mayan calendar predicted Bolon Yokte K'Uh (a god of the underworld) would turn up at the end of 2012 — supposedly marking the end of time. Maybe he did, but no one saw him, especially because he returned to the underworld, the inner earth which as I have documented and proved in Book One of *Who's Who in the Cosmic Zoo?* is inhabited by our resident 'Alien Presence' which many believe are Fallen Angels, aka Fallen Sons of Heaven that lost their original estate. [8] The Da'ath Wars have certainly been amped up since 2012, and here we are in 2020, and have already had multiple official Disclosures of UFOs from the Pentagon and Department of Defense. Just what or 'who' are they preparing the public for?

Nova KIC 9832227 which is part of a binary system, will pass in 2022. Some Orthodox Rabbis have said, it may herald the arrival of the Messiah based on Jewish Prophecy in both the Torah and the Zohar. Astronomers predict a nova will light up our night skies in 2022.

That is a lot of fuss being tied to a little extra light in the night sky. Even though the actual cataclysmic event is of interstellar proportions. At the heart of the matter is KIC 9832227, a dull star in the constellation of Cygnus, the Swan which is part of the greater constellations of Aquarius. Nova KIC 9832227 is about to blow.[9]

What is different about this one, is science knows when it is going to happen. And it will be noticeable to the naked eye. My assessment is that this is part of larger system of comets, asteroids, novae, and debris that is coming with the Nemesis-

CHAPTER 12: NIBIRU IN BIBLE PROPHECY

Nibiru system that has begun to intersect within our Solar System. Indeed, the red star has been prophesied as the harbinger of the Messiah which begins the Messianic Age.

The notion that a nova or the arrival of a red star for all to see, is not new to the return of gods, the messiah, or even the apocalypse, which remember in Greek means, 'revelation' or 'Disclosure.' The notion of a bright star leading the world into a new future is one well-grounded in Jewish and Christian lore. In the Christian Gospel of Matthew, the three wise men see the light of a guiding star as the fulfilment of prophecy — and are led to the baby Jesus.

Now a Jewish Rabbi says the same thing is about to happen again. (Or for the first time. Depending upon your faith.) Controversial ultra-Orthodox Rabbi Yosef Berger has been awaiting the coming of his religion's messiah all his life. He is the rabbi overseeing one of Judaism's most holy sites — the Tomb of King David on Mount Zion (in Jerusalem's old city).[9]

He wants Muslims expelled from Temple Mount and the Dome of the Rock demolished. He wants a Third (Jewish) Temple erected in its place. He has even crafted an elaborate copy of the Torah to personally place in the son-of-god's own hands. Now he says the impending nova will be a sign of fulfilled prophecy.

"The Rambam (a 12th Century rabbi who helped shape modern Jewish lore) brings this verse about a star appearing as proof that the Messiah will come one day," Rabbi Berger told Israeli media. "But he says it will come from Jacob, and not from Esau (the eldest son of the prophet Isaac). More specifically, from the tribe of Judah." He bases his argument on the Torah. Specifically, the Book of Numbers 24:17 as the basis of the 2022 prophecy:

> "I see him, but not now;
> I behold him, but not near —
> a star shall come out of Jacob,
> and a sceptre shall rise out of Israel;
> it shall crush the borderlands of Moab,
> and the territory of all the Shethites."

It's a prophecy essentially saying that a new Jewish leader will emerge in Israel who will lead it to world domination. This is NOT the true Messiah, but the Counterfeit Messiah who is the Antichrist. What's missing in his false belief, is that the true Messiah returns from the Heavens on the Mount of Olives in Jerusalem. He is saying, this Jewish Messiah will emerge in Israel. Regardless of where he comes from, the Bible Prophecy says, that he will broker a Peace deal after the great war against Israel, between Israel and her enemies, and declare himself god, and sit in the rebuilt 3rd temple, deceiving Israel, and unleash the greatest persecution of both Jews and Christians known to our human history.

It is my opinion as a Jewish – Christian, Messianic Jew, that this deception happens because the Jewish people are blinded by their rejection of the true

Messiah, who came first to deal with their sin issue, as a Lamb sacrificed to slaughter, to ransom them back from the Dark Lord of this world, who is Lucifer/Satan. Jewish people have been so heavily persecuted by so called Christians, that they reject Christ because of the Antisemitic Doctrines and Replacement Theologies of the Catholic Church. As a result, they are deceived into following after the counterfeit messiahs, which isn't a new concept to Jews, as they have historically turned Rabbis into Messiahs, which have all turned out to be false prophecies.

So which star will be that particular star? Could it be Nibiru? The Zohar, a collection of commentaries that define kabbalah, or Jewish mystical thought, outlines the astrology behind the prophecy. "The Zohar states explicitly that the Messianic process will be accompanied by several stars appearing. The Zohar goes into great depth, describing how many stars, and which colors they will be," Rabbi Berger says.

Prophetic nova-like stars are not restricted to the Torah. The New Testament's Book of 2 Peter 1:19 states:

> "We have also a more sure word of prophecy; whereunto ye do well that ye take heed, as unto a light that shineth in a dark place, until the day dawn, and the day star arise in your hearts:"

Then there's Revelations 22:16:

> "I Jesus have sent mine angel to testify unto you these things in the churches. I am the root and the offspring of David, and the bright and morning star."

The January 7 meeting of the American Astronomical Society was told by astronomers from Calvin College in Michigan that a nondescript star in the Cygnus constellation will likely become one of the brightest objects in the sky. By the way, KIC 9832227 is actually not one star. It's two. One is about the third of the size of our Sun. The other is roughly 40 per cent bigger. They're orbiting each other so close that their superheated atmospheres are actually touching each other.[9]

New observations reveal the stars are in an accelerating death spiral — constantly winding faster and closer. At the moment, that orbit is once every 11 hours. Whether KIC 9832227 foretells the first or second coming, therefore, is a matter of perspective. In my opinion, there is nothing new under the Sun, and this is cyclical. KIC 9832227 is actually two stars in the process of falling into each other. Soon the cores must tear each other apart. The fragments will then fall together, forming the heart of a new star.[9]

The Calvin College astronomers say they know when this will happen. It's an event that has been seen before. In 2008 a similar collision (V1309 Scorpii) produced what is called a red nova — a poorly understood phenomenon brighter than a standard nova but duller than a supernova. Observations of the lead-up to that cataclysmic blast have been compared to KIC 9832227 "Based on an updated fit to the exponential formula, we now estimate the time of merger to be the year 2022.2 with a random uncertainty of 0.6 years," the team reported. They are predicting this merger will occur March 15, 2022 — with an error of margin of roughly six months. Science also knows roughly what to expect: a steady doubling of the nova's red light every 19 or so days as the outer layers of the stars are blasted out into space. It is a process likely to last about six months.[9]

No matter how you look at it, it is clear from putting all the pieces together, one form of apocalypse or another is bound to happen. We are entering into the last days of the End Times Prophecies. How long that will last remains to be seen. But it is clear that the ancient battle over humankind and planet Earth will culminate in the return of the Gods of Nibiru who have a divine appointment on Earth to battle and claim what is rightfully theirs. Will the real Messiah please step up! Maranatha!

Notes:
1. Anderson Reed, *Shouting at the Wolf, A Guide to Identifying and Warding Off Evil in Everyday Life*, Library of the Mystic Arts, Citadel; First Edition, August 31, 1998.
2. Ella LeBain, *Who is God?* Book One, Chapter Four: *The Motherships of the Lord, The Cloudships and the Clouds of Heaven*, pp.71-88, Skypath Books, 2015.
3. Wikipedia, Giordano Bruno, https://en.wikipedia.org/wiki/Giordano_Bruno
4. Rafi Letzter, Thousands of Earthlike 'blanets' might circle the Milky Way's central black hole, https://www.space.com/black-hole-planets-blanets.html, August 06, 2020.
5. Sid Perkins, 'Planet Nine' may actually be a black hole, https://www.sciencemag.org/news/2019/09/planet-nine-may-actually-be-black-hole, September 27, 2019.
6. Book of Enoch, Chapters 77-88 - Bill's Bible Basics https://www.billkochman.com/Articles-Non-BBB/enoch-06.html
7. Ella LeBain, *Who are the Angels?* Book Three, Chapter, "Ascension or Rapture?" Skypath Books, 2016.
8. Ella LeBain, *Who's Who in the Cosmic Zoo?* Book One – Third Edition, Skypath Books, 2013.
9. Jamie Seidel, Jewish prophecy: *Will Nova KIC 9832227 in 2022 herald the arrival of the messiah?* January 25, 2017, News Corp Australia Network, https://www.adelaidenow.com.au/technology/science/jewish-prophecy-will-nova-kic-9832227-in-2022-herald-the-arrival-of-the-messiah/news-story/fee7e498804f8afe2695a8b235f69c02

CHAPTER THIRTEEN

WHO ARE THE ANUNNAKI?

"Knowledge makes a man unfit to be a slave."
~Frederick Douglass

In Sumerian, the word, *Annunaki* means, *Those who from Heaven to Earth came*, or *Those who come from the sky*. Thousands of ancient clay tablets translated contain information about a God-like race from the Planet Nibiru. 450,000 years ago, the immortal Annunaki arrived on Earth. Their purpose was to mine for gold to repair the atmosphere of Nibiru. Annunaki possess technology that was used to genetically alter life on Earth, creating the slave races of Grays (aliens) and humans.

According to the late Zecharia Sitchin1 who interpreted the clay tablets, the Annunaki created a slave race to mine gold for them which, in turn, was used as a heat shield to preserve their planet, Nibiru. This is why there is such a great value placed on gold. However, let's connect the dots here, if Nibiru comes around every 3,650 years, which creates cataclysms on Earth, in the way of comets, fires, Earthquakes, Pole Shifts and tsunamis, then perhaps the gold is not working too well? This is why the Lord of the Cosmos promises to create a New Heaven and a New Earth. (Revelation 21)

The greatest secret never told is how the church has been subservient to these alien masters. The Annunaki are the Archons of this world and are in legal agreements with the god of this world, who is Lucifer/Satan/Enlil. The Church has been infiltrated by the god of this world, who is Lucifer/Satan, who was Enlil. When the church learns to discern the Word and the Spirit rightly and understand the difference between the gods, and repent of their idolatry, then the God of gods will save them. The all-seeing eye… the ones who oversee the shadow governments… the Annunaki, this is Lucifer/Satan/Enlil's kingdom, not the Lord Yahuah, the God of Israel. The Annunaki is the race that has kept us working as slaves for millennia, both spiritually by demanding blood sacrifices on a regular basis, and economically by their oppression of slave labor along with their suppression of women, children, and animals.

The Annunaki according to the Sumerian Cuneiform Stone Tablets come from Nibiru. They return to Earth each time Nibiru does a fly by. They are at enmity with each other and with the God of Israel. The Sumerian Tablets, Enuma Elish, the

CHAPTER 13: WHO ARE THE ANUNNAKI?

Epic of Gilgamesh and the Old Testament Genesis accounts, are all part of the same story.

The Hebrew Bible has a cast of characters, that can only be discerned in the original Hebrew language but has been covered up in English translations with the words, 'God' and 'Lord.' I unpack these discernments and reveal there is in fact a cast of characters in the Old Testament revealing multiple gods at odds with each other, in *Who is God?* Book Two. This explains the contradictions in the scriptures. Many think it is the same god speaking, and it clearly is not.[2]

The Annunaki are still enslaving humankind today. The British Royal family and the Illuminati Elite worldwide trace their bloodlines back to ancient Egypt's Pharaohs. The Pharaohs at the time where human/Annunaki hybrids put into power as a go-between the Annunaki and the enslaved human race. The Elite continue to interbreed in an attempt to keep their Annunaki bloodlines strong. They consider themselves to be Annunaki, and not humans, which explains the Elite's utter contempt for the human race and their ability to kill humans with no remorse through genocides and ritual sacrifices. They feel their Annunaki reptilian bloodlines give them the "Divine Right to Rule" as they considered themselves to be gods, but in fact, were just an invading reptilian race of giants. Lucifer, considered by the Vatican, the UN, the royals and the Elite as the one true god, is actually just another Annunaki reptilian. The Illuminati and Annunaki bloodlines are one in the same.

Nibiru is said to be on a 3,650-year elliptical orbit. The last appearance of Nibiru may have accounted for all of the Great Flood stories told in virtually every religion. Depending on the closeness of Nibiru, it could cause an extinction level event where a reset of humanity is possible. That would explain how an unknown bronze coin could be discovered 114 feet into the ground.

According to the late Zecharia Sitchin, the Annunaki created a slave race to mine gold for them which, in turn, was used as a dust to preserve their planet, Nibiru. The greatest secret never told is how the church has been subservient to their masters... the all-seeing eye, the ones who oversee the shadow governments, the Illuminati are the Annunaki representatives on Earth. This is the race that has kept us working as economic slaves for millennia.[3]

The Sumerian tale is a family drama between two brothers, Enlil, and Enki. As I discerned and connected the dots in *Who is God?* Book Two, that Enlil became Lucifer-Satan, Baal who later was called, Allah. Enki became Yahuah, who sent His Son, Yahushua aka Yeshua aka Jesus Christ to save humankind and set all humans free from captivity from the shackles and strongholds of Enlil, who is Lucifer-Satan.[4]

This proves that not all Annunaki are malevolent. According to the Sumerian tale, one Annunaki in particular, Enki, loved the human Earth race, and intervened several times to save the humans that were genetically modified to be a slave race, used to mine gold for the Enlil's Annunaki tribe. It is expected and believed, that Enki will return to Earth, who will break the shackles of economic subservience to today's Rothschild-Annunaki lineage. This will be a paradigm shift scenario for all religions, as the TRUTH and DISCLOSURE of 'who' these beings were and are, will finally be exposed.

Zecharia Sitchin did not tell us when God created man and woman, why then are there so many blood types and Rh values? While it is possible that the Annunaki genetically manipulated our DNA to create a slave race, one must ask, why are there so many different races, languages, blood types and Rh values? Despite the Annunaki turning humankind into a slave race, it's clear that while we are all similar as humans, we differ in bloodlines, languages and genetic codings. This would indicate that there was more than one type of god that seeded this planet. The Bible is clear that there is the seed of Satan, i.e., the Nephilim seed, Reptilian or Serpent Seed, that coexists on Earth alongside and is in an ongoing battle with the seed of the Elohim, the gods involved in the Creation Story of Genesis, who serve the God of gods, who is Yahuah, the God of Abraham, Isaac, Jacob and the God of Israel.

This is what the basis on the End Game of the End Times is about, the final clash of two kingdoms culminating on and above the Earth over *who* gets to Lord it over the human race, and Planet Earth, and to some, that's even if there will be a human race left, after the alien agenda and alien presence on Earth is done with corrupting the image and likeness of God in the human race, and replacing Earth humans, with alien human hybrids and artificial intelligence. It is already happening.

As I postulated throughout this Book Series that the possibility that Earth is a Grand Experiment representing multiple races within the galaxies, that does not limit Earth humans to what the Annunaki slave masters initiated. There were more Extraterrestrials in our galactic neighborhood who also seeded this planet with various races, i.e., Lyran, Sirian, Pleaidian, Arcturian, Andromedan, as a galactic experiment to see how we would all get along. There are numerous extrabiblical Jewish texts, that were rejected by the Ecumenical Councils which sheds light on how our racial differences evolved from the geographical locations of our ancestors, as well as pointing us towards the multiple spiritual battles that ensued over the course of human history. The god of this world is intent to destroy these various human cultures and extraterrestrial experiments, that to date, seems to have survived as there are approximately 7.5 billion human beings living on planet Earth. See, the Books of Enoch, The First and Second Books of Adam and Eve, the Books of Jasher, the Book of the Giants, the War Scrolls, etc.

WHO IS ENKI?

Many get the gods confused, due to linguistics, and, well, the story is just so old. Falsehoods are spun over the ages, and like the game of telephone, the original story gets obscured and lost in mistranslations and misunderstandings. However, the following Tablet discerns that out of the two warring brothers, Enki and Enlil, it was Enki who was the compassionate and righteous one, especially in his decisions towards humankind. In the Bible, Yahuah is likened to Enki, whereas the Enlil character, is likened to Lucifer. Enlil later became Baal, which later on became known as Allah, all derivatives of Satan, also known as the god of *this* world.⁵ However, Enki, who sent his son, a messianic figure to save humankind, is not of this world. Jesus was quoted as saying, "I am not from this world. (John 17:16); "My kingdom is not of this world." (John 18-36)

In Tablet 11 of the Sumerian Cuneiform Stone Tablet, it says:

Enki is told to destroy the human race, which he loved, with a flood, and to swear an oath. But he rejects. Enki made his voice heard and spoke, "why should you make me swear an oath? And why should I use my power against my people?"

So, the other gods agreed to cause a flood themselves and forced Enki to swear an oath, promising not to interfere with their plan. When they left the Earth to be safe, Enki broke the oath and told humanity about the flood and saved them. When the others returned, they saw the boat and knew Enki had something to do with it. They were furious with him, and he responded:

"How could you bring about a flood without consideration? Charge the violation to the violator, charge the offense to the offender. But be compassionate lest mankind be cut off, be patient lest they be killed."

Not all of the Annunaki are malevolent. One Annunaki in particular, Enki, loved the human Earth race that was genetically modified to mine gold for the Annunaki.

The expectation amongst researcher anticipating it Enki returns to Earth, the shackles of economic subservience will finally be broken. This is a frightening scenario for all religions because the TRUTH will be exposed. They will try to deceive the church and the world, so that the story of Jesus and salvation will no longer apply. There is already confusion on *who* these gods are/were with respect to our ancient history. What many of these blog writers and researchers fail to understand about the Enki/Enlil story is that the Bible story continues their story.

As I've already connected the dots to the Sumerian and Genesis Creation accounts, that the same gods may already be recorded in the Hebrew Bible, as the confusion lies in the fact that the original Hebrew language was mistranslated into English to be all the same god, but the truth is in the details, and based on Hebrew Linguistics, there are a cast of characters of gods mentioned in the original Hebrew scriptures that most are oblivious too, because they've been so used to reading them as all the same. Plus, Jews have a major superstition of using and pronouncing the true names of the God of Israel, and the word, 'god'. 'Lord' and the phrase 'Ha-Shem' which literally means, 'the name' are not names, but titles, which obscures the correct discernment even deeper.

If Enki turns out to be Yahuah, the Father of Yeshua, (Jesus in English), then for these researchers to suggest and disseminate the false story that when or if the god Enki returns it will cancel out the Jesus story of salvation, are researchers who have failed to connect the dots between what many scholars, including myself believe to be the same story. And if Yahuah turns out to be a completely different group of Extraterrestrial or Ultraterrestrials, who sent His Son, Yeshua, to set the captives free from the bondage of the Annunaki gods, see, Isaiah 61:1; Luke 4:18, then they obviously failed to understand the purpose of the Messiah.

The Bible Prophecy in Isaiah 61:1, was fulfilled by Jesus Christ aka Yeshua HaMashiach, in Luke 4:18. This is why His last dying words were, "forgive them for they know not what they do," was about the spiritual, genetic, and generational bondage of humankind to the Annunaki slave masters and genetic manipulators. Besides the obvious, there is a much deeper reason for a Messiah, a Savior of humankind, and this is to be liberated from the bondages, ownership, and controls of the Annunaki.

Jesus said He was returning after there were great signs in the Heavens. The signs he was referring to are the Exoplanets of the Nemesis/Nibiru system. Every End Time Prophesy that relates to the Second Coming of Christ, and the Rapture of His Bride/Church, coincides with the passing of Nibiru. According to historical record, it was Jesus who gave St. John of Patmos the vision of Revelation of the End Times. John wrote down Wormwood was a fiery star that fell to Earth and destroyed a third of the waters and life on Earth and turning the waters bitter. This is by every Astronomer who understands Bible scripture conclusion: That this is one of the comets/asteroids that come out of the Nibiru system, and with it will bring the return of the Lord who is Jesus the Savior. So, these researchers should study their Bible, and connect the dots, the coming of Nibiru is the Second Coming of the Jesus Christ. Jesus Christ was the Son of God, which god? According to the Hebrew scriptures, His Father is Yahuah, the God of Israel. By anyone's guess, Enki could very well be Yahuah in another language.

BREAKING THE CURSE OF THE ANNUNAKI

In Book One of *Who's Who in the Cosmic Zoo?* in my Concluding Words,[5] I discussed the false belief that the Annunaki were the Creators of Humankind, which was a misinterpretation of the Sumerian Tablets that was presented through Zecharia Sitchin's epic book series, *The Earth Chronicles*, in his book, *Genesis Revisited*,[6] where he postulated that the Annunaki created humankind in what he called the first test tube babies, which was actually done in a pot of clay. However, what a lot of people have missed from this story, was that it clearly says, that when the Annunaki came to Earth, they used the woman, which was already on the Earth, to incubate a new race of humans, through mixing the genetics. Therefore, the Annunaki did not create humans, because the woman was already on the Earth. However, what they did do was genetically manipulate human DNA.

Therefore, we can conclude that the missing link was the hybridization of the human species by an extraterrestrial or an interdimensional race. Who or what fused our chromosomes? Today we only have 23 chromosomes compared to past hominids who had 24 chromosomes. This confirms my original thesis, that the 'Fall of Humankind' was genetic downgrading from the original race of humans that were created in the image and likeness of the Elohim gods, which carried all 12 strands of DNA and 24 Chromosomes. Then the Curses of Genesis 3 came upon the whole cast of characters, Satan, the Man, the Woman, and the Earth. This is why it says, 'all of creation moans and groans for the redemption of the Lord.' (Romans 8:22)

What the Sumerian Cuneiform Tablets tell us is who hybridized the human species, after the curses were doled out. They clearly created a slave race of human beings, that had just enough intelligence to understand and follow orders and mind the Earth for gold and other precious resources that the Annunaki needed. As a result, they set up temples, religions, and bloodlines to continue to serve them on the Earth until they returned to harvest their own. Because they were in rebellion to the God of gods, the Master Creator, they created competing religions and set themselves up as 'gods' on the Earth, to steal the worship from the Creator God. This underscores the present battle we find ourselves in Ephesians 6:12, "For our struggle is not against flesh and blood, but against the rulers, against the powers, against the world forces of this darkness, against the spiritual forces of wickedness in the heavenly places (who are the Aliens)."

The Lord Yahuah/Yeshua came to Earth to set the captives free. Free from the Annunaki slave masters who have ruled and oppressed humankind on this planet for millennia.

The history clearly details enslavement of humankind by the Annunaki. The history also involves a Savior/Messiah sent to save humankind. This is the consistent story in the major Earth religions, as well as being written into the star pictures of our Milky Way Galaxy.

Isaiah 61:1; Luke 4:18, "I have come to set the captives free."

Messiah/Savior promises to restore humankind into original glory bodies, which is all 12 strands of DNA breaking the genetic and spiritual curse of the Annunaki.

The Book of Revelation says the Pope is the False Prophet, who will be used to foster in the time of the Antichrist who will, but for a very short time, rule over a One World Government, that is being called the New World Order.

Why would an extraterrestrial, who is probably light years ahead of most humans, want to make contact with someone who is being controlled by Draconian forces that the Vatican represents? The truth is, today's Vatican worships Lucifer, who is the Reptilian Draconian who stakes claim to Earth, and is at war with the Elohim God of gods, who is Yahuah and His Son Yeshua, Jesus.

You see, what most Catholics do not understand is that the Spirit behind the Roman Catholic Church is what came out of the Roman Empire. This began with Constantine's Creed, otherwise known as the Doctrine of Anti-Semitism and Replacement Theology, by waging war against the God of Israel, who is the Father of Jesus Christ. This is our history. Rome became the new Babylon in the ancient world, as it practiced the ancient Babylonian Religion. It only switched out its two main gods to Jesus and Mary. Today's Vatican is Mystery Babylon. The Roman Empire stole the Babylonian religion which became the official religion of Rome. Then 350A.D., during the Emperor Constantine, Rome's religion was renamed, yet at its core, remained rooted in paganism which was condemned and punished by the God of Israel.

Babylon is Fallen:
> "Then I heard another voice from Heaven say: "<u>Come out of her, my people</u>, so that you will not share in her sins or contract any of her plagues. For her sins are piled up to Heaven, and God has remembered her iniquities."
> (Revelation 18:4-5 Berean Study Bible)

CREATED IN THE IMAGE OF GOD

Remember the original creation scripture is written in its plurality. "The gods created humans, male and female, in their image according to their likeness." (Genesis 1:27)

The **Image of God** in Hebrew: צֶלֶם אֱלֹהִים, Romanized: tzelem Elohim; Latin: Imago Dei; are beliefs and theological doctrine in Judaism, Christianity, and Sufism of Islam, which asserts that human beings are created **in the image** and likeness of **God**. Which God?

Why are there so many blood types and Rh values? The Sumerian record that the Anunnaki genetically manipulated our DNA to create a slave race. Then why did we end up so many different races, languages, blood types and Rh values?

I truly believe, as I stated in Book One of this book series, that Earth is a Grand Experiment that is being used to weed out good and evil humans, while all of us have karmic histories with different gods – Elohim, who planted their seeds within humankind, to represent different sectors of this galaxy, different alien races, which are all being homogenized through humankind. That is why we have such diversity, as well as karma, to work out on Earth.

Yet, because of a series of battles over humankind and over Earth, this is why we are not all the same, but must find a way to accept each other nonetheless, through our shared humanity. The other biggest common denominator that can unite us, is our connection to the one true God of gods, who is the Author and Finisher of all of our Faiths. (Hebrews 12:2)

Anunnaki creating a slave race of humans to serve them is only one part of our shared human history. We must incorporate the beliefs from around the world, that our galactic neighbors also seeded this planet with various types of human races from within our local galaxy. See, Book One of *Who's Who in the Cosmic Zoo?* Chapters on the Lyran Vine aka the Human Vine of Pleiadeans, Sirians from Sirius, Arcturians, Andromedans, Orion, etc. that held galactic history with each other in their battle against the Draconians, which was ultimately grounded on Earth. So, if we all can find that which unites us, becomes stronger than that which divides us, then suffice it to say, this galactic experiment to see how we would all get along, would be a success.

The Bible indeed has explanations as well as evidence within the ancient Jewish Scriptures, that stemmed from the early history of the different tribes of our ancestors that emerged after Noah's Flood. The Old Testament is an historical record evidence of warring gods who are extraterrestrial gods over their human experiments and the different core groups that began it. For example, the Jewish Race, which came from the Tribe of Judah, Jacob's sons. They were called to be set apart, to the God of Israel who is Yahuah. This is why they were given instructions, laws, called the Torah, and then the Messiah was prophesied to be born through this bloodline, bringing promises of salvation from captivity and from bondage to the other competing gods, i.e., warring extraterrestrials over "who" controls humankind on Earth. "Who" gets to rule over Governing the Affairs of Planet Earth?

This question is still being answered, evidenced in the present-day battles over our minds, bodies, and souls. The Bible Prophecies all point to the New Messianic Age to come, where the Kingdom of Heaven lands on Earth, and in its center is the Throne of the Messiah-King who rules over the Earth by His Eternal Light and Righteousness.

> "But for you who revere my name, the Sun of righteousness will rise *with healing in its rays*. And you will go out and frolic like well-fed calves."
>
> (Malachi 4:2)

> "They will see His face, and His name will be on their foreheads. There will be no more night in the city, and they will have no need for the light of a lamp or of the sun. *For the Lord God will shine on them*, and they will reign forever and ever."
>
> (Revelation 22:4-5)

According to an article from In5d.com titled: *What the Church Isn't Telling You About Nibiru and The Anunnaki*, they explain: Silently, the Vatican is releasing information about Nibiru entering our Solar System and extraterrestrial neighbors who have already met with Vatican officials, but they are not telling you everything.[8]

While the mainstream media remains conspicuously silent while distracting our attention with false flag events, disclosure information has in fact been slowly leaked by the Vatican, indicating the presence of extraterrestrials on, and visiting, our planet along with an incoming anomaly called Planet X and/or Nibiru.

As I laid out in *Covenants – Book Four*, the religions of the world need to recognize extraterrestrials, and find them in their respective literatures, because they are all mentioned in Ancient History. After much clinical research, it was concluded that it was the religion of Christianity that would have the hardest time accepting Disclosure of Extraterrestrials and Alien.[9]

Pierre Lena, a French astrophysicist, and member of the Pontifical Academy stated at a November 2009 Astrobiology Conference hosted by the Vatican, "Astrobiology is a mature science that says very interesting things that could change the vision humanity has of itself. The church cannot be indifferent to that."

Chris Impey, a University Distinguished Professor and Deputy Head of the Department at the University of Arizona and a keynote speaker at the Astrobiology Conference added, "The first discovery is only a few years away." Collectively they agree that full Disclosure is imminent.

COVERT OR OVERT VISITATIONS BY ANNUNAKI?

It is believed by many Planet X researchers that there will be covert visitations by the Annunaki who come from Nibiru, during the next passage of the Nibiru System. I have heard for years that they are coming to harvest their own. This is why I have postulated that there will be three (3) End Times Raptures, 1. The Tares are taken first, as Matthew 3:30, Allow both to grow together until the harvest; and in the time of the harvest I will say to the reapers, "First gather up the tares and bind them in bundles to burn them up; but gather the wheat into my barn."; 2. The Dead in Christ Rise First, 1 Thessalonians 4:16, "For the Lord Himself will descend from Heaven with a shout, with the voice of the archangel and with the trumpet of God, and the dead in Christ will rise first."; 3. Those who are left on the Earth at the End after the Last Trump are taken up, all before the Terrible Day of the Lord, when God's wrath is poured out. 1 Thessalonians 4:17, "After that, we who are alive and remain will be caught up together with them in the clouds to meet the Lord in the air. And so, we will always be with the Lord."

There is a belief amongst New Agers who anticipation the Ascension. I am going to get into that in more detail, to continue and conclude the ongoing discussion through this Book Series, on *Ascension vs. Rapture?* However, suffice it to say for now, that many believe the Aliens will lift them off the planet before the first impact event, which is when one of the asteroids from the Nibiru-Wormwood system hits Earth, causing one third of Earth's water to become bitter as prophesied in Revelation 8:11. What if the Raptures described in the Bible, depicted those who will be harvested by Annunaki, and will be seen as being abducted by aliens? And then those who live and believe on Christ, get raptured, ascend to Heaven at the very end? As I postulated in Book Three, *Who Are the Angels?*[8] and I analyzed the actual wording of the End Times scriptures, delineating that there was a total of ten (10) Raptures in the Bible, and that the End Times Raptures, according to the above referenced scriptures, are actually three (3) separate events.

According to Marshall Masters[11], when these deep impact events come upon the Earth, most people will respond with unbelief and say, "I just don't believe it!" Meanwhile, people are being abducted, while large planets and dwarf stars are visible in the skies, as this passing occurs, and while the Annunaki, who have been working with the Elite of this planet, Illuminati Globalists, will try to enslave humankind through the anticipated, but short lived, New World Order. Cognitive Dissonance will kick in, which plays right into the hand of the Annunaki Slave Masters, the inhabitants of Nibiru.

CHAPTER 13: WHO ARE THE ANUNNAKI?

Following the White Ash Event otherwise called the Winter of Ash, which according to Marshall Masters, we've already entered into this catastrophic timeline, that started with disturbances in the Ring of Fire, causing volcanic eruptions not limited to the Ring of Fire, but as a catalyst to initiate the eruptions of volcanoes all over the world, even those that have become dormant and inactive. This is due to the movement of magma at the Earth's Core, which is being magnetized by the incoming objects from the Nemesis-Nibiru System intersecting within our solar system.

The Mayans were looking for a Harbinger Event on 12.21.12, which was the return of Nibiru. We are seeing these harbingers in the way of increased seismic activity worldwide, and that is not limited to surface quakes, but deep inside the earthquakes that seem to be coming from the core. This is more indication that the Earth changes are being precipitated from the magnetism being felt by the Earth's inner core, as it is communicating with these incoming exoplanets and exomoons as well as their obvious drain on our Sun.

Physicist Dr. Claudia Albers is known for her discovery of solar cores, which she says are coming from the Planet X (Nibiru) incoming system and are draining the Sun to the point of causing the Sun to dim and die. She believes our Sun is already dead, and that's why the Artificial Sun Simulator was put into orbit between Earth and Sun, to keep the Earth warm, as the Sun is being drained and not showing Sunspots as usual. I happen to agree with her on this point, yes, there are solar cores literally sucking energy from the Sun, I have seen them myself, from SOHO (Solar and Heliospheric Observatory).

The Sun Simulator is real - see my chapter in this book. While I am not entirely convinced that our Sun is dead, the data I have seen indicates that it has dimmed. This is the reason for ice ages on Earth, and I believe that fact has been withheld from the public. It's my assessment of the information at hand stemming back to the 1990s when Al Gore came out with his iconic, *An Inconvenient Truth*, which was nothing, but a marketing scheme played upon the ignorant public, to twist the narrative into believing that the Earth was in a Global Warming phase, caused by human pollution, excessive use of fossil fuels, carbon emissions and cow farts.

This was cleverly done, as a type of sleight of hand, to get the public to believe that the Earth was warming, when in fact it was cooling, and all indicators showed that due to the lack of Sunspots, coupled with the fact that the Sun was being drained by multiple solar cores, and other exoplanets, and Nemesis itself, that the Earth was headed for another ice age. The Elite had a plan which was to launch the Solar Simulator, the Artificial Sun, to warm up the Earth, along with an elaborate plan to chemtrail the skies, cloud-seeding and various types of weather modification through HAARP, SMAC and HAMMER technologies. So, when monster storms would inevitably arise from these artificially induced storms, it could became plausible deniability, causing people to assume that it was due to Global Warming, thanks to the propaganda distributed far and wide by Gore, which has since been heavily debunked.

A lot of these schemes are part of an elaborate cover-up, as to not alarm the general public into panic and more importantly, survival. I don't think it comes as a surprise to any of my readers that it says on the Georgia Stones - which is literally written into stone - the intentions of the Elite and New World Order is to reduce the Earth's population to 500,000 from 7.5 billion. They are allowing the catastrophe to take the lives of most everyone on Earth, except for them and their chosen few. They have been terribly busy building underground cities, communities, and luxury doomsday bunkers where they plan on sticking out the coming apocalyptic catastrophes, that will literally rearrange land masses. Ocean levels will rise to take over land, and new land masses will emerge, such as what we see happening now, as the ice melts in Antarctica.

Together with global economic and political chaos, it's believed by researchers that the Annunaki will begin by making visits as their world approaches with the incoming Nemesis-Nibiru System, also called The Destroyer and the Red Dragon by the ancients, by contacting world leaders about warning them of the impending asteroid impact event. Marshall Masters thinks this asteroid is natural to our solar system, and orbits around our Sun while locked in a path between Mercury and Venus, which up until now, has presented no threat to Earth. However, the approaching Nibiru system will "bump" it into a trajectory by knocking it off its present orbit, and consequently bring it into contact with Earth. According to the Bible Prophecy in Revelation 8:11, a comet- like object or even an asteroid named *Wormwood* will hit the Earth. This could be one and the same event, and then again, it could be two separate impact events, that will have catastrophic effects on Earth and our solar system.

> "But the meek shall inherit the Earth; and shall delight themselves in the abundance of peace."
> (Psalm 37:11)

> Likewise, Jesus reiterated, "Blessed are the meek, for they will inherit the Earth."
> (Matthew 5:5)

Notes:
1. Zecharia Sitchin, *The Twelfth Planet: Book I of the Earth Chronicles (The Earth Chronicles)* Harper, reprint edition, 2007.
2. Ella LeBain, *Who is God?* Book Two, Skypath Books, 2015.
3. Gregg Prescott, M.S., *What the Church Isn't Telling You About Nibiru And The Annunaki*, August 24, 2016 | In5D.com
4. Ella LeBain, *Who is God?* Book Two, Chapter: *Who Are the Annunaki?* p.264, Skypath Books, 2015.
5. Ella LeBain, *Who's Who in the Cosmic Zoo?* Book One – Third Edition, Concluding Words, pp. 415-444, Skypath Books, 2013
6. Zecharia Sitchin, *Genesis Revisited, Is Modern Science Catching Up with Ancient Knowledge? (Earth Chronicles)* Avon Paperback – October 1, 1990.

7. Ella LeBain, *Who is God?* Book Two, Chapter: *Who is Allah?*, p.359, Skypath Books, 2015.
8. Gregg Prescott, M.S., *What the Church Isn't Telling You About Nibiru And the Anunnaki*, August 24, 2016 | In5D.com
9. Ella LeBain, *Covenants* – Book Four, Skypath Books, 2018.
10. Ella LeBain, *Who Are the Angels?* Book Three, Skypath Books, 2016
11. Marshall Masters, *Two Suns in the Sky, Who Lives, Who Dies?* Audio Transcript. CreateSpace, June 2017

CHAPTER FOURTEEN

SAVING EXTRATERRESTRIALS

"In order to understand more, it is imperative that we improve our knowledge, before choosing which side of the fence, we feel compelled to belong."
~J.P. Robinson

Guy Consolmagno, leading Astronomer for the Vatican stated, "Very soon the nations will look to aliens for their salvation." Consolmagno believes that humans are not the only intelligent beings created by God in the universe and added these nonhuman lifeforms are described in the Bible as the Nephilim. He also coauthored the book with another Jesuit Astronomer, Paul Mueller, Guy Consolmagno, *Would You Baptize an Extraterrestrial? . . . and Other Questions from the Astronomers' In-box at the Vatican Observatory*, in 2014.[1]

They postulated that it is their duty to lead aliens and extraterrestrial to salvation in Christ. Their thesis was structured to connect science with God, and God with the creation of extraterrestrials. They both do not claim to have met aliens, but as Vatican Astronomers and Jesuit Priests, they are preparing others to get ready for the implication of Alien Disclosure.

I have to admit, I don't agree with the Catholic Church on much, in fact, if you read the many essays, I offer on exposing the Church of Rome's heresies, as an enemy of the God, you will know I'm no fan of the Papacy, Catholics, or the Jesuit Priesthood. But these two Vatican Astronomers both intrigued and impressed me, with their open mindedness, and their spirits open to the possibility of Disclosure of Extraterrestrial life both on Earth and in the Universe, which is totally aligned with my interpretation of the Old and New Testaments Bibles.

As they say, even a broken clock is right twice a day, and the Vatican Astronomers are on the right track here. I do suspect, they know more than they are revealing, and that their book was a treatise on the End Times strategies of the Vatican. Nevertheless, the notion that it's the Vatican leading the way towards ET Disclosure, is both fascinating and disturbing at the same time, especially knowing that according to the Bible Prophecy, the Vatican is the Whore of Babylon, and will produce the End Times False Prophet, that will prop up the final Antichrist, aka Counterfeit Messiah, who will be Alien in human form, to lead the One World Government, One World Religion and One World Bank.

ET'S ARE ALREADY HERE

Dr. Christopher Corbally, Vice Director for the Vatican Observatory Research Group on Mount Graham until 2012, believes "our image of God must change if Disclosure of Alien life is soon revealed by scientists. A new truthful story needs to be written." When I read these types of comments coming from Vatican sources, my discernment tells me, that they are preparing the public for what they already know, that they already have had contacts with the Aliens.

Paola Harris translates Vatican Monsignor Corrado Balducci to explain the Vatican's acceptance of extra-terrestrials as our brothers endowed with life by God. Balducci said, "they have a body like ours (tangible) they make themselves seen to people on Earth...they have a soul with a body, otherwise they would be angels."[2]

The late Balducci not only believed in the presence of alien intelligences already interacting with our planet, but also believed that the Vatican has been aware of it. Balducci believed that extraterrestrial contact is real and the extraterrestrial encounters "are NOT demonic, they are NOT due to psychological impairment, they are NOT a case of entity attachment, but these encounters deserve to be studied carefully." Vatican spokesman to our galactic neighbors?

Balducci went on to state, "As God's power is limitless, it is not only possible but also likely that inhabited planets exist. I always wish to be the spokesman for these star peoples who also are part of God's glory, and I will continue to bring it to the attention of the Holy Mother Church."

Kind of riveting to think that someone from the Church of Rome, who has kept the truth hidden for millennia, wants to represent our planet as a spokesman with our galactic neighbors. Part of me is enthralled, the other part highly suspicious as to who the aliens are, they are in contact with, and why they should be trusted. As I wrote and exposed in *Covenants* – Book Four, that the aliens Hitler was in contact with also allegedly were in contact with Pope Pious II, who authorized his own concentration camps to get rid of who he deemed were undesirable humans, that made up Jews, gypsies, and homosexuals.[3] Why should we trust the Vatican to be the Ambassadors of Extraterrestrial contact on Earth?

Thus far, they have not proved to be right about anything to do with *who* the God of gods is – they worship Lucifer. They have been wrong about the Earth being flat, which they have since retracted. They were wrong about Earth's location in the universe, which they have admitted to these errors. They were wrong about the Inquisitions, murdering thousands of true believers of Christ, and now, they expect us to follow their lead with ET and Alien Disclosure? My discernment tells me that they should not be trusted. If the Vatican wants to lead Disclosure of Extraterrestrial Life, then first start with published and releasing all the thousands upon thousands of documents and books held under lock and key in their vast private Vatican Library. That would turn some heads around towards given them any credibility to lead on Alien Disclosure. Till then, my bets are on the facts that the Vatican has been meeting with the Grays and Tall White Nordic Aliens, the Aryan Race, that live, inside the Earth. They have been here for millennia, and they are the alien presence behind the Holocaust, and numerous wars against humankind.

> An excerpt from, EXO-VATICANA, Thomas Horn stated: "In a paper for the Interdisciplinary Encyclopedia of Religion and Science, Father Giuseppe Tanzella-Nitti—an Opus Dei theologian of the Pontifical University of the Holy Cross in Rome—explains just how we could actually be evangelized during contact with "spiritual aliens," as every believer in God would, he argues, greet an extraterrestrial civilization as an extraordinary experience and would be inclined to respect the alien and to recognize the common origin of our different species as originating from the same Creator. According to Giuseppe, this contact by non-terrestrial intelligence would then offer new possibilities "of better understanding the relationship between God and the whole of creation." Giuseppe states this would not immediately oblige the Christian "to renounce his own faith in God simply on the basis of the reception of new, unexpected information of a religious character from extraterrestrial civilizations," but that such a renunciation could come soon after as the new "religious content" originating from outside the Earth is confirmed as reasonable and credible. "Once the trustworthiness of the information has been verified" the believer would have to "reconcile such new information with the truth that he or she already knows and believes on the basis of the revelation of the One and Triune God, conducting a re-reading [of the Gospel] inclusive of the new data…" [4]

The Vatican has known about Nibiru for quite a while. In a 1997 interview with Art Bell on Coast-to-Coast AM, Father Malachi Martin was asked why the Vatican was heavily invested in the study of deep space at the Mount Graham Observatory. Martin replied, "Because the mentality amongst those who are at the highest levels of Vatican administration and geopolitics, know…what's going on in space, and what's approaching us, could be of great import in the next five years, ten years." [5]

They fear the Biblical creation story will be challenged. In Book One of *Who's Who in the Cosmic Zoo?*, I concluded that the big deception to come out with Disclosure of Alien Life, was that they were going to tell the public that humankind was created by Reptilians. This is a lie, which I discerned, in that, Reptilian Annunaki did not exactly create humankind, but rather genetically manipulated humankind, during the Genesis account of the 'Fall of Humankind.' I explained that due to the spiritual legal ground after the curses were pronounced on man, woman, the Earth and Satan, who stole back the rulership of planet Earth from the first Adam, made

CHAPTER 14: SAVING EXTRATERRESTRIALS

a deal with the incoming Annunaki who needed to replace their Igigi, a group of beings that allegedly rebelled against them, because they were being used as slaves, who disappeared, leaving the Annunaki to seek out another slave race, in which humans were tagged as 'it.' According to the Sumerian Mythology, Igigi was the term used to refer to the gods of Heaven who came down to Earth. Despite often being misunderstood as synonymous with the term 'Annunaki,' the myth tells us that the Igigi were allegedly younger gods and servants (slaves) of the Annunaki, until they rebelled and were replaced by the genetic manipulation/creation of humans.[6]

Originally humankind was created in the image and likeness of the Elohim gods, (Genesis 1:27) who were Human ETs from Lyra, where the Human Vine originated. But after the curse, the Annunaki were given permission to take the humans created on the Earth, who were created with all 12 strands of DNA, just like the Elohim gods, and genetically downgrade their DNA by disabling 10 strands of DNA, leaving humankind with two strands of DNA, giving them just enough intelligence to follow orders, and do the Anunnaki's dirty work, which was to mine the Earth for gold and other natural resources. Please keep my thesis in Book One in mind, as you read through these reports, and if you haven't read Book One of *Who's Who in the Cosmic Zoo?* you may want to go over the foundations I laid in the first four chapters and in my Concluding Words in order to connect the dots to what I'm asserting here.[7]

All in all, this is why the second Adam, who was Jesus Christ had to come to *set the captives free* in fulfilling the Bible Prophecy of Isaiah 61:1, promises to redeem humankind back for the Kingdom of Heaven. The promises of God when Jesus returns is to *restore* humankind back into their Glory Bodies which was the Original Blueprint that the Elohim created the Evadamic Race, i.e., Adam and Eve Creation story. The Glory Bodies are immortal, and the Bible Prophecy promises that we will live as the Angels, and never die. This suggests that the restoration into immortal incorporeal bodies was the original 12 strands of DNA, that the Elohim created Adam and Eve with.

Be that as it may, the Vatican has been sitting on the secrets of who the Annunaki are, and where and when will its home planet, called Nibiru in the Sumerian Tale, return to Earth again. This is why they own the world's most sophisticated telescopes, which they named, L.U.C.I.F.E.R.

Rev. Jose Gabriel Funes, an Astronomer and director of the Vatican Observatory, stated, "Just as there is a multitude of creatures on Earth, there could be other beings, even intelligent ones, | created by God. This does not contradict our faith, because we cannot put limits on God's creative freedom." [8]

Funes added, "God was made human in Jesus to save us. In this way, if other intelligent beings existed, it is not said that they would have need of redemption. They could remain in full friendship with their creator."

In 2005, a Vatican insider blew the whistle on *Secretum Omega* and *Nibiru*. This high-ranking Jesuit Vatican insider released footage of Nibiru along with documentation from the highest secrecy of the Vatican in a project called *Secretum*

Omega, which was a Top-Secret Astronomical Research venture with NASA, in which a high-powered infrared camera was secretly placed into orbit to monitor and track the elusive Planet X or Twelfth Planet, as Sitchin labeled it to be Nibiru.[9]

This insider stated: "What I can say is that it was built in the 1990s with the object of studying all anomalous celestial bodies approaching Earth, similar to what the CIA did with one of its "secret eyes," the twin to Hubble, called "SkyHole 12" (a.k.a. Keyhole 12).

Moreover, the Vatican Insider was informed during the meetings of the aliens with Pope Pius XII, during the 1930-1940s. As I pointed out in *Covenants* – Book Four, Pope Pius XII's (1876-1958) actions during the Holocaust remain controversial. For much of the war, he maintained a public front of indifference yet remained in an alliance with Adolph Hitler, by having their own death camps to rid the world of what he deemed were undesirables, which essentially included all those in opposition to the auspices of the Roman Catholic Church. This is why I concluded in *Covenants* – Book Four, that the aliens that both Hitler and Pope Pius XII alleged met with, were hellbent on destroying the Jewish People, because they were at war with the God of Israel and the God of the Jews, who is Yahuah.[10]

In the meetings, it was reported, the approach of a celestial body that was scheduled to enter into the Solar System in which resides an advanced very warlike alien race. It was during the analysis of certain data and information from the Alaska radio telescope that they discovered one remote deep space probe, part of a deep space exploration program called "SILOE", which was started in 1990 which had taken a photograph of a huge planet getting closer to the Solar System.

The information was received in Alaska during October 1995, which is when the Vatican Insider's problems started. There was a danger to this whistle-blower's life, due to the secrecy. He reported, his contact revealed that inside the Vatican, there were two factions who struggle over possession and control of this information, which was classified far beyond Top Secret. The insider said that this probe was created in Area 51, and has an electromagnetic impulse motor, and it was put in orbit by a space plane, similar to the Aurora. The probe did not have any calculations or pre-indications of the trajectory or the precise location of Nibiru, because its purpose was to approach that planet, correcting its direction to avoid impact and to return to this Solar System to a position close enough to transmit the data and images to the secret radio telescope located in Alaska." [11]

The insider went on to say, "The human race must surrender completely to the message of salvation and redemption of Christ, which St. Paul defined as Kerigma, which is Greek for preaching, a message that John Paul II tried to spread to all nations. Don't you think that the Pope knows how close these events are to us?" In other words, to preach the Gospel of Jesus Christ to all creation, which includes aliens. It is known as the Great Commission:

CHAPTER 14: SAVING EXTRATERRESTRIALS

> "Jesus said to them, 'Go into all the world and preach the gospel to *all creation*."
>
> (Mark 16:15)

NASA ACKNOWLEDGES NIBIRU

A NASA Press Release in 1992 stated the following: "Unexplained deviations in the orbits of Uranus and Neptune point to a large outer Solar System body of 4 to 8 Earth masses on a highly tilted orbit, beyond 7 billion miles from the Sun." [12]

Gerry Neugesbeuer, director of the Palomar Observatory for the California Institute of Technology has been studying a mysterious object, that is shrouded from the Sun's light, yet is tugging at the orbits of Uranus and Neptune. It is this unseen force that Astronomers suspect may be Planet X, which they say belongs in Earth's celestial neighborhood as its tenth planet, despite Zecharia Sitchin calling it the 12th Planet, because he counted Pluto as a Planet, which has since been downgraded as a planetoid. Sitchin also counted the Sun and Moon as celestial objects that belong here, which is why he labeled Nibiru as the 12th planet.

Last year, the infrared astronomical satellite (IRAS), circling in a polar orbit 560 miles from the Earth, detected heat from an object about 50 billion miles away that is now the subject of intense speculation. "All I can say is that we don't know what it is yet," says Gerry Neugesbeuer. I believe they know but have been keeping it secret. Scientists were hopeful that the one-way journeys of the Pioneer 10 and 11 space probes would help to locate this mysterious celestial body. In recent years, there have been so many articles published on Planet Nine, Planet X and Nibiru, that are too numerous to quote here.

Notes:
1. Guy Consolmagno, Paul Mueller, *Would You Baptize an Extraterrestrial? . . . and Other Questions from the Astronomers' In-box at the Vatican Observatory*, in 2014.
2. YouTube: http://youtube/ua26wcJ8Z80
3. Ella LeBain, *Covenants* – Book Four, Skypath Books, 2018.
4. Thomas Horn, Cris Putnam, *EXO-VATICANA, Petrus Romanus, Project L.U.C.I.F.E.R. And the Vatican's Astonishing Plan for the Arrival of an Alien Savior*, Defender Press, 2013.
5. https://archive.org/details/ArtBellAndMalachiMartin
6. Ella LeBain, *Who's Who in the Cosmic Zoo?* Book One, Annunaki p.110, 2012.
7. *Ibid, p.415*
8. Gregg Prescott, M.S., *What the Church Isn't Telling You About Nibiru And the Annunaki*, August 24, 2016 | In5D.com
9. *Ibid*
10. Ella LeBain, *Covenants* – Book Four, Skypath Books, 2018.
11. https://newspunch.com/exposed-the-church-isnt-telling-you-about-nibiru-and-the-anunnaki/
12. *Are We Alone in the Universe* - Zecharia Sitchin, UFOTV, 1992, 2003.

CHAPTER FIFTEEN

SALVATION FOR ALL CREATION

"For we know that *all* of creation groans and travails in pain,
together awaiting our Salvation."
Romans 8:22-23

This notion that salvation through Christ is for humans only, denies His sovereignty. He is not just some Messiah that has come and gone, He is The Messiah who was not just written in the Bible Prophecies but into the very Stars themselves! God's Divine Plan of Salvation was first written into the Stars before the scrolls on Earth. *The Gospel in the Stars* tells the story of redemption, a Kinsmen Redeemer hero King of the Kingdom of Heaven.

As I discerned in Book One of *Who's Who in the Cosmic Zoo?*, the Reptilians are a hierarchy and they're split into groups.[1] Some of which are in rebellion towards each other and others are moderates, just like our politics. What they all have in common is they were cursed along with the rest of Creation.

This is why scripture says, "For the creation was subjected to futility, not willingly, but because of him who subjected it, in hope that the creation itself will be set free from its bondage to corruption and obtain the freedom of the glory of the children of God. For we know that the whole creation has been groaning together in the pains of childbirth until now. And not only the creation, but we ourselves, who have the first fruits of the Spirit, groan inwardly as we wait eagerly for adoption as sons, the redemption of our bodies. For in this hope, we were saved. Now hope that is seen is not hope. For who hopes for what he sees? But if we hope for what we do not see, we wait for it with patience. (Romans 8:20-24) This includes the Reptilians, Dragons and Lizards.

The Serpent Race is ancient, and the Bible clearly says that the Almighty Lord created serpents to serve Him. Some fell, some were cursed, others were captured, and banished to the inner Earth. We tend to forget that they are part of Creation, that All of Creation waits for the Redemption from the Lord.

When the Lord returns, it is all going to change. We must prepare our hearts for disclosure where our Bible becomes like a science fiction story. Allow me to correct myself, it doesn't become it, it already is, we just wake up to that realization.[2]

Are we alone? The Search for Extraterrestrial intelligence is an exploration into the concept of searching for extraterrestrial intelligence in the universe with Dr. Seth Shostak of SETI and discussing the use of the Allen Telescope array. Dr. Shostak has long been a major player in SETI and predicts that we will very well find signs of alien civilizations within the next 20 years. Also discussed are new avenues for SETI such as biosignatures and techno-signatures, and how extraterrestrial civilizations might choose to try to communicate. [3]

CHRIST AND HIS KINGDOM ARE EXTRATERRESTRIAL

It should be no secret that the Lord Jesus Christ is Extraterrestrial, especially to my readers! The Bible says He presently dwells in the Third Heaven, where He remains until His return to Earth. By very definition, this qualifies Him as Extraterrestrial.

If that is not enough to convince you, then listen to the Words of Christ Himself:

> "I am not of this world." (John 17:16)

> "My Kingdom is not of this World." (John 18:36)

> "You are of this world; I am not of this world. New Living Translation Jesus continued, "You are from below; I am from above. You belong to this world; I do not." (John 8:23)

> "Are you not aware that I can call on My Father, and He will at once put at my disposal more than twelve legions of angels?" (Matthew 26:53)

OUR HOME IS IN HEAVEN

> Jesus said, "My kingdom is not of this world. If it were, my servants would fight to prevent my arrest by the Jewish leaders. But now my kingdom is from another place." (John 18:36)

> "But our citizenship is in Heaven. And we eagerly await a Savior from there, the Lord Jesus Christ." (Philippians 3:20)

Satan is the god of this world.

> "The time for judging this world has come, when satan, the ruler of this world, will be cast out." (John 12:31)

> "You, dear children, are from God and have overcome them, because the one who is in you is greater than the one who is in the world." (1 John 4:3)

With all the evidence written into the Scriptures and from a multitude of witnesses during the times of Christ through today, His Kingdom is clearly in a battle with the Kingdom of Darkness who rules from within this Earth and on the other planets in our Solar System.

When the Blood of Christ was shed on Calvary, which we all know was for the propitiation of the sins of humankind, it had another purpose and much deeper spiritual meaning, which has to do with a deposit on His Return to Earth.

The shedding of the blood liberated certain elements so that they might enter the Earth through the Spirit of Christ, which was found in His Blood. The Lifeforce is found in the Blood. (Leviticus 17:11, 14) This is why it had to be prophesied earlier, to prepare for this appointment on the timeline, which as I prove herein, was first written into the Heavenly Scroll, the Milky Way Constellations, before it was written on Earth.

> "And I will pour upon the house of David, and upon the inhabitants of Jerusalem, the spirit of Grace and of supplications: and they shall look upon me whom they have pierced.
> (Zechariah 12:10)

> "For dogs have compassed me: the assembly of the wicked have enclosed me; they pierced my hands and my feet."
> (Psalm 22:16)

> "But one of the soldiers with a spear pierced his side, and forthwith came there out blood and water.... They shall look on him whom they pierced."
> (John 19:34, 37)

Christ was pierced not only in His hands and feet, but in His side, which poured out His blood onto the ground. His Blood had certain elements in it which could not be released any other way. Blood carries the individual's frequency, and the Blood of Christ frequency is like no other anywhere in Creation. The crystalline formation of the blood carries light energy through the body, that is what Leviticus 17: 11, 14 means about, Life is in the Blood. But the discernment here is that the Blood is not Life, but the medium through which Life enters the body and is carried to all its parts. Therefore, when Christ's Blood fell upon the ground of the Earth at

Golgotha, it assured the World that He would return again. This was part of the plan, to assure His return, which creates a match in His vibration and frequency that is literally left behind on and into the Earth itself. This is partly why there was an Earthquake after His Blood fell to the ground, and the veil was split in two.

It served as a deposit with the frequencies and elements needed to fulfill His Return to Earth, which is to transform the Earth and all of its inhabitants of His Creation, into His Heavenly Kingdom which is scheduled to arrive on Earth at the end of this present Age.[4]

The Blood and Water from the side of Jesus signified the conjunction of the Christ with Humankind by His Godly Truth, both spiritual and natural, coming from the Divine Grace of His Loving-Kindness. It is the very foundation of Spiritual Legal Ground, as I laid out in Book One of *Who's Who in the Cosmic Zoo?*[5]

For millennia after the Crucifixion till today, it is a theological fact that the literal blood of Jesus authorizes complete and everlasting salvation to all those who accept Him. Our salvation is not just individual, but global, as His Blood was deposited on the Earth, thereby sealing the promise of His Return! The blood is a symbolic, as well as literal, physical manifestation of that promise.

> "Look, I am coming soon! My reward is with me, and I will give to each person according to what they have done."
> (Revelation 22:12)

THE ULTIMATE FULFILLMENT

The promise of a coming Eternal King, to emerge from the Bloodline of King David, was repeated over and over again to David, to his son Solomon, and again and again in the Psalms, and by the Prophets Isaiah, Zechariah, Jeremiah, Micah and Amos over a period of several hundred years.

The Prophesy about a coming Star out of Jacob (Israel) along with a Scepter was given to Moses in the Torah.

> "...there shall come a Star out of Jacob, and Scepter shall rise out of Israel."
> (Numbers 24:17)

The Zohar speaks of an ancient red-hot planet surrounded by seven revolving stars that will pass the Earth in the end of days that will appear as "Fire in the Sky," and then the Moshiach (Messiah) will be revealed. Jesus said there would be signs in the Heavens prior to His return. You are seeing them now.

> "There will be signs in the Sun and Moon and stars, and on the Earth dismay among the nations, bewildered by the roaring of the sea and the surging of the waves. Men will

CHAPTER 15: SALVATION FOR ALL CREATION

faint from fear and anxiety over what is coming upon the Earth, for the powers of the Heavens will be shaken."
(Luke 21:25-26)

The Zohar writes that during the Floods of Noah, God used two planets from the Pleiades to collide with one another and cause the floods on Earth. The Jewish Scriptures reveal that the Lord uses planetary bodies, and He rules over all of them, to judge the inhabitants of Earth. He has done it in the past, and according to today's Bible, He is scheduled to do it again.

The Lord is going to use these Exoplanets to wake up sleeping humanity and to turn people back to Him. The fact that the powers that be, have been going to great lengths and spending tons of money to keep this information from the general public should wake people up as to why.

But we have arrived at a time where it is becoming harder to hide Exoplanets. The abundance of Chemtrails in the sky, along with other types of technology such as Project Blue Beam, created to beam clouds onto the blue sky, to cover up and obscure these Exoplanets, are nothing but flawed technologies. Just as your computer systems and cell phones do not perform 100% of the time, neither do these technologies. Otherwise, why are so many people still able to view these Exoplanets and their brown dwarf star that they orbit?

People all over the globe are reporting through photographic and videographic evidence seeing these Exoplanets and their second Sun daily, despite the grand scale effort to cover them up.

There is cognitive dissonance which is a state of unbelief when new information or knowledge is disclosed, and then there are outright lies, mind control and, yes, magic spells being used to keep people in the dark, oppressed by spirits of unbelief and doubt. Meanwhile, you do not need to be a rocket scientist to figure out that something very unusual is happening in our skies and with our Sun.

The Sun controls climate changes, but the passing and presence of these Exoplanets, creates Pole Shifts, which is occurring like birth pangs. See, Matthew 24:8. These contractions are happening to the Earth's inner core, causing volcanic eruptions that are happening more frequently lately, as well as Earthquakes and climate changes in the way of extreme heat and cold. The Winter of 2019 was the coldest winter on record in the Northern Hemisphere with months of Polar Vortex and deep freezes with temps that plunged all the way to minus 50 degrees below zero. The Southern Hemisphere simultaneously experienced extreme heat that has gone up to 130 degrees Fahrenheit in Australia, causing the death of wildlife.

The fact that extreme temperatures are happening simultaneously is indicative of a Pole Shift, as well as the Earth's Core becoming unstable due to its magnetic push me, pull you effect, that's occurring with the presence of the Nemesis-Nibiru Exoplanets that are intersecting within our Solar neighborhood, subsequently affecting Earth's inner core and magnetic poles.

In fact, the Earth's Magnetic North Pole is shifting 40 miles per year and has been known to loop up to 50 miles in a single year, which signals a pole reversal. The Magnetic Pole is moving faster than at any time in human history, causing major problems for navigation and migratory wildlife.[6] We are at the precipice of a major Pole Shift, and the transformation from the Processional Age of Pisces into the Processional Age of Aquarius, that will be marked by the passing of Nibiru, the Flaming Red Star that has been predicted by the Ancients to change the Earth, and herald the Return of Christ to Earth.

However, before the Fulfillment of the final Prophecies, it is the Will of God that none should perish, but that everyone be saved through the Grace of God through Jesus Christ. This is not about religion, but about reconciliation with the Creator. The Lord is patient and wants all to come to Teshuva (Repentance) to Him, so that everyone can be saved. The Lord is merciful, and His Loving-Kindness is everlasting, but those who rebel and reject Him will perish, but those who turn to Him in repentance will be saved.

The recurring message throughout the Prophets in the Scriptures, is that the "Righteous will inherit the Earth, but the wicked will perish, be removed from the Earth and thrown into the fires as the Tares." (Matthew 13:30)

The Earth will be transformed. Those whose soul is safely tucked away under the Shadow of the Almighty, will be saved and will inherit the New Earth. That is the Gospel as written into the very Stars of Heaven as well as the Scriptures on Earth.

HEAVEN COMES TO EARTH

> "The Lord has established his throne in the Heavens,
> and His Kingdom rules over all."
> Psalms 103:19 ESV

The Divine Plan is not necessarily to escape Earth, but to bring Heaven to Earth. Therefore, Jesus instructed His disciples to pray, "Thy Kingdom come, they Will be done on Earth, as it is in Heaven." (Matthew 6:10) While at the same time He spoke about the Kingdom of God that has been established within us, however, the Heavenly Kingdom needed to be called in to manifest on Earth. What is the difference? The Kingdom of Heaven is *governed* by the Kingdom of God. The Kingdom of God is the presence of the Holy Spirit that was deposited into all His believers and into the faithful. It is the precursor that establishes the Way, the Truth, and the Life through Christ. The Holy Spirit teaches us how to govern ourselves, how to deal with our relationships, Heaven's Way, not in an Earthly Way, distinguishing ourselves between the clash of these two kingdoms on Earth. The Kingdom of God verses the Kingdom of Darkness.

CHAPTER 15: SALVATION FOR ALL CREATION

Those of us who are called to the Great Commission are here on Earth at this time, to make Heaven more crowded. The thing is, Heaven has a gate, and His name is Yeshua, Jesus the Christ. He is the Gate, the Gatekeeper, and the Door to Heaven. All who accept His Ultimate Sacrifice which was to purchase all of our souls back from being held captive by the Kingdom of Darkness, who is led by the god of this world, Lucifer/satan.

> "You were bought at a price; do not become slaves of human beings. God paid a high price for you, so don't be enslaved by the world; do not become slaves of men."
> (1 Corinthians 7:23)

The Kinsman Redeemer, as written into the very stars of Heaven itself, was the Original Divine Plan from the beginning of this Grand Experiment, that the Creator God would send a Kinsmen Redeemer whose blood was Divine and had the power to break all the curses, and ransom back all the souls that the Creator God created, who were held captive through various types of genetic manipulations, along with both spiritual and physical bondages, from being "held captive" from Lucifer/satan's kingdom. The Kinsmen Redeemer was sent to "set the captives free."

> "The Spirit of the LORD is upon me, because the LORD has anointed me; he has sent me to bring good news to the oppressed and to bind up the brokenhearted, *to proclaim freedom for the captives, and release from darkness for the prisoners.*"
> (Isaiah 61:1: Luke 4:18)

Notes:
1. Ella LeBain, *Who's Who in the Cosmic Zoo?* Book One – Third Edition, https://www.amazon.com/Whos-Who-Cosmic-Zoo-Third/dp/1629942065/, 2013.
2. Ella LeBain, *Who's Who in the Cosmic Zoo? A Spiritual Guide to ETs, Aliens, Gods & Angels*, Trafford, 2012.
3. *Are We Alone? The Search for Extraterrestrial Life: How Does SETI Search for Alien Civilizations*, https://www.etupdates.com/2019/02/23/how-does-seti-search-for-alien-civilizations/
4. George Hunt Williamson, *Secret Places of the Lion*, pp. 184-185; Warner Destiny Books, NYC, 1958.
5. Ella LeBain, *Who's Who in the Cosmic Zoo?* Book One, Chapter Two: *Exopolitics and Divine Jurisprudence, Spiritual Legal Ground*, p. 45-55; Trafford, 2012.
6. Brian Nelson, https://www.mnn.com/Earth-matters/climate-weather/stories/magnetic-north-shifting-by-40-miles-a-year-might-signal-pole-r, February 5, 2019.

CHAPTER SIXTEEN

NIBIRU AND MENTAL DISORDERS

"One must still have chaos in oneself to be able to give birth to a dancing star."
~Friedrich Nietzsche

According to The Kolbrin Bible, as Hercolubus the Greek name for Nibiru, approaches Earth in the Last Days, men will lose their minds. Mental Disorders and Mental Illnesses will increase as Nibiru (Hercolubus) gets closer to Earth. Are we there yet?? Because people are being unusually nasty to each other. Slandering others to pump yourself up is a mental disorder which comes from transgressing God's Laws. You can get into a lot of trouble over what you say about other people – legally, socially, and spiritually. When your mother told you, "if you can't say anything nice about someone, then don't say anything at all," she was actually teaching you what the Lord says. No doubt we are living in the end times, where darkness in people has increased and love grows cold.[1]

I have been observing the worst side of Humanity being expressed out of jealous competition on YouTube and Facebook. That is why I am a book nerd. I write long books and take the time to get it right, before hastily publishing with errors.

I have been attacked for sharing photos, too. People see what they want to see. Jesus said, it is not so important as to what you put into your body as to what comes out of it. Out of the abundance of the heart, the mouth speaks. At some point, the Bible tells us we will be judged by our own words and how we treat others.

Nibiru's closeness creates mental problems that includes those for researching it. They tend to get stuck on what they want to see, blinding them to what actually is, or could be.

I am getting ready to present research from my decades of research including what started my foundation study under Zecharia Sitchin, that is contrary to a lot of what YouTubers are putting out these days. As for data, I find that if you go with those mathematicians and Astronomers who are not associated with religious cults, you get a completely different assessment then the plethora of false predictions put out by Internet Sensationalism.

I sit back and watch people on Facebook and YouTube be wrong every time they set a date that does not come to pass. I think it presents a danger to the collective psyche to keep putting out false predictions and being wrong each time, as well as for those within Christian cults and Christian circles keep falsely predicting the Rapture on every Jewish holiday and Lunar Eclipse. It is antisemitic to abuse Judaism and Torah this way, and I wish those who keep doing it will be convicted by the Holy Spirit to repent of this egregious error. It is offensive to Jews, not to mention they are making constant fools of themselves with so many false predictions, that they lose credibility for the rest of Christians. Then nobody wants to take any believer seriously, because of so much fear mongering, false predictions, and anxiousness about the Rapture.

You will not find a single prediction of the Rapture in my work. No man knows the day or hour therefore people should stop trying to become soothsayers for how and when the Lord will do it.

Using Astrology to predict the Rapture has to be the ultimate perversion of the Word of God. And please, stop hiding behind Astronomy. When you take an astronomical event and use it to predict an outcome, that is the very definition of Astrology.

I am not mentioning anyone's name, but if the shoe fits, please wear it and repent so you do not receive the reward of the wicked, of those who deceive others.

Being kind is a mark that you live through the Kingdom of Heaven when you choose kindness over slander. Self-control is a fruit of the Spirit of God. Jesus said, by their fruits, you will know whose mine are.

END TIMES MADNESS

Interesting correlation from the Kolbrin Bible to exactly what the Bible Prophesy predicts as well.

Godlessness in the Last Days:

> "But understand this, that in the last days there will come times of difficulty. For people will be lovers of self, lovers of money, proud, arrogant, abusive, disobedient to their parents, ungrateful, unholy, heartless, unappeasable, slanderous, without self-control, brutal, not loving good, treacherous, reckless, swollen with conceit, lovers of pleasure rather than lovers of God, having the appearance of godliness, but denying its power. Avoid such people. For among them are those who creep into households and capture weak women, burdened with sins, and led astray by various passions, always learning and never able to arrive at a knowledge of the truth. Just as Jannes and Jambres opposed Moses, so these men also oppose the truth, men

CHAPTER 16: NIBIRU AND MENTAL DISORDERS

corrupted in mind and disqualified regarding the faith. But they will not get very far, for their folly will be plain to all, as was that of those two men."

(2 Timothy 3:1-9)

"Now the Spirit expressly states that in end times, some will abandon the faith to follow deceitful spirits and the teachings of demons,"

(1 Timothy 4:1)

"For the time will come when men will not tolerate sound doctrine, but with itching ears they will gather around themselves teachers to suit their own desires."

(2 Timothy 4:3)

"when they said to you, "In the last times there will be scoffers who will follow after their own ungodly desires."

(Jude 1:18)

"For the time will come when they will not endure sound doctrine; but after their own lusts shall they heap to themselves teachers, having itching ears;"

(2 Timothy 4:3)

"And Jacob called unto his sons, and said, gather yourselves together, that I may tell you that which shall befall you in the last days."

(Genesis 49:1)

"And it shall come to pass in the last days, that the mountain of the LORD'S house shall be established in the top of the mountains and shall be exalted above the hills; and all nations shall flow unto it."

(Isaiah 2:2)

Millennia ago, Egyptian, and Celtic authors recorded prophetic warnings for the future and their harbinger signs are now converging on these times. These predications are contained in The Kolbrin Bible, a secular wisdom text studied in the days of Jesus and lovingly preserved by generations of Celtic mystics in Great Britain.

Nearly as big as the King James Bible, this 3,600-year-old text warns of an imminent, Armageddon-like conflict with radical Islam, but this is not the greatest threat. The authors of the Kolbrin Bible predict an end to life as we know it by a celestial event – the return of a massive space object, in a long elliptical orbit around our Sun. Known to the Egyptians and Hebrews as the Destroyer, the Celts later called it the Frightener.

> Manuscripts 3:4 When blood drops upon the Earth, the Destroyer will appear, and mountains will open up and belch forth fire and ashes. Trees will be destroyed, and all living things engulfed. Waters will be swallowed up by the land, and seas will boil.
>
> Manuscripts 3:5 The Heavens will burn brightly and redly; there will be a copper hue over the face of the land, 'followed by a day of darkness. A new Moon will appear and break up and fall.
>
> Manuscripts 3:6 **The people will scatter in madness.** They will hear the trumpet and battle cry of the Destroyer and will seek refuge within dens in the Earth. Terror will eat away their hearts, and their courage will flow from them like water from a broken pitcher. They will be eaten in the flames of wrath and consumed by the breath of the Destroyer.
>
> Manuscripts 3:10 In those days, men will have the Great Book before them; wisdom will be revealed; the few will be gathered for the stand; it is the hour of trial. The dauntless ones will survive; the stouthearted will not go down to destruction.
>
> "*The Destroyer* will come against every town, and not a town will escape. The valley will be ruined, and the plateau destroyed, because the LORD has spoken."
> (Jeremiah 48:8)

The Destroyer is also known today as Wormwood, Nibiru, Planet X and Nemesis. There are also troubling prophetic correlations to the future predictions of Mother Shipton's *Fiery Dragon* and the *Red Comet* warning of Astronomer Carlos Ferrada.

The Kolbrin Bible contains passages that describe none other than the passing of Nibiru itself.

From the 'Book of Creation,' Chapter Three

> "…It is known, and the story comes down from ancient times that there was not one creation but two. A creation and a recreation. It is a fact known to the wise that the Earth was utterly destroyed once then reborn on a second wheel of

CHAPTER 16: NIBIRU AND MENTAL DISORDERS

creation. At the time of the great destruction of Earth God caused a [celestial] dragon from outer Heaven to come and encompass her about.

"The dragon was frightful to behold, it lashed its tail and breathed out fire and hot coals and a great catastrophe was inflicted upon mankind. The body of the dragon (Nibiru/Planet X) was wreathed in a cold bright light and beneath on the belly was a ruddy hued glow while behind it trailed a flowing tail of smoke.

"It spewed out cinders and hot stones and its breath was fowl and stenchful poisoning the nostrils of men. Its passage caused great thundering and lightning to rend the thick darkened sky, all Heaven and Earth being made hot. The seas were loosened from their cradles and rose up pouring across the land. There was an awful shrilling trumpeting which outpoured even the howling of the unleashed winds. Men stricken with terror went mad at the awful sight in the Heavens.

They were loosed from their senses and dashed about crazed not knowing what they did. The breath was sucked from their bodies and they were burned with a strange ash. Then it passed leaving Earth enwrapped with a dark and glowering mantle which was ruddily lit up inside. The bowels of the Earth were torn open and great writhing upheavals and a howling whirlwind rent the mountains apart.

"The wrath of the sky monster (Nibiru) was loosed in the Heavens. It lashed about in flaming fury roaring like a thousand thunders. It poured down fiery destruction amid a welter of THICK BLACK BLOOD (Red ash).

"So awesome was the fearfully aspected thing that the memory mercifully departed from Man. His thoughts were smothered under a cloud of forgetfulness…. in this manner the first Earth (Tiamat) was destroyed by calamity descending from the skies. Men and their dwelling places were gone. Only sky boulders and the red Earth remained where once they were. But amidst all the desolation a few survived for Man is not easily destroyed but crept out of their caves and came down the mountain sides.

> Their eyes were wild, and their limbs trembled. Their bodies shook and their calls lacked control. Their faces were twisted, and their skin hung loose on their bones. They were as maddened as wild beasts driven into an enclosure before flames. They knew no law being deprived of all the wisdom they once had and those who had guided them were gone…"

The Kolbrin Bible goes on to give us a warning of the red comet/planet's return, named the Destroyer:

> "When blood [the red ash] drops upon the Earth the DESTROYER (Nibiru) will appear and mountains will open up and belch forth fire and ashes. Trees will be destroyed, and all living things engulfed. Waters will be swallowed up by the land and the seas will boil. The Heavens will burn brightly and redly. There will be a copper hue over the face of the land followed by a day of darkness.
>
> A new Moon (one of Nibiru's satellites) will appear and break up and fall. The people will scatter in ***madness***. They will hear the trumpet and the battle cry of the DESTROYER and will seek refuge in the dens [underground bunkers] of the Earth. Terror will eat away their hearts and their courage will flow from them like water from a broken pitcher.
>
> They will be eaten up in the flames of wrath and consumed by the breath of the DESTROYER. In those days, men will have the great book before them, wisdom will be revealed. The few will be gathered for the stand. It is the hour of trial. The dauntless ones will survive. The stouthearted will not go down in destruction…"
>
> Manuscripts 3:3-6: "When ages pass, certain laws operate upon the stars in the Heavens. Their ways change; there is movement and restlessness, they are no longer constant, and a great light appears redly in the skies."
>
> "When blood drops upon the Earth, the Destroyer will appear, and mountains will open up and belch forth fire and ashes. Trees will be destroyed, and all living things engulfed. Waters will be swallowed up by the land, and seas will boil."

The people will scatter in madness. They will hear the trumpet and battle cry of The Destroyer and will seek refuge in the den in the Earth. Terror will eat away their hearts and their courage will flow from them like water from a broken pitcher. They will be eaten up in the flames of wrath and consumed by the breath of The Destroyer.

Thus, in the Days of Heavenly Wrath, which have gone, and thus it will be in the Days of Doom when it comes again. The times of its coming and going are known unto the wise. These are the signs and times which shall precede The Destroyer's return: A hundred and ten generations shall pass into the West and nations will rise and fall. Men will fly in the air as birds and swim in the seas as fishes. Men will talk peace one with another, hypocrisy and deceit shall have their day.

Manuscript 3:7, "*Women will be as men and men as women, passion will be a plaything of man.* A nation of soothsayers shall rise and fall, and their tongue shall be the speech learned. A nation of lawgivers shall rule the Earth and pass away into nothingness. One worship will pass into the four quarters of the Earth, talking peace and bringing war. A nation of the seas will be greater than any other but will be as an apple rotten at the core and will not endure. A nation of traders will destroy men with wonders, and it shall have its day. Then shall the high strive with the low, the North with the South, the East with the West, and the light with the darkness. Men shall be divided by their races and the children will be born as strangers among them.

Brother shall strive with brother and husband with wife. Fathers will no longer instruct their sons and their sons will be wayward. Women will become the common property of men and will no longer be held in regard and respect.

Then men will be ill at ease in their hearts, they will seek they know not what, and uncertainty and doubt will trouble them. They will possess great riches but be poor in spirit.

Kolbrin Bible 3:9, "Then will the Heavens tremble and the Earth move, men will quake in fear and while terror walks with them the Heralds of Doom will appear. They will come softly, as thieves to the tombs, men will not know them for what they are, men will be deceived, the hour of The Destroyer is at hand. In those days, men will have the Great

Book before them, wisdom will be revealed, the few will be gathered for the stand, it is the hour of trial. The dauntless ones will survive, the stout-hearted will not go down to destruction. Great God of All Ages, alike to all, who sets the trials of man, be merciful to our children in the Days of Doom. Man must suffer to be great but hasten not his progress unduly. In the great winnowing, be not too harsh on the lesser ones among men. Even the son of a thief has become Your scribe.

As the great salt-waters rise up in its train and roaring torrents pour towards the land, even the heroes among mortal men will be overcome with *madness*. As moths fly swiftly to their doom in the burning flame, so will these men rush to their own destruction. The flames going before will devour all the works of men, the waters following will sweep away whatever remains. The dew of death will fall softly, as grey carpet over the cleared land.

Men will cry out in their *madness*,

"O whatever Being there is, save us from this tall form of terror, save us from the grey dew of death."

Kolbrin Bible 4:4, "The flames going before will devour all the works of men, the waters following will sweep away whatever remains. The dew of death will fall softly, as a grey carpet over the cleared land."

With passages like these, I tend to put much weight in what the text has to say— a planetary encounter. Yes, I realize ancient manuscripts have certain built-in biases and exaggerations dependent upon the point of view of the author but, at the same time, you simply cannot just throw a precious document like this out with the bath water.

Therefore, it is reasonable to suggest the Destroyer mentioned in the Kolbrin, is in fact Nibiru, the celestial Sky Monster accompanied with a red meandering comet-like serpentine tail.[2]

"And they come to Jesus, and see him that was possessed with the devil, and had the legion, sitting, and clothed, and in his right mind: and they were afraid. And they began to pray him to depart out of their coasts."
(Mark 5:1-17)

However, the mystery of iniquity (and Satan's Mouthpiece, the Corporate Media) will insist that mental illness is a physical problem. Then they can offer their solutions.

The Bible says that those things which cause the problems men attribute to mental health will get worse in the end times. If the Bible is true, we should see an increase in both the number and severity of mental health problems. Many people have been seeing mental health professionals for years. Yet their conditions do not get any better. The mental health field spends as much time manufacturing and offering excuses for why their methods do not work as they do any other activity.

Yet, men still look to men for the solutions to their problems—even when it is obvious that other men have no answers.[3]

THE DECEPTION OF END TIMES MIND CONTROL

The story about mental health is a fairly recent one. The term mental hygiene, which was the forerunner of mental health, was not used until the middle of the 1800s. It wasn't popularized until late in the 1800s-early 1900s. Many of the supporting body of tales which surround mental health were not written until the 20th century. Some are still to being created and written down today. The solutions offered by mental health professionals are many, varied and ineffective.

The best they can do is sedate people and render them unable to hurt others or themselves. At least on occasion.

Google wants to monitor your mental health. There is a school of thought which posits that mental illness is a deeply spiritual problem, not a physical one.

> "And they came over unto the other side of the sea, into the country of the Gadarenes. And when he was come out of the ship, immediately there met him out of the tombs a man with an unclean spirit, Who had his dwelling among the tombs; and no man could bind him, no, not with chains: Because that he had been often bound with fetters and chains, and the chains had been plucked asunder by him, and the fetters broken in pieces: neither could any man tame him.
>
> And always, night and day, he was in the mountains, and in the tombs, crying, and cutting himself with stones.
>
> But when he saw Jesus afar off, he ran and worshipped him, and cried with a loud voice, and said, What have I to do with thee, Jesus, thou Son of the most high God? I adjure thee by God, that thou torment me not. For he said unto him, Come out of the man, thou unclean spirit.
>
> And he asked him, 'what is thy name?' And he answered, saying, 'my name is Legion: for we are many.'
>
> And he besought him much that he would not send them away out of the country. Now there was there nigh unto the mountains a great herd of swine feeding. And all

the devils besought him, saying, 'send us into the swine, that we may enter into them.'

And forthwith Jesus gave them leave. And the unclean spirits went out and entered into the swine: and the herd ran violently down a steep place into the sea, (they were about two thousand;) and were choked in the sea.

And they that fed the swine fled, and told it in the city, and in the country. And they went out to see what it was that was done.

And they come to Jesus, and see him that was possessed with the devil, and had the legion, sitting, and clothed, and in his right mind: and they were afraid. And they began to pray him to depart out of their coasts."

(Mark 5:1-17)

Often, celebrities with mental health issues speak openly about their problems. But many who are deceived believe when these tormented souls speak of their demons they are speaking metaphorically.

> "Yeah! Literally, it's like possession - all of a sudden, you're in, and because it's in front of a live audience, you just get this energy that just starts going…But there's also that thing - it is possession. it is Dr. Jekyll and Mr. Hyde, where you really can become this other force. In the old days you would be burned for it…"—Robin Williams told James Kaplan, US Weekly, 1999.[8]

MENTAL ILLNESS AND COVID-19 PANDEMIC

This chapter was written before the 2020 Covid-19 Shutdown but is being published in December of 2020. Therefore, I thought it fitting to add the following statistics about the rise in mental illnesses during 2020 as a result of the pandemic. There have been so many controversies along with conspiracy theories about the 2020 economic shutdowns, mask tyrannies, accompanied with all kinds of civil unrest, rioting, looting, arson, vandalism, and murders, which came as a direct result of enforced stay at home orders, social distancing, and restrictions.

Millions have lost their businesses and jobs, resulting in unemployment, causing anxiety and depressions levels to go through the roof. Living with uncertainty, along with the loss of loved ones, the fear of losing both jobs and relationships has resulted in a marked increase in mental illnesses.

I am reminded of the late Mother Theresa's word on the disastrous effect of loneliness and isolation in the world, which is a condition that has plagued far more people than the coronavirus has infected.

CHAPTER 16: NIBIRU AND MENTAL DISORDERS

> "The greatest disease in the West today is not TB or leprosy; it is being unwanted, unloved, and uncared for. We can cure physical diseases with medicine, but the only cure for loneliness, despair, and hopelessness is love. There are many in the world who are dying for a piece of bread but there are many more dying for a little love. The poverty in the West is a different kind of poverty -- it is not only a poverty of loneliness but also of spirituality. There's a hunger for love, as there is a hunger for God." (Mother Theresa, 1910-1997)

And to compound matters even worse, because of the shutdowns, health clinics, HMOs, doctor's offices have been operating with minimal staff due to the Covid-19 restrictions, which has prevented people from obtaining regular preventative and ongoing care through check-ups that includes testing outside of Covid-19. This has caused people with underlying health issues to be neglected from routine care, causing added depression, frustration, and anxieties.

During this unprecedented time of uncertainty and fear, mental health issues and substance use disorders among people with these conditions have exacerbated. In addition, epidemics have been shown to induce general stress across a population and may lead to new mental health and substance use issues. The COVID-19 pandemic and the resulting economic recession have negatively affected many people's mental health and created new barriers for people already suffering from mental illness and substance abuse.

In a KFF Tracking Poll conducted in mid-July, 53% of adults in the United States reported that their mental health has been negatively impacted due to worry and stress over the coronavirus. This is significantly higher than the 32% reported in March, the first time this question was included in KFF polling. Many adults are also reporting specific negative impacts on their mental health and wellbeing, such as difficulty sleeping (36%) or eating (32%), increases in alcohol consumption or substance use (12%), and worsening chronic conditions (12%), due to worry and stress over the coronavirus. As the pandemic wears on, ongoing and necessary public health measures expose many people to experiencing situations linked to poor mental health outcomes, such as isolation and job loss.[9]

More than one in three adults in the U.S. have reported symptoms of anxiety or depressive disorder during the pandemic (weekly average for May: 34.5%; weekly average for June: 36.5%; weekly average for July: 40.1%) (Figure 1). In comparison, from January to June 2019, more than one in ten (11%) adults reported symptoms of anxiety or depressive disorder. Additionally, a recent study found that 13.3% of adults reported new or increased substance use as a way to manage stress due to the coronavirus; and 10.7% of adults reported thoughts of suicide in the past 30 days.[9]

Therefore, we can conclude that the approach of Nibiru intersecting within our Solar System, is performing as expected, whether it be in the form of pandemics, or civil unrest, it is increasing mental disorders worldwide. What or who is the solution?

THE GOSPEL OF JESUS CHRIST

"Do not be conformed to this age (world), but be transformed by the renewing of the mind,"
(Romans 12:2)

"For there is one God, and there is one mediator between God and men, the man Christ Jesus, who gave himself as a ransom for all, which is the testimony given at the proper time."
(1Timothy 2:5-6)

"For God so loved the world, that he gave his only Son, that whoever believes in him should not perish but have eternal life. For God did not send his Son into the world to condemn the world, but in order that the world might be saved through him."
(John 3:16-17)

"All have sinned and fall short of the glory of God, and are justified by his grace as a gift, through the redemption that is in Christ Jesus, whom God put forward as a propitiation by his blood, to be received by faith."
(Romans 3:23-25)

"God shows his love for us in that while we were still sinners, Christ died for us."
(Romans 5:8)

"If you confess with your mouth that Jesus is Lord and believe in your heart that God raised him from the dead, you will be saved. For with the heart, one believes and is justified, and with the mouth one confesses and is saved."
(Romans 10:9-10)

"For I delivered to you as of first importance what I also received: that Christ died for our sins in accordance with the Scriptures, that he was buried, that he was raised on the third day in accordance with the Scriptures."
(1Corinthians 15:3-4)

> "Believe in the Lord Jesus Christ, and you will be saved, you and your household."
>
> (Acts 16:31)

Notes:
1. Ivan Petricevic, *The Kolbrin: A history-changing 3,600-year-old 'lost' Bible*, https://educateinspirechange.org/alternative-news/the-kolbrin-a-history-changing-3600-year-old-lost-Bible/, October 28, 20-17.
2. Greg Jenner, *Nibiru and the Kolbrin Bible – Correlations*, darkstar1.co.uk http://humansarefree.com/2011/09/nibiru-and-kolbrin-Bible-correlations.html
3. Jeremiah Jameson, Mondo Frazier, *Mental Health Problems in the End Times*, https://endtimesprophecyreport.com/2016/10/30/mental-health-problems-in-the-end-times/, October 30, 2016.
4. By Elizabeth Howell, What Is a Planet? https://www.space.com/25986-planet-definition.html, April 7, 2018.
5. Mike Wehner, *Pluto might become a planet again because Astronomers can't make up their mind*, https://bgr.com/2018/09/07/pluto-planet-argument-definition/, September 7, 2018.
6. Elizabeth Howell, Neil deGrasse Tyson Rejects Pluto Planethood Proposal on 'Colbert', https://www.space.com/36126-pluto-planethood-slammed-by-neil-degrasse-tyson.html, March 20, 2017
7. Tim Prudente, The Baltimore Sun (TNS), *Scientist leads effort to restore Pluto's planetary stature*, https://www.seattletimes.com/nation-world/scientist-leads-effort-to-restore-plutos-planetary-stature/, March 17, 2017
8. "Robin Williams," by James Kaplan, US Weekly, January 1999, p. 53.
9. The Implications of COVID-19 for Mental Health and Substance Use, https://www.kff.org/coronavirus-covid-19/issue-brief/the-implications-of-covid-19-for-mental-health-and-substance-use/

CHAPTER SEVENTEEN

REVIVED ROMAN EMPIRE

"A United Europe sounds like the Roman Empire!"
President George W. Bush, 2004

What is the Revived Roman Empire?

Rome at one time ruled nearly the entire known world, some use the term Revived Roman Empire to describe a coming world-wide government. The phrase does not appear in the Bible. However, the Bible does describe a powerful government that rises to power in the end times. According to some interpretations of Daniel and Revelation, the Revived Roman Empire is either a generic government system or a specific nation led by a ruler, centered in Rome, Turkey, or the Middle East. Ten lands that used to be the old Roman Empire are now all Muslim nations.

The Revived Roman Empire is often associated with the fourth beast in Daniel 7. The ten-horned beast is "terrifying and dreadful and exceedingly strong" (Daniel 7:7). It is believed to be a prophetic image of the Roman Empire (Daniel 7:19–24). As Daniel watched, a little horn grew out of the ten horns (Daniel 7:8), which is representative of the Antichrist, who is somehow connected with the Roman Empire. Since the Roman Empire lost its power and identity in the fifth century, we expect it to be "revived" in some sense in order to fulfill end-times prophecy.

The Revived Roman Empire is also associated with the fifth and final kingdom mentioned in Daniel 2. Daniel 2 describes Nebuchadnezzar's dream of an image made of various metals and Daniel's subsequent interpretation. The image had iron legs and feet made partially from iron and partially from clay. The iron legs represent the Roman Empire, and the dual-material feet represent the last world empire. That the feet are partially made of iron may suggest the last world empire is somehow associated with the Roman Empire. The ten toes may link to the ten horns in Daniel 7:20, perhaps a ten-nation coalition led by a ruler in Rome.

In Revelation 13, a ten-horned beast comes from the sea. The ten horns of this beast connect it to the fourth beast in Daniel 7. The Revelation 13 beast is described as blasphemous, tyrannical, and absolute in its rule. It has global power bestowed by Satan. Interpreting the symbolism as references to a coming political empire or ruler, rather than a figure of prior history, is more easily done with Revelation 13. [1]

CHAPTER 17: REVIVED ROMAN EMPIRE

If you listen in on the Vatican masses, which are all in Latin, you will hear them invoking Lucifer, in their praise and worship. Easter Mass is broadcast on their YouTube channel every year, as well as others. It is no secret that the Roman Catholic Church believes that Lucifer, who is the god of this world, also known as Satan, is the Light Bearer spoken of in the book of Isaiah 14:12, which was a major mistranslation by St. Jerome, who was commissioned by Pope Darius in 382AD to rewrite the Old Testament into the Latin language, known as the Vulgate Bible. St. Jerome did not understand the actual Hebrew Word/Name that the Lord used to describe this fallen and rebuked son of Heaven, which was Heylel, that literally translates to, "he who boasts in a foolish rage." When I researched this in depth, documented in Book Two, *Who is God?*,[2] I also found that it wasn't the fault of Pope Darius, who was plagued by violence and distracted by treachery when he went from being a Bishop in Rome to the Pope of Rome in 366AD. It is clear from the Darius Decree, Pope Darius knew exactly who Jesus Christ was, which is evidenced in His Decree:

> "The arrangement of the names of Christ, however, is manifold: Lord, because He is Spirit; Word, because He is God; Son, because He is the only-begotten son of the Father; Man, because He was born of the Virgin; Priest, because He offered Himself as a sacrifice; Shepherd, because He is a guardian; Worm, because He rose again; Mountain, because He is strong; Way, because there is a straight path through Him to life; Lamb, because He suffered; Corner-Stone, because instruction is His; Teacher, because He demonstrates how to live; Sun, because He is the illuminator; Truth, because He is from the Father; Life, because He is the creator; Bread because He is flesh; Samaritan, because He is the merciful protector; Christ, because He is anointed; Jesus, because He is a mediator; Vine, because we are redeemed by His blood; Lion, because he is king; Rock, because He is firm; Flower, because He is the chosen one; Prophet, because He has revealed what is to come."

Yet, St. Jerome in his ignorance or willful ignorance, (let God be the judge of that), inserted the Latin name, *Lucifer*, in place of the Hebrew name, *Heylel*, which has misguided Catholics for centuries to misunderstand Isaiah 14, and think that Lucifer is the Light of the World, which according to hundreds of other scriptures, too numerous to repeat here, states that it is Yeshua, who is Jesus in English, is the true Light of the World, the Sun/Son of Righteousness, the Messiah, the Prince of Peace, the Mighty Counselor, the King of kings and Lord of lords.

This is the crux of the quintessential Spiritual Battle of all time, which will culminate in the End Game or Armageddon. The Bible Prophecy states that the

Lord will return, the whole world will see Him with His armies of Angels (the Faithful Extraterrestrials), and He will end Armageddon, with His Breath and His Word. See, Revelations 19:15-16.

"And out of His mouth goes a sharp sword, that with it He should smite the nations" (Revelation 19:15). He is going to smite them with His Word. "The worlds were framed by the Word of God" (Hebrews 11:3).

If He can make the world and the whole universe just by speaking the Word, how much more can He smite the nations by His Word!

"And He shall rule them with a rod of iron." When Jesus died on the cross, He was God in the hands of men. But this time it is not going to be God in the hands of sinful, wicked men; it is going to be men in the hands of God, who is going to mete out the judgment they deserve.

"And He treads the winepress of the fierceness and wrath of Almighty God. And He hath on His vesture and on His thigh a name written, KING OF KINGS, AND LORD OF LORDS." (Revelation 19:15–16) This picture of the grand finale of the judgments of God, with Christ coming and casting the wicked into the great winepress of God's wrath, is even clearer back in chapter 14 of Revelation, where we read about the two great harvests.

"And the armies which were in Heaven followed Him upon white horses, clothed in fine linen, white and clean. And out of His mouth goes a sharp sword, that with it He should smite the nations: and He shall rule them with a rod of iron: and He treads the winepress of the fierceness and wrath of Almighty God. And He hath on His vesture and on His thigh a name written, KING OF KINGS, AND LORD OF LORDS." (Revelation 19:14–16)

Speculation assumes that the Vatican has these high-powered telescopes to monitor the biblical "Wormwood" aka Nibiru or Planet X:

> "And the third angel sounded, and there fell a great star from Heaven, burning as it were a lamp, and it fell upon the third part of the rivers, and upon the fountains of waters; And the name of the star is called Wormwood: and the third part of the waters became wormwood; and many men died of the waters, because they were made bitter."
>
> (Revelation 8:10, 11)

Another possibility is that the Vatican is monitoring extraterrestrial activity, tracking Wormwood and already made ET contact hundreds of years ago, which may explain the religious belief in God or the switch along the way to worship Lucifer in place of Jesus Christ who, in the religious texts, may have been influenced by the alien presence.

On February 11, 2013, just after Pope Benedict announced his resignation – an action unprecedented in Vatican History of the Papacy who are appointed for life and are only replaced when they die – the world saw a lightning bolt directly hit the

CHAPTER 17: REVIVED ROMAN EMPIRE

Papal Basilica Cathedral of St. Peter in the Vatican. Jesus said, "I watched Satan fall from Heaven like lightning." (Luke 10:18) Then, when the new Pope Jorge Bergoglio, aka Pope Francis, was elected, he released two doves (of peace), who were both attacked by a crow and a seagull. All this symbolism suggests that there will be no peace found within the Vatican, as the Church of Rome is coming under Judgement.

> "a woman sitting on a scarlet beast that was full of blasphemous names, and it had seven heads and ten horns. The woman was arrayed in purple and scarlet, and adorned with gold and jewels and pearls, holding in her hand a golden cup full of abominations and the impurities of her sexual immorality. And on her forehead was written a name of mystery: "Babylon the great, mother of prostitutes and of Earth's abominations." And I saw the woman, drunk with the blood of the saints, the blood of the martyrs of Jesus."
> (Revelation 17:3–6 ESV)

In the Old Testament, the city of Babylon gave birth to an empire that ruled the known world and imposed an alien-implanted religious worldview upon all that she conquered. Those she did not destroy, she subverted. Often, she was brutal; callous and arrogantly believed she would reign over the Earth forever. She made sure here bloodline would always fill her Earthly thrones. She spoke as a god and she thought herself secure. Her destruction is announced in Isaiah 47. In response to her wickedness, arrogance, self-indulgence, and brutality towards the people of God the Lord announces sudden doom:

> "But evil shall come upon you, which you will not know how to charm away; disaster shall fall upon you, for which you will not be able to atone; and ruin shall come upon you suddenly, of which you know nothing."
> (Isaiah 47:11 ESV)

History recorded that the Babylonian Empire fell suddenly and tragically to the Persians under Cyrus The Great in 539 BC. Later when Darius was King, the Babylonians revolted but were faced with the consequences of the same brutalities they had previously inflicted upon the Jews. The Babylonians strangled their wives and children to keep them from starving to death during the brutal siege of their capital city. When the city fell, Herodotus said the gates were pulled down and 3,000 of their leading citizens were impaled upon its walls. The once great city – the Queen of the world – was defeated, devastated, and despoiled. Just like God said.

Babylon appears in the biblical narrative again, about 630 years later. The former seat of empire is now a village surrounded and nearly swallowed by a sea of sand. Yet her name reappears in the New Testament as a symbol of the world at war with the people of God. Peter uses it as a sort of code. He ends his epistle to the churches of Pontus and Bithynia by saying:

> "She who is at Babylon, who is likewise chosen, sends you greetings, and so does Mark, my son. Greet one another with the kiss of love. Peace to all of you who are in Christ."
> (1 Peter 5:13–14 ESV)

She who is at Babylon? Peter was nowhere near Babylon when he wrote that letter; Peter was in Rome, but he uses the word Babylon as a symbolic way of referring to the new world culture at war with the *Covenant* community. All the false religions come out of Babylon. What Peter is actually saying is: *Rome is the new Babylon.* Rome is the new mistress who would seduce and undermine the people of God. The city had become a spirit.[3]

Revelations 17 references to Babylon is obviously symbolic:

> "a woman sitting on a scarlet beast that was full of blasphemous names, and it had seven heads and ten horns. The woman was arrayed in purple and scarlet, and adorned with gold and jewels and pearls, holding in her hand a golden cup full of abominations and the impurities of her sexual immorality. And on her forehead was written a name of mystery: "Babylon the great, mother of prostitutes and of Earth's abominations." And I saw the woman, drunk with the blood of the saints, the blood of the martyrs of Jesus."
> (Revelation 17:3–6 ESV)

The *Whore of Babylon* is the *spirit* of seductive culture, actively engaged in the deception and destruction of God's people. That she rides upon the Beast means she is propped up by the forces of enemies of Christ, who is the Anti-Christ, that embodies the Draconian Alien Presence on Earth, who rules the anti-Christian government. All the symbolism in this passage, a woman, who in scripture represents the Church. The Bride is the True Body of Christ, the Whore is the Antichrist. She is dressed in purple and scarlet, which are the colors of the Vatican. Just a look at the Pope's and Bishop's costumes. The jewels, pearls, golden cup full of abomination of sexual perversions, represent the countless pedophile abuses of Catholic priests all around the world.

The Boston Globe exposes the biggest pedophile ring in the history of the Catholic Church, and their statistics reported in 2002 that more than 80% of sexual abuses by priests were all male and were done by homosexual priests. While in defense of priests, the Globe reported "They abused mostly boys because they didn't have as much access to girls," "Some argue many boys were victimized because abusive priests had greater access to them."

The report found that 4,392 priests had been accused of sexually abusing minors between 1950 and 2002, or about 4% of the 109,694 priests in active ministry during that period. Of the 10,667 reported victims, 81% were male, the report said, and more than three-quarters of the victims were postpubescent, meaning the abuse did not meet the clinical definition of pedophilia.

Instead, most fell victim to ephebophiles, men who are sexually attracted to adolescent or postpubescent children.[4]

U.N. REPORT: VATICAN POLICIES ALLOWED PRIESTS TO RAPE CHILDREN

The United Nations heavily criticized the Vatican for a systematic adoption of policies allowing priests to rape and sexually abuse tens of thousands of children.

The devastating report published by the UN Committee on the Rights of a Child said the Vatican must "immediately remove" all known or suspected child abusers within the clergy.

It said the Holy See had "systematically placed preservation of the reputation of the church and the alleged offender over the protection of child victims."

In response, the Vatican said in a statement published on its website that some points made in the report were an "attempt to interfere with Catholic Church teaching."

The Vatican said it would examine the report thoroughly and reiterated its commitment to defending and protecting child rights in accordance with the UN guidelines and "the moral and religious values offered by Catholic doctrine."

The UN's conclusions come after an unprecedented hearing in Geneva on Jan. 16 in which Vatican representatives were questioned by the UN committee.

Its recommendations are non-binding, and the UN has given the Vatican until 2017 to report back. It criticized the institution for submitting its last report 14 years late.

"Well-known child sexual abusers have been transferred from parish to parish or to other countries in an attempt to cover-up such crimes," the report said.

It later added: "Due to a code of silence imposed on all members of the clergy under penalty of excommunication, cases of child sexual abuse have hardly ever been reported to the law enforcement authorities in the countries where such crimes occurred."

The UN report also denounced the Holy See for its attitudes toward homosexuality, contraception, and abortion.[5]

As it was in the past, so shall it be in the future. We can conclude, based on the many scandals, lawsuits, and sexual abuse cases that have and continue to plague the Vatican and Catholic priests worldwide, that invoking Lucifer in their masses, invites his demons to possess the Priesthood. But the problem with the Vatican, is it is not just about the world's biggest cult, by definition, it is about the Vatican's plans to rule the world through a One World Religion and One World Government. This is why the Pope is behind the Chrislam movement, by trying to sell the lie that Christianity and Islam worship the same god, which is easy for the Pope to get that one confused, because he worships Lucifer, who is Allah, yet Christians are supposed to worship Jesus Christ, not Lucifer, that's why they're called, CHRISTians! He also partners with the Co-Exist New Ager Movements, which is being purported by a number of New Age leaders, such as Rick Warren, the Dalai Lama, could not tell the difference between one god from the other. DISCERNMENT is the most important spiritual muscle that you can activate and exercise in these end times. If you're not grounded in the Word and Knowingness of *who* Jesus, Yeshua, Immanuel is, then you're easily going to be deceived by the many false messiahs, false prophets, false teachers, and more importantly, the false Spirit of Religion and Godlessness in the Last Days:

> "But understand this, that in the last days there will come times of difficulty. For people will be lovers of self, lovers of money, proud, arrogant, abusive, disobedient to their parents, ungrateful, unholy, heartless, unappeasable, slanderous, without self-control, brutal, not loving good, treacherous, reckless, swollen with conceit, lovers of pleasure rather than lovers of God, having the appearance of godliness, but denying its power. Avoid such people."
>
> (2 Timothy 3:1-3)

A religious spirit is the "leaven of the Pharisees and Sadducees" (Matthew 16:6) of which ... The solution to the Religious Spirit? Of course, is the Spirit of Jesus Christ, which is sent to us through the Holy Spirit.

> "Come to Me, all who are weary and heavy-laden, and I will give you rest. "Take My yoke upon you and learn from Me, for I am gentle and humble in heart, and YOU WILL FIND REST FOR YOUR SOULS. "For My yoke is easy and My burden is light."
>
> (Matthew 11:28-30)

THE SPIRIT OF RELIGION

When ritual controls our relationships, this is religion. Ritual minus relationship equals religion. The spirit of religion is an agent of Satan assigned to prevent change and to maintain the status quo by using religious devices. There are many spirits of religion as we see in the many types of non-Christian and Christian religions both personal and corporate in the world today.

It is a spirit that is assigned to distract people from the truth by making us think that we are okay. The spirit of religion is the meanest, foulest, most ruthless spirit that you will ever confront. It comes as an impostor, masquerading as the real thing, cutting off believers from relationships with God and blocking the work of the Holy Spirit.

The fruit of the spirit compared to the fruit of the religious spirit. The word of God says we can recognize a tree by the fruit that it bears (Matthew 12:33-37).

The Bible describes two kingdoms operating in the world. The kingdom of God seeks to accomplish the will of the Father on Earth (Matthew 6:10), which is to bring righteousness, peace, and joy in the Holy Spirit, (Romans 14:17).

In opposition to God's kingdom, Satan has established the kingdom of this world. Satan's goal is to resist the progress of God's kingdom and hinder God's plan of redemption. These kingdoms are locked in a conflict that will not end until Jesus returns. Satan's goal is to counterfeit God's kingdom by offering the world a dead imitation of the true kingdom to hinder the progress of God's work by vaccinating the world against God's love. Satan's counterfeit kingdom is called religion. God's goal is not to create people who are religious. His goal is to bring men and women into a relationship with Himself and combat the spirit of religion through love.

WARNING SIGNS OF A RELIGIOUS SPIRIT

Most of us are subject to religious spirits to some degree. Paul exhorts us to "test ourselves to see if we are in the faith," (2 Corinthians 13:5).

Go through the warning signs below and check the ways you behave because you might have a religious spirit in that area of your life. Remember religious spirits are seven times stronger than other types of spirits as demonstrated in Daniel 3:19. Remember that religious spirits hold people in spiritual slavery. The battle against this spiritual slavery can only be won by unconditional love, in living the Christian life focused on Jesus and the Word of God and being filled with the Holy Spirit.

Essentially there are five types of Religious Spirits: a counterfeit gift of discernment, this is someone who listens to familiar spirits, a spirit of Jezebel, who wars against the Holy Spirit and God's prophets, self-righteousness, which is something the Lord hates, because no one is righteous, except for the Lord, and our righteousness comes from Him, in order to honor the Lord and give Him the Glory, not us; the martyr syndrome, this syndrome is almost always partnered with those who adopt victim consciousness, which denies the victory of Jesus Christ, and those who are narcissistic, think they don't need the Lord, are in rebellion to His Holy Spirit, these types can be secular, new age, elitist types who think they are better than everyone else.

THE COUNTERFEIT GIFT OF DISCERNMENT

A religious spirit gives birth to a counterfeit gift of discernment of spirits. This counterfeit gift thrives on seeing what is wrong with others rather than seeing what God is doing to help them along. Its wisdom is rooted in the Tree of Knowledge of Good and Evil, and though the truth may be accurate, it is ministered in a spirit that kills. Suspicion, which is motivated by rejection, territorial preservation, or general insecurity, causes this counterfeit gift. However, the true gift of discernment can only function with love. In 2 Corinthians 11:13-15, Paul warns them about those who minister in a religious spirit that they try to place a yoke of legalism on the believers. Religions that are based on works will tend to become violent easily, especially when confronted with those who live by faith. This includes "Christian Religions" where a doctrine of works has supplanted the cross of Christ. Those who are driven by religious spirits may try to destroy the true Christian by onslaughts of slander.

THE SPIRIT OF JEZEBEL

The Spirit of Jezebel is a combination of the religious spirit and the spirit of witchcraft that is the spirit of manipulation and control. This spirit is often found in deeply wounded people. Every trial in our lives will either make us bitter or better, and the finished work on the cross will heal every spiritual wound if we will turn to it. Those who are deeply wounded who do not go to the cross can open themselves up to this evil spirit. The Jezebel spirit is one of the enemy's most potent forms of the religious spirit that seeks to keep the church and the world from being prepared for the Lord. This spirit attacks the prophetic ministry through which the Lord wants to give timely, strategic direction to His people. Jezebel knows that by removing the true prophets, the people will be vulnerable to her false prophets, which always leads to idolatry and spiritual adultery. When there is a void of the true prophetic word, the people will be much more prone to the deception of the enemy.

A religious spirit produces religious pride so that God will not communicate with those who are proud. In Matthew 15:14, Jesus called the religious leaders of the day "blind guides." A primary strategy of the religious spirit is to get the church devoted to sacrifice in a way that perverts the command for us to take up our crosses daily. This perversion will have us putting more faith in our sacrifices than in the Lord's sacrifice. It will also use sacrifices and offerings to try to get God to manifest Himself. This is a form of the terrible delusion that we can somehow purchase the grace and presence of God with our good works.

SELF-RIGHTEOUSNESS

We do not crucify ourselves for the sake of righteousness, purification, and spiritual maturity or to get the Lord to manifest Himself; this is nothing less than conjuring. We are "crucified with Christ" by faith, (Galatians 2:20). If we crucify ourselves, it will only result in self-righteousness. This is pride in its most base form because it gives the appearance of wisdom and righteousness. In Colossians 2:18-23 Paul warns us about pride about how well we are doing compared to others, not on God's glory. This results in our putting confidence in discipline and personal sacrifice rather than in the Lord and His sacrifice. It is the motivation behind everything that we do that is important – driven by the Holy Spirit or a religious spirit. A religious spirit motivates through fear, guilt or pride and ambition. The Holy Spirit motivates through love for Jesus the Son of God.

In Colossians 2:18-19, Paul explains that people who delight in self-abasement will often be given to worshiping angels and taking improper stands on visions they have seen. A religious spirit wants us to worship anything or anyone but Jesus. God does not give us revelations to prove our ministry or so people will respect us more. The fruit of true revelation will be humility, not pride. A religious spirit will always feed our pride, whereas true spiritual maturity will always lead to increasing humility.

THE MARTYR SYNDROME

To be a true martyr for the faith is one of the greatest honors that we can receive in this life. When this is perverted, it is a tragic form of deception. When a religious spirit is combined with the martyr syndrome, it is almost impossible for that person to be delivered from his deception. At that point, any rejection or correction is perceived as the price he must bear to "stand for the truth" which drives him even farther from the truth and any possibility of correction. The martyr spirit can also be a form of the spirit of suicide. It is sometimes much easier to "die for the Lord' than it is to live for Him. Those who have a perverted understanding of the cross often glory more in death than they do in life. The goal of the cross is the resurrection, not the grave.

In Jude 12-16 we see the false prophets as "grumblers and fault-finders."

The world is becoming increasingly repulsed by religion. However, when Jesus is lifted up, everyone will be drawn to Him. Because the whole creation was created through Him and for Him, we all have a huge Jesus-size hole in our soul. Nothing else will ever satisfy us or bring us peace, except a genuine relationship with Him.[6]

Until the Vatican repents of their Lucifer worship, and both the Constantine and Nicean Creeds, which are Anti-Christ, Anti the God of Israel, Anti-Semitic, and full of the evils and falsehoods of Replacement Theology, then, their Priests will be plagued by evil spirits and their cult will be manifested demons. This concludes my discernment that the Vatican is clearly not the best representative for contact with Extraterrestrials, because as history proves, they will mistake the bad Aliens for the good ones, as the Spiritual Legal Ground is ripe for satan to have his foothold in their church government, as he always has. In other words, do not trust the Vatican or the Pope, especially Pope Francis, who behaves like the Antichrist.

If you are a Catholic, or you were born into a Catholic family, do not worry. As long as you repent to Jesus, you will be saved. But you have to repent from idolatry, because Jesus is the only one who ultimately has the authority to save your soul, deliver you from your generational iniquities, and ancestral transgressions, and break the curses of Catholicism over your life. Mother Mary is powerless to do so. Those who think she has this power, are greatly mistaken. I tested it once, which is the basis of my testimony in my future book, *CinderElla's Shadow*, my testimony that Jesus Christ is the Lord of lords, and Mother Mary was not given that title.

You see, what Constantine did was switch out the gods and goddesses of the Ancient Babylonian religion of Rome to incorporate Christianity into their already existing religion of Mithraism. They made Jesus the Sun god, and they switched Semiramis out for Mother Mary. Jezebel is the Whore of Babylon, and the Lord speaks directly to her in Revelations, 2:20-22:

> "To the Church in Thyatira: But I have this against you: You tolerate that woman Jezebel, who calls herself a prophetess. By her teaching she misleads My servants to be sexually immoral and to eat food sacrificed to idols. Even though I have given her time to repent of her immorality, she is unwilling. Behold, I will cast her onto a bed of sickness, and those who commit adultery with her will suffer great tribulation unless they repent of her deeds."
> (Revelations 2:20-22)

The Church of Thyatira originally located in Turkey was known as the Church that tolerated sin. It does not matter the location, it is about the Jezebel spirit ruling the church, because the Jezebel spirit rules over all false beliefs, false prophets, and is at war with Yahuah/Yeshua/Jesus, the Gods of Israel. The battle is real over this ancient demon.

Contrary to popular belief, the Jezebel spirit is a demon, and all demons are male in nature. Jezebel is not limited to attaching to women. Jezebel spirit was named that because of the evil queen Jezebel who tried to bring down the kingdom of Israel through her marriage to King Ahab. She was eventually thrown out the window by King Jehu, but never received a proper burial because she was eaten by dogs. This is why she became a *spirit*, but the demon that rules the Jezebel spirit existed long before the evil queen Jezebel. Therefore, all kinds of sexual immorality come under the influence and rulership of this ancient demon. As does, the Religious Spirits that exalt themselves through spiritual pride and spiritual arrogance against the knowledge of God, and the Truth in God's Words, and more importantly, wars against the Holy Spirit, the Spirit of Christ, and the last bastion of Grace of Planet Earth.

> "And I heard another voice from Heaven saying, "Come out of *her*, my people, lest you share in her sins, and lest you receive of her plagues."
> (Revelations 18:4)

We can conclude, as the Vatican's spiritual stronghold and partnership with the alien presence, that obeys Lucifer-Satan, the god of this world, who is the Commander of the Draconian Empire inside the Earth, rules the Vatican and the Catholic Church today. This is why he attacks God's people via the front door through governmental persecutions around the world, and attacks God's people via the backdoor through seductive, idolatrous cultures globally, while turning a blind eye to the sexual immoralities in the Church as the Vatican quietly settles multi-million dollar lawsuits worldwide, with the victims of pedophilia, through sexually perverted the youth, who turn to the Church and the Priesthood for safety and trust, yet are betrayed by both pedophile and homosexual priests, whose sexual lusts are unbridled, while covering up their abuses to the world, today's Pope Francis just lets them all go unrepentant. The Lord of lords takes this seriously and promises to avenge. All the enemies of God's people will finally receive retribution for misleading God's people into all kinds of falsehoods, abuses, and perversions. The Lord has His day, and that day is coming.[7]

The harlot is sometimes thought of as being a specific church or organization. The harlot will lose all her power and will then be completely destroyed. Revelations 17. In the Bible, the Body of Christ is referred to as the Bride, who are the redeemed and the ones made righteous through their faith in Christ. The Whore is always referred to as the false church. Both are references to women. One is good, and the other is evil.

THE NEPHILIM GIANTS IN STASIS

The giants in stasis, are not 200 who landed on Mount Hermon. They are said by all sources to have the ability to change their appearance through shape shifting. This is how they were able to mate with human women. The giants in stasis however are arraigned in a circle with their heads pointing to the middle. The fallen ones cannot be killed by humans, as they are supernatural. The giants are in small groups all over the world, which are called *hives,* and no one can come within a certain distance, because they can fall apart. These giants are arranged in a circle, to bolster their power. They are cannibals. They are as strong as they look, weighing approximately 2,000 lbs. or more. They are very smart and agile. Thought they can move very quickly they can never go to Heaven. They were not supposed to exist in the first place therefore the Creator did not designate a place for them in the afterlife. They are miscreants. [8]

They and their fathers will be put down when the Lord returns. According to the Book of Enoch, they are bound to the spiritual realms of the fourth density, condemned to be Watchers of humankind, and serve as evil spirits to torment us until the Lord returns to judge them. They are half human, half Annunaki, fallen angels.

The fathers of these hybrid children are not made in the human image. They are massive, gigantic. They are the ones that left their first abode as described in Jude 6. They are waiting for the day when they shall be judged, by the Heavenly Father's children, and the Heavenly Father Himself, with Jesus, who sits on the right hand of the Heavenly Father. They are some that have been wide awake for years, living in captivity behind titanium bars cared for by the Illuminati Elite in underground bases inside the Earth. Allegedly, according to anonymous sources, there are shamans who speak to them. They are inside a portal. No one crosses that line. They just talk to them and ask questions. Though they can be killed with solid copper bullets, they must be beheaded to prevent them from regenerating.

HEAVENLY CHARIOTS

Who and What were the wheels in Ezekiel 1? For those who have not read Book Two, *Who is God?* please refer to the Chapter Two: *Ancient Technology & Biblical Astronauts*, pp. 21-47, where I analyze the original Hebrew words used for aerospace technology and Divine Spaceships. [9]

Ezekiel's vision of the four wheels dramatically illustrates the omnipresence and omniscience of God. These wheels were associated with the *four living creatures* (Ezekiel 1:4), who were later described (Ezekiel 10:5, 20) as cherubim, angelic beings appointed as guardians of the holiness of God. Each wheel was actually two in one, with one apparently set inside the other at right angles which enabled the *living creatures* to move in any direction instantly without having to turn, like a flash of lightning. These wheels had the appearance of chrysolite, which may have been a topaz or other semiprecious stone. The outer rim of the wheels was described as high and awesome with the outer edge of the rims inset with *eyes*. (Ezekiel 1:14-18)

The Spirit of the living creatures were in the wheels (Ezekiel 1:20-21). As a result, the creatures were able to move any direction the wheels moved. Most biblical scholars hold to the idea that the Spirit of God gave direction to the wheels through direct knowledge of and access to the will of God. The mobility of the wheels suggests the omnipresence of God; the eyes, His omniscience; and the elevated position, His omnipotence. However, if we take the Biblical account literally, these Four Living Creatures were aliens, as I proved in great detail in *Who Are the Angels?* Book Three, see, Chapter Two: *The Angelic Government*, subsection *Who Are the Living Creatures?* [34-37].[10]

This vision appeared to Ezekiel as a powerful imagery of movement and action demonstrating the characteristics of God's Divine Nature. It presents God as being on a chariot-like throne, that also serves as starship, cloaked in His glory cloud, that is both supreme and immanent, existing in and extending into all of the created universe. As such, the whole revelation by God in this vision to Ezekiel, the cherubim, the chariot, the Spirit, and the wheels, emphasized their unity and coordination.

As terrifying as this vision was, it vividly displayed the majesty and Glory of God (Ezekiel 1:28), who visited Ezekiel and the children of Israel while in the midst of their Babylonian exile. It reminded them of His holiness and power as the Lord of all creation. The message was clear: though His people were in exile and their nation was about to be destroyed, God was still on His Throne and able to handle every situation. The lesson for us is that through His awe-inspiring providence, the God of gods moves through and often in the affairs of all nations to work out His own unseen Divine plan, always at work, intricately designed, never wrong, and never late. (Romans 8:28)

CHAPTER 17: REVIVED ROMAN EMPIRE

Notes:
1. The Revived Roman Empire – What is it?, https://www.compellingtruth.org/Revived-Roman-Empire.html
2. Ella LeBain, *Who is God?* Book Two, Chapter Twenty: The God of this World, pp. 311-318, Skypath Books, 2015.
3. Paul Carter, https://ca.thegospelcoalition.org/columns/ad-fontes/w11.ho-is-the-whore-of-babylon-and-why-does-it-matter/, June 14, 2018.
4. Kevin Cullen, More than 80 percent of Victims since 1950 were Male, Report says, https://archive.boston.com/globe/spotlight/abuse/stories5/022804_victims.htm, February 28, 20014.
5. The Associated Press contributed to this report. https://www.nbcnews.com/news/world/u-n-report-vatican-policies-allowed-priests-rape-children-n22531
6. Dr. C. Peter Wagner, *Freedom from the Religious Spirit*, Chosen Books (May 10, 2005).
7. *The Book of the Watchers*, Chapters 1-36, in the *Books of Enoch*, translated by Dr. Richard Laurence in 1821.
8. *What The Church Isn't Telling You About Nibiru And The Anunnaki*, In5D.com
https://newspunch.com/exposed-the-church-isnt-telling-you-about-nibiru-and-the-anunnaki/
9. Ella LeBain, *Who is God?* Book Two, Chapter Two: *Ancient Technology & Biblical Astronauts*, pp. 21-47, Skypath Books, 2015.
10. Ella LeBain, *Who Are the Angels?* Book Three, see, Chapter Two: *The Angelic Government*, subsection *Who Are the Living Creatures?* pp34-37, Skypath Books, 2016.

CHAPTER EIGHTEEN

NEMESIS & THE DEATH OF OUR SUN

"Immediately after the tribulation of those days the Sun will be darkened, and the Moon will not give its light, and the Stars will fall from Heaven, and the powers of the Heavens will be shaken."
Matthew 24:29

"And there will be signs in Sun and Moon and Stars, and on the Earth distress of nations in perplexity because of the roaring of the sea and the waves,"
Luke 21:25

There has been a rash of articles from 2018-2019 about the intention of some scientists, in China and around the world, to dim the Sun. All of these stories reeked of propaganda manipulations, almost false flag type stuff. Please explain to me how and why they want or even can dim the Sun? Those of us who know the truth, about the Sun Simulator, are not fooled by their dimming propaganda to cover up the fact that they there have been a Sun Simulator in place since around 2011.

The reason they put the Sun Simulator up was to cover the fact that the Sun is actually dying and dimming by itself. In or around 2011 the Artificial Sun Simulator, Flashlight Prism Sun lensing system, was installed to orbit our Earth between us and the Sun. It follows the Sun. It is followed by a jet. We have all seen it.

As a child of the 1960s and 70s, I remember growing up when our Sun was golden yellow. Now it is bright blinding white, like a flashlight. Now Sunglasses are not enough, now I absolutely have to drive with my visors down in the afternoon, because the Sun is so bright, it is literally blinding, obscuring everything, which can be dangerous while driving. I wonder how many accidents are caused by this bright blinding flashlight Sun? Here in Colorado, we are already a mile high into the altitude, 5,830 feet above sea level, and it is crazy bright on us especially when it is at an angle setting over the mountains.

So, all these lies spawning lately that they want to dim the Sun, then turn off the Flashlight Artificial Sun! Or at least tone it down! This past growing season, we have seen plant life literally fry from the scorching brightness of the Sun, despite the fact that the plants got plenty of water and were not dehydrated.

Something is not quite right with all of these articles. Here is my sense: it is kind of like picking up a message on your voicemail. It could be sitting for hours, even days, yet you first learn of it only when you retrieve the message. Similarly, the Lamestream Media, which has ignored just about everything on the Sun Simulator and constant daily chemtrailing, of multiple chemicals into our atmosphere, but has now been given new marching orders by the powers that be, who want to control the narrative, so they put out a twisted form of Disclosure through lies, propaganda and mind control, to prepare for another chemical being added to our upper atmosphere, trying to confuse those of us who know the opposite of what they claim, is truth.

The Sun is dying, it is dimming and cannot be fixed, except by God. That's why they installed the Sun Simulator to thwart the climate change, which was destined to send Earth into a mini-ice age, so they heated it up, through a combination of chemtrailing chemicals and the installation of an artificial Sun. That serves as a lensing system to obscure and cover up through cloaking technology the multiple objects around our Sun that are draining it from plasma and energy, like vampires. Suffice it to say, it is not a flawless system, depending on time of day and angles of light, it can reflect, refract, illuminate, or obscure the objects.

OUR SUN IS REPLACED
And according to the End Times Prophecies, God is not going to fix the Sun, but replace it with Himself!

> "But for you who revere my name, the *Sun of Righteousness* will arise with healing in its wings (rays). And you will go out and frolic like well-fed calves."
>
> (Malachi 4:2)

> "The Sun will no more be your light by day, nor will the brightness of the Moon shine on you, for the LORD will be your everlasting LIGHT, and your God will be your glory. Your Sun will never set again, and your Moon will wane no more; the LORD will be your everlasting LIGHT, and your days of sorrow will end."
>
> (Isaiah 60:19-20)

> "For *the LORD God is a Sun and shield*: the LORD will give grace and glory: no good thing will he withhold from them that walk uprightly."
>
> (Psalm 84:11)

> "Arise, shine, for your light has come, and the glory of the LORD rises upon you."
>
> (Isaiah 60:1)

> "But I will restore your health and heal your wounds, declares the LORD, because they call you an outcast, Zion, for whom no one cares."
>
> (Jeremiah 30:17)

> "The Sun shall be turned to darkness and the Moon to blood, before the day of the LORD comes, that great and magnificent day."
>
> (Acts 2:20)

Before the Sun gets replaced by the Son of Glory, who is Yeshua/Jesus Christ returning with His Kingdom of Heaven in the New Jerusalem, the Holy City that descends out of the Heavens and lands on a scorched Earth, as prophesied in Revelation 21, there will be a lot of trouble on the Earth. I do believe we are living in the last days of the End of this present Age. A time where Jesus warned about how the love of most will grow cold and wickedness will grow.

END TIMES PROPHECIES

> "Now as He sat on the Mount of Olives, the disciples came to Him privately, saying, "Tell us, when will these things be? And what will be the sign of Your coming, and of the end of the age?" And Jesus answered and said to them: "Take heed that no one deceives you. For many will come in My name, saying, "I am the Christ," and will deceive many."

> 'Immediately after the tribulation of those days the Sun will be darkened, and the Moon will not give its light; the stars will fall from Heaven, and the powers of the Heavens will be shaken. Then the sign of the Son of Man will appear in Heaven, and then all the tribes of the Earth will mourn, and they will see the Son of Man coming on the clouds of Heaven with power and great glory. And He will send His angels with a great sound of a trumpet, and they will gather together His elect from the four winds, from one end of Heaven to the other. "Now learn this parable from the fig tree: When its branch has already become tender and puts forth leaves, you know that summer is near. So also, when you see all these things, know that it is near—at the doors! Assuredly, I say to you, this generation will by no means pass

away till all these things take place. Heaven and Earth will pass away, but My words will by no means pass away.

"But of that day and hour no one knows, not even the angels of Heaven, but My Father only. But as the days of Noah were, so also will the coming of the Son of Man be. For as in the days before the flood, they were eating and drinking, marrying and giving in marriage, until the day that Noah entered the ark, and did not know until the flood came and took them all away, so also will the coming of the Son of Man be."

"Watch therefore, for you do not know what hour your Lord is coming. But know this, that if the master of the house had known what hour the thief would come, he would have watched and not allowed his house to be broken into. Therefore, you also be ready, for the Son of Man is coming at an hour you do not expect. "Who then is a faithful and wise servant, whom his master made ruler over his household, to give them food in due season? Blessed is that servant whom his master, when he comes, will find so doing. Assuredly, I say to you that he will make him ruler over all his goods. But if that evil servant says in his heart, "My master is delaying his coming," and begins to beat his fellow servants, and to eat and drink with the drunkards, the master of that servant will come on a day when he is not looking for him and at an hour that he is not aware of, and will cut him in two and appoint him his portion with the hypocrites. There shall be weeping and gnashing of teeth."
(Matthew 24:3-5; 29-39; 42-51)

THE EVENT
"The Event" from what I understand is supposed to be a Full Circumference Mass Coronal Ejection, a kind of solar sneeze of our Sun. The ancients referred to this event as the Solar Flash. It is what happens when stars die, they supernova, they expand out by flashing and belching solar rays of light. When this happens, it will end all life on Earth as we know it.

This is why the Bible Prophecy through the Revelation of Jesus Christ says, "Then I saw *"a new Heaven and a new Earth,"* for *the first Heaven and the first Earth had passed away*, and there was no longer any sea. (Revelation 21:1)

"For, behold, *I create new Heavens and a new Earth*: and the former shall not be remembered, nor come into mind.
(Isaiah 65:17-19)

"For *as the new Heavens and the new Earth, which I will make*, shall remain before me, saith the LORD, so shall your seed and your name remain."
(Isaiah 66:22)

"And I saw a great white throne, and him that sat on it, *from whose face the Earth and the Heaven fled away*; and there was found no place for them."
(Revelation 20:11)

"that the creation itself will be set free from its bondage to decay and brought into the glorious freedom of the children of God."
(Romans 8:21)

"But the day of the Lord will come like a thief. The Heavens will disappear with a roar, the elements will be dissolved in the fire, and the Earth and its works will not be found."
(2 Peter 3:10)

"But in keeping with God's promise, we are looking forward to a new Heaven and a new Earth, where righteousness dwells."
(2 Peter 3:13)

"The sky receded like a scroll being rolled up, and every mountain and island was moved from its place."
(Revelations 6:14)

"And He that sat upon the Throne said, *Behold, I make all things new*. And he said unto me, Write: for these words are true and faithful."
(Revelation 21:5)

THE UNSHAKABLE KINGDOM

"At that time, His voice shook the Earth, but now He has promised, "*Once more I will shake not only the Earth, but Heaven as well.*" The words, "Once more," signify the removal of what can be shaken— that is, created things— *so that the unshakable may remain*. Therefore, since we are receiving an *unshakable kingdom*, let us be filled with gratitude, and so worship God acceptably with reverence and awe, for our God is a consuming fire." [1]

(Hebrews 12:26-29-Berean Study Bible)

WHY IS THE SUN SO BRIGHT?

"And in those days the Sun shall be seen, and he shall journey in the evening on the extremity of the great chariot in the west. And shall shine *more brightly* than accords with the order of light. And many chiefs of the stars shall transgress the order (prescribed). And these shall alter their orbits and tasks, and not appear at the seasons prescribed to them."

(3 Enoch Chapter 80:5-6)

These are part of the *signs and wonders* Jesus Christ predicted would precede His return.

The approaching dwarf star, known as Nemesis, is part of a system of seven planets with 10 Moons. This is what the Revelations seven-headed Dragon with 10 crowns is referring to. Seven planetary Kingdoms that govern 10 Moons aka kingdoms.

The Truth is indeed Stranger than Fiction, and what has occurred in the past decade, is, the installation of a Solar Simulator into the Earth's upper ionosphere, which travels with the path of our Sun daily. Why would they do such a thing? Several reasons, one, they are trying to retard climate changes, warming, cooling patterns are all governed by the Sun. Secondly, it serves to obfuscate the objects in our Space around the Sun and around the Earth, i.e., spaceships around the Sun, Exoplanets, Exomoons, Solar Cores, and even our own Secret Space Technologies, the Solar Simulator being one of them.

The Solar Simulator blocks objects passing in front of the Sun from our viewpoint on Earth. But not necessarily from the viewpoint of Space, as we are seeing photos released from the International Space Station, through the European Space Agency, that reveal multiple objects around our Sun. The Helioviewer uploads photos from SOHO daily and are public domain.

The Sun Simulator is a giant prism-like flashlight that is situated between Earth and the Sun. It serves as an enormous flashlight. Think of what happens when a flashlight is shone in your eyes, you are blinded by the light. You cannot see past it. Everything you are seeing is now obscured, you lose clarity and focus.

Likewise, this is exactly what the Sun Simulator aka giant flashlight is doing by covering up objects and Exoplanets around the Sun, as it sits between us and the Sun. Its set into orbit to follow the path of the Sun. Every now and then, it separates, and we see what appears to be two Suns. One is our Sun, the other, sometimes is the Sun Simulator, aka prism.

This gigantic flashlight is designed to obscure Eclipses from the Exoplanets which seems to pass over our Sun much more frequently than our Moon eclipses the Sun, which happens like clockwork every six months. However, there's eclipses over our Sun weekly.

The Sun Simulator blocks the object passing in front of the Sun from our viewing. These are part of the signs and wonders Jesus Christ predicted would precede His return.

> The sight of the Sun is Life, and Life is Love. ~Chernihovsky

> The idea of more light, more beauty, more love possesses an irresistible magic.
> ~Ernest Dimnet

> I will tell you how the Sun rose, a ribbon at a time. ~ Emily Dickinson

> Light is Sweet. (Ecclesiastes 11:7)

THE SOLAR FLASH

> *All winged things are flying. They live when you have shone on them.*
> Poem to the Sun ~ Ancient Egypt.

These words in the Bible are important visual cues about the fluorescent bodies of angels, i.e., Extraterrestrials.

> "The appearance of the living creatures was like burning coals of fire or like torches. Fire moved back and forth among the creatures; it was bright, and lightning flashed out of it. The creatures sped back and forth like flashes of lightning" .... "Spread out above the heads of the living creatures was what looked something like a vault, sparkling like crystal, and awesome. Under the vault their wings were stretched out one toward the other, and each had two wings covering its body."
> (Ezekiel 1:13-14, 22-23)

Chosen people may look like that after solar flash, just as is described in the Bible.

Luke 17:24-25, "For the Son of Man in his day will be like the lightning, which flashes and lights up the sky from one end to the other. But first he must suffer many things and be rejected by this generation."

Isaiah 4:3-5, "Those who are left in Zion, who remain in Jerusalem, will be called holy, all who are recorded among the living in Jerusalem. The Lord will wash away the filth of the women of Zion; he will cleanse the bloodstains from Jerusalem by a spirit of judgment and a spirit of fire. Then the Lord will create over all of Mount Zion and over those who assemble there a cloud of smoke by day and a glow of flaming fire by night; over everything the glory will be a canopy."

There are so many fires and explosions in the world. Massive flooding, tornados, cyclones, hailstorms, landslides, lightning, and meteors.

THE SOLAR WIND IS HERE

Earth is caught in a stream of solar wind flowing faster than 500 km/s (1.1 million mph). This is causing geomagnetic unrest and intermittent auroras around the Arctic Circle. Solar wind speeds should remain elevated for the next 48 to 72 hours as our planet surfs two additional streams of solar wind in the offing. We are still deep within the Solar Minimum.

Solar Minimum means the Earth has less protection from outer space disturbance as our magnetic field lessens due to minimum Sun activity.

From the Coronal hole that facing Earth, we are under a massive solar wind attack. But during solar minimum, this natural heating mechanism subsides. Earth's upper atmosphere cools and, to some degree, can collapse.

Unique space weather effects become stronger during solar minimum. For example, the number of galactic cosmic rays that reach Earth's upper atmosphere increases during solar minimum. Galactic cosmic rays are high energy particles accelerated toward the Solar System by distant supernova explosions and other violent events in the galaxy.

Lamp of the World, light of the Universe. ~ Sylvester.

THE SOLAR SNEEZE AKA SOLAR FLASH

This will be followed by a few days of darkness before the Sun reignites to a higher frequency essentially moving itself and ALL life within its influence into the 4th dimension.

The solar sneeze should also cause an electromagnetic pulse destroying most, if not all, of the global electronic infrastructure along with the artificial intelligence infection at the root of our planetary issues created to control humans through a digital society and failed New World Order.

I have heard it called by many names; ascension, Revelations, evolution, graduation, quantum shift, galactic super wave or Leveling-Up. The Event, 200 years of Earth Purification – 2017 to 2217. Many will die by 2029-2046 during the passing of the Nemesis-Nibiru System interloping within our Solar System space, which is initiating shifts and massive changes. What if the ascension and the rapture are all precipitated by the Solar Sneeze? If the Sun is going to be replaced by the Son of Glory, then perhaps these events are somewhat related or at the very least, connected? Just something to ponder.

Imagine, says Plato, that the real world is like some people inside a cave. These people are chained and made to look at a wall since their childhood. Behind these people is the entrance of the cave and, outside the entrance, the real world. Between these chained people and the real world, there is a big fire. Therefore, the chained people cannot see the real world, but only their shadows/ representations as reflected by the big fire on the wall. Wouldn't they believe that these shadows are the real world, for they have seen only that? Imagine now, he continues, that someone manages to unchain himself and turn the head towards the outside world. He sees the truth, he gets to know the real side of things, but when he tells the other people, they do not believe him. This unchained man, for Plato, is the philosopher. That is why, Plato thinks that knowledge is all about recollection: we all have seen the truth in another parallel world, but only a few get to remember it.[2]

> "The final mystery is oneself. When one has weighed the Sun in the balance, and measured the steps of the Moon, and mapped out the seven Heavens star by star, there still remains oneself. Who can calculate the orbit of his own soul?"
> ~ Oscar Wilde

Love is the Sun. It does not choose the plants and animals upon which it shines. It does not warm only the worthy. It sends its light to all beings and all worlds.

> "Love your enemies and pray for those who persecute you, that you may be sons of your Father in Heaven. He causes His Sun to rise on the evil and the good and sends rain on the righteous and the unrighteous."
> (Matthew 5:44-45)

Notes:
1. Hebrews 12:26-29-Berean Study Bible, Biblehub.com
2. Plato, *Allegory of the Cave—Book VII of the Republic, Life Lessons on How to Think for Yourself*, https://www.mayooshin.com/plato-allegory-of-the-cave/, January 2018

CHAPTER NINETEEN

SOLAR SIMULATOR

"The Moon will be as bright as the Sun,
and the Sun will be seven times brighter – like the light of seven days in one!
*So, it will be when the LORD begins to heal his people,
and cure the wounds he gave them."*
Isaiah 30:26

The Solar Simulator, White Sun, Flashlight Sun, Prism Sun, probably started around 2011. The legends of vampires being affected by the Sun is probably about the end of days. There is also a correlation to when the Old Testament Prophet Daniel was put in the oven. King Nebuchadnezzar turned the heat up seven times, and that is when the Son of Man showed up. The fact that we now have a Sun Simulator that is placed between the Earth and the Sun is more evidence that the days are leading up to the coming of the Son of Man and the Sons of God returning to Earth in a divine appointment to complete their unfinished business with the LORD of Lords and KING of Kings, in a final battle we all know as Armageddon, which will actually be a Space War, the real Star Wars.

The artificial Sun Simulator is mounted on the nose of an anti-gravity, magnetic-propulsion space plane without wings. Most vehicles are white although some are yellow with faces. The tankers/carriers are huge and transport the others which vary in size and shape. Some are painted with the same snake-like pattern the UN paints their vehicles. They all blow CLOUDS to cloak themselves and others, otherwise known as Predator's Cloaking.

The Solar Simulator is a sophisticated piece of technology that was patented #US3325238A on June 13, 1967 by G. GEIER 3,325,238 and classified under the name: SOLAR SIMULATOR, originally with the US Patent office on June 4, 1963.[1]

United States Patent 3,325,238 SOLAR SIMULATOR: Inventor George Geier, Teaneck, NJ., assignor to Keulfel & Esser Company, Hoboken, N.J., a corporation of New Jersey Filed June 4, 1963, Ser. No. 285,4?6 2 Claims. (Cl. 35027) This invention relates to systems, methods, means, and devices for obtaining parallel radiant energy rays or light rays of uniform illumination intensity and uniform energy distribution over a region or space from radiant energy emitted by a radiant energy source, and refers, more particularly, to obtaining parallel radiant energy rays of uniform illumination intensity and uniform energy distribution and having a complete spectral range and which simulates solar radiation in outer space over a region or space on Earth.

Until very recently man had no particular need for simulating solar radiation in the sense of obtaining radiation from a radiant energy source which would give parallel light rays or radiant energy rays of uniform illumination and energy intensity having a complete spectral range and simulating the spectral range of Sun light. However, with the recent advances of the various disciplines of science, there has arisen a need for practical means for simulating solar radiation and consequently the need has arisen for systems, methods, means, and devices for obtaining from a radiant energy or light source, radiation simulating solar radiation particularly as it occurs in outer space.

THE CLAIM: 1. In a SOLAR SIMULATOR comprising a radiant energy source, combination of REFRACTIVE and REFLECTIVE means for respectively projecting inner zone rays and collecting outer zone rays from said source and focusing all said rays at a point in a focal plane and means for COLLIMATING said rays EMANATING from said focal point.

Definition of a collimator is a device that narrows a beam of particles or waves. To narrow can mean either to cause the directions of motion to become more aligned in a specific direction or to cause the spatial cross section of the beam to become smaller – a device for producing a parallel beam of rays or radiation. A Collimator is any a device for producing a parallel beam of rays or radiation. Today's Collimator is a small, fixed telescope used for adjusting the line of sight of an astronomical telescope.

The improvement comprising COLLIMATING means providing a large, uniformly illuminated area, said collimating means comprising:

(A) A first CASSEGRAIN COLLIMATING SYSTEM comprising a primary and a secondary REFLECTOR each having an axial APERTURE there through for intercepting and COLLIMATING said outer zone rays to form an annular outer illuminating zone, and passing said inner zone rays, said first system being axially disposed in alignment with and in the path of said emanating rays;

(B) A second CASSEGRAIN COLLIMATING SYSTEM having an outer diameter substantially equal to the greater aperture of said REFLECTORS of said first system and comprising a primary and a secondary REFLECTOR each having an axial aperture there through for intercepting and COLLIMATING an outer portion of said inner zone rays to form an annular intermediate illuminating zone contiguous with said outer illuminating zone, and passing the remaining inner portion of said inner zone rays, said second system being disposed at a greater distance from said focal point than said first system and in axial alignment therewith in the path of said emanating rays, and;

(C) A REFRACTIVE COLLIMATING SYSTEM having an outer diameter substantially equal to the greater aperture of said reflectors of said second system for intercepting and COLLIMATING said remaining inner portion of said inner zone rays to form an inner illuminating zone, contiguous with said intermediate zone, said refractive system being disposed at a greater distance from said focal point than said second system and in axial alignment therewith in the path of said emanating rays.[1]

ARTIFICIAL SUN AND PROJECT BLUEBEAM

All you have to do is go outside during midday, take some photos with your camera phone or any digital camera, and you will see many different anomalies, such as the petal effects, that seem to create a flower pattern in reds and pinks around the sun. Or, depending on time of day and the angle of the sun, you may just get a complete white out, and depending on the day, you may even see what appears to be two suns setting in the sky. One of them is the Solar Simulator that sometimes separates from its alignment to the real sun, and it causes a lot of confusion in analyzing photos, as it appears to be two separate light sources. I am a member of multiple Facebook Groups that are devoted to posting these types of photos daily, and without a doubt, there is something artificial covering the sun. My astronomer friends are convinced its fake sun simulators, established over the sun which is programmed to Project Bluebeam.

Project Bluebeam was created decades ago to 'beam' light onto the Blue sky for the purpose of obscuring its objects, which includes celestial objects and UFOs. I kid you not. Anyone who has researched this knows about Project Blue Beam. It's become popular on the religious fringe as it has experimented with counterfeit religious imagery beamed onto the sky to create a false flag Messiah scenario and/or manipulate religious imagery such as crosses, Christ figures, angels made of clouds, so earth humans think it's a sign.

Blue beam is somehow involved with the sun simulators to intentionally cover up planets, its moons and spacecraft around the sun. Chemtrails have been amped up since 2007 to create clouds to obfuscate views from earth. It is no coincidence that Chemtrails are done in the afternoons to crisscross the skies with artificial clouds to obscure the sunsets from revealing the exoplanets from the sun's reflective light.

This is not good. Our government is bereft in warning people. Everyone else who knows what is going on is telling people to get off the coastlines.

Astronomer Terral Croft is probably the most resourceful and credible researcher who is tracking the exoplanets and measuring earthquakes. He used to work for the government as a scientist.

He collects and tracks data daily on the earthquakes of the inner core and the movement of Nemesis and Nibiru systems. I highly recommend Terral as the best scientific resource for what is going on around our sun. He calls it the black star. [2]

Many people have commented on the phenomenon that is nothing more than the effects of Project Blue Beam that is being used as mechanism to prevent us from seeing the approach of Nibiru that most icy contents on Earth are seeing. The powers that have been using a combination of Chemtrails, the Sun Simulator that has a lensing system that works to obscure the incoming Nemesis-Nibiru system.

Between Project Blue Beam, the Sun Simulator's lensing system and the chemtrailing of Earth's skies worldwide, is being implemented by the elite in control of these technologies, as tool, to not only obscure the appearance of the Nemesis-Nibiru planets, Moons, and dwarf star, but also to prevent us from seeing UFOs that travel dialing along the trajectory of Earth's skies.

There are multiple reasons for chemtrailing the skies, one of which is the intention to keep humankind in an increasingly dense and limiting sensory perception as this approaching Nemesis-Nibiru system intersects and approaches closer to Earth. Chemtrails contain a toxic cocktail of barium, glass fibers coated with Nano Aluminum known as chaff, cadmium, chromium, nickel, dehydrated animal blood, mold spores, fungal mycotoxins, etc. We all need to stay tuned to this ongoing 'global' event and always question by focusing our discerning their fruits, that not everything is of extraterrestrial fruit, but is caused by the intentional manipulation by surface humans to confuse and control us.

Top Geophysicist with the U.S. Geological Survey Kenneth Hudnut stated, "At this point, we know that one GPS station moved (8 feet), and we have seen a map from GSI (Geospatial Information Authority) in Japan showing the pattern of shift over a large area is consistent with about that much shift of the land mass." National Institute of Geophysics and Volcanology in Italy, that estimated that "the 8.9-magnitude quake shifted the planet on its axis by nearly 4 inches (10 centimeters)." Astronomers concur that there has not been a shift in the Earth's rotational axis, but that there have been subtle polar shifts over the last 10 years. This is a change in what is called the figure axis.[3] The Inuit People conclude that Earth has actually changed its tilt.[4]

Contrary to the environmental narrative by conservationists and scientists, the Elders believe polar bear populations are increasing not decreasing, which causes bears to wander into the Inuit neighborhoods. They say as more bears find their way into Inuit communities, the more they are being traumatized by scientists, who tranquilize the bears to put radio collars around the bears' necks to track them for their research. The scientists also do other research which disturbs their natural life, and that requires them to spend most of their lives in almost total isolation and silence.[5]

Project Blue Beam certainly isn't a new idea or concept. In reality, it was thought up by powers, authorities, and rulers of the world since ancient times. Project Blue Beam is about using imagery through holograph projections to manipulate the minds of the public who witness it. What, you might ask, would the government or the powers that be, have use for this type of technology?

A staged alien invasion using holographic imagery that terrorizes the public, to justify their protection and ultimate control. The Hegelian Principle is about creating a false flag problem, that would cause the public to want more security, justifying more government and/or military control. In other words, giving up your freedoms, in exchange for security, usually ends up meaning you lose both.

Blue Beam has the ability to project images of planets, UFOs, even armadas. This is done to control the thoughts and sociopolitical management of surface humans. It can also project false messiahs, religious figures coming back in the sky, in order to manipulate certain cultures and nations to falsely believe their Mahdi, Messiah, Buddha, or Space Brotherhood of Ascended Masters have arrived to save them. In this case, Project Blue Beam is being used to project a false reality, by playing on the religious beliefs of humans.

Recently, Project Blue Beam is being used to obscure, obfuscate, and distort eclipses of exoplanets from the Nemesis-Nibiru system that works with a lensing system that was put in place through a sophisticated and patented Sun Simulator. This acts like a prism when used with an accompanied lensing system used to create the effect of lens flares across the sky, which confuses people as to what exactly they are seeing. It is also caused a great deal of dissension amongst researchers for quite some time, who were debunking each other's photos and videos, because of lens flares, and lens arrays.

Then a Nibiru researcher and YouTube channel known as Jeff P,[6] discerned that there was an intentional lensing system put into place, that was creating lens arrays, which are different than lens flares. Lens flares are lights that show up on photos and videos from the capture of light into the camera lens. However, lens arrays are quite different. It has nothing to do with anyone's individual camera equipment used to photograph the sky, but rather is caused by the prism Sun aka Sun Simulator and accompanied lensing system put into orbit in the upper layers of our atmosphere, that is programmed with the changing angles of light, to create refractions of light, reflections of light, and even to illuminate light. These lens arrays show up around the sky and have been captured in weather cams all around the planet, that look like huge red and blue planets. What is actually happening, is this lensing system is being used to obfuscate eclipses happening on a daily basis over our own Sun, by the

Nemesis-Nibiru system. Part of the lensing system actually works as a Cloaking Device.[6]

Another object approaching our Solar System shrouded in red iron oxide. The Red Dragon maybe imminent in a late August Fly-By, according to the YouTube channel, "Soul Evolution". A geological Pole Shift may occur in late August.

Most people think that planets cannot get too close to Earth without their gravity having a deleterious effect on the Earth. This is not true. Gravity is a much lesser force that Electromagnetism. As much as 10 to the 40th powerless.[7]

The Electromagnetic forces from our binary twin Nemesis will be the cause of the predicted Geophysical Pole Shift and not gravity from Nibiru or any other planet in that system.

Question: What force in the universe is the strongest or the most powerful?

Answer: There are four fundamental interactions that we know of. These four interactions account for all observed forces. These interactions are: Strong Nuclear, Electromagnetic, Weak Nuclear and Gravitational.

The strongest of these four is the Strong Nuclear. However, since this interaction operates at a very short range inside the nucleus (as little as 1 fm - which is a femto-meter or 10-15 meters) it is not the sort of force that we experience in everyday lives. This interaction acts between the nucleons (particles in the atomic nucleus - protons and neutrons which are made of quarks) and it is responsible for keeping the nucleus together. This interaction is mediated by so called gluons.

WHAT IS PROJECT BLUE BEAM?

Project Blue Beam was created decades ago to 'beam' light onto the Blue sky for the purpose of obscuring its objects, which includes celestial objects and UFOs. It's become popular on the religious fringe as it has experimented with counterfeit religious imagery beamed onto the sky to create a false flag Messiah scenario and/or manipulate religious imagery such as crosses, Christ figures, angels made of clouds, so Earth humans think it's a sign. This is why many fundamentalists on the religious fringe think every real UFO is Project Blue Beam. Whether this was intentional mind control, it nevertheless served to blind countless Conspiracy-oriented Funda-Mental Christians on the Fringe to deny the reality of UFOs and Exoplanets, who think it is all a hologram.

Project Blue Beam is a combination of advanced technologies which include satellite systems in outer space and terrestrial technologies like HAARP, SMAAC and HAMMER. The High-Altitude Meteorological Manipulation Energy Research (HAMMER) project is linked to HAARP, the High Frequency Active Auroral Research Program. Another separate program also used for weather warfare is called the Solar Magnetic Amplification and Causative Configurator or SMACC. Before the revelations of HAMMER and SMACC and their harnessing of solar energy for weather warfare, I was aware that the Large Hadron Collider at CERN had a side project called, S.A.T.A.N. SATAN is an acronym for the Solar Axion Telescopic AntenNa. It is believed that CERN used the SATAN antenna to create quantum Vortex experiments using solar axions. We've got one of the world's most sophisticated telescopes, named LUCIFER and an antenna named SATAN, is it any wonder why the Bible says, that Lucifer/Satan is the prince of the power of the kingdom of the air (space)? (Ephesians 2:2)

HAARP, SMAAC and HAMMER work in conjunction to harness solar plasma and use it for weather warfare. These technologies combined are capable of creating Earthquakes, tidal waves, fires and life-like holographic images that can be beamed up or down all over the planet. Each culture is meant to see an image of their god or deity corresponding to their specific religion. They have developed what they call, God's voice technology, which works telepathically, using a form of mind control, that will mislead unsuspecting masses into the false belief that they are having direct communication with their god. The so-called telepathic communication will be the most convincing part of the event. The God voice technology has been discussed at great length by DARPA scientist. The Hive Mind technology which also has been made public will convince the uneducated masses that their version of what they believe is god has now arrived.[7]

The Broadcasting of the God voice to the masses is planned to subdue the public so people allow and accept the enactment of the New World Order system. Before the arrival of their so-called "gods," the following events must take place:

1. Cataclysms such as massive Earthquakes, volcanic eruptions, floods, fires, hurricanes and meteors or fireballs must be seen and experienced.

2. World wars, nuclear wars, civil wars, and constant violence.

3. And this is the most important of all, the implanting of the RFID Chip into every human around the world. Without the implant of the RFID Chip, Lucifer will not be successful in overriding and gaining full control of the spirit, mind, and soul.

The idea planned is that the RFID chip would eliminate everyone's' free will choice which provide the Luciferians/Illuminati/New World Order with ultimate power over the spirit and mind of humankind through technology. That technology is now capable of thinking on its own and we know it as the (AI) artificial intelligence. Humanity would literally be in a real-life Matrix situation with an additional program, that being demonic energy possession, which will be traveling through every part of this Matrix. Every thought is compromised, and your free will is terminated.

WHEN ARTIFICIAL INTELLIGENCE RUNS AMOK

The Elite will use frequency waves directed to RFID chips just as it is used for cell phones through cell phone towers (5G's) now operational. This will provide you with memories, ideas, images, thoughts, that are remotely transmitted to your conscious and subconscious mind. Society will be under the Hive Mind. You will not be able to tell the difference between fantasy and reality nor will you be able to tell the difference between what is truthful or false. Forever being removed from God the creator.

The Illuminati (Luciferians) must announce their intent before completing their rituals. Whether you believe in this or not is no concern to them. They warn you by way of movies, music, art, announcements in newspapers and during public program announcements. They discuss it during public conferences, in government documents made available to the public, in their written books and personal interviews. The occultist does provide you with a warning of what is ahead, and they do not hold back on information concerning the spiritual aspect.

Through films like The Ring, 13 Ghosts, The Omen, Hell Raiser, The Exorcist, Legion, The Prophecy and Flatliners, they have provided you with the announcement. The physical aspect is shown in films and documentaries like What the Bleep do we Know, The Secrets of Quantum Physics, CERN Documentaries, Scientist Lectures, Resident Evil, Mad Max, Hunger Games, Blade Runner, The Dark Tower, Stargate, Star Trek, Star Wars, I Robot, AI, Terminator, Lucy, The Ghost in a Shell and so on.

We need to understand these so-called elites (Royals, Billionaires, Trillionaires, Satanists) have access to real history, real archeology, real future technology and have the means to keep all secret information hidden.[8]

The late Canadian journalist Serge Monast, a writer and poet who began investigating the New World Order secret plans in the 1990s, exposed NASA's Project Blue Beam. Serge was a member of the Social Credit Party of Canada, a Conservative populist political party that was created in the 1930s and dissolved in 1993. Monast exposed how NASA and the United Nations were working together with advanced black program technologies to simulate the Second Coming of Christ.[9]

Monast believed this would allow the shadow elite to fool the citizens of the world and provide them with the necessary breakdown of religious society in order to implement the New Age Religion that will become the One World Religion established by the New World Order. The plan is to have the Anti-Christ at the head of the New World Religion and give him all authority to govern according to its doctrine.

In 1954, the US military were interested in weather modification and documents surfaced outlining Project Stormfury for Hurricane modification. According to a declassified 258-page document entitled, *Weather and Climate Modification, Problems and Prospects*. There were brief summaries of the interests of the US military services in weather modification.

CHAPTER 19: SOLAR SIMULATOR

The document stated: "A portion of the United States Navy program in weather modification is devoted to the logistic and flight support of the joint N O A A Navy research program on hurricane modification (Project Stormfury). In addition, it supports studies of cloud and fog dissipation and the development of pyrotechnic seeding devices. The Office of Naval Research has for many years supported several basic research programs in atmospheric electricity, some of which are relevant to the possibility of artificially modifying lightning from thunderstorms."[10]

Interesting that the CIA recently declassified a paper on the Pole Shift. The 57-page booklet called "The Adam and Eve Story" authored by "Chan Thomas," was classified by the CIA at some point after its 1963 publication. Part of it was declassified in 2013.[11]

What stargazers today need to do is learn how to discern the sky and its astrophotography for objects, that includes UFOs; how to tell the difference between light refraction, light reflection, and when an object has become illuminated and exposed by the light; and how to distinguish them from photographic lens flares and sun dogs.

By definition, "Atmospheric phenomena: Halos, Sundogs and Light Pillars. Sundogs, light pillars, and other kinds of halos seen in the sky are atmospheric phenomena that occur when light, usually from a natural source, but also sometimes from ARTIFICIAL sources, is reflected or refracted by ice crystals in the atmosphere."

There is something artificial covering the Sun. All of my Astronomer friends are convinced it is fake Sun Simulators, established over the Sun which is programmed to Project Blue Beam.

According to my research, which includes reviewing hundreds, if not thousands of hours of weather cams, videos of the sky and video footage from SOHO, there is most certainly a 'system' in place to obfuscate Exoplanets and spacecraft, whether that belongs to the Secret Space Program or otherwise, and the holographic technology that is known as Project Blue Beam is somehow involved to work with the Sun Simulators to intentionally cover up eclipses of Exoplanets, its Moons and spacecraft around the Sun. Chemtrails have been amped up since 2007 to create clouds to obfuscate views from Earth. It is no coincidence that Chemtrails are done in the afternoons to crisscross the skies with artificial clouds to obscure the Sunsets from revealing the Exoplanets from the Sun's reflective light.

Our government is bereft in warning people of the true threat of this incoming Nemesis-Nibiru System intersecting within our Solar System, and crossing Earth's orbit around our Sun. Those who know what is really going on are telling people to get off the coastlines.

Notes:
1. United States Patent 3,325,238 SOLAR SIMULATOR, https://patents.google.com/patent/US3325238A/en
2. Astronomer Terral Croft, http://terral03.com

3. https://watchers.news/2015/03/08/inuit-elders-tell-nasa-Earth-axis-shifted/
4. New documentary recounts bizarre climate changes seen by Inuit elders. Originally published, October 19, 2010, Updated, May 2, 2018, https://www.theglobeandmail.com/arts/film/new-documentary-recounts-bizarre-climate-changes-seen-by-inuit-elders/article1215305/
5. Jeff P – Nibiru Watcher YouTube Channel, https://www.youtube.com/channel/UCi62JvN-lUn7hVL3ffofADA, *The Very Best Of Jeff P Proof of Planet X and Sun Simulator*, https://youtu.be/AK-EjmP-8cQ
6. https://www.usatoday.com/story/tech/2015/09/21/breakthrough-cloaking-technology-grabs-militarys-attention/72544510/
7. *What force in the universe is the strongest or the most powerful?*, www.physlink.com/education/askexperts/ae268.cfm
8. Blue Beam Project by NASA, http://politicalavengernews.com/index.php/2018/03/31/blue-beam-project-nasa/
9. Serge Monast, 1994, http://educate-yourself.org/cn/projectbluebeam25jul05.shtml
10. https://www.groundzeromedia.org/9-13-18-hammer-smacc-weaponizing-natures-fury-w-christopher-fontenot/
11. Massive Pole Shifts are Cyclic According to Declassified CIA document, https://www.Exoplanets.org/massive-pole-shifts-are-cyclic-according-to-declassified-cia-document/

CHAPTER TWENTY

INTERLOPERS IN OUR SOLAR SYSTEM

And therefore, as a stranger give it welcome.
There are more things in Heaven and Earth, Horatio,
Then are dreamt of in your philosophy.
(Hamlet Act 1, scene 5, 159–167, William Shakespeare)

OUR SUN'S BINARY TWIN
The Earth is currently in the grip of two magnetic fields – the Sun's dominant field and the approaching field of Planet X.[2]

Where it is true that the intense electromagnetic fields generated by high power lines can affect human health, such as those living under high tension wires developing cancer at a higher rate than normal, this does not result due to a changing planetary magnetic field. The intensity of the field, for those creatures crawling about on the surface, remains the same regardless of which direction the magnetic particles flow. Certainly, confusion reigns in those migrating birds and whales using the field for direction. Havoc will certainly be wreaked in man's technology.

The compass has already become erratic, with alternate guidance systems being used in planes as a consequence. The degree to which man's technologies rely on a general N/S direction holding steady will become apparent as satellite and navigation systems fail. Submarines, which are essentially blind under the waves, are particularly susceptible. Planes with poor visibility and satellites without manual controls also are essentially blind and rely primarily on the compass for positional stability. Electromagnetic motors are subject to electromagnetic pulse, which can sometimes emerge when magnetic fields are suddenly switched around, creating crowded particles in crunch points during the switch. These pulses are known to destroy electronic equipment, when they occur, and little of mankind's technology is shielded from these pulses as they are rare and unforeseen in the normal day-to-day operation of equipment.[1]

These are the primary consequences of clashing or merging magnetic fields, and the contortions that occur before they are merged or joined to the extent of causing the lean to the left and three days of darkness foretold. By this time, the Earth will be in a severe wobble, caused by the daily tugging on the magnetic poles that present to the influence of Planet X – the wobble of today, exacerbated. In this regard, the steady approach of Planet X, coming ever closer so the influence of its magnetic field is ever stronger, does equate to more violent Earth changes.

The Earth torque is one example, where the magnetic S Pole is tugged back each day, creating more stress on the plates in the Pacific and allowing the Atlantic to widen. The wobble is another such example, creating more erratic and violent weather, storms, and weather extremes. These will continue to increase, not because the fields are merging and clashing, but because Planet X continues to close the distance between itself and Earth, moving toward the hapless Earth during its passage, and thus the tugging on Earth will increase in its violence.

What the clashing and merging fields mean for Earth is that it now has more than a torque and a wobble, it has a bully in the neighborhood that will eventually supplant the Sun's influence. Supplanting the Sun's magnetic dominance, combined with the increasing 270° roll that will place the N Pole of Planet X pointing directly at Earth, is what allows the Pole Shift drama to commence. These merged fields are natural for two planets forced close together, in such a setting. [2]

NIBIRU AND THE SECRET SPACE PROGRAM

In recent years, we have had multiple whistleblowers come forward sharing their experiences and eye-witness accounts of a secret space program that has been going on since the end of World War II in America. One witness was not exactly a whistleblower, by definition; he was just doing research and accidentally hacked into the secret space program's Solar Warden. It consists of deep spaceships that patrol this Solar System. But why? I realize some of you might be thinking, how can that be? Please see my Chapters on the Secret Space Programs and Solar Warden.

I'm so convinced that this has been going on for over 70 years, and without spending pages on proving that to you, as there's a plethora of information on the Disclosure of the *Secret Space Program*, available by several credible authors, such as Dr. Michael Salla[3], who has a 5-book series called the Secret Space Program Book Series, that includes research from multiple whistleblowers detailing this 70-year clandestine program that most know, as the UFO Coverup, which I want to refer you to. I am going to be touching upon the Secret Space Program in how it relates to the U.S. Space Force in keeping with my theme, that the *Truth is Stranger than Fiction*, while discerning Solar Warden later in this book. But suffice it to say, that it is real, and it is no longer the kind of conspiracy it once was but has become a reality that has been going on beneath our noses and our feet for decades.

Notes:
1. Kojima's Blog, *EMP Caused by Planet X*, http://poleshift.ning.com/profiles/blogs/emp-caused-by-planet-x, March 18, 2017
2. Nancy Leider, *ZetaTalk: Magnetic Clash*, http://www.zetatalk.com/index/zeta300.htm, July 1, 2006.
3. Michael Salla, PhD., The Exopolitics Institute, https://www.exopolitics.org

CHAPTER TWENTY-ONE

POLE SHIFT

"Therefore, we will not fear,
Even though the Earth be removed,
And though the mountains be carried into the [b]midst of the sea;
Though its waters roar and be troubled,
Though the mountains shake with its swelling."
Psalm 46:2-3

At the time of this publication, the magnetic north pole of Earth has shifted approximately 40 degrees and has moved from the Canadian Arctic and has now crossed continents to Siberia. June 21, 2020, record high temperatures of 100 degrees F were recorded in Siberia.[1]

Psalm 46 is known as the Psalm for Protection, which is an affirmation of God being our refuge and an ever-present help in times of trouble. It also describes a pole shift in verses 2 and 3 above. That's exactly what goes on during axis pole shifts, resulting in displacement of the oceans, as water takes over the land, the mountains end up in the water, everything is shaken and the ground from underneath you, is removed and moved. But those who trust in the Lord are saved, maybe not so much in their physical life, but certainly in their afterlife, which is not promised to the wicked.

Insiders say the Earth's poles are shifting 42 miles per day due to Planet X. Extinction level event Nibiru is here. Predicting waves three miles high from the coasts around the world as it passes.

Sun is pulling it in. If it is behind the Sun, there it is, it is going to orbit around the Sun then go back out. Sun pulls it in like a magnet does not collide with the Sun but disrupt everything. See, Wendell Stevens, Pleiadeans about Nemesis System.

It is its own system, seven planets and moons, with debris around it, made of iron, average debris is 72 lbs., millions of degrees hot -- A third of the Earth is burned.

Nibiru is five times the mass of Earth. 70 lbs. of iron rocks come down upon the Earth like Hail everything will be destroyed. Earth's tectonic plates are being pulled like a magnet as Nibiru passes seven planets and one brown dwarf sun.

Insider said it was fuzzy dust ball then got bigger and bigger, then the CIA pulled the plug. Sea levels rise as high as six hundred as the Earth crust shifts.

PROPHET ISAIAH ON POLE SHIFT

Interesting that the Old Testament Prophet Isaiah has a name in Hebrew, Yeshayahu, that means, *The LORD Yahuah is our Salvation*. Old Testaments Prophets were messengers of Yahuah. Isaiah Chapter 24 spoke about an Earth wobble, that is clearly outstanding and different from the Earth's natural wobble on its axis, that is much more dramatic, devastating, and rapid than what naturally occurs. A pole shift was predicted – Earth rotates 180 degrees, and literally goes upside down.

> "Behold, the LORD makes the earth empty, and makes it waste, and turns it upside down, and scatters abroad the inhabitants thereof."
>
> (Isaiah 24:1)

Then in verse 18 Isaiah forewarns of "fearful noises" that will be heard while the Earth is wobbling and turning upside, as this all happens, the sky opens up.

> "And it shall come to pass, [that] he who flees from the noise of the fear shall fall into the pit; and he that cometh up out of the midst of the pit shall be taken in the snare: for the windows from on high are open, and the foundations of the earth do shake."
>
> (Isaiah 24:18)

Then Isaiah tells us that this Earth wobble becomes excessive:

> "The earth is utterly broken down, the earth is clean dissolved, the earth is moved exceedingly. The earth shall reel to and fro like a drunkard and shall be removed like a cottage; and the transgression thereof shall be heavy upon it; and it shall fall, and not rise again."
>
> (Isaiah 24:19-20)

Isaiah foresees world leaders, who are called the kings of the earth, will hide inside the Earth during this event. We know that underground military bases have been built all around the world, as well as whole luxury communities and buildings that are built upside down underground exist all around the planet, hidden inside mountain ranges and underground.

> "And it shall come to pass in that day, [that] the LORD shall punish the host of the high ones [that are] on high, and the kings of the earth upon the earth. And they shall be gathered together, [as] prisoners are gathered in the pit, and shall be shut up in the prison, and after many days shall they be visited."
>
> (Isaiah 24:21-22)

When does Isaiah tell us this will it happen?

> "Then the moon shall be confounded, and the sun ashamed, when the LORD of hosts shall reign in mount Zion, and in Jerusalem, and before his ancients gloriously."
>
> (Isaiah 24:23)

Isaiah's prophecy corroborates with St. John's Revelation of Jesus Christ, when the Sixth Seal of Cosmic Disturbances is opened, during the time when the LORD returns to Zion.

> "I looked when He opened the sixth seal, and behold, there was a great earthquake; and the sun became black as sackcloth of hair, and the moon became like blood. And the stars of heaven fell to the earth, as a fig tree drops its late figs when it is shaken by a mighty wind. Then the sky receded as a scroll when it is rolled up, and every mountain and island was moved out of its place. And the kings of the earth, the great men, the rich men, the commanders, the mighty men, every slave and every free man, hid themselves in the caves and in the rocks of the mountains, 16and said to the mountains and rocks, "Fall on us and hide us from the face of Him who sits on the throne and from the wrath of the Lamb! For the great day of His wrath has come, and who is able to stand?"
>
> (Revelation 6:12-17)

WHO SURVIVES ON EARTH?

Edgar Cayce talked about this event. Great lakes end up in Gulf of Mexico. Al Bielek described when he went to the future during the Philadelphia Experiment. Andy Basiago saw Washington DC was under a hundred feet of water during Project Pegasus. See, Book One of *Who's Who in the Cosmic Zoo?* Chapter: Life on Mars – Project Pegasus.[2]

Insider saw Blazing hot ball of fire coming through our Solar System. Asteroids five hundred miles across.

Meteorites in the tail of this Solar System. Insider says meteorites extend Millions of miles in the tail and Nibiru will come 14-20 million miles to Earth. It is going to flip us upside down. Earth Pole Shift. See Isaiah 24, the Lord literally turns Earth upside down, which is Biblical language for a Pole Shift. Crust shifts, but core stays the same.

> "For the windows of Heaven are open, and the foundations of the Earth are shaken. The Earth is utterly broken apart, the Earth is split open, the Earth is shaken violently. The Earth staggers like a drunkard and sways like a shack. Earth's rebellion weighs it down, and it falls, never to rise again. In that day, the LORD will punish the host of Heaven above and the kings of the Earth below. They will be gathered together like prisoners in a pit. They will be confined to a dungeon and punished after many days. The Moon will be confounded and the Sun ashamed; for the LORD of Hosts will reign on Mount Zion and in Jerusalem, and before His elders with great glory."
>
> (Isaiah 24:18-23)

RED IRON OXIDE DUST CONTAINS CYANIDE

Warning: the red iron oxide dust is Poisonous because it contains Cyanide! It attaches to the iron within this protein. The binding of cyanide to this enzyme prevents transport of electrons from cytochrome c to oxygen. As a result, the electron transport chain is disrupted, meaning that the cell can no longer aerobically produce ATP for energy. Tissues that depend highly on aerobic respiration, such as the central nervous system and the heart, are particularly affected. This is an example of histotoxic hypoxia. The most hazardous compound is hydrogen cyanide, which is a gas that kills by inhalation.

Red dust, a red Moon, red skies at Sunrise and Sunset and even a halo around Venus have been observed and, on occasion, reported in the media - all in the last 10 years; what's causing this and why is it happening now?

Accounts of red dust during past upheaval on Earth are brought to mind, such as rivers turning red during the Jewish exodus and prophecies of future events, as in Revelations where red is a theme throughout.

For Example:

> "And I beheld when he had opened the sixth seal, and, lo, there was a great Earthquake; and the Sun became black as sackcloth of hair, and the Moon became as blood."
>
> (Revelations 6:12)

Red dust comes from the massive tail that trails behind Planet X during its travels. The tail, which contains Moons, boulders, gravel, dust, and gases is the reason why Planet X is described by witnesses during a pole shift as a dragon, a mythical creature with a long, twisting tail. The tail originates from when the Solar System itself was forming, but mainly from trash picked up after the asteroid belt was created; each passage causing more and more debris to be captured in Planet X's gravitational and magnetic pull. The red dust is of course the reason why Planet X appears to be red.

The dust is charged and so does not drift down to the surface of Planet X due to gravity but rather is influenced by the planet's magnetic field which keeps the dust suspended in the air. Continued light dusting, red skies and red Moons can be expected up until the last weeks when the heavy red dust occurs, turning the rivers and ponds red and giving the water a bitter taste. The blood red dust of legend, turning ponds and rivers blood red, occurs closer to the time of the pole shift as the iron oxide dust is charged, and thus clings to the magnetic giant – Planet X – which is at the center of the magnetic attraction. At the far ends of the tail lie debris not so heavily charged, and this does not have the preponderance of the blood red colored iron oxide dust. As with other atmospheric changes – Moon and Sun halos and neon clouds and an increase in upper atmosphere lightning – the increased dust from the tail will be explained away by the media, which is controlled by the elite who are desperate to forestall rioting and panic in the people.

The Shift is increasing so fast that Earth's north magnetic pole is now shifting 34 miles per year! That is almost 3 miles a month! (2.8 miles) That is almost 500 feet a day. This will continue to increase and speed up. But it is already moving so fast that in about four years it will reach its weakest magnetic position (40 degrees mark). Remember that means at that point the Pole Shift is imminent.[3]

NIBIRU AND POLE SHIFT

An unnamed CIA whistleblower, whose testimony has made its rounds among YouTubers, claims that while he is not an Astronomer, he did, in fact, see Nibiru through the Hubble Telescope and described it as a red-orange hot star with a huge debris tail sending iron oxide dust into space around its enormous tail. He said Astronomers have been tracking it since the 1930s through 1960s and mentioned it was reported in Popular Science and Astronomy magazines. I looked up his claims, and sure enough, I found articles referencing Planet X and the Death Star in past Popular Science magazines that published articles referencing Planet X: August 1932, p.21; October 1955, p. 174; July 1972, p. 42; May 1985, p.99 on the Death Star.

The Whistleblower said the government has known about it for decades, but when they realized it was going to flip the planet, they kept it quiet and took steps to cover it up so as not to alarm the public. He said it was going to cause a Pole Shift. Its effects would be:

- Sea levels rise 500-700 ft.
- Earth crustal movement
- Magnitude 12+ Earthquakes
- Massive Tornadoes
- Volcanic Eruptions
- EMP (Electro-Magnetic Pulse)
- Tsunamis of 100-1000 ft.

This would be followed by the debris tail of Nibiru, and its thousands of asteroids. The entire power grid will go down, no internet or cell phone services will be available, and all bridges, tunnels and roads would be destroyed caused by 5k shock waves around the Earth.

A 700 ft. tidal surge will take over coastlines flooded by pole meltdown, tsunamis, and hurricanes.

Nibiru according to Nostradamus: "The Divine word will be struck from the sky, one who cannot proceed any further, the secret closed up with the revelation, such that they will march over and ahead." (Quatrain 2-27)

EARTHQUAKES IN THE BIBLE

"And there was a great Earthquake, such a mighty and great Earthquake as had not occurred since men were on the Earth."
(Revelation 16:18)

"This Earthquake may cause the splitting of the Mount of Olives in two when Jesus descends to Earth."
(Zechariah 14:4)

"You will be punished by the LORD of hosts with thunder and Earthquake and great noise, with storm and tempest and the flame of devouring fire."
(Isaiah 29:6)

"shall go into the holes of the rocks, and into the caves of the Earth, from the terror of the LORD and the glory of His majesty, when He arises to shake the Earth mightily."
(Isaiah 2:19)

"There will be famines, pestilences, and Earthquakes in various places."
(Matthew 24:7)

The 8.2 and 7.1 magnitude Earthquakes that have just struck days apart are two of the biggest deepest Earthquakes we have ever seen. "This is unprecedented, both the sides of the planet moving like this." "Phenomenal beyond measure." [4]

There are a number of further warnings about the sudden and unexpected fulfillment of Bible prophecies:

> "But the day of the Lord will come as a thief in the night."
> (2 Peter 3:10)

> "Remember therefore how you have received and heard; hold fast and repent. Therefore, if you will not watch, I will come upon you as a thief, and you will not know what hour I will come upon you."
> (Revelation 3:3)

> "Behold, I am coming as a thief. Blessed is he who watches, and keeps his garments, lest he walk naked and they see his shame."
> (Revelation 16:15)

> "Therefore, you also be ready, for the Son of Man is coming at an hour you do not expect."
> (Matthew 24:44)

> "But you, brethren, are not in darkness, so that this Day should overtake you as a thief. ... Therefore, let us not sleep, as others do, but let us watch and be sober."
> (1 Thessalonians 5:4-6)

The fall of the Babylonian system extant at the end time will happen quickly and suddenly: "Her plagues will come in one day—death and mourning and famine. ... In one hour, your judgment has come. ... In one hour, such great riches came to nothing. In one hour, she is made desolate."
(Revelation 18:8, 10, 17, 19)

In the last book and the last chapter of the Bible, Christ stated, "Surely I am coming quickly."
(Revelation 22:20)

The ancient prophet Isaiah issued a similar warning for the end times. In a message that applied first to ancient Jerusalem and then to Jerusalem prior to the return of Christ, the prophet said,

> "You will be punished by the LORD of hosts with thunder and Earthquake and great noise, with storm and tempest and the flame of devouring fire."
>
> (Isaiah 29:6)

During the "day of the Lord" (Isaiah 2:12), a time of judgment upon the nations when Christ returns, Isaiah said the people "shall go into the holes of the rocks, and into the caves of the Earth, from the terror of the LORD and the glory of His majesty, when He arises to shake the Earth mightily." (Isaiah 2:19, compare verse 21)

An Earthquake Analogy:

To encourage Christians to remain faithful to God, the writer of the book of Hebrews referred to a prophecy by the Old Testament prophet Haggai, where God said, "Once more (it is a little while) I will shake Heaven and Earth, the sea and dry land; and I will shake all nations, and they shall come to the Desire of All Nations [Christ], and I will fill this temple with glory." (Haggai 2:6-7)

Although this prophecy in Haggai was directed toward the temple that was being rebuilt by the Jews, the author of the book of Hebrews used this concept of *shaking* to encourage us not to refuse God's instruction as we shall "not escape if we turn away from Him who speaks from Heaven." (Hebrews 12:25)

Continuing, the passage in Hebrews states, "But now He has promised, saying, 'Yet once more [just before Christ's return] I shake [margin: will shake] not only the Earth, but also Heaven.' Now this, 'Yet once more,' indicates the removal of those things that are being shaken, as of things that are made [physical things], that the things which cannot be shaken may remain." (verses 26-27)

The point of this comparison is that we should firmly focus on "receiving a kingdom [the coming Kingdom of God] which cannot be shaken" by serving God "with reverence and godly fear" (Hebrews 12:28). This coming Kingdom will stand forever because it cannot be shaken or destroyed. (Daniel 2:44) This is the Kingdom Christ will establish at His coming and the end of the age.[5]

POLE SHIFT IN THE BIBLE

> "This is what the Sovereign Lord says to Tyre: *Will not the coastlands tremble at the sound of your fall*, when the wounded groan and the slaughter takes place in you? Then all the princes of the coast will step down from their thrones and lay aside their robes and take off their embroidered garments. Clothed with terror, they will sit on the ground,

trembling every moment, appalled at you. Then they will take up a lament concerning you and say to you:

"How you are destroyed, city of renown, peopled by men of the sea! You were a power on the seas, you, and your citizens; you put your terror on all who lived there. Now the coastlands tremble on the day of your fall; *the islands in the sea are terrified at your collapse.*' "This is what the Sovereign Lord says: When I make you a desolate city, like cities no longer inhabited, and when I bring the ocean depths over you and its vast waters cover you, then I will bring you down with those who go down to the pit, to the people of long ago. *I will make you dwell in the Earth below, as in ancient ruins, with those who go down to the pit, and you will not return or take your place in the land of the living.* I will bring you to a horrible end and you will be no more. You will be sought, but you will never again be found, declares the Sovereign Lord."

(Ezekiel 26:15-21)

"The shorelands will quake when your sailors cry out. All who handle the oars will abandon their ships; the mariners and all the sailors will stand on the shore. They will raise their voice and cry bitterly over you; they will sprinkle dust on their heads and roll in ashes. They will shave their heads because of you and will put on sackcloth. They will weep over you with anguish of soul and with bitter mourning. As they wail and mourn over you, they will take up a lament concerning you:

"Who was ever silenced like Tyre, surrounded by the sea?" When your merchandise went out on the seas, you satisfied many nations; with your great wealth and your wares you enriched the kings of the Earth. Now you are shattered by the sea in the depths of the waters; your wares and all your company have gone down with you. All who live in the coastlands are appalled at you; their kings shudder with horror and their faces are distorted with fear. The merchants among the nations scoff at you; you have come to a horrible end and will be no more.'"

(Ezekiel 27:28-36)

BIBLE PROPHECY MATTERS

All of us need to take the warnings of the Bible seriously. We must be careful not to make the mistake of rejecting any of its teachings because they do not agree with our human reasoning—our opinions, our ideas, and our beliefs.

Bible prophecy shows that earthquakes, along with other major calamities, will strike the Earth in the not-too-distant future. Repentance and reconciliation to the Lord, will save souls for the Kingdom of Heaven which is coming to Earth in the Age to Come.

God takes no pleasure in the suffering humans bring on themselves (2 Peter 3:9). Instead, He wants us to choose the way of joy and abundance that comes with a reconciled relationship with Him, as He offers to those who seek Him and live His way of life. (Isaiah 2:2-4; 11:9)

POLE SHIFT AND PROJECT BLUE BEAM

Pole shifting has already occurred on the polar caps of the North Pole, which the Inuit Peoples have discerned from their navigation system by stars and landmarks across their land.[6]

The Inuit Tribe are indigenous people who live in Alaska, the Canadian Arctic, Greenland, and Siberia. The Inuit Elders wrote to the National Space and Aeronautics Administration (NASA), telling them that the Earth's axis has indeed shifted. The Elders do not believe that CO2 emissions from humans are the cause of today's climate changes. The Inuit Elders declared, "the sky has changed."

The Inuit elders noticed climate change in the disappearing sea ice, melting glaciers, deterioration of sealskin and burns on seals. They say these changes are attributed to changes in the sky. The tribal elders claim that the Sun no longer rises where it used to rise, which indicates a shifting of the poles and Earth's axis. They observed the days heat up more quickly and last longer. The stars and Moon are also in different places in the sky, and this affects the temperatures. Inuits rely on the placement of the Moon and stars for their survival as they live in total darkness during part of the year.

The elders say they can no longer predict the weather like they used to in the past. They observe that warmer winds are changing the snowbanks, making their ability to navigate over land more difficult.[6]

BEHOLD A PALE GREEN HORSE

Revelation 6:7-8 is typically translated as: "And I looked, and behold, a pale horse! And its rider's name was Death, and Hades followed him. And they were given authority over a fourth of the Earth, to kill with sword and with famine and with pestilence and by wild beasts of the Earth."

This is a mistranslation. In the original Hebrew text, the color of the horse is not *pale* but rather *chloros*, the Greek word for *green*, sharing the same root word as *chlorophyll*, the molecule of photosynthesis for all green plants on our planet."[7]

Environmentalism is the new Green Religion, Gaia Worship, Green New Deal, etc. All efforts in futility and idolatry.

The Elite and the wealthy of the world are busy building underground luxury bunkers, communities to simulate day and night, 26 feet below the surface. But what happens if and when there's underground earthquakes that will destroy them? Or, what happens when their power source is cut off? Who are they going to turn to?

This is what the Ancient Prophet Isaiah had to say about them:

> "On that day, the LORD will punish the host of Heaven, in Heaven, and **the kings of the Earth, on the Earth**. They will be herded together like prisoners bound in a dungeon; **they will be shut up in prison and be punished after many days**. Then the Moon will be confounded and the Sun ashamed, for the LORD of hosts reigns on Mount Zion and in Jerusalem, and his glory will be before his elders."
> (Isaiah 34:21-23)

The whole Global Warming propaganda was to get people to believe the Earth was warming in order to justify the elite using a Sun Simulator to warm up the Earth. If it warmed too much, they have the plausible deniability to say it was due to Global Warming, which during warming trends, during summers, it can appear that way. Climate change encompasses both extremes, until the wobble settles down, which is not happening anytime soon.

However, the deep and disturbing truth here, is, if they turned off the Simulator, or it got damaged, the Earth would cool. The Sun is dying, the Sun is dimming, we are missing a Sunspot Cycle which loops every 11 years. Not happening now. Anyone who seriously studies Earth science, must take into consideration that the Elite are manipulating the narrative and the data in order to cover up the truth about the Nemesis-Nibiru System and the real situation with our Sun.

The Global Warming propaganda campaign was perpetuated by Al Gore, to Cover Up the real inconvenient truth that the Earth is cooling, the Sun is dying, and there's another Solar System intersecting within our Solar System, which is not only going to flip our Earth and complete its Pole Shift, but there's also going to be cataclysmic events that will lead to a new Earth. See, Revelations 21:1;4.

The Earth is being transformed. Nothing we can do about it. There is another Solar System intersecting with ours. We are seeing it every day, all over our planet.

The Quintessential Ultimate End Times Prophecy is when the Creator LORD, God Almighty, creates a new Heavens and a New Earth.

CHAPTER 21: POLE SHIFT

A New Heaven and a New Earth:

> "Then I saw "a new Heaven and a new Earth," for the first Heaven and the first Earth had passed away, and there was no longer any sea. I saw the Holy City, the new Jerusalem, coming down out of Heaven from God, prepared as a bride beautifully dressed for her husband. And I heard a loud voice from the throne saying, "Look! God's dwelling place is now among the people, and he will dwell with them. They will be his people, and God himself will be with them and be their God. 'He will wipe every tear from their eyes. There will be no more death' or mourning or crying or pain, for the old order of things has passed away."
>
> (Revelations 21:1-4)

THE NORTH MAGNETIC POLE HAS MOVED TO SIBERIA

I want to end this Chapter the way it began, with the recent news of the north magnetic pole that is rapidly moving eastward to Siberia, Russia, which sounds alarming. Here is what NBC4's Ben Gelber explained on February 15, 2019, and what that means:

> "The term "polar wander" accounts for the constant gradual movement of Earth's magnetic poles in relation to the planet's axis of rotation. This is different from the North Pole we study about in geography class, which resides beneath the Arctic Ocean at 90 degrees north, mostly in a fixed position, and currently 4 degrees north of the North Magnetic Pole.
>
> The Earth does wobble a bit on its spin axis, causing the geographic North and South Poles drift slightly, which scientists think is correlated to melting ice sheets and glaciers, which shift the balance of mass. But the quickening pace of the normally wandering north magnetic pole has caught the attention of NOAA and academic researchers.
>
> "Over the last 100 years or so, it has migrated from northern Canada towards Siberia," said Wendy Panero, an Ohio State University professor in the School of Earth Sciences.
>
> This has raised concerns, because navigation systems used in aviation and to locate your iPhone rely on magnetic compasses. The magnetic heading of airport runways that

appears in big white numbers has to be redone as the magnetic field moves.

Scientists and mathematicians have known about the erratic motion of the north magnetic pole for hundreds of years, but the recent accelerating drift -- about 30-35 miles per year since 2000 -- has been the subject of intensive study, with the help of satellites and surface-based data.

Earth's magnetic field is created by wavy motions that give rise to convection currents -- rising and sinking liquid iron -- in the fluid outer core of Earth, a 1,500-mile layer between the surface and a solid inner core beneath Earth's mantle. The other issue is the protective role of the magnetic field from strong solar radiation when the geomagnetic poles switch positions during a period of several thousand years. Earth's magnetic field has undergone a reversal every 200,000 to 300,000 years throughout most of the past 20 million years. However, the last occurrence was around 780,000 years ago. A weaker magnetic field triggered by a field reversal results in an uptick in cosmic background microwave radiation reaching Earth, with potentially harmful effects to life on Earth, not to mention modern power grids.

Ohio State University scientist, Lonnie Thompson, at the Byrd Polar and Climate Research Center, said, "If we have the collapse of Earth's magnetic field, we will have intense cosmic radiation for a short period of time. It is kind of our shield to all those intense cosmic rays coming in."

NOAA researchers just finished updating mapping of the North Magnetic Pole, which is incorporated into smartphones and navigation systems, considering the recent eastward acceleration in the past few years.[8]

Pole Shift is increasing so fast that Earth's North magnetic pole is now shifting 34 miles per year! That is almost three miles a month! (2.8 miles) That is almost 500 feet a day... This will continue to increase and speed up. But it is already moving so fast that in about four years it will reach its weakest magnetic position (40 degrees mark). Remember, that means at that point the Pole Shift is imminent.

CHAPTER 21: POLE SHIFT

Notes:
1. Jonathan Erdman, *It Just Hit 100 Degrees Fahrenheit in Siberia, the Hottest Temperature on Record So Far North in the Arctic*, https://weather.com/news/climate/news/2020-06-21-siberia-russia-100-degrees-heat-record-arctic, June 21, 2020.
2. Ella LeBain, *Who's Who in the Cosmic Zoo? Book One – Third Edition*, Chapter: Life on Mars, pp. 310-313, Skypath Books, 2013.
3. Seth Borenstein / AP Updated: Earth's North Pole Shifts About 34 Miles Per Year, Report Finds http://time.com/5520537/Earth-north-pole-shifts/, February 5, 2019.
4. Dutchsinse, *8/24/2018 --Large M7.1 Earthquake strikes South America -- Deep EQ at 610km -- NEW UNREST POSSIBLE*, https://youtube/NMq-3ipwpE4
5. André van Belkum, Earthquakes in Bible Prophecy, https://lifehopeandtruth.com/prophecy/end-times/Earthquakes-in-prophecy/
6. *Inuit Elders tell NASA Earth Axis Shifted*, https://watchers.news/2015/03/08/inuit-elders-tell-nasa-Earth-axis-shifted/, March 8, 2015,
7. Mike Adams, *Behold, a pale horse? How the "green" environmental movement may be the Biblical Fourth Seal of an End Times global death cult*, https://www.naturalnews.com/2019-02-10-behold-a-pale-horse-how-the-green-environmental-movement-may-be-the-biblical-fourth-seal-end-times.html, February 10, 2019
8. Ben Gelber, *Earth's North Magnetic Pole moving east faster – NBC4's Ben Gelber explains what that means*, February 15, 2019, https://www.nbc4i.com/news/local-news/Earth-s-north-magnetic-pole-rapidly-moving-east-nbc4-s-ben-gelber-explains-what-that-means/1785001457

CHAPTER TWENTY-TWO

TIME TRAVEL IN THE BIBLE

"Time is the father of truth . . . its mother is our mind."
Giordano Bruno, Cosmologist (1548-1600)

"There's something about the future, every time you look at it, it changes."
Feres in *Next* (2007 Film)

TIME DOESN'T ACTUALLY EXIST

Time does not exist. Clocks exist. Time is just an agreed upon construct. We have taken distance (one rotation of the Earth, and one orbit of the Sun), divided it up into segments, then given those segments labels, while it has its uses, we have been programmed to live our lives by this construct as if it were real. We have confused our shared construct with something that is tangible and thus have become its slave.

The scripture in the Old Testament repeated verbatim in the New Testament that defines the Lord as a Time Traveler.

> "With the Lord a day is like a thousand years, and a thousand years are like a day."
>
> (2Peter 3:8)

The original corroborating scripture came from, Psalms 90:4, "A thousand years in your sight are like a day that has just gone by, or like a watch in the night," which 2 Peter 3:8 copied verbatim.

A day = a thousand years
A thousand years = a day

This defines time-travel and more importantly, it defines the Lord as a time-traveler! Using this language established in the Word of God, we can attach the same equation to the six days of Creation as told in the Book of Genesis. That the world and all of its creatures including humans, were not created in six literal Earth days, but six of Heaven's days, which is how the LORD of Heaven views days, as a thousand years in His time. Making the creation story, from the Lord's perspective, 6,000 Earth years, which are six days in Heaven.

In the Book of Joshua Chapter 10, it is recorded that the Sun stood still over Gibeon and the Moon stood still in the Valley of Aijalon. While that certainly qualifies as a Supernatural event, it does not specify time-travel, however it is close. There are two corroborating scriptures, one in the Old Testament and one in the New Testament which is more specific to time-travel.

> "At that time Joshua spoke to the LORD in the day when the LORD gave the Amorites over to the sons of Israel, and he said in the sign of Israel, "Sun, stand still at Gibeon, and Moon, in the Valley of Aijalon. And the Sun stood still, and the Moon stopped, until the nation took vengeance on their enemies. Is this not written in the Book of Jasher? The Sun stopped in the midst of Heaven and did not hurry to set for about a whole day. There has been no day like it before or since, when the LORD listened to the voice of a man, because the LORD found for Israel."
>
> (Joshua 10:12-14)

Examples of time-travel in the Bible:

> "I was watching the night visions, and behold, *One* like the Son of Man, Coming with the *Clouds of Heaven*! He came to the Ancient of Days, and they brought Him near before Him. Then to Him was given dominion and glory and a kingdom, that all people, nations, and languages should serve Him. His dominion *is* an everlasting dominion, which shall not pass away, and His Kingdom *the one* which shall not be destroyed."
>
> (Daniel 7:13-14)

> "I John, both your brother and your companion in the tribulation and kingdom and patience of Jesus Christ, was on the island that is called Patmos for the word of God and the testimony of Jesus Christ. I was in the Spirit on the Lord's Day, and I heard behind me a loud voice, as of a trumpet saying, "I am the Alpha and the Omega, the First and the Last," and, "What you see, write in the book and

> send it to the seven churches which are in Asia: to Ephesus, to Smyrna, to Pergamos, to Thyatira, to Sardis, to Philadelphia and to Laodicea."
>
> (Revelation 1:9-11)

Ezekiel 1:4 is a good example of space travel, but not specific to time-travel.

> "And I looked, and, behold, a whirlwind came out of the north, a great cloud, and a fire enfolding itself, and a brightness was about it, and out of the midst thereof as the color of amber, out of the midst of the fire."
>
> (Ezekiel 1:4)

TIME DILATION

Time dilation explains why two working clocks will report different times after different accelerations. For example, at the ISS time goes slower, lagging 0.007 seconds behind for every six months. For GPS satellites to work, they must adjust for similar bending of spacetime to coordinate with systems on Earth.[1]

Time dilation is a difference in the elapsed time measured by two clocks, either due to them having a velocity relative to each other, or by there being a gravitational potential difference between their locations. After compensating for varying signal delays due to the changing distance between an observer and a moving clock (i.e., Doppler effect), the observer will measure the moving clock was ticking slower than a clock that is at rest in the observer's own reference frame. A clock that is close to a massive body (and which therefore is at lower gravitational potential) will record less elapsed time than a clock situated further from the said massive body (and which is at a higher gravitational potential).

These predictions of the theory of relativity have been repeatedly confirmed by experiment, and they are of practical concern, for instance in the operation of satellite navigation systems such as GPS and Galileo.[1,2] Time dilation has also been the subject of science fiction works, as it technically provides the means for forward time-travel.[3]

Disclosure of the Secret Space Program is happening as more if the program's "assets" aka experiencers are waking up and remembering their experiences and owning their memories.

According to the witness of Peggy Bradley, "the technology the SSP programs have gotten from ET's [in exchange for human cyborg slaves] is beyond the belief structure of Earth based humans."

This is just the tip of the iceberg:
- Time-travel through man-made time dilation portals and clock works devices or through naturally opening energetic portals that open on planets these are activated with SSP Asset psychic abilities;

- Soul transference between bodies, with soul trap technologies operated by electromagnetic stasis fields
- Regeneration that can bring you back from the dead, regeneration tanks with different green and blue goo's, holographic medical pods, and portable Meditech units;
- Personal cloaking devices, smart suits with electromagnetic fields to shield again energy weapons and projectile bullets;
- Devices that allow SSP Assets to alter their DNA and genetic body structures on the molecular level to temporarily transform into other ET's and have their genetic stalk in your body's system. This can be achieved with drug serums or shape shifting devices.
- Personal portal tech like star gate platforms or naturally opening teleportation portals that are opened with psychic abilities;
- Directed Energy Weapons that cause internal bleeding or cancer these weapons have invisible laser frequency fields;
- Warp drives and crystalline based engines or plasma cores to travel beyond the speed of light or for inter-dimensional travel;
- Access to the wormhole system of the galaxy or the cosmic web of space teleportation energy transfer hubs;
- Replicators for food or any other 3D material you want;
- Zero Point energy and clean energy that can remove pollution and terraform planets or moon systems.

The tech in Star Trek is behind what is available to the more advanced groups like the Planetary Corporations and Nacht Waffen.[4]

Revelation 8:11 mentions Wormwood. "The name of the star is Wormwood. A third of the waters became wormwood, and many men died from the water, because it was made bitter." In my opinion, wormwood is a comet like asteroid that hits Earth causing a third of Earth's oceans to become bitter because it is made from iron oxide, which comes out of the Nemesis-Nibiru system. However, there is a possibility that Wormwood may come to our planet *through* a wormhole. Others have wondered if Wormwood is in and of itself a wormhole? We know based on the prophecy that it is the name of a star. And we know that it means bitterness, which is what it causes our waters to turn bitter from its toxic substance and energies. This is why the Nemesis-Nibiru system has been called the Destroyer because each time it passes through our Solar System, it creates destruction in its path.

What if Planet Nine, which has recently been hypothesize by Astronomers as not a planet but a moving black hole, is the portal, that Nibiru enters into our space through? Something to ponder.

Harvard University undergraduate Amir Siraj and theoretical astrophysicist Avi Loeb suggests that a new telescope currently being constructed in Chile could hold the key to discovering whether there is, in fact, a black hole located in our own solar system.

That would be the find of the century, arguably way more fundamental than the discovery of any "Planet Nine." [1]

"Planet Nine" is speculated to be a super-Earth—a planet about five to 15 times larger than Earth—that was first theorized in 2016. Its existence would explain why objects in the Kuiper Belt—a doughnut-shaped region of the outer solar system beyond the orbit of Neptune, and home to Pluto—are clustered and aligned in a particular way. If there is an as-yet-undetected planet in the Kuiper Belt, it is cold, dark, and way too small to see.

"Planet Nine" is considered unlikely to exist because it would be difficult for the solar system to collect enough material at such a distance from the Sun to form a super-Earth-sized planet. Which has led astronomers to think-up some even crazier-sounding theories about what, exactly, the observed "Planet Nine effect" could actually be. One of the theories is that it is that there is a primordial black hole in our solar system. Gulp.[5]

Notes:
1. Ashby, Neil (2003). "Relativity in the Global Positioning System" (PDF). Living Reviews in Relativity. 6 (1): 16. Bibcode:2003LRR....6....1A. doi:10.12942/lrr-2003-1. PMC 5253894. PMID 28163638.
2. Miller, Arthur I. (1981). Albert Einstein's Special Theory of Relativity: Emergence (1905) and Early Interpretation (1905–1911). Reading, Massachusetts: Addison–Wesley. ISBN 978-0-201-04679-3.
3. Darrigol, Olivier (2005). *The Genesis of the Theory of Relativity*. Einstein, 1905–2005 (PDF). Séminaire Poincaré. 1. pp. 1–22. doi:10.1007/3-7643-7436-5_1. ISBN 978-3-7643-7435-8.
4. Penny Bradley and Elena Ka from the Private Secret Space Program Facebook Group
5. Jamie Carter, *Is 'Planet Nine' Actually A Black Hole In The Solar System? There's Only One Way To Find Out*, https://www.forbes.com/sites/jamiecartereurope/2020/06/03/is-planet-nine-actually-a-black-hole-in-the-solar-system-theres-only-one-way-to-find-out/#7f924493ef40

CHAPTER TWENTY-THREE

CALENDAR SHENANIGANS

The purpose of Time is, so that everything doesn't happen all at once.
~Albert Einstein, Nobel Prize for Physics, 1921

JEWISH CALENDAR VS. GREGORIAN CALENDAR

September 18, 2020 was the Jewish New Year stardate – 5781. That means there's only 219 years to the year 6,000. If a day is like a thousand years unto the Lord, then that would be the end of a full work week, which would indicate that the seventh day would be the Sabbath, the day of Rest.

The Jewish viewpoint is that Time is not a train of cars hitched one to another. A year is not dragged along by the year preceding. The present is not hitched tightly to the past. The future is not enslaved to the present. Rather, every year arrives fresh from its Creator, a year that never was before and could never have been known before its arrival.

That is why we call Rosh Hashanah "the birth of the world" in our prayers. The past has returned to its place, never to return. With the blowing of the shofar, the entirety of Creation is renewed. From this point on, even the past exists only by virtue of the present.

What if, due to Calendar Shenanigans and manipulating time, our counting of the days and years, ages and millennia are not where we think we are, but perhaps we are further along to the end of this age than we think.

According to a new controversial theory, everything around us is intricately planned, and each and everyone's destiny has already been decided. The new theory suggests that time does not PASS and that everything is ever-present. In fact, time is not linear as we have been thinking all along, and everything around us is ever present. [1]

The researchers indicate that time should be regarded as a dimension of space-time, as relativity theory holds — so it does not pass by us in some way, because space-time does not. Instead, time is part of the uniform larger fabric of the universe, not something moving around inside it. According to a scientist, everything that has happened and everything that will happen is, in fact, occurring at this very moment as time is positioned in space.

The new theory proposed by Dr. Bradford Skow, Professor of Philosophy at the Massachusetts Institute of Technology (MIT), indicates that if we were to look down on the universe, we would actually observe time and events spreading out in all directions. [2]

So, what does this actually mean? Well, it suggests that time as we know it is incorrect, in other words, it is not linear as we have been thinking all along. In fact, everything around us is ever present.

The new theory is detailed in Dr. Skow's book, *Objective Beginning*, where he writes: "When you ask people, 'Tell me about the passage of time,' they usually make a metaphor. They say time flows like a river, or we move through time like a ship sailing through the sea."

In *Objective Becoming*, Skow aims to convince readers that things could hardly be otherwise. To do so, he spends much of the book considering competing ideas about time — the ones that assume time does pass or move by us in some way. "I was interested in seeing what kind of view of the universe you would have if you took these metaphors about the passage of time very, very seriously," Skow says.[2]

Skow believes in a so-called block universe – a theory which states that the past, present and future exist simultaneously.

In other words, this means that once an event has occurred, it continues to exist somewhere in space-time.

The new 'controversial' theory is backed up by Albert Einstein's theory of relativity which indicates that space and time are in fact part of an intricate, four-dimensional structure where everything that has occurred has its own coordinates in space-time.

Skow further details: "The block universe theory says you're spread out in time, something like the way you're spread out in space. We're not located at a single time."

Skow agrees that while things change and we see time as if it were passing, he also believes that we are in 'scattered conditions,' and that different parts of time may be dotted around the infinite universe.

"If you walk into a cocktail party and say, 'I don't believe that time passes,' everyone is going to think you're completely insane," says Brad Skow. He does not believe time passes, at least not in the way most have been taught to believe. Skow believes, "that time flows like a river, or we move through time the way a ship sails on the sea."[3]

TIME IS A DIMENSION OF SPACE-TIME

In the first place, Skow says, time should be regarded as a dimension of space-time, as Einstein's theory of relativity holds — so it does not pass by us in some way, because space-time does not. Instead, time is part of the uniform larger fabric of the universe, not something moving around inside it.

Skow's bombshell book, *Objective Becoming*, published by Oxford University Press, details this view, which philosophers call the "block universe theory of time."[3]

CHAPTER 23: CALENDAR SHENANIGANS

In one sense, the block universe theory seems unthreatening to our intuitions: When Skow says time does not pass, he does not believe that nothing ever happens. Events occur, people age, and so on. "Things change," he agrees. However, Skow believes that events do not sail past us and vanish forever; they just continue to exist in different parts of space-time. Many may find this view of time intuitive.

Skow's view of time may be unusual, but that does not mean it is wrong. How else can we explain having memories that tend to live on forever, this confirms his theory that perhaps they still exist, just in another dimension of space-time. He says, we exist in a "temporally scattered" condition. Let's face it, some days are better than others in many of our lives. There are many who feel scattered, fragmented. Memories can flutter in and out of people's minds, causing them to be distracted from being in present time, which causes them to live in the past. Likewise, many who are anxious about the future, are also not living in the present time, as their anxiety creates a distraction to appreciate and be fully present in the presence of now.

"The block universe theory says you're spread out in time, something like the way you're spread out in space," Skow says. "We're not located at a single time."

Both viewpoints are right, but just layers of one another. For instance, it does not matter how many layers make up an onion, if the outer layers are peeled back, it still remains an onion. In relationships, love is timeless. Good relationships share timelessness together. That is where it does not matter if you are apart for a period of time, when you come back together, the experience of feeling like no time has passed, exists, which comes from strong bonds. Therefore, we can add to this discussion, that our emotional and spiritual experiences often dictate how we perceive time. Timelessness is a state of feeling that you are either outside of time, or that time does not exist, only the love you feel.

Likewise, when people suffer trauma, deep sorrow, grief, and pain, it can alter our perception of time, which feels like it is dragging on slowly due to the heaviness of the experience and its emotions. This is why people who experience the pain of trauma and loss, must take the time, however long that takes them, to heal their fragmentation and feel whole again. When people are having good times, time appears to go by fast. When people suffer, time feels like it is slower than molasses. Patience could never be achieved, if the density of the third dimension did not cause things to feel slow, so during those time periods, we exercise patience, and then time seems to shift again. Therefore, we can safely say, that our experiences and how we feel about them, alter time and most certainly our perception of time.

The popular view known as "presentism," which posits that only events and objects in the present can be said to exist, Skow thinks defies the physics of space-time. Skow is more impressed with the "moving spotlight" theory, which may allow the past and future to exist on a par with the present. Only one moment at a time is absolutely present, and that moment keeps changing, as if a spotlight were moving over it. This is also consistent with relativity, Skow thinks — but it still treats the present as being too distinct, as if the present were cut from a different cloth than the rest of the universal fabric, which it clearly is not, as the present is made up of the effects of the past and the preparation and anticipation of the future.

All we really have are a series of moments strung together. "I think the theory is fantastic," Skow writes of the moving spotlight idea. "That is, I think it is a fantasy. But I also have a tremendous amount of sympathy for it." After all, the moving spotlight idea does address our sense that there must be something special about the present.

"The best argument for the moving spotlight theory focuses on the seemingly incredible nature of what the block universe theory is saying about our experience in time," Skow adds. Still, he says, that argument ultimately "rests on a big confusion about what the block universe theory is saying. Even the block universe theory agrees that ... the only experiences I'm having are the ones I'm having now in this room." The experiences you had a year ago or 10 years ago are still just as real, Skow asserts; they are just "inaccessible" because you are now in a different part of space-time.

That may take a chunk of, well, time to digest. But by treating the past, present, and future as materially identical, the theory is consistent with the laws of physics as we understand them.

Our life is but a series of defining moments, strung together by the passing of time. When we fully surrender to this present moment, it is not the moment itself that defines us, but how we choose to live in it.

> "Life is a series of moments and moments are always changing, just like thoughts, negative and positive. And though it may be human nature to dwell, like many natural things it is senseless, senseless to allow a single thought to inhabit a mind because thoughts are like guests or fair-weather friends. As soon as they arrive, they can leave, and even the ones that take a long time to emerge fully can disappear in an instant. Moments are precious; sometimes they linger and other times they're fleeting, and yet so much could be done in them; you could change a mind, you could save a life and you could even fall in love."
>
> — Cecelia Ahern, How to Fall in Love

ENDING TIME MANIPULATIONS

Daylight Savings Time is another indicator of Time Manipulation and Calendar Shenanigans. There are more than 24 time zones. If each time zone were one hour apart, there would be 24 in the world. However, the International Date Line (IDL) creates three more. Also, several time zones are only 30 or 45 minutes apart, increasing the total number of standard time zones even further.

They need to stop manipulating Time. The amount of Calendar Shenanigans that has occurred on our timeline, which are probably more that I am presenting here in Book Five - *The HEAVENS*, is enough to make you doubt where we actually are in our trip around the Sun. Objects appear closer than you think. Did you know that at any given time, time around the world is divided into over two dozen different time zones at once? And that does not include those with broken clocks, and those whose clocks are inaccurate or off. Remember, *even a broken clock is right twice a day*. No matter what time zone it is in.

My husband and daughter deliberately have their clocks set ahead 5 minutes in their cars, so they will not be late to appointments. My chiropractor sets his clock 5 minutes behind to keep his appointments on schedule because people like me are always a few minutes late stuck in traffic, which works to take anxiety away for being late.

Did you know, that if you have multiple clocks in your house, and they are not all in sync, it has been known to confuse the spirits. Why? Because every hour of every day is ruled by a different angel and when evil spirits come along that are already in a state of confusion because they come from the kingdom of darkness, they cannot always tell time. So, it has been said, if you keep one clock in the house slightly off the actual time, it can confuse spirits, which can sometimes work in your favor. Yes, some see this a silly superstition, but those who are awake to the spiritual realm, understand. I am not going to get into the nitty gritty of Kabbalah, but anyone who has researched and studied it, knows that every hour of every day is ruled by different ruling and overseeing angels. Some may think these are fallen angels, and as I have discerned in much greater detail in *Who Are the Angels?* Book Three of this Book Series, that the fallen angels network with the evil spirits who are bound to the astral realm of Earth, until the Day of the Lord, when He releases them and orders them to their permanent place. [4] The other point I make is that the Lord changed the names of the original angels to confuse humans because they were invoking them and casting spells, which He felt was a misuse of their power. So, the fallen angels adopted their names, as counterfeits and are the ones who the ones being invoked. So, to confound the fallen and evil spirits, is not such a bad idea.

Of course, affirming Psalm 91 over yourself, is a prayer of assurance that only the Faithful Angels who obey the command of the Lord of Hosts, the Lord of all Angels and Spirits, will send His Angels to be in charge of you, to keep you in all of your ways, is part of a Covenant between you and the Lord, that is sealed with you choosing to put your faith and trust and Him. In this way, His Angels supersede and overcome anything the fallen angels and their evil spirits can do.

I agree with doing away with these ridiculous time changes.

It is a fact that heart attacks are up 25% on the Monday after Spring Forward, but down on the Monday after Fall Behind. That extra hour of sleep is important to our body clocks, so here is how you can avoid losing sleep over Spring Forward. Go to sleep at your normal time Saturday night and do not move your clocks forward until AFTER you wake up. The only hour you will lose will be off your day Sunday.

What year is it really?
- Gregorian calendar – 2020
- Jewish calendar - 5781
- Chinese calendar - 4717
- Arabic (Hijri) calendar - 1443
- Persian calendar - 1399
- Armenian calendar - 1470
- Buddhist calendar - 2563
- Korean calendar - 4353

It certainly appears we have the Tower of Babel of calendrical systems as well, to add to our language and cultural differences. The Jewish calendar is considered the most accurate, however, some say it may be missing approximately 150 to 240 years. There are scholars who actually believe we are in or approaching the beginning of the seventh millennium. If they are correct, then Judaic prophecies about the seventh millennium may become more chronologically aligned and coincidental with Christian and Islamic prophecies about the end times. Things could get interesting. Remember a day is like a thousand years to the Lord, so if we're at the end of the 6th day, or the year 6,000 in the Jewish calendar, then on the 7th day, we rest in the Lord.

TIME JUMPING DURING APPOINTED TIMES

Time travel is real. Can we jump from time to time, even in our minds, just as we jump from point to point on a line, only it is not linear, but moves in a spiral. Can we ever go back even as we move forward, somehow moving in both directions simultaneously?

Jorian Polis Schutz, Publisher of Deuteronomy Press, publisher of the Jewish Circular Calendars, known in Hebrew as the Misaviv seems to think we can by following the Misaviv is a Hebrew calendar, based on the Torah, that shapes time as a circle, overlaid with the Gregorian Calendar, featuring original artwork from 13 artists and innovative calendar designers placing Shabbat (the Sabbath) at the center of the week. It contains all Jewish holidays, also known as the Six Annual Feasts of the Lord, along with some important holidays celebrated in the U.S.A, and a collection of important yarzeits (memorial dates that honor the dead), included in the overlay.

Schutz says, when we take our calendar as a circle and our journey as a spiral in time, we have a different frame to approach such questions. There is no wide gap between past and future if particular coils are compressed. When we are coming around the bend towards a repetition of past events, we make choices that bring us backwards, or forwards, or maybe a combination of both. This is experienced in annual events, like birthdays, which are called, Solar Returns, when the Sun returns back to the exact same degree and minute it was at one's time of birth. It is an astronomical event that spirals through one's lifetime, which is explored deeply in Astrology. Nevertheless, people all over the world acknowledge their birthdays, which are likened to one's personal new year, one's special mark in time between past and future.

Annual feast and fast days also bring the same repetition, while we can remember our previous years and compare how we felt then to the present holiday. In fact, all the holidays have the same effect on our consciousness, such as Christmas and New Year's.

Who does not feel that time has warped and brought us back to where we just have been? This is because time does expand and contract for us — especially when we are involved with cleansing, transforming, and evolving. Evolution is the purpose of the Spiral of Time. That we return back to the same place in space-time, but we are one notch up from the last year. In this way, we can conclude that the purpose of Time is growth.

Shutz says, in truth, we are always rehearsing the future and recapitulating the past. According to the Jewish Calendar, it's a tradition that from Wednesday (yom revi'i) to Friday (yom sh'lishi) of every week, many are warmed by the glow of the approaching Shabbat, whereas from Sunday (yom rishon) to Tuesday (yom shlishi) of each week, we still bask in the light from the previous Shabbat.

Every holiday, holy day, or festival is a sort of wormhole in space-time, a pinhole for transcendent light. Through our experience of these days, it is as if we exist in all generations: which is repeated in the prayers and blessings for each holiday, *bayamim hahem bazman hazeh* ("on those days, in this time"). It is as if the cumulative history of the day—and its future—is stored for us in a supernal cloud, to which we are granted access. In this way, we can also deduce that, that holiday or special day, is a portal in time that has the potential to change events, if we choose to direct our energies in a positive and thoughtful way.

The same can be said of the negative days of the year, which are memorialized such as Yom HaShoah, which is Holocaust Day, or Tisha B'Av "the ninth of Av," an annual fast day in Judaism that memorializes a day where curses came down on Jews through multiple disasters in Jewish history. Primarily, the destruction of both Solomon's Temple by the Neo-Babylonian Empire and the Second Temple by the Roman Empire in Jerusalem. For this reason, Tisha B'Av is considered the saddest day in the Jewish calendar, which has spawned the belief that the 'day' itself is destined for tragedy. Tisha B'Av usually falls some time towards the end July or early August in the Gregorian calendar.

The observance of the day includes five prohibitions, the most notable of which is a 25-hour fast. The Book of Lamentations, which mourns the destruction of Jerusalem, is read in the synagogue followed by the recitation of kinnot, liturgical dirges that lament the loss of the Temples and Jerusalem. As the day has become associated with remembrance of other major calamities which have befallen the Jewish people, some kinnot also recall events such as the murder of the Ten Martyrs by the Romans, massacres in numerous medieval Jewish communities during the Crusades, and the Holocaust.

When we add our personally significant days to the year—like birthdays or yahrzeits (memorials) for our loved ones — these, too, may become portals of past and future worlds, that hold prophetic insight with the opportunity for renewed retrospect. We cannot resurrect those who have departed this life, but we can bring ourselves to a place and time where dialogue with their spirits is real. When we think of a yahrzeit, a memorial of the dead, as a supernal landline, and the intervening years as a superfluous spiral cord, that we can twirl in our hands, we are therefore transformed into something new and healed from our grief at the same time.

In healing from PTSD, Post Traumatic Stress Disorder, therapists will work on guiding the patient to the point in time where the memory is held prior to the trauma, when they felt whole. This is a strategy to get the memory to return back to that space-time and to retain the memory of wholeness and wellness. How often do we hear the words, 'if we can turn back time,' we would do things differently, is often the impetus to use time travel to correct the past. In various types of therapy, it can be done, as well as the miracle healings that come through the outpouring of the Holy Spirit through belief in Yeshua-ti, our Messiah and Salvation, Yeshua, who is Jesus, whose very name means Salvation. He has promised to restore people back to their wholeness state before the years that the locusts have eaten. See, Joel 2:25.

"May we return forwards even as we return back, and may we remember—no matter where and when we are—that all our blessings continually flow from the Creator of all Life who is the Source of All Time."[5]

TIME AS A CIRCLE

The Earth is not flat, and neither is our timeline. Both are spherical. Time moves in a spiral. We walk in circles, literally. Every step we take on the surface of God's Round Circular Spherical Earth is part of a much greater circle that ultimately encompasses the globe.

It is easy to forget this fact because more often we seem to move in a world of lines. We go down boulevards and up elevators; we have destinations; we can measure our progress on a simple chart or stock graph. Lines give us a feeling of directedness, empowerment, accomplishment. On the contrary, the roundness of circles can seem idle, indecisive, self-defeating. Expressions like "walking in circles," "round and round we go," "here we are again," imply that moving in circles is unproductive and monotonous.

The same dynamic applies to time. We experience time moving forward inexorably; we arrange history as a timeline; we use rectilinear calendars with the days stacked upon each other and progress marked in a well-ordered series of a rows. At the end of the week, a magical corner is turned without turning, and we appear again at the front.

There is no crime in this arrangement. It corresponds to the way we generally read, line after line. But ultimately such forms are arbitrary, obscuring our involvement with cycles. For time moves just as much in circles as in lines, and the Jewish tradition firmly upholds this—even in the way we read. The word sefer (ספר) now refers to any book, but it once only meant "scroll," that ancient frame for our holy texts, which sends out a flat section like a linear emissary from the rolls and then soon after "gathers it in" to the forever-furl of its brother pole. It is a powerful metaphor for our own lives, which seem to emanate from and return to a place beyond time, and which at times feel as if they are superimposed on primordial rolls, with prophetic letters shining through.

There are circles to be found explicitly in Torah. The manna, which sustains the people in the desert for 40 years, is round (Ex. 16:14). The etrog we hold together with the lulav on Sukkot unites circle and line in one gesture. The unique prayer power of Honi the Circle Drawer (Honi HaMa'agel) is set apart from his generation by a circle drawn in the dirt (Ta'anit 23a). But there are also more hidden references. What is the deeper meaning of the first two rivers going out of Eden, which are said to encompass (ha'sovev) the land? What are those ma'aglei tzedek, (circles of righteousness?) that King David sings about in Psalm 13? What is the significance of the stones that Ya'akov places in a circle around his head before he dreams of ascending and descending beings (28:11)? And why does the Torah tell us just as we exit Egypt that God does not lead us on the direct route but rather "curves" (vayaseiv) the people towards the desert path (Exodus 13:17-18)?

THE SHAPE OF TIME

The Hebrew calendar is an incomparable marvel of circles within circles, spirals within spirals of temporal rhythm and symmetry, of balance between solar and lunar. Each month (chodesh) follows the cycle of the Moon, beginning and ending with its renewal. The seven-day shabbat cycle is extended into the weeks (sefirat ha'omer), into the years (shmitah), and into the weeks of years (yovel). We journey along with the patriarchs and matriarchs and with the children of Israel in a spiraling journey of ascent; we return to the same place/time, and we remember, but we are changed. The beginning of the solar year, Tishrei, is opposite the beginning of the lunar year, Nisan; Sukkot is opposite Pesach, one marking the beginning of the rainy season, and one marking its end. On Simchat Torah we move from the final to the first Torah portion, proving that the cycle does not pause or finish, and we learn that the final letter and first letter, seemingly enclosing the "white space" of the world, spell the word lev (לב), heart.

On this first year of a new sabbatical cycle, we welcome you to join us in this new evocation of the holy imagination, for the purpose of a new circular inhabitation of Jewish time-space. And we invite you to consider submitting sketches and concepts for, God willing, next year's calendar.

The circle is necessary. It reminds us of our equidistance to the center, no matter where we stand, and the equivalence of every arc and angle in this world. The intention of MISAVIV, God willing, and its potential toledot (offspring), is not to overthrow the rectilinear regime on time, but to show that another possibility, and indeed many other possibilities could exist. And that these could be just as practical guides through the year as their counterparts, while representing a beautiful, holy window to cyclicality, spirality, and to the great circle of which we are all a part, and all will become a part, as our Sages teach:

In the future the Holy One will make a circle (machol) for the saintly ones, and He will sit between them in the garden of Eden, and each one will show with his finger, as it is written (Isaiah 25:9), "And it shall be said in that day: 'Behold, this is our Lord, for whom we waited, that He might save us; this is God, for whom we waited, we will be glad and rejoice in His salvation."

The promise of the New Jerusalem, the Holy City that comes out of the Heavens that lays over the Earth, says that when the Moshiach-Nagid (Messiah King) returns to Earth, he will sit in the middle of the New Jerusalem, on His Throne, and rule over the all the nations of the world.

In the 5781 Misaviv, Schutz describes the shape of time, based on the Sefer Yetzirah. A depiction of the 22 cosmic fire letters of the Hebrew Aleph-Bet that combines 231 unique pairs or double the if you count reverse pairs. These are the 231 gates, paths or portal that manifest the world, including space-time.[5] The spiral of time leads us to the great homecoming, (the Yovel) and the final Sabbath year (shmitah), that takes us above the petty tyrannies of exiled time through the Roman Gregorian manipulated calendar, that has very little connection to the actual Astronomy that matches the true shape of time, which is the spiral. Even our Sun moves in a spiral in its journey and trajectory within our galaxy. Likewise, galaxies are shaped as spirals which also journey within our Universe and Multiverse.

TIME ZONES AND PARALLEL UNIVERSES

According to Jewish Cosmology, known as Kabbalah, there are days and months that are deemed positive cosmic energy and there are days and times considered to be negative energies. According to the Zohar, "All the celestial treasures and hidden mysteries which were not revealed to succeeding generations will be revealed in the Age of Aquarius." It is believed that in the age to come we will have a new understanding not only of our familiar universe, but of that which lies beyond the range observation in the realm of the metaphysical and spiritual realms, the non-space domain.[6]

The Force is a Jewish concept, which is known in the Zohar as the raw cosmic energy that separates the light and darkness. See, my related Chapter on the *Real Star Wars, Jediism v Judaism* further in this book.

CHAPTER 23: CALENDAR SHENANIGANS

Notes:
1. A New Theory On Time Indicates Present, And Future Simultaneously Exist, http://www.thescinewsreporter.com/2018/09/a-new-theory-on-time-indicates-present.html
2. Peter Dizikes | MIT News Office, *Does time pass? Philosopher Brad Skow's new book says it does — but not in the way you may think.* January 28, 2015
3. Brad Skow, PhD., *Objective Becoming*, Oxford University Press, Oxford, UK, 2015.
4. Ella LeBain, *Who Are the Angels* Book Three, Skypath Books, 2016.
5. Jorian Polis Schutz, Publisher, Deuteronomy Press, 5779, 5781.
6. Philip S. Berg, PhD., *Time Zones: Your Key to Control*, Research Centre of Kabbalah, Jerusalem, Israel, 1990.

CHAPTER TWENTY-FOUR

THE BATTLE FOR TIME

"Every square inch in the universe is either claimed by God,
or countered by Satan." ~ C.S. Lewis

That includes Time and Space-time!

Everyday people wrestle over schedules and how they want to spend their time. To many, a tight schedule is considered the norm, making their weekends and days off also full of activities, leaving little time for resting and doing nothing.

There is an ancient Middle Eastern saying that goes like this, "never criticize a man for being late, because you never know what the will of God is." We see this played out in the plethora of testimonies during disasters, accidents, and tragedies, when being on time proved to be catastrophic, yet those who were delayed for all kinds of reasons were saved from essentially being in the wrong place and at the wrong time. There are so many testimonies of salvation on September 11, 2001, from those who were scheduled to be in their offices in the World Trade Center that morning, but were delayed for various reasons, thereby saving their lives.

Then again, there is the flip side, where 'timing is everything,' proving that being in the right place at the right time can be fortuitous and beneficial for opportunity, favor, and good fortune.

TIME IS MONEY

The fact that we charge for our time, tells us time is a valuable commodity. The field of psychology has a mindset, that if you do not pay for your time, then you do not appreciate or value your therapy. Then again, there are people who give freely of their time and attention, without expecting payment in return. Love is supposed to be free, but not in our present world, where people pay to be loved through having someone listen to them in a safe place of acceptance, or where others pay for sex which is defined as prostitution.

Then on the flip side, you have children who would do anything to spend time with their parents, but because their parents are too busy, they pay for babysitters or daycare professionals to entertain their children for them. Therefore, when it comes to time and the people we pay for their time, we must assess what our priorities are, and if it is worth it? Obviously, parents have to provide for their children, and cannot be with them 24/7, which justifies employing others to look after their pride and joy, but not without a price. And I'm not just talking about money, but the guilt many parents feel for doing that, the anxiety they experience for giving up control and having to trust a professional they're paying, that can cause all kinds of dysfunctional families, because raising children is a hands-on, be present, kind of job.

Of course, parents deserve regular breaks to get restored, but parenting is not just limited to taking care of children's immediate needs and teaching them to become independent and responsible adults. Rather parenting is about building close family bonds and relationships that last a lifetime and throughout eternity. Those who pay others to raise and bond with their children, pay the price of not enjoying close bonds, and may have to work harder in later years, when children grow distant, because that's simply how they were trained, to give Mommy and Daddy lots of space to work and nurture careers, while their children grow up to feel neglected, unloved and estranged. Therefore, the philosophy, "time is money," may not be the best philosophy for building strong family bonds.

To many children, time is the love and attention they crave and need to grow up with. Parents who do not pay attention to their children enough cause these kids to grow up with attention deficit disorder, which essentially means that no one took the time to be patient and pay attention to not only their needs, but their thoughts and to validate 'who' they are and their potential of 'who' they could be. For this reason, children grow up lost and confused and, therefore, cannot focus on one thing too long, because they did not have a role model to show them how. Their parents were too much in a hurry or they were not psychologically and mentally healthy and could not relate to their children, because they might have grown old, but have not grown up.

Many of us feel like time has sped up or like we are running out of time. As a result, we work extra hard, doing "double time," and hope we can "catch up" with time. But then time becomes elusive, and we are chasing good times, which become fleeting in between spending our time hard at work.

Steve Jobs, the creator of the Apple iPhone died a billionaire, with a fortune of $7 billion, at the age of 56 from pancreatic cancer. Here were some of his last words:

> "At this moment, lying on the bed, sick and remembering all my life, I realize that all my recognition and wealth that I have is meaningless in the face of imminent death. You can hire someone to drive a car for you, make money for you – but you cannot rent someone to carry the disease for you. As we get older, we are smarter, and we slowly realize that the watch is worth $30 or $300 – both of which show the same time. Whether we drive a car worth $150,000, or a car worth $2,000 – the road and distance are the same, we reach the same destination. If we drink a bottle of wine worth $300 or wine worth $10, we're still drunk."

Five Undeniable Facts
1. Do not educate your children to be rich. Educate them to be happy. — So, when they grow up, they will know the value of things, not the price.
2. Eat your food as medicine, otherwise you will need to eat your medicine as food.
3. Whoever loves you will never leave you, even if he has 100 reasons to give up. He will always find one reason to hold on.
4. There is a big difference between being human and human being.
5. If you want to go fast – go alone! But if you want to go far – go together.

SECRET TIME TRAVEL PROJECTS

The following list documents the origins of the US Time Travel Technology Program, according to one of its participants and whistleblowers, Andrew D. Basiago,[1] who is presently a practicing attorney in Washington State. His story was featured in Book One of *Who's Who in the Cosmic Zoo?* – Third Edition, under the Chapter: Life on Mars.[2]

1. Remote viewing was developed by the US military in the 1960s years before it was supposedly developed at Stanford Research Institute [SRI] in 1972.
2. Spinning to induce out-of-body experiences so as to travel on the astral plane is an ancient occult practice.
3. The so-called Montauk chair was reverse engineered from the pilot's seat aboard a crashed ET craft, by which the ET pilot piloted the craft psychically to avoid collisions in space in light of the speed of the craft.
4. The teleporter, which opens up a portal tunnel as an interstitial chasm in time-space through which the teleportee passes from Point A to Point B in several seconds, was invented by Nikola Tesla.
5. The chronovisor was accidentally discovered by Vatican musicologists Father Pellegrino Ernetti and Father Augustino Gemelli when they were studying the harmonic patterns in Gregorian chants at the Catholic University of Milan in the 1940s and found the microphone, they were developing could pick up the sounds of past events.
6. The advanced chronovisor was developed from a TV-like screen into a standing cubical hologram of moving, multi-colored light by US defense contractors under DARPA after the Vatican gave the chronovisor technology to the US government for further development.
7. The plasma confinement chamber was invented by Dr. Stirling Colgate, president, and dean of physics at the New Mexico Institute of Science and Technology [NMIST].
8. The jump room or "aeronautical repositioning chamber" [ARC] was developed in a joint venture between Parsons and Lockheed, possibly

after being reverse-engineered from an extraterrestrial device or as a result of ET-human liaison in which the device was given to the US government by one of the Grey ET species.[3]

> "The time-space age has begun."
> — Andrew D. Basiago

PROJECT PEGASUS

Seattle-based lawyer Andrew Basiago became a whistleblower in 2004, when he spilled the beans on a secret space government program, *Project Pegasus*, in which he has repeatedly claimed he participated in as a child from 7 to 12 years old, as a test subject in which he worked on teleportation and time travel.[1]

"They taught children with adults to test if there were differences in the impact of time travel on people of different ages. And it turned out that children had an advantage over adults in terms of adapting to the tension of movement between the past, present, and future".

Many scientists think that time travel is impossible because nothing moves faster than the speed of light. Nevertheless, Andrew Basiago found many supporters, including the conspiracy theorist Dr. Alfred Webre, who specializes in exopolitics – the secret presence of aliens on Earth and their interference in the politics of different countries.

According to Webre, teleportation and time travel technologies have been available to the US government for more than 40 years, but so far it is wary of openly using them and is accumulating experiences. Webre stated:

> "Long-distance teleportation is an inexpensive and environmentally friendly means of transportation. The Department of Defense has had it for 40 years and former Secretary of Defense Donald Rumsfeld once used his share of moving soldiers onto the battlefield".

Basiago said that during his participation in the Project Pegasus program, he traveled through time eight times and the time machine was mainly built on the basis of a teleport, plans for which were found in the New York apartment of mechanical engineer Nikola Tesla after his death in January 1943.[3]

"The machine consisted of two gray elliptical arrows about 8 feet high, separated by about 10 feet, between which was a shimmering curtain of what Tesla called "Radiant Energy." This is a form of energy that Tesla discovered, and which is latent and penetrating into the universe, and among its properties is the ability to bend space-time."

Basiago stated that through this field, the project participants jumped into the vortal tunnel, and when it closed, they were at their destination. "It felt like the tunnel was either moving at a tremendous speed or not moving at all since the universe was around us."

According to Basiago, he did not just fall into the past, but once got into a historical photo which captures President Abraham Lincoln's visit to Gettysburg in 1863. Basiago made this trip in 1972 and the picture was captured from a camera that was installed in East Hanover, New Jersey.

"I was dressed in vintage clothes and looked like an ordinary boy. I was wearing huge men's shoes, and it was only about 100 steps from the place of my arrival to the place in the picture. I was filmed very close to Lincoln."

Then, Basiago claims he went to the Ford Theater on the night of Lincoln's murder 5 to 6 times, but he never got to the moment of the murder. Once he managed to hear a shot and witnessed a huge commotion in the crowd. He was very scared at that moment.

"Each of my visits to the past was a little different as if I were being sent to different alternate realities located on adjacent timelines. As these visits accumulated, I bumped into myself twice on two different visits.

One day, young Andrew Basiago witnessed a time travel failure that happened to another young boy. The child's legs were a few seconds late from the time the main body moved. "He lay and writhed in pain, and he had stumps for legs." He reported when the settings of the car were adjusted, that did not happen again. In total, Basiago reported, about a hundred people participated in Project Pegasus. [3]

TIMES TO PRAY

Did you know that God prepares our day during the night? Why is that? It is because God is the master and author of time. Understand that the Hebrew day actually begins, not ends, at Sunset. In Hebrew time, which was something established by God, the day starts at night. And much of our daylight then is prepared through prayer at night. Prayer during daylight hours has its place and value, but prayer during nighttime hours sets the table for the energies of that which manifests during daylight hours.

Many people I know find themselves waking up at strange hours in the middle of the night and feeling the unction to pray. Some people wake up almost every night at 1:11, 2:22, 3:33, 4:44, 5:55 AM. Many people talk about these numbers as having some mystical meanings. And many books have been written about Angel Numbers, but here is another take on those who are on Night Watch. Embedded in ancient Hebraic traditions are the four prayer watches, segments of time during the night, for covering different needs in prayer. This originated back in Ancient Israel when every hour was ruled over by different Angels. To this day, these practices are incorporated into Kabbalistic teachings, which have, unfortunately, been corrupted with occult teachings and the practice of witchcraft and magic. Nevertheless, what really matters, is *who* you are praying to in the wee hours of the morning.

CHAPTER 24: THE BATTLE FOR TIME

God is the Author and Lord of Time. Ancient Israel believed that the Lord God prepared our day during the night. This is why, in the Jewish Calendar and all Hebraic traditions, the day begins at Sunset, not Sunrise. All Jewish holidays begin at Sunset, even Shabbat, which begins at Sunset on Friday night. This began in Hebrew time when the Lord established the day to begin at night. Just as babies are prepared in the darkness of the womb for 9 nine months before being born into the Light, so does our day begin in the darkness of the nighttime hours—through our resting, sleeping, dreaming, and praying. While praying during daylight hours has its place and value, nighttime prayer can be even more meaningful.

As a night owl, I get most of my writing and work done during the night hours when everyone in my house is sound asleep. No phones to answer, no children to tend to, no chores to do and no errands to run. Silence is golden is true, but after working nights for most of my life, I can almost "feel" when the hours of the night after midnight shift. There is a most definitive, marked shift when it becomes 2 AM., and an even more pronounced energy shift at 3 AM. The 3 AM. hour has long been known to be the witching hour, when the ghosts and goblins run amok. And coincidentally, more alien abductions have taken place during the hours of 3-4 AM. than any other time period.

For years, due to my PTSD, I could not sleep during the night, so I eventually gave into my insomnia and chose to work through the night instead. Things really do change overnight in my house, in the truest sense of the words. I find myself most productive between the hours of 1-5 AM. With that said, I have become accustomed to the hourly energetic shifts. And yes, I do take breaks, and I am often given the unction to pray. After coming across this ancient material, I thought it was fitting to include it here, for my readers, who I know, are spiritually minded, and for those of you who wake up during the early morning hours with heavy hearts.

Here is why, sometimes, when you awaken, from a dream, or with a bad feeling, and you want to pray, you may not have the words to describe what you are feeling. Turn to the Psalms and read them out loud. Jeremiah 1:12 states, "The Lord is watching over His Word to perform it." That implies that Angels are assigned to the Word of God, and when God's Angels intervene, after they hear you vocalize God's Word, you put them into action, to sweep away the weird feelings, which are often demon spirits trying to get a hold of you during the witching hour, 3-4 a.m.

Also, depending on where you live, and the time of year, as the age old saying goes, "it's always darkest before the dawn," has meaning with spiritual warfare, as when the light comes, it literally dissipates the spirits of darkness by exposing them into the light. So, take heart to pray, when you are called to wake up during these hours.

1. First Watch is 6-9 PM.: Enter this watch, and you will see His time of visitation.
2. Second Watch is 9-12 AM.: Enter this watch, and your life will be ordered and not scattered.
3. Third Watch is 12-3 AM.: Enter this watch, and your faith will remain strong during a trial of faith. You will keep your covenant with God and not break it.
4. Fourth Watch is 3-6 AM.: Enter this watch, and your future will be established. [4]

This information about prayer watches is quoted from Chuck Pierce's book *Reordering Your Day*.

Notes:
1. Andrew D. Basiago, Project Pegasus. https://www.projectpegasus.net/
2. Ella LeBain, *Who's Who in the Cosmic Zoo?* – Book One - Third Edition, Chapter: Life on Mars, pp. 310-313, Skypath Books, 2013.
3. Andrew D. Basiago, *A lawyer who claims he traveled in time and even got into an old photo*, https://infinityexplorers.com/andrew-basiago-a-lawyer-who-claims-he-has-traveled-in-time-and-even-got-into-an-old-photo, September 11, 2020
4. Chuck Pierce, *Reordering Your Day: Understanding and Embracing the Four Prayer Watches*, Glory of Zion International, (November 10, 2011).

CHAPTER TWENTY-FIVE

FROM SECRET SPACE PROGRAM TO SPACE FORCE

"Maybe your purpose on this planet isn't on this planet?"
https://www.space.com/space-force-recruitment-video.html

THE 20 AND BACK PROGRAM

Whistleblower Jason Rice recently came out as serving in the Secret Space Command, which President Trump has now renamed, or more accurately, added a 6th arm of the Military, calling it the Space Force. March of 2018, President Donald Trump announced his intention of creating a new branch of the U.S. military, the "Space Force," during a National Space Council meeting. He directed the Department of Defense and the Pentagon to immediately begin establishing the force, which will focus on outer space. "Our destiny beyond the Earth is not only a matter of national identity but a matter of national security," Trump said. 1

Zohar Stargate TV UFO & Ancient Mysteries Networking TV, (UAMN TV) published the following interview with Whistleblower Jason Rice on their YouTube Channel on March 25, 2019.

Jason Rice, SSP Veteran, US Army Veteran and Engineer, provided Disclosure of the Secret Space Program, revealing that the Secret Space Program is not so secret anymore. He was a member of the Secret Space Program. He stated that he is coming forward now to disclose that there are thousands of other people who are also involved with this program that is known as the "20 and Back." On a personal level, he felt that there were many who he worked with, who gave up their lives for humanity through this secret program, that have not come back, and need to be shared with the public now.

He was asked to give a brief summary of his life: "I had what I thought of as a normal life, I started waking up to the memories that were buried, but the amount of them increased, enough that I was starting to piece together over a couple of years, which allowed him to come forward with the events in the Secret Space Program.

"I was on a particular mission in 1999 on Planet 1 - Area 26, our mission at that time, was to gather intelligence of the local populations so that we could develop a better assistance plan to help them prepare for their defense of their planet."

Q: Was there a battle going on? In our Solar System?
"Yes, there are many battles going on."

Q: Do our governments know about it?
"Yes, there are leadership there are some who know about the SSP activities, however, the vast scope of the SSP means that there are extreme cases of compartmentalization, and nobody, no matter who they are or what they claim, knows all of it."
"You asked about battles going on currently: The unit that I was a part of, was one leg of a false flag. The same command structure was also in charge of ordering an alien invasion fleet, they were in charge of our fleet, that was there to help the populations, so they could speed up planetary control from a century or more to less than ten years."

Q: What kind of aliens?
"The Draco primarily and their boss the AI were the ones calling the shots for the unit that I was in."

Q: Do these guys have any kind of religious or spirituality?
"They have their own. Mostly centered around AI and Transhumanism."

Q: What is their opinion of humans?
"That we are animals, feed animals, slave animals, they don't have a very high regard for us."

Q: When we talk about Alliance? Is that real? Do we have ET support?
"Yes, we do have benevolent ETs that are on our side. However, what they are doing, is they are doing their best to stop intrusions from other ETs, the mess that we have got on this planet, we're going to have to take care of that."

CHAPTER 25: FROM SECRET SPACE PROGRAM TO SPACE FORCE

Q: Did Hitler get involved in a Secret Space Program?
"He was highly involved in Germany's Secret Space Program. He was getting assistance from the Draco. The Mars Germans I served with, I had the blessing of becoming friends with a few of them and was able to find out information that we don't have privilege to, coming from the USA."

Q: Do you have information from the Pyramids.
"My information starts around the 1930's time frame."

Q: Where are we with this battle?
"I can tell you that the IDARF (Interplanetary Defense and Reaction Forces) carrier fleet still have units out on deployment. The false flag nature of that unit is franchising evil is setting up despot control over planetary populations. They do that through mass casualty war. The other battles that are going on amongst ETs races for control of territory or ideological, I do not know.

Q: Sounds like Star Wars
"Yes, the Truth is Stranger than Fiction. You have got a rebel alliance against the Draco Empire. The Draco Empire spreads across many solar systems."

Q: What kind of technology do we have?
"As humans technology we have at our disposals, is nanite-enhancement therapies. The enhancement therapies that they gave to us as soldiers were primarily designed to help give us best chances to stay alive, that included anti-radiation protocols. These are micro machines. They are non-living. But they help aid in the healing process. They help in glandular control either increasing the particular output, if you need to stay awake longer, or curb an appetite or any number of things.

Q: What kind of distances are we traveling?
"Thousands of light years. There are humans from Earth all over this galaxy and others."

Q: Are humans involved on other planets that have not experienced Earth?
"The aliens that I encountered appeared to be human. Just like other humanoids here on Earth, they have different

color hair, light and dark skin, eye color, from blue to green to brown to hazel.

Q: Mankind's greatest achievement was the Moon Landing was it all set up? If we have this technology, why are they wasting money to go to the Moon. Did we go to the Moon?

"Yeah, we did go to the Moon. They had sets, set up in the event that they were not able to get good footage that is what the sets were for. Mankind's movement into space with the Apollo Program was a coverup for hiding Black Budget money, i.e., the financing that goes into the Secret Space Program. And to get the public involved for getting exciting about Space for political gain."

Q: Did they have craft watching?

"Yes, they had company watching, they had craft watching them. We have a base there. The Lunar Command Center. The Germans put their first base there in the 1940s. We took over that base in 1991, as part of the agreement that started IDARF, the unit that I was in. Paying for the lease of the moon base, required that Earth provide 25,000 soldiers for use where and when the IDARF command structure was needed. That is how it came about.

How many UFOs belong to us? And How many belong to ETs?

"Earth has a defense grid. The UFOs that are viewed within our skies have permission to be here. Whether they are human or whether they are ET. The majority that are seen, are most certainly human, and only a small percentage are ET."

Q: The ET that we do see, are they part of the Alliance or are they part of the Dark Forces?

"There are still a lot of dark forces at play here on Earth and they have control over certain geographical areas. So that the ETs that come in through that portal or through that door through those geographies are allowed in. Over here they would be shot down. It is a mess. We have a mess here on our planet. It is something that we have got to get straightened out.

CHAPTER 25: FROM SECRET SPACE PROGRAM TO SPACE FORCE

Q: Will we get Disclosure?
"Yes, we will have ET Disclosure, it is not a matter of if, but when. They're intentions initially think that the time frame when this was first an issue was 1947-48, the social morae's, i.e., Eisenhower, was a different generation, their thinking was the public was just not ready for it, so they felt that they needed to keep it under wraps, so that they could get a better handle on it. Unfortunately, there were far too many nefarious groups that wanted that technology that subverted the control of it and were able to gain control over a period of decades."

Q: This community is hostile towards the SSP, how do your colleagues feel about you coming forward?
"We weren't told about the memory wipe compartmentalization when we volunteered. That was not part of the deal. Among the things that they told us, were we were going to have a nice nest egg, when we graduated and/or retired. They did not say anything about memory wipes. They never told us they were going to tell our families we were dead. So that they would assure our commitment. So those who are in, that have found out those very disturbing truths, I would say, that they would want, I know I would have, if I was back in, like year 15 of my 20, and someone would have said, 'Major they're going to wipe your memory,' I would say, whoa, wait a minute, time out."

Q: I have no doubt we can wipe memories. Can you understand why people find it hard to believe?
"I have no tangible proof, I understand that, and I'm asking people to take my word for it. There are circumstantial proofs that I have been able to find and locate. I think once the technology was acquired for mind wipe that that was probably on par, if not more important for human development and technology then the Manhattan Project, because they were able to assure secrecy, you can't tell something that you don't remember.

Q: Would the ETs continue to communicate with you, since you have been out?
"No, I've had no contact with them since getting out.

Q: Are you disappointed that they have not contacted you, to give you reassurances?

325

"No, I actually have, my faith is very strong, so needing a reassurance from ETs, no. I am of the mind that there was help in getting into the mess we have here. They were ET races were directly involved in creating the mess that we have got. My thoughts are, thank you from keeping our environment safe and clear from meddling ETs. We need to clean this up. Humans need to be involved with cleaning this up, not the ETs races.

Q: What about Disclosure net 5 years?
"Yes. I see it happening. The existing Disclosure plan that is happening now, involves slow release of technology, February of 2018 they released information on transparent metal. Came out as Patent. They are working on it. They have had that for decades. The DARPA last year released neuro-interface between a pilot and 3 drones. Another technology that was mature when I went in 1997 well developed again. They're releasing of technologies through universities and corporations so get to pay for them twice."

Q: Is it consciousness or technology that gets us to that tipping point?
"I believe it's consciousness that's going to get us there. And demand from the people. Those responsible for 80 years of secrecy are the same ones that came up with this drip-drip disclosure plan. I do not want them deciding that. They are the ones that kept us in the dark and lied to us for 80 years. I want the people, the civilians, the citizens to decide what we are going to do with technology integration. In order to do that, we need a number of people that come forward to put a plan together. This is how we are going to obsolete certain job industries, like the power company grid, there's thousands of workers there, that are now out of work if you drop new technology that allows power technology at each house. What are we going to do with all those workers? Well, let's get them to work on the next generation of power. Well, there is some training involved, so let's get them involved in determining their future. How do they want to get trained how do they want to work? Same thing for the medical industry. Same thing for transportation, same thing for ecology. Same principles apply." [2]

CHAPTER 25: FROM SECRET SPACE PROGRAM TO SPACE FORCE

Jason Rice on *Fade to Black with Jimmy Church*, October 3, 2018. [3]

Q: Have you met Corey Goode?
"I have briefly."

Q: Had Corey not come forward; would you have come forward?
"That is a great question, I had been motivated to come forward for a number of reasons. Had Corey 3 not, I still would have been motivated to come forward, whether it was one platform or social media platform, I would have come forward.

Q: Had you discussed the SSP with friends and family before 2016?
"The elements I discussed with family had to do with events of my childhood when I would leave the house. The other events and memories that came back to me, when they started coming back, I didn't have all the pieces put together. There was a string of bizarre events that were happening, that I didn't have enough of a complete picture yet, until much later. I think a lot of that has to do with personal spiritual growth, individual vibrations, and the passage of time, of where our Earth is at in our celestial orbit. The energies that Mother Gaia is going through at this time, contributed to the awakenings that I was going through."

Q: This is a heavy subject for the community that we are trying to digest at the same time. What caused these memories to start coming back? I understand you had memories that were in real time throughout your childhood, so you knew something was happening. But as far as the SSP, Mars, Draco, the Moon and underground bases and training facilities, and MILAB, when did those memories come flooding back? Was Corey a catalyst for that?

"When I first got out of the US Army, I was having issues with PTSD, waking up in the middle of the night fighting, it was confusing because I had no frame of reference. So, I sought out professional help, to get the help I needed, and I was able to center myself again to take care of myself my family and my profession, went on about life. There were a number of events over the years that happened, I had outpatient surgery done, I was in the waiting room, after anesthesia, the nurses were joking with me, I was talking about being in space and helping humanity under anesthesia, but I didn't remember any of that. This was in 2002.

Q: The nurses asked you about the SSP?
"Yes, they mentioned to me, that I was talking about being in space in some classified program and helping humanity. They told me what I said, after coming out of the anesthesia. I do not recall any of it.

Q: Did this help your memories come back?
"No as what I refer to as parting gifts from these USAPS, let's keep in mind too, every person that goes into the "20 and Back SSP Program" has their minds tinkered with, make no mistake about that. That leaves an impression that is different for each person. So having those parting gifts that they leave with you, some of them include, suggestions, commands, triggers for seeking out destructive behavior, or for changing your behavior if you start thinking or going into too deeply in trying to seek information to discover what's happened. So, my programming held up and worked, and every time I thought, gosh, that's strange, then something would happen, and I would be completely distracted, and far too busy with other things than to be thinking about any strange conspiracy."

Q: Then naturally you would have to ask yourself are these memories that have been implanted? Have I been programmed? Are these real memories from real events? And that is a bad situation, and the third thing can be that you are crazy and lost your mind.
"Sure. I got my training in college as a scientist, as an engineer. Going all the way to the standard scientific process you go through, am I losing my mind? Why am I having these thoughts or memories? What is going on? One of the things that I can tell you is that the memories are crystal clear when they first come back. It is like the difference between watching rabbit ears television black and white and watching a high-definition LED TV. The memories I have gotten to recognize are vastly different, than just a normal dream of walking around my yard, it is not the same.
Of course, I ask myself, what in the world is going on? As a scientist, there is another component, another piece in the puzzle. The awakening process is both invigorating, terrifying, enlightening and ultimately a very spiritual experience, because you realize all the things and possibilities that you've done, so trying to impart that information to other people is challenging to say the least. Because it is such a departure from our normal frame of reference. Because

how can I explain this? It is like trying to explain the concept of Time Dilation or Timelines. You can get into an entire 5-hour conversation trying to explain how time works. In this consensus reality, that information has not been released. So, some of the memories are still hidden and this is an ongoing dynamic process, at least for me. I have memories that are still coming back, more pieces of puzzles. If I knew and have an understanding of time manipulation, which I'm sure I did, because as an engineer I would have drilled down into it and figured out what it was about, then I would explain it better. So as the memories start coming back, more and more, it was a tremendously spiritual experience after realizing what had happened, with all the pain and trauma, you could go crazy and end up being a homeless person on the street and end up battling spiritual forces.

Q: Any other crew members you worked with in the program that you are in contact with today?
"Yes. I don't want to violate the free will of the others. And being red pilled. Whether or not another remembers, it is not my responsibility to wake them up. Because each person determines their own speed, their own determination of when or if they wake up.

Q: Do any of you talk about it?
"No.

Q: Do you anticipate others reaching out to you with similar experiences?
"There are Thousands from every country on the planet, except for those that do not have a central bank.

Q: So, there's 6 countries, no one from the 20 and Back Program from N. Korea.
"Of course, this is all back in 1991 when I first went into the program. I don't have the central bank list memorized.

Q: One thing that is interesting was your knowledge of history on your Gaia interview?
"Part of the conventionally available information of history was one of the subtle signs of my awakening process, was an interest in this topic or that topic, was an insatiable need to find out more about that. For e.g., operation paperclip, operation castle 1234, those types of we can find out on the

internet, a lot of the history that was not published was conveyed to me by one of my fellow soldiers, Yohan.

Q: Information from Mars, how was it taught to you?
"Information that is not taught to us as ground pounders in the program that I was in. The SSP and the Cabal have insured their secrecy and privacy through compartmentalization. As was mentioned on Gaia's Cosmic Disclosure, the Manhattan Project had over 100,000 people working on it, but they did not know that. So, through compartmentalization they were able to keep the full picture away from so many people. I think a more modern-day example would be if you believe in the Q-Anon phenomenon. Q said that there are less than 10 I think that there are 7 who know the full plan of what is going on. Rest assured that there are people working for those goals through compartmentalization the air traffic control helped with the planes that were on an operation with the pilots who had a mission, but they do not know the full picture. So, through compartmentalization they're able to keep it to a very small number of people that know the full picture."

Q: Insider, whistleblower, is a footprint in social media, background, there is a lack of social media presence of you. Is Jason Rice your real name?
"Yes, it is my real name. There is a very specific reason why I do not have a website and that has everything to do with I have come out to talk about my story. I am not interested in attention for me. I have had a couple of years of time where I could have started posting, but I am concerned about the safety and privacy of my family. So, I made the choice not to start blasting the internet with the things that I have to say."

Q: The community and media journalists out there are going to look into your background, has that already started?
"I would be surprised if it hasn't. I do want to say that I have, plans for a website in the works, I expect to have out in a month or so. Again, I am a family man I have worked in the engineering field. This has come out to respect and honor those that I served with, as well as to raise the consciousness and awareness of everybody that I can, about what has been going on. Good, bad, ugly, and different, whatever it is. I have come out to do that, respecting my privacy and my family's privacy. You only must look at what has been happening in Washington, to see what people can do to a family. To recognize, hey look this is a real life we are talking

about. Real family. Real people. I did not want it to turn into a circus at my house. I very specifically did not go public, until things were ready to be aired on Gaia. A large reason for that, is I was concerned for my safety.

The people that run these programs, have absolutely zero respect for the life of us sheeple, if you will. They have killed Presidents, they have killed generals, they have killed thousands of other people in the blink of an eye. What is some engineer in some city to them? Nothing. So, my coming public has now closed off one avenue in which I do not think they will come after me directly. They have other tools at their disposal. The next will be character assassination, the next will be financial ruin, there is a whole list, you can go through their playbook, it is not hard to find. Coming out at one time, was the best possible way I can assure that myself and my family are safe.

Q: Do you have plans have you thought about sitting down with Corey or other whistleblowers or taking notes or figuring out ways to move forward.

"I have nothing to fear by sitting down with other whistleblowers. I am not concerned about encroachment on what I have to say. Because this is my truth, my experiences. Everyone else in the SSP through the compartmentalization, will have their own experiences. Now there is some things that we will not share if we were not in the same programs. Obviously if we are in the same ones, we will have similarities and a lot to talk about. I welcome that. Especially in this day and age I can't be more grateful enough for the mainstreaming of the whole scientific process and taking responsibility for information for knowing what's going, it's forced people to ask the hard questions, and look into satanic ritual abuse, child and human trafficking, all the ugliness that exists out there, because it's made people more aware of what's happening around us. So, this particularly, is a great time, there is not a better time than now for the whistleblowers to work together for not being afraid of encroachment or having a fear that their voices will not be heard. That is not what this is about. This is about finding out information because Disclosure is going to happen. I refer to it as Disclosure D Day. Because eventually there is going to be dump and I fully expect and demand full Disclosure. Do not hold anything back. There is going to be that dump day - D Day. It is going to happen. I think there has been a change in what used to be considered tinfoil hat conspiracy theories. A lot people are waking up and going, wow, they did that? Oh my gosh, like how? What a better time for it to happen.

Q: After you left the program and were raising your family, did you have a handler, did you have anybody that stayed with you after you left the program and went back into civilian life?

"No, I did not. Not that I was aware of. There have been several incidences, where I suspect that I was getting closer to recall, and I suspect that there was a re-abduction and re-wiping that occurred. Those are only suspicions at this point based on circumstantial evidence.

Q: Has any reached out to you since the episodes of Gaia have aired?
No.

Q: The announcement by President Trump about the Space Force – we are all intrigued by creating a new arm force in the Pentagon. Is this part of the SSP?

"I suspect so, is that remember last year all those launches that occurred globally in a short period of time. I doubt seriously that there was anything in them. My feelings are that was a cover for the already SSP assets in place, so they can say, yeah, we put that in place. I worked at Kennedy Space Center (KSC), you ask any space engineer, how long does it take to design prep and build a satellite? They don't' keep these things just laying around in the warehouse. They are too expensive, far too classified, too many security risks to just keep them lying around. These are planned years in advance, so, Trump has not been in office that long for them to have built and go through the entire process to build a satellite, there just wasn't enough time. I think Trump's mentioning of the new Space Force was a part of the unmasking of the drip-drip disclosure, to get to the point where he can say, hey look, we've got these assets in space, hey look we've got this technology in space. I have some other suspicions that we can get into, we can talk about several things, that has to do with the Mars Germans.

'Mars Germans' refers to the Nazis who allegedly established Mars colonies. They are also on the Moon. [3]

CHAPTER 25: FROM SECRET SPACE PROGRAM TO SPACE FORCE

Notes:
1. Trump Orders Pentagon to 'Immediately' Establish 'Space Force', https://www.thedailybeast.com/trump-orders-pentagon-to-immediately-establish-space-force, June 2018.
2. Whistleblower Jason Rice Interview, Zohar Stargate TV aka UFO & Ancient Mysteries Networking TV, (UAMN TV), YouTube Channel, March 25, 2019.
3. Jason Rice on *Fade to Black with Jimmy Church*, October 3, 2018

CHAPTER TWENTY-SIX

WHO OWNS SPACE?

*"The mind is its own place,
and in itself can make a heaven of hell,
a hell of heaven."*
~ John Milton, *Paradise Lost*

The Exopolitics of Space determines the rules and policies space explorers agree to live by in our present Space Age and the emerging private aerospace industry. As humans continue exploring the vastness of space though, the question of ownership and creating laws for Earth's space businesses becomes more relevant.

It may be hard to imagine any group or any one individual owning the vastness of outer space, but with nascent rise of Space businesses, the question becomes important, to establish rules of engagement and perhaps a set of cosmic real estate laws to abide by as Space becomes a popular an area for exploration.

In 1967, the United Nations Declaration called *The Outer Space Treaty*, stated that no nation on Earth can claim the Moon, asteroids, or other celestial bodies as their own. It was essentially written and established to prevent a nuclear standoff between nations in Space, the real Star Wars. But today, this treaty no longer holds ground, as private aerospace companies, rather than nations, look to make a profit in the cosmos. For example, there are aerospace industries that will mainly focus on Space Tourism, while others will provide the necessary taxi service to and from the International Space Station, and even more Space Stations established around the Moon and Mars.

Needless to say, there are several gray areas, like who is responsible for satellite damage? And, who gets to clean up space debris caused by an awful of lot of refuse and retired space technologies?

Legal experts are working to create new laws like Antarctic mining and sea drilling bans, while providing international rules for businesses, which apply to those who want to mine asteroids. It is still uncertain, but suffice it to say, the future will likely be a mad grab for space in Outer Space, hopefully not likened to the mad grab of Earth humans hoarding toilet paper during panic attacks.

Space law is the body of law governing space-related activities, encompassing both international and domestic agreements, rules, and principles. Parameters of space law include space exploration, liability for damage, weapons use, rescue efforts, environmental preservation, information sharing, new technologies, and ethics.[2] Other fields of law, such as administrative law, intellectual property law, arms control law, insurance law, environmental law, criminal law, and commercial law, are also integrated within space law.

The origins of space law date back to 1919, with international law recognizing each country's sovereignty over the airspace directly above their territory, later reinforced at the Chicago Convention in 1944. The onset of domestic space programs during the Cold War propelled the official creation of international space policy (i.e., the International Geophysical Year) initiated by the International Council of Scientific Unions. The Soviet Union's 1957 launch of the world's first artificial satellite, Sputnik 1, directly spurred the United States Congress to pass the Space Act, thus creating the National Aeronautics and Space Administration (NASA). Because space exploration required crossing transnational boundaries, it was during this era where space law became a field independent from traditional aerospace law.

Since the Cold War, the Treaty on Principles Governing the Activities of States in the Exploration and Use of Outer Space, including the Moon and Other Celestial Bodies (the "Outer Space Treaty") and the International Telecommunications Union have served as the constitutional legal framework and set of principles and procedures constituting space law.[7][8] Further, the United Nations Committee on the Peaceful Uses of Outer Space (COPUOS), along with its Legal and Scientific and Technical Subcommittees, are responsible for debating issues of international space law and policy. The United Nations Office for Outer Space Affairs (UNOOSA) serves as the secretariat of the Committee and is promoting Access to Space for All through a wide range of conferences and capacity-building programs. Challenges that space law will continue to face in the future are fourfold—spanning across dimensions of domestic compliance, international cooperation, ethics, and the advent of scientific innovations. Furthermore, specific guidelines on the definition of airspace have yet to be universally determined.

The US flag was planted on the Moon in 1969, two years after the Outer Space Treaty was created.

When space crops up in conversation, ownership does not immediately spring to mind. But as the human race continues to advance in this field, and with commercial space enterprises just around the corner, questions about power politics and their interaction with space exploration must be asked and answered.

Neil Armstrong famously planted a US flag on the Moon in 1969. This gesture may have implied territorial ownership but was purely symbolic because of the 1967 Outer Space Treaty.

129 countries, including China, Russia, the UK, and the US, have committed to this treaty, which is overseen by the United Nations Office for Outer Space Affairs.

It sets out important principles, such as the concept that space should be considered the province of all mankind, that outer space is free for the exploration and use by all states, and that the Moon and other celestial bodies cannot be claimed by a sovereign nation state. Additionally, the Moon and celestial bodies are to be used purely for peaceful purposes, and weapons will not be placed in orbit or in space.

"This is frequently referred to as the outer space constitution," says Dr. Jill Stuart, a visiting fellow at the London School of Economics and Editor of the journal Space Policy. She spoke to BBC News at the British Science Festival in Bradford.

WHERE IS OUTER SPACE?

This treaty has worked so far, but there are some potential pitfalls - as Dr. Stuart explains.

"There is no official definition of outer space, but it's something on which a United Nations working group is currently consulting member states. I suspect we will settle for a physical demarcation at the Karman Line, which is about 100km up, but it is also an option to go for a functional definition. This is where laws are defined based on the function of a space object rather than where it is in space."

A physical demarcation results in a lot of paperwork for commercial spaceflight companies, such as Virgin Galactic, which is developing a sub-orbital tourist space plane. It means Virgin has to abide by both international aviation laws as well as space laws, despite only being "in space" for five or six minutes. A sensible compromise has to be reached.

MINING THE MOON

Commentators agree that the Outer Space Treaty is an excellent foundation for international space law, but it makes no reference to commercial space activities, such as the exploitation of space resources; presumably because this was not foreseen back in 1967.

"International law is ambiguous about private companies setting up mining operations in space. There is a strong case for revisiting the Outer Space Treaty to bring it up to date," argued Ian Crawford, a professor of planetary science at Birkbeck College, University of London.

There is an argument that in the future, when assets are developed in space, it is more cost-effective to use raw materials mined from space rather than transporting them from Earth.

As well as the Outer Space Treaty, there are four other treaties governing space law. According to the Liability Convention, anything that goes into space must be registered with its launching state and becomes sovereign territory.

"If you were to target another country's satellites, you will create a lot of space debris, which could impact other satellites," said Dr Steer. This is where ambiguity arises over who is responsible for clearing up the mess. Dr Steer added that some satellites have dual use. Their technologies can be used in the military as well as the civilian context - and this makes the issues around the Outer Space Treaty quite complex.

What if other intelligent life is encountered, with their own set of rules? Whose laws would take precedence. This topic perhaps throws up more questions than solutions. "We're at a point in time where it's ever-more pressing to re-evaluate our current legal infrastructure that governs outer space," Dr Stuart concluded.[2]

THE REAL STAR WARS

I have been saying this all along, *the Truth is Stranger than Fiction*! I still say it, and have a running commentary on this very theme on my Public Facebook Page: *Who's Who in the Cosmic Zoo?* http://www.Facebook.com/whoswhointhecosmiczoo

More and more we are seeing that our sci-fi movies no longer necessarily depict *futuristic concepts*, but concepts that have been around for millennia. As King Solomon said, "What has been will be again, and what has been done will be done again; there is nothing new under the sun. Is there a case where one can say, "Look, this is new"? It has already existed in the ages before us." (Ecclesiastes 1:9-10) We are spiraling back to these events, as the real Star Wars comes full circle for its final Battle to be fought in the Heavens, or the skies and space above and around the Earth.

Space Force Boss Says One Of Russia's Killer Satellites Fired A Projectile In Orbit

A very recent test involved small orbital inspectors that had previously been observed shadowing an American spy satellite. The head of U.S. Space Force, the U.S. military's newest branch, had said publicly for the first time that the Russian government has conducted two on-orbit anti-satellite weapon tests in the past three years. These revelations come less than six months after the U.S. military expressed concern about a Russian "inspector" satellite that appeared to be shadowing an American KH-11 spy satellite.

Chief of Space Operations General John "Jay" Raymond, Space Force's top uniformed officer, revealed the two apparent tests in an interview with Time for a profile of the new service. Raymond is also presently head of the joint-service U.S. Space Command (SPACECOM). That piece, which was published on July 23, 2020, offers a detailed look at where Space Force is now and where it hopes to go in the future and is worth reading in full.

USAF Chief of Space Operations General John "Jay" Raymond told Time on July 15, 2020, "Russia is developing on-orbit capabilities that seek to exploit our reliance on space-based systems." He explained that a satellite identified as Cosmos 2543 launched a projectile that could be used to destroy another craft in space.

The Kremlin describes Cosmos 2543, also sometimes written Kosmos 2543, as a "space apparatus inspector," one of a number in orbit now, which are ostensibly intended to do just what the name says, inspect other satellites. On the face of it, this offers Russian officials a way to investigate problems with or assess damage to other space-based assets on-orbit.

However, given their small size and high degrees of maneuverability, there have been long-standing concerns that these orbital inspectors could double as spies or even "killer satellites" capable of getting close to and then disrupting or destroying other space-based platforms by any of a number of means, including electronic warfare jamming or a directed energy weapon, such as a high-powered microwave beam. They could also potentially manipulate a satellite in a way that would disable it or launch kinetic attacks, either smashing into the target themselves or launching projectiles, the latter being something that Space Force now says the Kremlin has been actively testing.

Defense Intelligence Agency depicted a number of ways in which one satellite might attack another, including by acting as a kinetic kill vehicle.[10]

The destruction of various satellites as "high flying assets" could easily be a key feature of any future large-scale conflict, especially during its opening stages. The U.S. military, among others, is heavily reliant on space-based systems for a wide array of functions, including early warning, intelligence gathering, communications and data-sharing, navigation, and weapons guidance, and more.

It is worth noting that another inspector satellite, Cosmos 2542, which the Russian government launched in November 2019, had actually deployed Cosmos 2543 while in orbit the following month. Time says the U.S. military has dubbed these "nesting doll" satellites, a reference to traditional Russian matryoshka dolls.

The Russian Ministry of Defense had previously announced a test involving Cosmos 2543 on July 15. However, it said that this simply involved the satellite maneuvering close to another orbital inspector, Cosmos 2535, and gathering imagery and other information in line with its publicly stated inspection mission. At that time, U.S. Space Command did note that it had detected something separating from Cosmos 2543, which it labeled Object 45915, but recorded its type as "to be determined."

The assertion that these satellites are actually part of a space-based anti-satellite weapon system is even more significant given that Cosmos 2542 had moved into a position to shadow a U.S. KH-11 spy satellite, publicly identified only as USA 245, in January. The month before, USA 245 had shifted its own orbit, potentially to avoid hitting the smaller Cosmos 2543, which Space Force later said also appeared to be following the American satellite.

"This is unusual and disturbing behavior and has the potential to create a dangerous situation in space," General Raymond said at the time in a statement to Business Insider. "The United States finds these recent activities to be concerning and do not reflect the behavior of a responsible spacefaring nation."

It is worth noting that this was hardly the first instances of satellites shadowing each other in space and that U.S. government has conducted similar activities in the past, as well. However, it appears now that the U.S. military had a particular concern about what was going on in January based on existing intelligence.

General Raymond also told Time that the space-based anti-satellite weapon test on July 15 was similar to something the U.S. military had observed in 2017. He did not elaborate on that previous incident, but it seems very possible that is related to another set of "nesting doll" satellites that emerged that year. The Kremlin launched the first of these, Cosmos 2519, in June 2017. Two months later, that satellite deployed another, dubbed Cosmos 2521. These were joined by a third, Cosmos 2523, which also appeared to have been deployed on-orbit from other these other satellites, in November 2017.

If that trio is what Raymond was referring to, it is unclear if Cosmos 2523 is what the U.S. military believes was actually an anti-satellite interceptor or if one of these satellites launched yet another object at some point.

"This is further evidence of Russia's continuing efforts to develop and test space-based systems, and consistent with the Kremlin's published military doctrine to employ weapons that hold U.S. and allied space assets at risk," the head of Space Force said in a separate statement following the publishing of Time's piece.

"This event highlights Russia's hypocritical advocacy of outer space arms control, with which Moscow aims to restrict the capabilities of the United States while clearly having no intention of halting its own counter-space program – both ground-based anti-satellite capabilities and what would appear to be actual in-orbit anti-satellite weaponry," Assistant Secretary of State for International Security and Nonproliferation Christopher Ford also said.

Russia, as well as China and others around the world, has been actively developing various anti-satellite systems, including ground-based and air-launched interceptors, and directed energy weapons. In April 2020, Space Force publicly accused Russia of testing an unspecified "direct-ascent anti-satellite missile," or DS-ASAT.

"Russia's DA-ASAT test provides yet another example that the threats to U.S. and allied space systems are real, serious and growing," General Raymond said in a statement after that test. "The United States is ready and committed to deterring aggression and defending the Nation, our allies and U.S. interests from hostile acts in space."

These public statements, together with the disclosure of the earlier assessment that an on-orbit anti-satellite weapon test had taken place in 2017, also follow comments from various senior U.S. military officials about the need to declassify more about what it is doing in space and the threats that it faces, as well as how it might be able to respond. In 2019, then-Secretary of the Air Force Heather Wilson warned that the United States might have to demonstrate its own counter-space capabilities in the future in order to deter potential adversaries.

"We've got some education to do for the average Americans on just how reliant their lives are on space capabilities," General Raymond told Time.

The disclosure that Russia has been testing on-orbit anti-satellite weapons certainly appears to be part of a campaign to increase public awareness of the very real threats to space-based systems that exist now and are continuing to emerge. [10]

SPACE FORCE JUST RECEIVED ITS FIRST NEW OFFENSIVE WEAPON

This is just one of two acknowledged US counterspace capabilities, but there are more in the classified realm. U.S. Space Force has begun operating a new offensive weapon system, an upgraded version of a ground-based satellite communications jamming system, for the first time in its short history. The first iteration of the Counter Communications System entered U.S. Air Force service in 2004 and the program has now gotten transferred to the newest branch of the American military.

The Space Force declared it had reached initial operational capability with the Counter Communications System Block 10.2, or CCS B10.2, on Mar. 9. The Harris Corporation, which merged with L3 Technologies last year to form L3Harris Technologies, had received the contract from the Air Force to develop this upgraded variant of the system in 2014.

The National Defense Authorization Act for the 2020 Fiscal Year, which Congress passed, and President Donald Trump signed in December 2019, officially established Space Force as a separate service within the Department of the Air Force. Units and assets previously assigned to Air Force Space Command now form the core of the new service, which is still very much in the process of standing up.

"CCS is the only offensive system in the United States Space Force arsenal," Lieutenant Colonel Steve Brogan, the Combat Systems branch materiel leader within the Space Force's Space and Missile Systems Center's (SMC) Special Programs Directorate, said in an official news piece about the system in January 2020. "This upgrade puts the 'force' in Space Force and is critical for Space as a warfighting domain."

The Air Force introduced CCS in 2004 and fielded the Block 10.1 upgrade in 2014. Last year, Harris also received a $72 million contract to begin developing the new iteration, the Block 10.3 variant. As of September 2019, the Air Force reportedly had at least seven complete CCS packages, which are intended to be rapidly deployable.

"CCS has had incremental upgrades since the early 2000's, which have incorporated new techniques, frequency bands, technology refreshes, and lessons learned from previous block upgrades," Major Seth Horner, the CCS B10.2 Program Manager, also said in January. "This specific upgrade includes new software capabilities to counter new adversary targets and threats."

The various versions of the system all include a number of trailer-mounted dishes and associated equipment. Details on exactly how the CCS functions are limited, but it is understood to be a jamming system that disrupts transmissions from enemy communications satellites. This could give U.S. forces valuable advantages on the battlefield by disrupting the ability of enemy units to rapidly communicate and share information using satellite-based systems.

The Air Force has described the effects as "reversible" in the past, meaning that when the jammer shuts off, the target satellite would go back to functioning are normal. The service had specifically chosen a non-kinetic anti-satellite weapon in order to avoid having to physically destroy hostile assets in space, which would create debris that could pose risks to friendly systems in orbit. This is notably different from Russian and Chinese anti-satellite efforts, among others, which include ground-based and air-launched kinetic interceptors, as well as possible dual-use "killer satellites." There are also reports that both countries are working on ground-based anti-satellite directed energy weapons.

"We are not going down that path," U.S. Air Force General John Hyten, then-head of Air Force Space Command, told lawmakers in 2016, referring to the development of capabilities that would totally destroy enemy satellites and create debris. Hyten is now Vice Chairman of the Joint Chiefs of Staff.

Hyten's comments would not necessarily preclude the development of offensive space capabilities that could create non-reversible effects and effectively destroy a target in space without totally disintegrating it. This could still include kinetic attacks, such as using a directed energy weapon, possibly mounted on another satellite, to destroy certain components without creating any substantial amount of debris.

Regardless, Space Force reaching initial operational capability with a new offensive weapon system for the first time is certainly a notable milestone for the still extremely young service. Beyond that, it is also a rare public announcement regarding the development of an American anti-satellite capability of any kind.

The Counter Communications System has long been the only such capability the United States publicly acknowledges having, though it is understood that it has other offensive counter-space capabilities. Space Force Vice Commander Lieutenant General David Thompson told lawmakers the new service was working on new capabilities "to protect and defend" America's space-based assets during a hearing on Mar. 4.

"We began prototyping, and demonstrating, and preparing for what I'll call abilities to protect and defend our assets, and we did that extensively in the budget in [the 2020 Fiscal Year]," he said. "In [the 2021 Fiscal Year], we are now taking steps to extend that across the fleet, as well as look at other capabilities to be able to continue to defend those assets that we have and defend adversary use of space."

Members of the 4th Space Operations Squadron, an element of the 50th Space Wing, load a trailer-mounted Low-Profile Antenna, used to support deployed satellite communications activities, on a U.S. Air Force C-5 Galaxy airlifter during an exercise in 2018.

In February 2020, the Air Force mentioned another counter-space system, called Bounty Hunter, in a news story about recent tests that the 17th Test Squadron had conducted. "The 17th TS' [Test Squadron] Operating Location-Charlie [OL-C] conducts testing on the base's higher classification systems," that official news item said.

"Recent tests we've done have been on the Bounty Hunter and the Counter Communications System, which are both electronic warfare systems located at Peterson [Air Force Base]," Technical Sergeant Tricia Benson, the OL-C Flight Chief and Project Manager, said in an official interview. There was no further information on Bounty Hunter or its capabilities.

In its first-ever budget request, for the 2021 Fiscal Year, Space Force asked for nearly $54.7 million in funds for research and development of "counterspace systems," including almost $50.5 million for continued work on the CCS program and just under $2 million for upgrades to Bounty Hunter. The remaining funds would support the development of a new and improved command and control architecture for counterspace capabilities. Details about all of these projects in the publicly available budget documents were extremely limited.

Other U.S. military officials have alluded to the still-classified capabilities recently, as well. Former Secretary of the Air Force ominously stated last year that it might be necessary to conduct a show of force using unspecified systems to deter potential adversaries, such as Russia or China, from attacking American satellites.

"We have a capability called a Counter Communications System that is built to deny an adversary the use of space communications," General Hyten had also said in an interview with CBS News' "60 Minutes" back in 2015. "All I can say is it's a capability that exists on the ground and it does not create debris in any way."

"The only two things you told me about the U.S. ability to fight in space, are the ability to maneuver your satellites and to jam other satellites. Is that it?" CBS News' David Martin then asked the general.

"That's not it, but that's all I can tell you," Hyten responded.

Ensuring the United States has the ability to conduct a conflict in space, including as part of a larger crisis on Earth, and what such a conflict might actually look like are certainly among the most pressing issues facing Space Force, something the War Zone has discussed in detail in the past. The fledgling service has now publicly acquired its first new counter-space system and it will likely take over other existing top-secret offensive space capabilities, and the development of future ones, soon, if it hasn't already.[11]

USAF SECY WARNS FORCE JUSTIFIED TO DETER SPACE ATTACKS

USAF Secretary Gives Ominous Warning that Show of Force Needed to Deter Space Attacks: Outgoing Secretary of the Air Force Heather Wilson says the United States may need to conduct a show of force to deter opponents, such as Russia and China, from attacking U.S. military satellites in space. But it was not entirely clear whether this would involve demonstrating the ability to attack an adversary's own space assets only or if it would even show an ability or willingness to attack hostile space-based systems at all. This cuts to a larger issue, that no one really knows for sure what a war in space would actually look like, though the United States is certainly working to answer that question.

Wilson offered her view on the future of potential conflict in space to reporters on the sidelines of the Space Foundation's 35th annual Space Symposium, which began on Apr. 8, 2019. Other Air Force and U.S. officials have also used the event as an opportunity to highlight the very real and growing threats to vital satellites that support critical military capabilities, such as long-distance communications, navigation, weapon guidance, early warning, ballistic missile defense, and intelligence gathering. Defending those assets has been an increasingly hot topic in recent years and President Donald Trump's administration has gone so far as to propose the creation of a sixth military branch, the Space Force, to focus on these issues. [12]

"There may come a point where we demonstrate some of our capabilities so that our adversaries understand they cannot deny us the use of space without consequence," Wilson said, according to The Daily Beast.

"That capability needs to be one that's understood by your adversary," she added, according to Air Force Magazine. "They need to know there are certain things we can do, at least at some broad level, and the final element of deterrence is uncertainty. How confident are they that they know everything we can do? Because there's a risk calculation in the mind of an adversary."

Wilson was not the only one to warn that the United States might retaliate forcefully in response to an attack on U.S. military space-based systems. "It's not enough to stand in the ring and take punches. You have to have the will and capability to punch back," U.S. Air Force Chief of Staff David Goldfein also said at the Space Symposium.

USAF

Secretary of the Air Force Heather Wilson and Air Force Chief of Staff General David Goldfein, to her immediate right, at the Space Foundation's 35th annual Space Symposium.

As *The War Zone* has pointed out on multiple occasions, the threats to U.S. military space assets are very real. The U.S. military's historical advantages in space-based capabilities and its increasing reliance thereon have long made them attractive targets to potential adversaries, such as Russia and China.

Most recently, there have been reports that China may be working on ground-based laser weapons to blind or otherwise disrupt optical systems on U.S. satellites, including ones tasked with missile warning and intelligence gathering missions. On Apr. 9, 2019, at the Space Symposium, Acting Secretary of Defense Patrick Shanahan said that the United States is of the opinion that the Chinese will have an operational anti-satellite laser capability by 2020. A report from the Defense Intelligence Agency in February 2019 said that China may be able to actually damage or destroy a satellite with a laser system by the mid-2020s.

In December 2018, Russia announced that they had put a ground-based laser system, called Peresvet, into service. The Russians claim this system could have an anti-satellite role and also say they are developing an aerial anti-satellite laser system.

This is to say nothing of the various anti-satellite interceptors and potential small "killer satellites" Russia and China have also deployed. In March 2019, India demonstrated its ability to shoot down a satellite, which shows that these capabilities are only proliferating.

Secretary Wilson declined to say what specific offensive capabilities the United States might employ as deterrents against attacks on U.S. military satellites, or if it had any such systems in place now. In February 2019, the Air Force did wrap up a four-month study into the issue, according to Air Force Magazine.

"We looked at all of our missions in space, from missile warning to communications and intelligence collection," Wilson said at the Space Symposium. "We took the best estimates of the threat and presumed a thinking adversary who would respond to the actions that we take."

A general overview of the potential threats to US satellites posed by offensive killer satellites. Ground-based threats down on Earth can also include jamming, laser and other directed-energy attacks, and hard-kill interceptors.

This process involved thousands of wargames and at least 25 tabletop exercises involving not only Air Force personnel, but individuals from other military branches, the National Reconnaissance Office (NRO), and the Defense Advanced Research Projects Agency (DARPA). These exercises also included simulated means of mitigating the threats, including the use of disaggregated constellations of hundreds or potentially thousands of small commercial satellites to make it more difficult for an opponent to swat down an entire network. Air Force Secretary Wilson has suggested that the results did not reflect favorably on this idea, which the Office of the Secretary of Defense and the newly minted Space Development Agency have been keen to promote.

While we don't know what responses the U.S. military might already have or be considering, we do know that it has demonstrated its ability to shoot down a satellite with an air-launched interceptor in 1985 and with a surface-launched missile in 2008. The latter event was known as Operation Burnt Frost. In 1997, it also conducted a test of a ground-based laser weapon that may have at least blinded or otherwise temporarily disrupted the optics on a target satellite.

The U.S. military also has a congressional mandate at present to be developing space-based missile defense systems. Many of the concepts on the table now would have an inherent anti-satellite capability. But officially, the United States does not have an anti-satellite system deployed and it has never described any past developments and experiments as offensive in any way, despite their obvious potential applications.

Of course, neither Wilson nor Goldfein specifically said that any potential retaliatory attacks would have to be limited to targets in space, or necessary include them at all. An American response to an attack in space could include conventional strikes, or even potentially nuclear ones if the damage done to U.S. space-based systems were to have sufficiently severe impacts on the ability of the United States to continue functioning as normal.

There could be non-kinetic options, as well, including offensive cyber-attacks. Either kinetic or non-kinetic options could be focused specifically on disabling an opponent's ability to control their own space-based systems.

CHAPTER 26: WHO OWNS SPACE?

This all leads to the bigger question of what conflict, or even in isolated attack, looks like in space and what a proportional response would be if one were to erupt. What kind of attack prompts what level of retaliation? Is an opponent temporarily disrupting a satellite's operation enough to warrant a kinetic response?

What if an adversary destroys a top-secret satellite the very existence of which is classified? What if that was a weapon you had not publicly acknowledged you already had? What if you cannot tell where the attack came from or who is responsible?

US ARMY

This 1960s US Army concept area showing troops fighting on the Moon with futuristic space guns will not be what an actual modern conflict in space looks like any time soon.

These are hardly new questions.

"It's really difficult to go ahead and justify how you might attack somebody's homeland if they've taken out a satellite that you don't even admit exists," Douglas Loverro, then-Deputy Assistant Secretary of Defense for Space Policy, said during a talk in 2016. "Is jamming an attack? Is a laser an attack? Does it have to be a kinetic hit on a satellite to be an attack?"

"None of us actually know how a war – if and when it extends to space – will actually evolve, where and what phase will it happen, when will it happen in the conflict, how will it be engineered, we don't know any of that," he continued. "It's difficult politically, it's difficult emotionally. Probably people are going to die on the ground where nobody's going to die in space."

Actually, destroying something in space in a retaliatory strike may not even be a realistic option. Blowing up a satellite can only create a cloud of extremely fast flying debris that is impossible to control and could threaten other friendly military and civilian space assets, as well as enemy targets.

The world is grappling with that particular issue right now in the aftermath of India's anti-satellite interceptor test. The debris field from that event presents a serious threat to objects in low earth orbit and may not fully dissipate for up to 18 months. [12]

WEAPONS IN SPACE

SpaceX Chief Operating Officer and President Gwynne Shotwell, told an 'We would launch a weapon to defend the U.S.' [13]

During an appearance on at the Air Force Association's annual symposium in Maryland, September of 2018, Shotwell was asked a question she said she had never heard before: "Would SpaceX launch military weapons?" Shotwell replied somewhat surprised, "I've never been asked that question." Her response: "If it's for the defense of this country, yes, I think we would." [13]

The room packed with Air Force service members and military contractors burst into applause. They seemed impressed that SpaceX is one of the world's coolest companies and also a staunch patriot, that if push came to shove, they would defend the United States of America with Space Technology Defense Systems.

It seems that it is inevitable that with all the high-flying assets that a defense system has to be prominent, which was initially launched by the United States under President Ronald Reagan's Star Wars Program, Strategic Defense Initiative, (SDI).

Notes:
1. "What Is Space Law?". Legal Career Path.
2. "Space Law". United Nations Office for Outer Space Affairs.
3. Gabrynowicz, Joanne Irene. "Space law: Its Cold War origins and challenges in the era of globalization". Suffolk University Law Review – via Law Journal Library.
4. Robert Wickramatunga, *The Outer Space Treaty*. http://www.unoosa.org.
5. Henri, Yvon. "Orbit Spectrum Allocation Procedures ITU Registration Mechanism" (PDF). *International Telecommunications Union*.
6. "Committee on the Peaceful Uses of Outer Space and its Subcommittees". *United Nations Office for Outer Space Affairs*.
7. Jacques Arnould (2011-09-15). Icarus' Second Chance: The Basis and Perspectives of Space Ethics. ISBN 9783709107126.
8. Gorove, Stephen (1979). "*The Geostationary Orbit: Issues of Law and Policy*". The American Journal of International Law. 73 (3): 444–461. doi:10.2307/2201144. JSTOR 2201144.
9. Yasmin Ali, *Who Owns Outer Space?* September 25, 2015, https://www.bbc.com/news/science-environment-34324443
10. Joseph Trevithick, *Space Force Boss Says One Of Russia's Killer Satellites Fired A Projectile In Orbit*, posted in THE WAR ZONE, https://www.thedrive.com/the-war-zone/35057/space-force-boss-says-russia-has-been-testing-its-killer-satellites-in-orbit
11. Joseph Trevithick, *Space Force Just Received its First New Offensive Weapon*, posted in THE WAR ZONE, https://www.thedrive.com/the-war-zone/32570/space-force-just-received-its-first-new-offensive-weapon
12. Joseph Trevithick, *USAF Secretary Gives Ominous Warning that Show of Force Needed to Deter Space Attacks*, posted in THE WAR ZONE, https://www.thedrive.com/the-war-zone/27396/usaf-secretary-gives-ominous-warning-that-show-of-force-needed-to-deter-space-attacks
13. Sandra Erwin, *SpaceX President Gwynne Shotwell: We would launch a weapon to defend the U.S.*'https://spacenews.com/spacex-president-gwynne-shotwell-we-would-launch-a-weapon-to-defend-the-u-s/, September 17, 2018.

CHAPTER TWENTY-SEVEN

THE NEW SPACE RACE

"...the commercial world will absolutely be the leader for everything in space."
~Robert Bigelow – Bigelow Aerospace

The new space race is no longer between the U.S. and Russia, it is between the privately owned aerospace companies which are owned and operated by billionaires, Elon Musk, Robert Bigelow, and Jeff Bezos. They compete on expandable spacecraft, large, lightweight structures that inflate in space, reusable rockets that return to Earth intact after each journey, and technologies that are dramatically changing how humans can live and work in zero gravity.

Entrepreneur Billionaire Robert T. Bigelow is the Founder and President of Bigelow Aerospace, LLC. Headquartered in Las Vegas, Nevada, Bigelow Aerospace is a general contracting, research and development company that concentrates on achieving economic breakthroughs in the costs associated with the design, development, and construction of habitable space structures for private enterprise and government use.

Since 1999, Mr. Bigelow has personally provided all financial support totaling over $350 million to date. In addition, Mr. Bigelow provides the daily strategic leadership at Bigelow Aerospace in its design, development, and testing of expandable habitat architectures where Bigelow Aerospace employs approximately 150 employees at its Las Vegas facility. Mr. Bigelow has successfully launched two subscale spacecraft called Genesis I & II into orbit.

He developed the Bigelow Expandable Activity Module (BEAM). There is a similar BEAM attached to the Tranquility module of the International Space Station. Moreover, Mr. Bigelow serves as the program manager of the B330 spacecraft – Bigelow Aerospace's main habitation system for LEO and beyond LEO destinations.

Robert Bigelow is an experienced general contractor, designer, developer, financier, buyer, and manager of many large real estate projects in the US. Mr. Bigelow holds the exclusive licensing rights to commercialize expandable habitat technology originally conceived but abandoned by NASA in the 1990's. Over the last seventeen years, Mr. Bigelow has earned over twenty patents, launched three prototype spacecraft, partnered with NASA on several contracts, built the necessary facilities to design and fabricate expandable habitat technology, and has advocated for a sustainable commercial space economy.

Elon Musk, Owner and Engineer of SpaceX is now partnering with NASA to transport Astronauts to the ISS. SpaceX also has recently acquired the contract to train Astronauts for NASA. His spacesuits are more advanced and streamline than the older version. Recently, on June 30, 2020, SpaceX launched a Satellite GPS III SPACE VEHICLE 03 MISSION for the U.S. Space Force. [1] See, https://www.spacex.com/launches/

PRIVATE AEROSPACE PARTNERS WITH NASA

Billionaire Aerospace Engineer Robert Bigelow, is a lifelong Ufologist and Las Vegas Real Estate Entrepreneur, maybe considered an unconventional figure in the aerospace world, but because of his obsession with aliens and UFOs he began engineering his own rockets, Bigelow Aerospace. In 2016, he partnered with NASA to prove his high-flying technology is ready to support humans in space. Bigelow says, "Flying to the International Space Station is a really big deal. It's an experimental spacecraft, which is a big deal. We don't know how it's going to behave for sure." Bigelow designed and developed, "the Bigelow Expandable Activity Module" (BEAM) as an experimental expandable Space Station module to be used on the International Space Station, that launched April 8, 2016.

In steps Tesla Billionaire, Aerospace Engineer, and owner of "SpaceX" – Elon Musk who engineered the SpaceX Falcon 9 full thrust rocket to transport BEAM, which left Earth on a mission for NASA, that successfully carried nearly 7,000 pounds of cargo to the International Space Station – including food, supplies and the Bigelow's expandable spacecraft. Suffice to say, there is a working relationship and partnership between Bigelow, Musk and NASA.

Recently, Elon Musk won the contract from NASA to transport NASA Astronauts back and forth to the International Space Station, by making history, with their first successful launch of the SpaceX-Crew Dragon Spacecraft that transported NASA astronauts, Bob Behnken and Doug Hurley using the Falcon 9 Rocket to the International Space Station as part of the agency's Commercial Crew Program. The SpaceX Crew Dragon spacecraft successfully docked at the ISS on May 31, 2020. This began a new age of human spaceflight in the United States, and potentially ended NASA's dependency on the Russian Cosmonauts to transport NAS Astronauts back and forth to the space station, which is how they were traveling since NASA ended their shuttle program, back in 2011.

Bigelow's inflatable was packed inside the Falcon 9 rocket's cargo capsule, known as Dragon, which made it to the space station and was successfully expanded May 28, 2016, and was monitored in the ISS for 2 years. In October of 2017, they decided that BEAM would stay attached to the ISS until 2020, with options for two further one-year extensions. They are using the module to store up to 130 cargo

CHAPTER 30: THE NEW SPACE RACE

transfer bags in an effort to free up additional space aboard the ISS. BEAM made history as the first expandable structure for humans to utilize in space.

Robert Bigelow sat down with CBS Correspondent, Lara Logan, for a 60 Minutes Interview in 2017, and he tells her, unequivocally, without a doubt, that Aliens are already living amongst us. [2]

Under 80-foot ceilings at Bigelow Aerospace in Las Vegas, Robert Bigelow shared with CBS his next generation plans for outer space. He calls it "the Olympus," a mansion for the skies of outer space. It is so large that the rocket it needs powerful enough to launch it into space, is still years away.

The CBS transcript:[2]

> Lara Logan: Is there anything this big that astronauts are working in today?
>
> Robert Bigelow: No, nothing even remotely close.
>
> Bigelow said he can turn this into anything a client wants -- a first-ever orbiting hotel, hospital, or inflatable research facility. The B330 is smaller than the Olympus.
>
> Lara Logan: This can function on its own?
>
> Robert Bigelow: Well, it can function as a standalone destination because it has all the facilities that it would take to keep people alive.
>
> For countries hoping to make their mark in space, Bigelow said this offers an affordable way in. Private industry, he believes, is becoming more dominant in developing space.
>
> Robert Bigelow: NASA and government still has a role because it is still in a sort of embryonic stage. And there will come a time when it's not necessary at all. And the commercial world will absolutely be the leader for everything in space.
>
> Lara Logan: What about the idea of national pride and what we do as nations. Neil Armstrong walking on the moon. He did not do that for a private company. He did that for the United States of America, and it meant something.
>
> Robert Bigelow: You bet it did. And that created a period of inspiration that has not been matched ever since. So now we're looking for a new era that says, "All right, how can we now shape this to where it's more accessible for a lot more folks at a lot less cost and still have safety and reliability at the same time?" "...the commercial world will absolutely be the leader for everything in space."

With no formal training in science or engineering, Robert Bigelow created an aerospace company with scientists and engineers that is achieved what no one else in the industry has done. His expandable spacecraft are the first and only alternative to the metal structures that have housed every astronaut in space for over half a century. For Bigelow, it all began with growing up in a time of nuclear tests. As a young boy, he would watch the skies over Nevada light up with the bursts of atomic bombs. This also inspired his research and obsession with UFOs, as Area 51 was also part of his neighborhood.

Robert Bigelow: Witnessing those explosions in the 50s and 60s, you weren't aware of the ultimate ramifications of those kinds of things but there was a real strong feeling of energy and a secretiveness and so forth and it was cool.

Bigelow watches Neil Armstrong takes the first steps on the moon, a moment in history he said still inspires him.

Armstrong: "That's one small step for man…"

Robert Bigelow: The approach wasn't lightning fast…

But on this canyon road just outside Las Vegas, Robert Bigelow's story takes a turn that some may find, to put it lightly, improbable. He told us this is where his grandparents had a close encounter with a UFO.

Robert Bigelow: It really sped up and came right into their face and filled up the entire windshield of the car. And it took off at a right angle and shot off into the distance. The story sparked his obsession and explained the alien looking out from the side of Bigelow Aerospace. And it made for the kind of conversation you do not ordinarily have with an accomplished CEO.

Lara Logan: Do you believe in aliens?

Robert Bigelow: I am absolutely convinced. That is all there is to it.

Lara Logan: Do you also believe that UFOs have come to Earth?

Robert Bigelow: There has been and is an existing presence, an ET presence. And I spent millions and millions and millions -- I probably spent more as an individual than anybody else in the United States has ever spent on this subject.

Lara Logan: Is it risky for you to say in public that you believe in UFOs and aliens?

Robert Bigelow: I don't give a damn. I don't care.

Lara Logan: You don't worry that some people will say, "Did you hear that guy, he sounds like he's crazy"?

Robert Bigelow: I don't care.

Lara Logan: Why not?

Robert Bigelow: It's not going to make a difference. It's not going to change reality of what I know.

CHAPTER 30: THE NEW SPACE RACE

Lara Logan: Do you imagine that in our space travels we will encounter other forms of intelligent life?

Robert Bigelow: You don't have to go anywhere.

Lara Logan: You can find it here? Where exactly?

Robert Bigelow: It's just like right under people's noses. Oh my gosh. Wow.

The FAA confirmed to us that for years, it referred reports of UFOs and other unexplained phenomena to a company Bigelow owns. Robert Bigelow was the secret owner of the infamous Skinwalker Ranch, which seemed to be a hotbed for UFOs and other strange paranormal activities.

He bought the idea for his inflatable technology from the space agency. NASA had been working on this since the early 1960s -- but when Congress killed its program in 2000, Bigelow seized the opportunity and invested tens of millions of dollars to advance NASA's original idea.

It took him just six years to launch the first expandable spacecraft into orbit. His second followed a year later. But those were never meant for humans.

At that time, Bigelow was still trying to prove his inflatable structures would survive in space. A decade later, with both still circling the Earth intact, NASA felt the technology was ready to be tested for humans. Bigelow still monitors these spacecrafts from his own mission control in Las Vegas.

Robert Bigelow made his fortune with a low-budget chain of rental apartments called Budget Suites of America.

Lara Logan: And made you how much money?

Robert Bigelow: Enough to support the aerospace indulgence and Bigelow Aerospace.

Lara Logan: Well, "aerospace indulgence" is a big term. So, do you want to put a figure on that?

Robert Bigelow: We are approaching 290 million.

Lara Logan: Of your own money?

Robert Bigelow: Oh, yeah.

Lara Logan: So, for a man who has had extraordinary success in the business world--

Robert Bigelow: Some, some.

Lara Logan: This space business is financially the worst investment you've ever made?

Robert Bigelow: It's atrocious. I mean, we are not in control of our own destiny.

Lara Logan: Who is in control of your destiny?

Robert Bigelow: We are hostage to what happens with transportation, space transportation.

Thanks to innovators like Elon Musk and Jeff Bezos, Bigelow told us reusable rockets -- like this one you see sticking a perfect landing after launching his inflatable -- are making routine transportation a real possibility in the next decade.

When Robert Bigelow's inflatable structure was added by NASA to the International Space Station, humans had never been inside one in space. This test would be the first time. And astronaut Jeff Williams, the first American to experience it.

The interior only has sensors, and for two years, NASA will use them to monitor how this holds up to solar radiation and extreme temperatures.

Lara Logan: How far are you from us right now?

Astronaut Jeff Williams from the ISS: Well currently we are over Libya as I recall, but when we started this conversation, I think we were probably over the Atlantic Ocean.

From Bigelow's Mission Control, NASA connected CBS with Jeff Williams on the space station. They spoke to him about Bigelow's inflatable room, he was orbiting the Earth at nearly five miles a second.

Jeff Williams: I was very excited to be part of it and to be part of something new.

Lara Logan: And how is it holding up?

Jeff Williams: Oh, it is holding up well. It was a little bit cool, and that was expected. Cooler than the air here, but not overly cold, and had that new car smell that you would expect. And it was very quiet, too.

Lara Logan: What do you think about men like Mr. Bigelow, people with deep pockets, private citizens getting involved in your world, in the world of space?

Jeff Williams: Oh, I would say anybody that has the means to do exploration and is willing to jump in the game, more power to them. Private enterprise is trying to open up economic doors to space exploration and everybody will benefit from that.[1]

SSP, SPACE FORCE & PRIVATE INDUSTRY
Connecting the dots to the Secret Space Program, Space Force and Private Aerospace Industries:

The US Government (USG) uses a series of alphabet soup names to confuse, hide, or designate often classified programs. On December 16, 2017, the New York Times disclosed that the Pentagon had secretly funded research into UFOs through the Advanced Aerospace Threat Identification Program, or (AATIP). In an instant, UFOs were no longer relegated to societies nihilistically curious, and for the first time in decades, droves of the mainstream public suddenly found themselves peering skyward with wonder. Adding to the chaos, an entirely different program moniker emerged: the Advanced Aerospace Weapons Systems Applications Program, or (AAWSAP). For over two years, no one has been able to adequately explain whether AAWSAP and AATIP were two separate programs, or the same intuitive under two separate names.

Bigelow Aerospace Advanced Space Studies (BAASS) received a contract from the Defense Intelligence Agency (DIA) regarding the Advanced Aerospace Weapon System Applications (AAWSA) program, there is no mention of UFOs or Unidentified Aerial Phenomenon (UAPs) in that solicitation/contract.

The DoD will usher in a new age of space technology and field new systems in order to deter, and if necessary, degrade, deny, disrupt, destroy, and manipulate adversary capabilities to protect U.S. interests, assets, and way of life... This new age will unlock growth in the U.S. industrial base, expand the commercial space economy and strengthen partnerships with our allies. It is also opening up a new level of the commercial space race, i.e., space tourism and space taxis, something Virgin Galactic and Space X are competing for. [1]

The U.S. Space Force has now actually gone to space. This is the first time a satellite has flown into space under the auspices of the U.S. Space Force. So far, Covid-19 has yet to substantially slow activity at the military's spaceports.[3]

COMMERCIAL AEROSPACE
Private Aerospace companies, such as SpaceX owned by Elon Musk, are now playing important roles in reducing the cost of space travel for NASA. A new era of human spaceflight began as American astronauts once again launched on an American rocket from American soil to the International Space Station as part of NASA's Commercial Crew Program. NASA astronauts Robert Behnken and Douglas Hurley lifted off from the Kennedy Space Center in Florida at 3:22 p.m. EDT May 30, 2020, on SpaceX's Crew Dragon spacecraft. It was the first time that people headed to orbit in an American rocket launching from the United States since the previous space shuttle mission, in 2011. SpaceX's mission successfully splashed down in the Gulf of Mexico with NASA astronauts Bob Behnken and Doug Hurley on August 2, 2020, after returning from a 63-day mission to the International Space Station.

Now, there seems to be at least 1-2 rockets launched each month from Cape Canaveral's Kennedy Space Center launch pads, making a revived new era of the Space Age, which despite shutdowns on Earth during the pandemic of 2020, stayed on schedule in their activities on space.

Due SpaceX's success, they are now one of the main ways that US astronauts travel to and from the space station. After the Space Shuttle program retired in 2011, American Astronauts relied on the Russian Cosmonauts on the Soyuz spacecraft to be transported back and forth to the International Space State. Under new NASA leadership through the Trump Administration, NASA has new breath and new life by relying on American Private Aerospace companies, instead of foreign entities for space travel. However, with that said, it is important to note, that if Russia were American's enemy, then why are they partners in space? Why would America have trusted Russians to transport their astronauts back and forth to the ISS for nine years, if they were enemies on Earth? The truth is, they are not enemies on earth or in space. The Russians and Americans have been working together on joint secret space programs since WW2.

But SpaceX is not the only private aerospace company who build spaceships for Nasa's commercial crew program. Boeing is also in play with its Starliner spaceship to fly astronauts to the space station in its Starliner spaceship.

Boeing's Crew Space Transportation (CST)-100 Starliner spacecraft is being developed in collaboration with NASA's Commercial Crew Program. Nasa astronauts Nicole Mann and Michael Fincke, and Boeing astronaut Christopher Ferguson, will form the first crew for Starliner. The Starliner was designed to accommodate seven passengers, or a mix of crew and cargo, for missions to low-Earth orbit. For NASA service missions to the International Space Station, it will carry up to four NASA-sponsored crew members and time-critical scientific research. The Starliner has an innovative, weldless structure and is reusable up to 10 times with a six-month turnaround time. It also features wireless internet and tablet technology for crew interfaces.[4]

Notes:
1. SpaceX -- FALCON 9 -- FALCON HEAVY – DRAGON – STARSHIP -- HUMAN SPACEFLIGHT -- STARLINK MISSION, https://www.spacex.com/launches/
2. Lara Logan, CBS 60 Minutes, *Bigelow Aerospace founder says commercial world will lead in space: NASA and Las Vegas entrepreneur have partnered on a technology that could change how humans live and work in space*, May 28, 2017, https://www.cbsnews.com/news/bigelow-aerospace-founder-says-commercial-world-will-lead-in-space/
3. Eric Berger, *The U.S. Space Force has now actually gone to space: So far, Covid-19 has yet to substantially slow activity at the military's spaceports*, https://arstechnica.com/science/2020/03/the-first-launch-for-the-us-space-force-is-set-for-today-from-florida/, 3/26/2020.
4. Boeing CST-100 Starliner Spacecraft, http://www.boeing.com/space/starliner/

CHAPTER TWENTY-EIGHT

NASA's BOMBSHELL ON THE MOON

"The politics of the future will be cosmic, or interplanetary."
—General Douglas MacArthur,
(New York Times, Oct 8, 1955)

In greater skill the paths of heaven to ride.
—Gordon Alchin

I want to begin this chapter by reprinting my dream/vision that I had in 2009 which I shared in Book One of *Who's Who in the Cosmic Zoo?* I was shown a vision in a semi-sleeping state early one morning, of an explosion that took place on the moon, that was so great, it was seen from earth, and then I saw, what looked like a dozen or so stars surrounding the moon, which I later realized weren't stars at all, but spaceships. It was such a foreboding event, one which shifted everything on earth. I awoke with a frightened gasp, and I heard an Angel's voice tell me, to write it down, so I did. Little did I know that NASA sent a Mission to the Moon I 2009 to drop a 2-ton kinetic bomb on the Moon.[1] But I know about it now.

Many people who have paid attention to this mission, think NASA bombed the moon in order to destroy an Alien base. I saw in my vision, spaceships all around the Moon, which revealed that there was some kind of skirmish or battle, after an act of aggression, which was a deliberate explosion. This was known as the LCROSS Mission.

There are multiple sets of images and reports, which reveal there are alien structures on the surface of the moon. It's believed by researchers, that NASA launched a 2-ton kinetic weapon to destroy them, which was covered-up in the usual NASA way, saying NASA's LCROSS mission, that they literally BOMBED the surface of the moon for alleged 'Scientific' purposes.

Despite the fact that bombing the Moon was strictly prohibited, NASA released a Centaur kinetic weapon, which impacted the Moon. The official LCROSS mission's objective was to explore the presence of water ice in a permanently shadowed crater near a lunar polar region. The mission was launched together with the Lunar Reconnaissance Orbiter (LRO) on June 18, 2009, as part of the shared Lunar Precursor Robotic Program, the first American mission to the Moon in over ten years.

Bombing the Moon breaks all kinds of international space laws. Why would NASA go against their very own standards? According to many, the true purpose behind the 2009 LCROSS 'Moon bombing' was way more perplexing than NASA was willing to accept.

One of the greatest enigmas regarding UFOs and Alien life is whether governments and Space Agencies around the world are covering up such information. While seeing UFOs on Earth and videos from space are not something new, in the last couple of years, a lot of attention has been drawn to Earth's moon. There, on the surface of Earth's natural surface lay numerous Alien Bases. The fact that many believe NASA and governments around the world have covered up information on these alien bases has become a widely accepted ideology in the last decade among Ufologists and believers. [2]

SPACE POLICIES

Since the advent of the Space Age, from the 1960s and 70s, space treaties were established, that became the foundation for US Military Space Policy. According to the book *The Paths of Heaven: The Evolution of Airpower Theory*, that was a collaborative effort by *The School of Advanced Airpower Studies*, they outlined the following outer space treaties:[3]

1. The Outer Space Treaty (OST) which dates back to 1967, clearly states that international law applies BEYOND the atmosphere. The treaty of 1967 reemphasized standing international laws and initiated new space-related laws: Free Access to space and celestial bodies for peaceful intent, prohibitions on national appropriations of space or celestial bodies, prohibitions on putting any weapons of mass destruction in space or on celestial bodies.
2. The Antiballistic Missile (ABM) treaty of 1972 (which was signed between the USA and the USSR) banned the development, testing, and employment of space-based ABMs.
3. The Convention on Registration (1974) requires parties to maintain a registry of objects launched into space and report orbital parameters and general function of those objects to the UN.
4. And most importantly, the Environmental Modification Convention signed in 1980 which prohibits the hostile use of environmental modification.

Apart of the above-mentioned treaties, during a Convention on International Law in 1977, they concluded, that the military was prohibited towards any type of hostilities nor were they allowed the use of environmental modification techniques, which set out a number of prohibitions also with respect to outer space and celestial bodies.[4] (Source: William Elliott Butler, *Perestroika and International Law*)

"What is not stated in international law is probably more important than what is; legal interpretation follows the convention that if the law does not explicitly prohibit something, it implicitly allows it. One must also understand that treaties are just mutual agreements between signatories—they hold in peacetime but not necessarily in wartime. Given these considerations, weapons of mass destruction, ABMs, and environmental modification weapons are all prohibited, but many conventional weapons (including ASATs) and tests of those weapons are allowed in space. The appropriation of space or any celestial body is illegal. Military maneuvers, bases, or installations on celestial bodies (the Moon) are also illegal; however, military maneuvers, bases, and installations in space (artificial satellites) constitute legal uses of that realm. Unreported space vehicles are prohibited but reporting vague functional specifications and changing orbital parameters after launch are not prohibited.

In addition to international law, several domestic laws affect how the military might use space, although they certainly do not inhibit the use of space. They include the DEBLOIS 537 Communications Act of 1934, whereby the president can commandeer private communications assets in times of crisis, and the Commercial Space Launch Act of 1984, which provides commercial customers access to military space launch facilities (at a price). Taken as a whole, international and domestic law limits but does not preclude the conventional weaponization of space."[4]

Despite clearly violating International Outer Space Treaties, NASA nevertheless altered the lunar surface after exploding a 2-ton kinetic weapon, which created another lunar crater that is now 5-mile wide at the South Pole of the Moon.

Ufologists believe based on leaked Apollo images that show alien structures on the surface of the moon, that NASA's LCROSS mission had a more militaristic objective rather than scientific one. Many believe that the 2-ton kinetic weapon that was detonated on the Moon's South Pole was aimed at an Alien Base located there.[2]

Perhaps this was a game changer, and explains why NASA hasn't been back to the Moon since? They know the Moon is filled with minerals, it has water making a perfect launchpad for Space Exploration of our solar system, making our journey

CHAPTER 32: NASA's BOMBSHELL ON THE MOON

to Mars and beyond, quicker, and perhaps easier to navigate. On the 50th Anniversary of Apollo 11, NASA announced its intention and goals for just that, to launch their Artemis Mission, which would not only place the first female astronaut on the Moon, but it would serve to set up an outpost on the Moon, to be used to launch rockets to Mars, so they could set up colonies, cities and bases on both the Moon and Mars.

There are some who believe what NASA bombed wasn't an alien base, but ancient structures that have been abandoned by ancient aliens from our forgotten history, when they traveled between the spheres, Earth, Moon, Mars, elsewhere. Perhaps they felt that disclosing these ancient structures would upset what we have been programmed to believe about our past, along with covering up what the Apollo Mission found on the Moon in the first place.

Then again, it could have been motivated by pure ego, as NASA along with many government military advisors, scientists and historic societies who cannot handle the fact they were not the first humans to step foot on the moon, but that our human ancestors or ancient aliens were far more advanced than we are today.[4]

A UFO hunter claims to have obtained new smoking gun evidence that NASA dropped a nuclear bomb on an alien Moon base on October 9, 2009. According to UFO blogger Scott Waring, he has stumbled on "100 percent" proof of conspiracy theory rumors that on October 9, 2009, NASA nuked an alien Moon base deemed a national security threat.[5]

Rumors that the U.S. government secretly launched a nuclear strike to destroy a threatening Moon base have been around in the online UFO community since 2012. But skeptics have dismissed evidence presented by UFO conspiracy theorists as bogus and fanciful.

Now Waring claims to have obtained "100 percent" proof that will convince even the most dyed-in-the-wool skeptic. According to Waring, he had watched the live internet transmission of the alleged Moon bombing on October 9, 2009, but had not realized the significance of the event at the time. But years later, while reviewing live video footage showing NASA employees at the Ames Research Center during the alleged nuking of the alien lunar base, he realized suddenly that the footage contains evidence that a nuclear bombing of the lunar surface actually took place on October 9, 2009.

Scott C. Waring of ufosightingsdaily.com reported: "Up till now, NASA dropping the nuke on a moon base to destroy the alien base found was always theory. Today I have proved that its 100% fact. See video [6], https://youtu.be/avmDnwsOGDw

Waring claims he was able to match the faces of two NASA employees in the live video footage with faces in a famous photo that first surfaced online in March 2012, allegedly showing two young NASA scientists reviewing a photo of the Cabeus crater hours before the alleged nuclear bombing by NASA.

According to UFO hunters, the photo that the NASA scientists were reviewing shows an artificial structure in the Cabeus crater and proves that NASA's "bombing" of the crater during the LCROSS mission on October 9, 2009, had actually targeted alien structures on the lunar surface.[5]

Ware observed and reported: "I matched the two faces from the live video footage at AMES with those of the two AMES employees in the moon base photo. Anthony Colaprete and Karen Gundy-Burlet are the ones in the HD photo with the moon base and in the live nuking of the moon video. LCROSS Satellite launched into Space July 18, 2009. See, NASA Bombing the Moon LIVE on Fox News, htttps://youtu.be/avmDnwsOGDw.[6]

"The two scientists can be seen scratching their heads with puzzled expressions after the bombing in October 2009. He claims they were puzzled because the nuclear explosion failed to destroy the targeted alien Moon base building. At 3:59 until 4:04 into the video you and see the two of them rubbing their heads, as if wondering WTF just happened? Since the explosion did not destroy the building at all nor did it create an upraised cloud of dust. It's as if the base had a shield around it stopping the blast from harming it."[7]

"The base...and I'm saying this for every country in the world to know that they may be able to claim it...is in the crater Cabius. America gave the order to destroy it because it was seen by Chinese and Russian satellites. Apparently, it is still there and DGAF what nukes America throws at it. That's...what alien technology can do. This base is occupied...not abandoned. It protected itself from the nuke and the world needs to know that America would rather destroy it, than let it fall into the hands of Russia or China." [6, 7]

Claims that NASA "nuked" the Moon to destroy an alien base on the lunar surface first emerged in March 2012, when some UFO hunters claimed that a top-secret photo that a NASA employee with the Ames Research Center left in front of him on his desk during the LCROSS mission showed alien lunar bases in the Cabeus crater. A reporter allegedly noticed the photo in front of the NASA scientist and snapped a photograph.

Although the alleged photograph first surfaced online in 2012, alien hunters claimed it was taken in 2009 and that it shows two NASA scientists looking at photos of alien bases located in the Cabeus crater hours before the "bombing."

The NASA mission that Waring and UFO conspiracy theorists claim was launched to drop a nuclear bomb on the Moon to destroy alien bases in the Cabeus crater was actually the 2009 LCROSS mission during which NASA spacecraft "bombed" the surface of the Moon — as media reports termed it — by crashing into it. The kinetic energy released by the high-velocity impact of NASA's craft on the lunar surface generated an explosion.

NASA claimed at the time that the purpose of the project was to confirm the presence or absence of water in the Cabeus crater at the Moon's South Pole. But after a photo emerged online in 2012 allegedly showing NASA scientists reviewing a photo claimed to show an artificial structure on the Moon, UFO conspiracy theorists announced that the LCROSS mission was actually a secret mission to destroy an alien base on the Moon by dropping a small nuclear bomb over it.

According to Waring and fellow UFO hunters, the photo snapped by the unnamed reporter shows two NASA scientists — allegedly Anthony Colaprete and Karen Gundy-Burlet — reviewing photos on October 9, 2009, before the "bombing" of the Cabeus crater.

Skeptics dismissed claims that the photo was snapped on October 9, 2009, pointing out that it first emerged online in March 2012. But now UFO blogger Waring claims to have found "smoking gun" evidence that the photo was snapped on October 9, 2009. According to Waring, he has been able to match the faces of the NASA scientists in the photo with the faces in the live video footage allegedly showing the NASA scientists at Ames Research Center during the "bombing" on October 9, 2009.

"Now that's a huge cat left out of NASA's bag. [This is] the smoking gun evidence that NASA knows about the presence of extraterrestrial bases on the Moon."

Waring claims at 3:59-4:04 in the recording of the live video (see first YouTube above) the two scientists can be seen scratching their heads with puzzled expressions after the bombing October 9, 2009. He claims they were puzzled because the nuclear explosion failed to destroy the targeted alien Moon base building.

According to Waring, the aliens occupying the building used advanced technology to shield their base from the nuclear attack. "The explosion didn't destroy the building at all nor did it create an upraised cloud of dust," Waring writes, "as if the base had a shield around it stopping the blast from harming it."

Waring claims the U.S. government tried to destroy the base only because the Chinese and Russians had spotted it on satellite and the U.S. was afraid that both countries could come into possession of advanced alien technology. But he also claims the alleged alien base was being occupied by the technologically advanced aliens, and that the aliens designed the base to withstand a nuclear attack.

"The base... and I'm saying this for every country in the world to know that they may be able to claim it... is in the crater Cabeus," Waring writes. "America gave the order to destroy it because it was seen by Chinese and Russian satellites... This base is occupied... not abandoned. It protected itself from the nuke and the world needs to know that America would rather destroy it, than let it fall into the hands of Russia or China." [5,6]

MINING THE MOON & ASTEROIDS

On April 6, 2020, President Trump signed an executive order that supports moon mining along with tapping into asteroid's resources. The U.S. sees a clear path to the use of moon and asteroid resources. It seems pretty clear, by anyone who connect these dots, that the space policies of the 60s and 70s, do not really count anymore. We are going to stake our claim on whatever Space holds for our own purposes. If it protects the survival of humankind, then so be it. We have been taught a lot of misnomers about our history as humans on the earth, moon and mars, which all the evidence points to the fact that humans have already colonized the Moon and Mars, and God knows where else? Europa, one of the Moons of Jupiter, is inhabited. See, Book One of *Who's Who in the Cosmic Zoo?* [8]

President Trump's Executive Order Public Law 114-90 established U.S. policy on the exploitation of off-Earth resources. That policy states that it is following the 1967 Outer Space Treaty, which allows the use of such resources. He believes that obtaining the water ice and other lunar resources from mining the moon, will help the United States establish a long-term human presence on the moon as the intention and goals are now with NASA's next big Lunar Missions under the Artemis. The White House believes, it is all there for the taking – with no digging required. Space mining on the moon and beyond may be solar powered.

The United States, like the other major spacefaring nations, has not signed the 1979 Moon Treaty, which stipulated that non-scientific use of space resources be governed by an international regulatory framework. Then in 2015, Congress passed a law explicitly allowing American companies and citizens to use the moon and asteroid resources. So, it is perfectly 'legal.'

Trump's executive order makes things even more official, stressing that the United States does not view space as a "global-commons" and sees a clear path to off-Earth mining, without the need for further international treaty-level agreements.[9] "Outer space is a legally and physically unique domain of human activity, and the United States does not view it as a global commons," the executive order states.

The executive order, called "Encouraging International Support for the Recovery and Use of Space Resources," has been in the works for about a year, a senior administration official said during a teleconference with reporters today. The order was prompted, at least in part, by a desire to clarify the United States' position as it negotiates with international partners to help advance NASA's Artemis program for crewed lunar exploration. Engagement with international partners remains important, the official said.

Artemis aims to land two astronauts on the moon in 2024 and to establish a sustainable human presence on and around Earth's nearest neighbor by 2028. Lunar resources, especially the water ice thought to be plentiful on the permanently shadowed floors of polar craters, are key to Artemis' grand ambitions, NASA officials have said.

The moon is not the final destination for these ambitions, by the way. Artemis is designed to help NASA and its partners learn how to support astronauts in deep space for long stretches, lessons that will be key to putting boots on Mars, which NASA wants to do in the 2030s.

"As America prepares to return humans to the moon and journey on to Mars, this executive order establishes U.S. policy toward the recovery and use of space resources, such as water and certain minerals, in order to encourage the commercial development of space," Scott Pace, deputy assistant to the president and executive secretary of the U.S. National Space Council, said in a statement today.

CHAPTER 32: NASA's BOMBSHELL ON THE MOON

President Trump has shown considerable interest in shaping U.S. space policy. In December 2017, for example, he signed Space Policy Directive-1, which laid the groundwork for the Artemis campaign. Two other directives have aimed to streamline commercial space regulation and the protocols for space traffic control. And Space Policy Directive-4, which the president signed in February 2019, called for the creation of the Space Force, the first new U.S. military branch since the Air Force was stood up in 1947.[9]

HUMAN SPACE EXPLORATION PROGRAMS

INFRASTRUCTURE & TECHNOLOGY

Executive Order on Encouraging International Support for the Recovery and Use of Space Resources:

By the authority vested in me as President by the Constitution and the laws of the United States of America, including title IV of the U.S. Commercial Space Launch Competitiveness Act (Public Law 114-90), it is hereby ordered as follows:

Section 1. Policy. Space Policy Directive-1 of December 11, 2017 (Reinvigorating America's Human Space Exploration Program), provides that commercial partners will participate in an "innovative and sustainable program" headed by the United States to "lead the return of humans to the Moon for long-term exploration and utilization, followed by human missions to Mars and other destinations." Successful long-term exploration and scientific discovery of the Moon, Mars, and other celestial bodies will require partnership with commercial entities to recover and use resources, including water and certain minerals, in outer space.

Uncertainty regarding the right to recover and use space resources, including the extension of the right to commercial recovery and use of lunar resources, however, has discouraged some commercial entities from participating in this enterprise. Questions as to whether the 1979 Agreement Governing the Activities of States on the Moon and Other Celestial Bodies (the "Moon Agreement") establishes the legal framework for nation states concerning the recovery and use of space resources have deepened this uncertainty, particularly because the United States has neither signed nor ratified the Moon Agreement. In fact, only 18 countries have ratified the Moon Agreement, including just 17 of the 95 Member States of the United Nations Committee on the

Peaceful Uses of Outer Space. Moreover, differences between the Moon Agreement and the 1967 Treaty on Principles Governing the Activities of States in the Exploration and Use of Outer Space, Including the Moon and Other Celestial Bodies — which the United States and 108 other countries have joined — also contribute to uncertainty regarding the right to recover and use space resources.

Americans should have the right to engage in commercial exploration, recovery, and use of resources in outer space, consistent with applicable law. Outer space is a legally and physically unique domain of human activity, and the United States does not view it as a global-commons. Accordingly, it shall be the policy of the United States to encourage international support for the public and private recovery and use of resources in outer space, consistent with applicable law.

Sec. 2. The Moon Agreement. The United States is not a party to the Moon Agreement. Further, the United States does not consider the Moon Agreement to be an effective or necessary instrument to guide nation states regarding the promotion of commercial participation in the long-term exploration, scientific discovery, and use of the Moon, Mars, or other celestial bodies. Accordingly, the Secretary of State shall object to any attempt by any other state or international organization to treat the Moon Agreement as reflecting or otherwise expressing customary international law.

Sec. 3. Encouraging International Support for the Recovery and Use of Space Resources. The Secretary of State, in consultation with the Secretary of Commerce, the Secretary of Transportation, the Administrator of the National Aeronautics and Space Administration, and the head of any other executive department or agency the Secretary of State determines to be appropriate, shall take all appropriate actions to encourage international support for the public and private recovery and use of resources in outer space, consistent with the policy set forth in section 1 of this order. In carrying out this section, the Secretary of State shall seek to negotiate joint statements and bilateral and multilateral arrangements with foreign states regarding safe and sustainable operations for the public and private recovery and use of space resources.

Sec. 4. Report on Efforts to Encourage International Support for the Recovery and Use of Space Resources. No later than 180 days after the date of this order, the Secretary of State shall report to the President, through the Chair of the National Space Council and the Assistant to the President for National Security Affairs, regarding activities carried out under section 3 of this order.

Sec. 5. General Provisions. (a) Nothing in this order shall be construed to impair or otherwise affect:

(i) the authority granted by law to an executive department or agency, or the head thereof; or

(ii) the functions of the Director of the Office of Management and Budget relating to budgetary, administrative, or legislative proposals.

(b) This order shall be implemented consistent with applicable law and subject to the availability of appropriations.

(c) This order is not intended to, and does not, create any right or benefit, substantive or procedural, enforceable at law or in equity by any party against the United States, its departments, agencies, or entities, its officers, employees, or agents, or any other person.

Signed by DONALD J. TRUMP
THE WHITE HOUSE, April 6, 2020.[10]

President Donald J. Trump is Encouraging International Support for the Recovery and Use of Space Resources

> "After braving the vast unknown and discovering the new world, our forefathers did not only merely sail home — and, in some cases, never to return. They stayed, they explored, they built, they guided, and through that pioneering spirit, they imagined all of the possibilities that few dared to dream." — President Donald J. Trump

The Executive Order on "Encouraging International Support for the Recovery and Use of Space Resources" addresses U.S. policy regarding the recovery and use of resources in outer space, including the Moon and other celestial bodies. Supportive policy regarding the recovery and use of space resources is important to the creation of a stable and predictable investment environment for commercial space innovators and entrepreneurs, and it is vital to the long-term sustainability of human exploration and development of the Moon, Mars, and other destinations.

As part of further implementing Space Policy Directive 1, **"Reinvigorating America's Human Space Exploration Program,"** President Trump underscores our commitment to the 1967 Outer Space Treaty, which has provided a foundational set of rules for the successful use of outer space for more than fifty years. The Executive Order also affirms Congress' intent that Americans should have the right to engage in commercial exploration, recovery, and use of resources in outer space, consistent with applicable law. Outer space is a legally and physically unique domain of human activity, and the United States does not view space as a global-commons.

- This Executive Order directs the Secretary of State to lead a U.S. Government effort to develop joint statements, bilateral agreements, and multilateral instruments with like-minded foreign states to enable safe and sustainable operations for the commercial recovery and use of space resources, and to object to any attempt to treat the 1979 Moon Agreement as expressing customary international law.
- Nevertheless, in seeking international support, the United States may draw on legal precedents and examples from other domains to promote the recovery and use of space resources.
- American industry and the industries of like-minded countries will benefit from the establishment of stable international practices by which private citizens, companies and the economy will benefit from expanding the economic sphere of human activity beyond the Earth.[11]

Notes:
1. Ella LeBain, *Who's Who in the Cosmic Zoo?* Book One – Third Edition, Chapter: Solarians, pp.403. Skypath Books, 2013.
2. Janice Friedman, Ancient Aliens Blog, *NASA Dropped A 2-Ton Kinetic Missile on the Moon*, https://www.ancient-code.com/nasa-dropped-a-2-ton-kinetic-missile-on-the-moon-what-did-they-destroy/
3. Col Phillip S. Meilinger, USAF, Editor, *The Paths of Heaven: The Evolution of Airpower Theory* – by The School of Advanced Airpower Studies, Air University Press, Maxwell Air Force Base, Alabama, 1997.
4. William Elliott Butler, *Perestroika and International Law*, Springer; 1990 edition (January 31, 1990)
5. 'NASA Dropped Nuclear Bomb on Alien Moon Base On October 9, 2009': UFO Blogger Claims 'Smoking Gun' Evidence, : https://www.inquisitr.com/2988781/nasa-dropped-nuclear-bomb-on-alien-moon-base-on-october-9-2009-ufo-blogger-claims-smoking-gun-evidence-video/#ixzz6IzmnxHO3, April 12, 2016.
6. Fox News Live Report of NASA Bombing the Moon, https://youtu.be/avmDnwsOGDw, October 9, 2009.
7. Scott C. Waring, *CONFIRMED! NASA Did Drop the Nuke In Oct 2009 On An Alien Base*, https://www.ufosightingsdaily.com/2016/04/confirmed-nasa-did-drop-nuke-in-oct.html, April 2016, Video, UFO Sighting News.

8. Ella LeBain, *Who's Who in the Cosmic Zoo?* Book One – Third Edition, Chapter: Solarians, pp.403. Skypath Books, 2013.
9. Mike Wall, *Trump signs executive order to support moon mining, tap asteroid resources, The U.S. sees a clear path to the use of moon and asteroid* resources, https://www.space.com/trump-moon-mining-space-resources-executive-order.html, April 6, 2020
10. whitehouse.gov/presidential-actions/executive-order-encouraging-international-support-recovery-use-space-resources/
11. whitehouse.gov/wp-content/uploads/2020/04/Fact-Sheet-on-EO-Encouraging-International-Support-for-the-Recovery-and-Use-of-Space-Resources.pdf

CHAPTER TWENTY-NINE

PRESIDENTIAL SPACE POLICIES

"I don't laugh at anybody anymore, when they say they've seen UFOs, I've seen one myself."
~ Former U.S. President Jimmy Carter, Washington Post Interview, 1975

The Space Age really began after WW2 with President Eisenhower's agreement with the Grays. Here are some quotes strung together from America's Presidential Visions for Space Exploration, and how their viewpoints shaped space policies.

President John F. Kennedy declared in his historic message to a joint session of the Congress, on May 25, 1961, a bold new plan for NASA and the nation: To send an American to the moon, and to return him safely, by the close of the decade. Kennedy's speech, which came just six weeks after cosmonaut Yuri Gagarin became the first person to reach outer space, had a huge impact on NASA and space exploration. It jump-started the agency's Apollo program, a full-bore race to the moon that succeeded on July 20, 1969, when Neil Armstrong's boot crunched down into the gray lunar dirt.

Kennedy, of course, is not the only leader who had a vision for the nation's space program. Since NASA's founding in 1958, every president from Eisenhower to Obama has left his mark. Take a look at how each U.S. commander-in-chief helped shape and steer American activities in space.

SPACE COMMANDERS-IN-CHIEF: FROM IKE TO TRUMP

DWIGHT EISENHOWER (1953-1961)

President Dwight Eisenhower was president when the Soviet Union launched the world's first artificial satellite, Sputnik I, in October 1957. This seminal event shocked the United States, started the Cold War space race between the two superpowers and helped lead to the creation of NASA in 1958. However, Eisenhower did not get too swept up the short-term goals of the space race. He valued the measured development of unmanned, scientific missions that could have big commercial or military payoffs down the road.

For example, even before Sputnik, Eisenhower had authorized a ballistic missile and scientific satellite program to be developed as part of the International Geophysical Year project of 1957-58. The United States' first successful satellite, Explorer I, blasted off Jan. 31, 1958. By 1960, the nation had launched and retrieved film from a spy satellite called Discoverer 14.

JOHN F. KENNEDY (1961-1963)

President John F. Kennedy effectively charted NASA's course for the rest of the 1960s with his famous speech before Congress on May 25, 1961. Dr. Wernher von Braun explained the Saturn Launch System to President John F. Kennedy.

The Soviets had launched Sputnik I in 1957, and cosmonaut Yuri Gagarin had become the first person in space on April 12, 1961, just six weeks before the speech. On top of those space race defeats, the U.S. plan to topple the Soviet-backed regime of Cuban leader Fidel Castro — the so-called Bay of Pigs invasion — had failed miserably in April 1961.

Kennedy and his advisers figured they needed a way to beat the Soviets, to re-establish American prestige and demonstrate the country's international leadership. So, they came up with an ambitious plan to land an astronaut on the moon by the end of the 1960s, which Kennedy laid out in his speech.

The Apollo program roared to life as a result, and NASA embarked on a crash mission to put a man on the moon. The agency succeeded, of course, in 1969. By the end of Apollo in 1972, the United States had spent about $25 billion on the program — well over $100 billion in today's dollars.

LYNDON JOHNSON (1963-1969)

President Lyndon Johnson was instrumental in both ratcheting up and scaling back the United States' space race with the Soviet Union. As Senate majority leader in the late 1950s, he had helped raise the alarm regarding Sputnik, stressing that the satellite launch had initiated a race for "control of space." Later, Kennedy put Johnson, his vice president, in personal charge of the nation's space program. When Johnson became commander-in-chief after Kennedy's assassination, he continued to support the goals of the Apollo program.

However, the high costs of Johnson's Great Society programs and the Vietnam War forced the president to cut NASA's budget. To avoid ceding control of space to the Soviets (as some historians have argued), his administration proposed a treaty that would outlaw nuclear weapons in space and bar national sovereignty over celestial objects. The result was 1967's Outer Space Treaty (OST), which forms the basis of international space law to this day. The OST has been ratified by all of the major space-faring nations, including Russia and its forerunner, the Soviet Union.

RICHARD NIXON (1969-1974)

President Nixon Was Prepared for Apollo Disaster. All of NASA's manned moon landings occurred during President Richard Nixon's presidency. However, the wheels of the Apollo program had been set in motion during the Kennedy and Johnson years. So, Nixon's most lasting mark on American space activities is probably the space shuttle program.

By the late 1960s, NASA managers had begun drawing up ambitious plans to set up a manned moon base by 1980 and to send astronauts to Mars by 1983. Nixon nixed these ideas, however. In 1972, he approved the development of the space shuttle, which would be NASA's workhorse space vehicle for three decades, starting in 1981. In 1972, Nixon signed off on a five-year cooperative program between NASA and the Soviet space agency. This deal resulted in 1975's Apollo-Soyuz Test Project, a joint space mission between the two superpowers.

GERALD FORD (1974-1977)

President Gerald Ford was in office for less than 2 1/2 years, so he did not have much time to shape American space policy. He did, however, continue to support development of the space shuttle program, despite calls in some quarters to shelve it during the tough economic times of the mid-1970s. Ford also signed off on the creation of the Office of Science and Technology Policy (OSTP) in 1976. The OSTP advises the president about how science and technology may affect domestic and international affairs.

JIMMY CARTER (1977-1981)

President Jimmy Carter did not articulate big, ambitious spaceflight goals during his one term in office. However, his administration did break some ground in the area of military space policy. While Carter wanted to restrict the use of space weapons, he signed a 1978 directive that stressed the importance of space systems to national survival, as well as the administration's willingness to keep developing an antisatellite capability.

The 1978 document helped establish a key plank of American space policy: the right of self-defense in space. And it helped the United States military view space as an arena in which wars could be fought, not just a place to put hardware that could coordinate and enhance actions on the ground.

RONALD REAGAN (1981-1989)

President Ronald Reagan offered strong support for NASA's space shuttle program. After the shuttle Challenger exploded in 1986, he delivered a moving speech to the nation, insisting that the tragedy would not halt America's drive to explore space. Reagan said, "The future doesn't belong to the fainthearted; it belongs to the brave."

Consistent with his belief in the power of the free market, Reagan wanted to increase and streamline private-sector involvement in space. He issued a policy statement to that effect in 1982. And two years later, his administration set up the Office of Commercial Space Transportation, which to this day regulates commercial launch and re-entry operations.

Reagan also believed strongly in ramping up the nation's space-defense capabilities. In 1983, he proposed the ambitious Strategic Defense Initiative (SDI), which would have used a network of missiles and lasers in space and on the ground to protect the United States against nuclear ballistic missile attacks.

Many observers at the time viewed SDI as unrealistic, famously branding the program "Star Wars" to emphasize its supposed sci-fi nature. SDI was never fully developed or deployed, though pieces of it have helped pave the way for some current missile-defense technology and strategies. Even Reagan knew that the *Truth is Stranger than Fiction*. One of the reasons, I believe they tried to assassinate him, which he bounced back from, but was later plagued with Alzheimer's disease of the brain.

President Reagan was a brilliant and charismatic leader and articulate communicator. He tried several times in his hand-written speeches to disclose the presence of aliens to us but was assigned Colin Powell to edit his speeches before he read them to the public on the teleprompters. So much of what President Reagan wanted to disclose to us, was censored. Eventually when they could not kill him, they censored his brain.

On April 13, 2009, the US National Archives of Administration disclosed nearly 250,000 pages of documents during the Reagan administration, including its Personal Journal. On June 11, 1985, the President wrote:

> "We had lunch with 5 top scientists in the field of Aerospace. It was fascinating. Space is really the last frontier, and some of the developments there in Astronomy are science fiction unless they are real. Learned that our transfer capacity is such that we could put 300 people in orbit."

Was he referring to the Secret Space Program using the alien tech?

GEORGE H. W. BUSH (1989-1993)

President George H.W. Bush (the first Bush in office) supported space development and exploration, ordering a bump in NASA's budget in tough economic times. His administration also commissioned a report on the future of NASA, which came to be known as the Augustine report when it was published in 1990.

Bush had big dreams for the American space program. On July 20, 1989 — the 20th anniversary of the first manned moon landing — he announced a bold plan called the Space Exploration Initiative. SEI called for the construction of a space station called Freedom, an eventual permanent presence on the moon and, by 2019, a manned mission to Mars.

These ambitious goals were estimated to cost at least $500 billion over the ensuing 20 to 30 years. Many in Congress balked at the high price tag, and the initiative was never implemented.

BILL CLINTON (1993-2001)

Construction of the International Space Station began in late 1998, in the middle of President Bill Clinton's second term as president. And in 1996, he announced a new national space policy. According to the policy, the United States' chief space goals going forward were to "enhance knowledge of the Earth, the solar system and the universe through human and robotic exploration" and to "strengthen and maintain the national security of the United States."

This latter sentiment was consistent with other space policy statements from previous administrations. However, some scholars argue that the 1996 document opened the door to the development of space weapons by the United States, though the policy states that any potential "control" actions would be "consistent with treaty obligations."

GEORGE W. BUSH (2001-2009)

President George W. Bush issued his own space policy statement in 2006, which further encouraged private enterprise in space. It also asserted national self-defense rights more aggressively than previous administrations had, claiming that the United States can deny any hostile party access to space if it so chooses.

Bush also dramatically shaped NASA's direction and future, laying out a new Vision for Space Exploration in 2004. The Vision was a bold plan, calling for a manned return to the moon by 2020 to help prepare for future human trips to Mars and beyond. It also instructed NASA to complete the International Space Station and retire the space shuttle fleet by 2010.

To help achieve these goals, NASA embarked upon the Constellation program, which sought to develop a new crewed spacecraft called Orion, a lunar lander named Altair and two new rockets: the Ares I for manned missions and the Ares V for cargo. But it was not to be Bush's successor, President Barack Obama, axed Constellation in 2010.

BARACK OBAMA (2009-2017)

President Barack Obama waves farewell after speaking at the NASA Kennedy Space Center in Cape Canaveral, Fla. on Thursday, April 15, 2010. Obama visited Kennedy to deliver remarks on the bold new course the administration is charting to maintain U.S. lead in Space.

In 2009, President Barack Obama called for a review of American human spaceflight plans by an expert panel, which came to be known as the Augustine Commission (not to be confused with the similarly named report President George H.W. Bush ordered two decades earlier).

A year later, Obama announced his administration's space policy, which represented a radical departure from the path NASA had been on. The new policy canceled George W. Bush's Constellation program, which the Augustine Commission had found to be significantly behind schedule and over budget. (Obama did support continued development of the Orion spacecraft for use as a possible escape vehicle at the space station, however.)

In place of Constellation, Obama's policy directed NASA to focus on getting humans to an asteroid by 2025 and then on to Mars by the mid-2030s. This entails, in part, developing a new heavy-lift rocket, with design completion desired by 2015.

The new policy also seeks to jump-start commercial spaceflight capabilities. Obama's plan relies on Russian Soyuz vehicles to ferry NASA astronauts to the space station in the short term after the space shuttles retire in 2011.

But over the long haul, Obama wants this burden shouldered by private American spaceships that have yet to be built. So, Obama promised NASA an extra $6 billion over five years, which the agency would use to help companies develop these new craft.

DONALD TRUMP (2017-2021)

President Trump Signs 2017 NASA Authorization Act. President Donald Trump has directed NASA to return astronauts to the moon in preparation for future crewed missions to Mars and other locations across our solar system. The directive, which has no set timetable of funding, was unveiled Dec. 11, 2017, when Trump signed Space Policy Directive 1.

In March 2019, the Trump administration unveiled a loftier target: land the first woman and next man on the moon by 2024. That plan, called the Artemis program, calls for the creation of a small space station in orbit around the moon and extensive cooperation with private companies to build the moon landers, habitats and other gear astronauts would need on the lunar surface. Its ultimate goal is to establish human colonies on Mars. The Artemis Program wants to use the Moon as a launch pad to Mars, shortening its travel time from Earth launches. Now private Aerospace Companies like Space-X is taking on the goal of colonizing Mars and having shuttle service.

NASA would us its Orion space capsules and Space Launch System mega rocket as core components of the Artemis program but would also use commercial rockets and vehicles to achieve the goal.

Meanwhile, the Trump administration also officially created the U.S. Space Force in 2019. The Space Force is the new 6th branch of the military service aimed at organizing and overseeing U.S. military "high-flying" assets and operations in space.[1]

Notes:
1. Mike Wall, *Presidential Visions for Space Exploration: From Ike to Trump*, https://www.space.com/11751-nasa-american-presidential-visions-space-exploration.html, February 05, 2020.

CHAPTER THIRTY

THE GRAYS & THE U.S. GOVERNMENT

"In order to understand more, it is imperative that we improve our knowledge, before choosing which side of the fence, we feel compelled to belong."
— J.P. Robinson

There is an ongoing repetitive account throughout history that consistently points to the U.S. Government entered into a secret agreement with The Grays in 1954 under President Eisenhower. It has been reported that a small group of Grey alien representatives arrived at Edwards Air Force Base, to contact President Dwight Eisenhower. According to Timothy Good, a former U.S. Government consultant claimed Eisenhower had three secret meetings with aliens. The 34th President of the United States met the extraterrestrials at a remote air base in New Mexico in 1954 at the Holloman Air Force base and there were many witnesses. Eisenhower met with the Grays on three occasions, resulting in entering a legal contract with them, which has since been referred to as the Edwards Agreement or, as Dr. Dan Burisch called it, the Tau IX Treaty for the Preservation of Humanity.

Eisenhower, who was a former 5 Star General who commanded the Allied Forces in Europe during World War II, was elected U.S. President from 1953 to 1961, was known to hold strong beliefs in Extraterrestrial life on other planets and pushed the U.S. Space Program.

His meetings with the Gray aliens took place while officials were told that he was on vacation in Palm Springs, California, in February 1954. The initial meeting was supposed to be with aliens who were 'Nordic' in appearance, but the agreement was eventually signed with the Grays aliens. See, Book One, Chapter on the Grays, in *Who's Who in the Cosmic Zoo?* that details the 35 different types of Grays.[1]

According to classified documents released by the Ministry of Defense in 2010, Winston Churchill ordered a UFO sighting to be kept secret. The UFO was seen over the East Coast of England by an RAF reconnaissance plane returning from a mission in France or Germany towards the end of the war. Churchill is said to have discussed how to deal with UFO sightings with Eisenhower. Their alliance and their friendship were no secret. They saw eye to eye on a lot of things, so it's certainly plausible to accept that knowing that Nazis were heavily involved in UFOs, that the two Allies who were Victorious in the Second World War, would have such discussions.

President Eisenhower met with extra-terrestrials on three separate occasions at New Mexico air base. It has been widely reported that Eisenhower and FBI officials organized the meetings with the extraterrestrial by using telepathic messages.[2]

TAU IX TREATY FOR THE PRESERVATION OF HUMANITY

The deal specified that the US Government would let certain people be abducted on a regular basis as the Greys solemnly promised they would bring them back safely and only after erasing their memory.

But, as many abductees could confirm, the situation was slightly different. Often, those who were kidnapped by aliens would recall the incident, in all its dreadful details.

In exchange for allowing the Greys to have human guinea pigs, the Government got their hands-on advanced alien knowledge and technology. Another 'clause in the contract' was the exchange program of ambassadors from both sides. However, it is unclear how many were involved in this exchange program, but the most renowned are Krill, the reptilian from the Draco constellation and J-Rod, the Grey from Zeta Reticulum.

The Edwards Agreement was the pinnacle of a series of events that began in 1947, when the wreckage of an alien spacecraft was recovered near Roswell, NM. Inside the wreckage, several alien bodies were found. This incident undoubtedly raised awareness towards the UFO phenomenon and undeniably gave rise to modern ufology.

In 1949, another crash took place in New Mexico but this time however, there was one survivor left. He was dubbed EBE, which is short for extraterrestrial biological entity and the details he spilled out allowed the US Government to make contact with his people.

In 1951, the Greys were reached using a device built with the help of the EBE. One year later, extraterrestrial communications were well underway, paving the road for the 1954 treaty.

On the night of February 20 to 21, 1954, while enjoying his vacation in Palm Springs, California, President Dwight Eisenhower disappeared for a short while. It is thought he was secretly escorted to nearby Edwards Air Force base for an encounter with the Grey aliens. His official statement was that he had to undergo a dental emergency and so he visited a local dentist.

While the meeting did go as planned, there were some unexpected consequences. The landing was kept secret in order to see how people would react when confronted with a technologically advanced alien race. The hundreds of soldiers present at Edwards AFB during first contact received no briefing prior to the event; they were the unaware test audience.

This approach would later prove to be disastrous, as a high percentage of those present began suffering from various psychological ailments, ranging from dysfunctional behavior to psychosis, criminal intentions, and suicide.

Several confessions have surfaced throughout the years, from people who say they witnessed the event firsthand. One of them is Gerald Light, a famous metaphysical community leader at the time. Light says his involvement was gauging the effect this breakthrough would have over the general public.

My dear friends: I have just returned from Muroc [Edwards Air Force Base]. The report is true — devastatingly true! During my two days' visit I saw five separate and distinct types of aircraft being studied and handled by our Air Force officials — with the assistance and permission of the Etherians! I have no words to express my reactions. It has finally happened. It is now a matter of history, wrote Light.

If Light is to be believed, several other important figures were present at the meeting. Among them was Eisenhower's chief economic advisor, Dr. Edwin Nourse, who offered his expertise on the potential economic outcome of a first contact with intelligent alien life. Light also said several trustworthy religious leaders had attended the meeting, their roles being obvious.

Several other sources support Light's claims. Whistleblowers have put forward their credentials, describing two sets of meetings involving different extraterrestrial groups.

According to former Naval Intelligence officer William Cooper, large objects traveling towards Earth had been seen in 1953. First believed to be asteroids, it was later determined they were spaceships. Two distinct missions, Project Sigma and Project Plato managed to establish contact with the extraterrestrials, through binary code.

Cooper distinguishes between two alien races: the Nordic ones, friendly towards humanity and the Greys, who had different goals with humans. According to him, the Nordics contributed to the signing of a non-aggression treaty between humanity and the Greys. The Nordics did not offer technology but rather the chance for spiritual advancement from our current dimension to the next.

The treaty told that aliens would not intervene in our affairs, and neither would we interfere with theirs. Their presence on Earth would be kept as a secret. They could regularly abduct humans for medical and scientific purposes if they would keep them safe from harm. The humans would then be returned to their abduction point, with no memory of the event.

Their demands were that humans break apart its nuclear weapons arsenal. They reminded of humanity's self-destructive potential and condemned the fact that we were killing each other, harming the planet, and wasting its natural resources. The committee interpreted their demands in a very suspicious way, believing that nuclear disarming was not in the best interest of the United States since it would leave the

CHAPTER 36: THE GRAYS & THE U.S. GOVERNMENT

world defenseless in the face of an alien threat. Eisenhower ultimately rejected their proposition.

Another witness to the meetings was John Lear, son of the creator of the Lear Jet, William Lear. According to him, the Nordics offered a hand in extinguishing the Greys' threat, but President Eisenhower refused it because there was no technology involved.

Did this meeting really happen? If so, it would appear to confirm the current world-wide UFO phenomenon. However, more than 60 years have passed since the alleged First Contact, and official disclosure is still missing.[3]

Eisenhower's First Contact was reported by U.S. Navy Intelligence Researcher and Whistleblower Milton William Cooper who served the Navy's Intelligence briefing team for the Commander of the Pacific Fleet between 1970-73, allowing him access to classified documents. His groundbreaking book, which established him as an insider Whistleblower, *Behold A Pale Horse*, is the source of most conspiracy theories, which does not mean it isn't true. He described the background and nature of the 'First Contact' with extraterrestrials as follows:

> "In 1953 Astronomers discovered large objects in space which were moving toward the Earth. It was first believed that they were asteroids. Later evidence proved that the objects could only be Spaceships. Project Sigma intercepted alien radio communications. When the objects reached the Earth, they took up an extremely high orbit around the Equator. There were several huge ships, and their actual intent was unknown. Project Sigma, and a new project, Plato, through radio communications using the computer binary language, was able to arrange a landing that resulted in face-to-face contact with alien beings from another planet. Project Plato was tasked with establishing diplomatic relations with this race of space aliens. In the meantime, a race of human looking aliens contacted the U.S. Government. This alien group warned us against the aliens that were orbiting the Equator and offered to help us with our spiritual development. They demanded that we dismantle and destroy our nuclear weapons as the major condition. They refused to exchange technology citing that we were spiritually unable to handle the technology which we then possessed. They believed that we would use any new technology to destroy each other. This race stated that we were on a path of self-destruction, and we must stop killing each other, stop polluting the Earth, stop raping the Earth's natural resources, and learn to live in harmony. These terms were met with extreme suspicion, especially the major condition of nuclear disarmament. It was believed that meeting that condition would leave us helpless in the face of

an obvious alien threat. We also had nothing in history to help with the decision. Nuclear disarmament was not considered to be within the best interest of the United States. The overtures were rejected." [4]

William Cooper said the humanoid extraterrestrial race was not willing to enter technology exchanges with the U.S. Government if it was going towards weapons development. The Tall Blond Nordic type Human Extraterrestrial wanted the U.S. Government to focused on spiritual development instead. Eisenhower rejected their officer and ended up entering into a type of Faustian Contract with the Grays, who have since breached and betrayed the Americans multiple times. One of the reasons, full Disclosure will unlikely be admitted by U.S. because they would have to admit their mistakes, and the disclose the alien threat. [4]

Another Whistleblower, former Lockheed L-1011 Captain John Lear who flew over 150 test aircraft and held 18 world speed records, during the late 1960's, 1970's. He was contract pilot in the early 1980's for the CIA. According to Lear, there was a warning from another ET race before Eisenhower signed the agreement with the Grays. He claimed they visited Muroc/Edward and the following occurred:

According to John Lear who told Art Bell on his original Coast to Coast Radio Broadcast in 2003, that in 1954, President Eisenhower met with a representative of another alien species at Muroc Test Center, which is now called Edwards Airforce Base. This alien suggested that they could help us get rid of the Grays, but Eisenhower turned down their offer because they offered no technology. This was the tall Blond Human looking Alien, who wanted the U.S. to focus on their spiritual development. [5]

So, instead, Eisenhower made a deal with the insidious demonic Grays in exchange for allowing a 'limited number of humans' on a short list, to be abducted by the Grays for their purpose of genetic experimentation, or so they said, and in turn, the Grays would share Eisenhower and the U.S. Government aerospace technology, which is how the Secret Space Program got its wings, literally. The Grays taught U.S. officials anti-gravity technology, which allowed them to develop several different types of interstellar craft.

Another Whistleblower, Master Sergeant Robert Dean came forward, who had access to top secret documents, just like Cooper, while working in the intelligence division for the Supreme Commander of one of the Major U.S military commands. Dean had a 27-year distinguished military career, he served at the Supreme Headquarters Allied Powers Europe where he witnessed these documents while serving under the Supreme Allied Commander of Europe. Dean claimed:

"The group at the time, there were just four that they knew of for certain and the Greys were one of those groups. There was a group that looked exactly like we do. There was a human group that looked so much like us that that really drove the admirals and the generals crazy because they determined that these people, and they had seen them repeatedly, they had had contact with them, there had been abductions, there had been contacts… Two other groups, there was a very large group, I say large, they were 6-8 maybe sometimes 9 feet tall and they were humanoid, but they were very pale, very white, didn't have any hair on their bodies at all. And then there was another group that had sort of a reptilian quality to them. We had encountered them, military people, and police officers all over the world have run into these guys. They had vertical pupils in their eyes and their skin seemed to have a quality very much like what you find on the stomach of a lizard. So those were the four they knew of in 1964." [5]

US PRESIDENTS & ALIENS

President Eisenhower was not the only one to encounter the aliens. This story is widely known amongst Ufologists and Researchers, when President Richard Nixon took his long-time golfing buddy, Jackie Gleason (from the famed TV Show, Honeymooners) to show him the Gray alien they were keeping at one of their secret military bases. Jackie Gleason was known as jolly hefty man, but it has been reported by his wife and others, that after he saw the Gray Alien, he was so shocked and that he did not eat for a month. Gleason was known for his interest in aliens and UFOs. He was outspoken about it, and regularly appeared on *The Long John Nebel Late Night Radio Program*, who was known for his reporting on the Paranormal and UFOs. Gleason was well known for his massive collection on UFO literature from the 1950s and 60s. He lived in a round house, dubbed, 'The Mothership,' because of his fear of ghosts. He believed the round corners confused them.

Gleason's second wife, Beverly McKittrick reported the story when Gleason saw proof that aliens exist on February 19, 1973, when his golfing buddy, President Nixon invited him to have a looksie where the dead, pickled alien bodies were allegedly being held at Homestead Air Force Base, Florida.[6] The base was destroyed during the Category 5 Hurricane Andrew on August 24, 1992, along with all its alien counterparts and was rebuilt and renamed, Homestead Air Reserve Base.

Notes:
1. Ella LeBain, *Who's Who in the Cosmic Zoo?* Book One – Third Edition, Chapter: The Grays, pp. 248-289, Tate, 2013.
2. Daily Mail, *President Eisenhower had Three Secret Meetings with Aliens, former Pentagon Consultant claims*, https://www.dailymail.co.uk/news/article-2100947/Eisenhower-secret-meetings-aliens-pentagon-consultant-claims.html, February 13, 2012.
3. Ana Ionita, *TAU IX Treaty—The US Government Has a Secret Pact with the Greys*, 3/26/16, http://ufoholic.com/tau-ix-treaty-the-us-government-has-a-secret-pact-with-the-greys/
4. William Cooper, "Origin, Identity, and Purpose of MJ-12," http://www.geocities.com/Area51/Shadowlands/6583/maji007.html William Cooper, *A Covenant With Death*, http://www.alienshift.com/id40.html taken from his book, Behold a Pale Horse, Light Technology Publishing, 1991, p. 202-203.
5. John Lear, *Disclosure Briefing*, Coast to Coast Radio with Art Bell, November 2003 http://www.coasttocoastam.com/shows/2003/11/02.html
6. Nick Redfern, *A President, a Comedian, and Pickled Aliens*, https://mysteriousuniverse.org/2017/03/a-president-a-comedian-and-pickled-aliens/, March 3, 2017.

CHAPTER THIRTY-ONE

SECRET GOVERNMENT SPACE PROGRAMS

"We can do anything you can imagine.
We can take E.T. home."
~Ben Rich, Director of Lockheed's Skunk Works, 1975-1991

The United States has a history of government agencies existing in secret for years. The National Security Agency (NSA) was founded in 1952; its existence was hidden until the mid-1960's. Even more secretive is the National Reconnaissance Office, which was founded in 1960 but remained completely secret for 30 years. Just because the mainstream media does not tell you about these agencies, does not mean they do not exist.

DECLASSIFIED: AMERICA'S SECRET FLYING SAUCER
In the 1950s, a small team of engineers set to work on a secret program called Project 1794—a supersonic craft designed to shoot down Soviet bombers. Now a trove of declassified documents reveals the audacious mission to build a flying saucer.

In September 2012, Michael Rhodes, a technician at the National Declassification Center (NDC) in College Park, Md., donned white cotton gloves, entered a climate-controlled room, and opened a cardboard file box. It was time for the report inside—" Project 1794 Final Development Summary Report 2 April—30 May 1956"—to become public.

Popular Mechanics mentioned the Air Force's "vertical-rising, high-speed" craft in 1956 and published a photo in 1960. In the decades since the program was canceled in 1961, aviation buffs and UFO researchers have unearthed technical papers written near the end of America's flying saucer experiment, but the document that originally convinced the government to invest in a military flying disc has languished in the NDC under the SECRET designation. This recently discovered report describes in previously unknown detail how aviation engineers tried to harness what were then cutting-edge aerodynamic concepts to make their improbable creation fly. Although Avro's saucer never completed a successful flight, some of the most sophisticated aircraft flying today adopted many of the same technologies.[1]

The Avrocar I Continuation and Terrain Test Program June 1960 to June 1961 Avro-US Army – Great vintage footage showing the development of the USAF-US Army Avrocar VZ-9-AV flying disc from June 1960 to June 1961.[2]

We are talking about Special Access Programs (SAP). From these we have unacknowledged and waived SAPs. These programs do not exist publicly, but they do indeed exist. They are better known as 'deep black programs.' A 1997 US Senate report described them as "so sensitive that they are exempt from standard reporting requirements to the Congress."

President Obama was sworn in office in 2013, when The Washington Post revealed that the "black-budget" documents report a staggering 52.6 billion dollars that was set aside for operations in the fiscal year 2013. Although it is great to have this type of documentation in the public domain proving the existence of these black budget programs, the numbers seem to be too low according to statements made by some very prominent people. There is a lot of evidence to suggest that these programs are not using billions of dollars, but trillions of dollars that are unaccounted for.

"It is ironic that the U.S. would begin a devastating war, allegedly in search of weapons of mass destruction when the most worrisome developments in this field are occurring in your own backyard. It is ironic that the U.S. should be fighting monstrously expensive wars allegedly to bring democracy to those countries, when it itself can no longer claim to be called a democracy when trillions, and I mean thousands of billions of dollars have been spent on projects which both congress and the commander in chief no nothing about." (Paul Hellyer, Former Canadian Defense Minister, 2008)[3]

We are talking about large amounts of unaccounted-for money going into programs we know nothing about. There have been several congressional inquiries that have noted billions, and even trillions of dollars that have gone missing from the Federal Reserve System.

On July 16, 2001, in front of the house appropriations committee, Secretary of Defense Donald Rumsfeld stated: "The financial systems of the department of defense are so snarled up that we can't account for some $2.6 trillion in transactions that exist, if that's believable. We do not really hear about black budget programs, or about people who have investigated them. However, the topic was discussed in 2010 by Washington Post journalists Dana Priest and William Arkin. Their investigation lasted two years and concluded that America's classified world has: Become so large, so unwieldy and so secretive that no one knows how much money it costs, how many people it employs, how many programs exist within it or exactly how many agencies do the same work."

It was only 2013 that the Central Intelligence Agency (CIA) finally admitted to the existence of Area 51. Although it did not 'officially' exist before the CIA made this admission, it was pretty clear that something secretive was going on in the Nevada desert. That secretive something would be the testing of secret aircraft and technology that the public has absolutely no idea about. Take for example the U.S. air strike against Libya in 1996. An f-111 jet was used, which had been operational since 1983, but its existence was still kept secret for a number of years after.[3]

These programs are referred to as Special Access Programs (SAP), and they are funded from what is known as the 'Black Budget.' From these we have unacknowledged and waived SAPs. These programs do not exist publicly, but they do indeed exist. They are better known as 'deep black programs.' A 1997 US Senate report described them as "so sensitive that they are exempt from standard reporting requirements to the Congress."[3]

UNDERSEA MILITARY BASES

In 1994, when I was living on a St. Petersburg beach off the Gulf of Mexico, I witnessed an Underwater Submerged Object (USO) shaped like a silver metallic disc the size of a football field, come out of the ocean, right before my eyes, life itself up into the air, shedding sea foam and water, and then in a New York second, project itself into space that turned into a star, only to disappear into what appeared to be a portal in space. That night the local MUFON received 350 calls reporting a USO out of the Gulf of Mexico. I was not alone in my witness however; it was unforgettable and life changing. Seeing is believing, and that sighting so close to me, inspired me to dig deep and investigate into what is really going on, which has culminated into this book series of *Who's Who in the Cosmic Zoo?* Book One has chapters on what is going on underground, the ancient Inner Earth civilizations, along with secret bases networked and tunneled throughout the US and around the World. [4] [See, Chapter: Dulce Underground, pp. 185-202].

The research into under-ocean military bases began decades ago. For example, in 1968 the Stanford Research Institute discussed the construction of dozens of undersea bases. The study was titled *Feasibility of Manned In-Botton Bases* which was published in the 2015. Here is an excerpt from UNDERGROUND BASES: Subterranean Military Facilities and the Cities Beneath Our Feet:[5]

> "The North Atlantic seems to be very significant and is possibly (the site of) the largest sea base in European waters. Other reported underwater bases are in South America waters – in the areas of Puerto Rico and Brazil – in Antarctica and other deep unobserved areas of ocean." – Tony Dodd, former British police officer.

> "Could there also be secret military facilities under our ocean floors? In 1969, the Stanford Research Institute in Menlo

Park, California, published a report titled The Feasibility of manned in-bottom bases. The report states, "The construction of thirty in-bottom bases within the ocean floors is technically and economically feasible … The cost of such a base program would be about $2.7 billion".

Now keep in mind that was in 1969. So, given the multi-trillion-dollar black budgets numerous researchers claim the US Government and its agencies have access to annually, who is to say undersea bases were not financed and built decades ago?

Interestingly, the Stanford Research Institute's 1969 report also states that deep submergence vehicles would need to be developed to build undersea bases. The following year it was announced Lockheed had launched deep sea vehicles with the necessary capabilities to do just that.

According to a lecture that independent researcher Dr. Richard Sauder gave at the Xcon 2004 conference, there could easily be US-built undersea bases in the Persian Gulf, the North Sea, *and the Gulf of Mexico*. Dr. Sauder, author of the 1996 book *Underground Bases and Tunnels: What is the Government Trying to Hide?* also spoke of the US Navy's undersea test and research center off the coast of Andros Island, in the Bahamas. He speculated that this facility, which is known as AUTEC (Atlantic Undersea Test and Evaluation Centre), could be a front for an undersea complex of secret bases.

On May 1, 2008, UK newspaper The Telegraph ran an article about China's underwater sea bases. The article, which contained satellite imagery of base openings on Hainan Island, China, states there's "a network of underground tunnels at the Sanya base on the southern tip of Hainan island". The article also states that the tunnels allow Chinese submarines to travel out into the ocean from the base completely undetected. "Of even greater concern to the Pentagon," the article continues, "are massive tunnel entrances, estimated to be 60ft high, built into hillsides around the base … While it has been known that China might be developing an underground base at Sanya, the pictures provide the first proof of the base's existence and the rapid progress made".

CHAPTER 37: SECRET GOVERNMENT SPACE PROGRAMS

Although likely to be more complex and costly to build than bases beneath land, undersea bases would obviously provide even more secure hideaways for the Splinter Civilization to go about their business. It's unlikely China and the US would be the only countries building undersea bases for their naval advancements and other purposes." [5]

In 1987 Deputy Director of Engineering and Construction for the U.S. Army Corps of Engineers, Lloyd A. Duscha, gave a speech at an engineering conference titled "Underground Facilities for Defense – Experience and Lessons." In the first paragraph of his speech, he states the following: "After World War II, political and economic factors changed the underground construction picture and caused a renewed interest to "think underground." As a result of this interest, the Corps of Engineers became involved in the design and construction of some very complex and interesting military projects. Although the conference program indicates the topic to be "Underground Facilities for Defense – Experience and Lessons," I must deviate a little because several of the most interesting facilities that have been designed and constructed by the Corps are classified.[6] [pp. 109-113]

He then went into a discussion of the Corps' involvement in the 1960's in the construction of the large and elaborate NORAD base buried deep beneath Cheyenne Mountain in Colorado. This is just a public statement, but you will not find a more significant public admission of secret, underground bases than this one. Such speeches are not the only evidence available, however. There also exist documents obtained by researchers through the Freedom of Information Act (FOIA) that shed more light on the subject, and clearly outline plans for the contraction of underground facilities.

Another great example of in-bottom bases deep underneath the ocean floor comes from William B. McLean, who was the inventor of the Sidewinder air-to-air missile and former Technical Director of the China Lake Naval, Naval Ordnance Test Center (NOTS). He was also the Technical Director of the U.S. Naval Undersea Warfare Center in San Diego. McLean made some comments to John Newbauer, who at the time was the Editor-in-Chief of Astronautics and Aeronautics, stating that these plants and projects were already in development. ("A Bedrock View of Ocean Engineering," Interview of William B. Mclean by A/A Editor-in-Chief John Newbauer, Astronautics and Aeronautics (April 1969: 30-36.) [5,6]

One of the most credible and prominent researchers on Underground Military Facilities is Richard Sauder, Ph.D., author of *Hidden In Plain Sight*, details the science and engineering behind these massive underground projects. Here is an excerpt from the book:

> This 'racetrack' facility – also called the 'Nautilus Concept' – that can dock three submarines at a time, with an adjoining sister facility that also can handle multiple submarines. The

385

picture is virtually self-explanatory. Large submarines are hundreds of feet long, so the dimensions of a facility such as shown here would have to be very large. The central docking area might be more than a thousand feet long and easily more than a hundred feet in diameter. The living quarters would obviously have to accommodate hundreds of crew members in some degree of creature comfort.
There are also known underground facilities in existence. Take for example the Swedish underground military facility at Musko. It is a large naval base built underneath a mountain. The hospital alone within this facility holds over 1,000 beds. Musko engineers blasted out 1,500,000 cubic meters of stone in order to build it.

As it happens, after giving a public talk a couple of years ago, I was approached by a man who had been a uniformed member of the United States Navy. We chatted for a while and when he mentioned that he had spent some time at China Lake my ears perked up. I asked him if there was an underground facility at China Lake. He said that indeed there is, and that it is impressively large and deep. I asked him if he had ever been in it, and he said that he had, though not to the deepest levels. I asked him how deep the deepest part extended. He looked at me soberly and said very quietly, "It goes one mile deep." I then asked him what the underground base contains. He replied, 'Weapons.' I responded, "What sort of weaponry?" And he answered without pausing, "Weapons more powerful than nuclear weapons."[7]

There are documents available which expose a deep underground command center that was to be built far below regions such as Washington, D.C. and China Lake, California during the Cold War. Documents show that in 1964 the military was considering building a huge underground cavity 4,000 feet deep beneath China Lake, and it's well known that the United States and the Soviet Union created a vast infrastructure to support a complex of offensive and defensive weapons during the Cold War. This infrastructure included sites and facilities for developing, testing, storing, and manufacturing weapons. There was also a host of communication and command centers." [6,7]

The very first TOP SECRET memo on the subject was issued by Robert McNamara on November 7th, 1963, from the office of the Secretary of Defense. A second memo was issued on the same day concerning a proposed Deep Underground National Command Center that would be approximately 3,500 feet beneath Washington. The memo also mentioned elevator shafts below the State Department and White House that would descend to 3,500 feet, with high speed, horizontal tunnel transport to the main facility. And this was way back in the 60's. Imagine what technological feats we are capable of now.[6]

TR-3B

Aviation journalist Bill Sweetman, who working within the Pentagon, estimated that 150 special access programs existed that weren't even acknowledged. These programs are not known about by the highest members of government and the highest-ranking officials in the military. He determined that most of these programs were dominated by private contractors (Lockheed Martin, Boeing, etc.) and that he had no idea as to how these programs were funded. These programs include advanced spaceships craft like TR-3 Manta.

Dwight Eisenhower, former five stars U.S. general and President of the United States also warned us about secrecy and the acquisition of unwarranted influence within the "Department of Defense" with his farewell speech:

In the council of government, we must guard against the acquisition of unwarranted influence whether sought or unsought, by the military industrial complex. The potential disaster of the rise of misplaced power exists and will persist. We must never let the weight of this combination endanger our liberties or democratic processes.

The only other president that blew the whistle on secrecy beyond the government was President John F. Kennedy referring to the military industrial complex:

It appears some of these black budget programs deal with UFOs. Documents from the NSA prove that UFOs and extraterrestrials are of high interest to the agency and over whelming evidence suggest that these black budget programs deal with matters beyond our world. Garry McKinnon has also shed light on this fact, as have thousands of previously classified documents and statements from high level government and military personnel. [2]

> "There are objects in our atmosphere which are technically miles in advance of anything we can deploy, that we have no means of stopping them coming here ... [and] that there is a serious possibility that we are being visited and have been visited for many years by people from outer space, from other civilizations. That it behooves us in case some of these people in the future or now should turn hostile, to find out who they are, where they come from, and what they want.

> This should be the subject of rigorous scientific investigation and not the subject of 'rubbishing' by tabloid newspapers." – Lord Admiral Hill-Norton, Former Chief of Defense Staff, 5 Star Admiral of the Royal Navy, Chairman of the NATO Military Committee

In 2002, the U.S. government charged and arrested Gary McKinnon, a UK Hacker, for performing what is said to be the biggest military computer hack of all time. He was charged with causing $700,000 worth of damage to government computer systems. McKinnon has been facing extradition to the United States. McKinnon was a threat to the secrecy of the cover-up of the Secret Space Program, that he serendipitously stumbled upon, which he described as Solar Warden. McKinnon became a political scapegoat whose actions became justification for heightened national security, but it was the secrecy within the extraterrestrial military industrialized complex that was threatened, not necessarily the U.S. Government.

Ten years later, in October of 2012, it was decided that McKinnon would not be extradited to the UK, and in December of the same year it was determined that McKinnon would face no further charges in the UK. The insiders decided that it was drawing more attention to his findings that defeated their whole purpose of secrecy.

McKinnon found a list of non-terrestrials 'off-world' officers of rank. He could not tell if they represented the Air Force, Navy, or Army or perhaps another agency, that today we are calling the Space Force. He found multiple photos of UFOs and lists of fleet-to-fleet transfers of materials from ship to ship.[8]

"I was convinced...that certain secretive parts of the American government intelligence agencies did have access to crashed extra-terrestrial technology which could...save us in the form of a free, clean, pollution-free energy." Gary McKinnon, Sources: BBC News article [9, 10]

THE NEW SPACE RACE
Who is Lieutenant General Steven L. Kwast?

According to his official USAF biography, Lt. Gen. Kwast graduated from the United States Air Force Academy with a degree in astronautical engineering and holds a master's degree in public policy from Harvard's Kennedy School of Government. Kwast previously served as Commander of the 47th Operations Group at Laughlin Air Force Base and the 4th Fighter Wing at Seymour Johnson AFB. Kwast boasts more than 3,300 flight hours in the F-15E, T-6, T-37, and T-38 and over 650 combat hours.

Lt. Gen. Kwast most recently served as Commander of the Air Education and Training Command (AETC) at Joint Base San Antonio (JBSA) but retired in August. According to some reports, Kwast was prematurely relieved of his duties at JBSA and blacklisted for promotion after speaking out on space-related issues despite a

CHAPTER 37: SECRET GOVERNMENT SPACE PROGRAMS

service-wide gag order. Kwast declined to comment on the reports and retired on September 1, 2019.

Despite the controversy surrounding his removal from his post at AETC, some defense analysts and Lt. Gen. Kwast's own supporters within the Armed Forces were suggesting prior to his retirement that he should be appointed as Commander of the Pentagon's budding Space Force. Kwast has published several op-eds in recent years pushing for the U.S. military to take on a greater role in space in order to ensure American economic dominance and what he sees as the continued proliferation of American values.

Lieutenant General Steven L. Kwast fired Kwast delivered a lecture at Hillsdale College in Washington, D.C. on November 20, 2019, titled "The Urgent Need for a U.S. Space Force." Kwast's wide-ranging speech described the power of new technologies to revolutionize humankind, referencing the competitive advantage the discovery of fire offered to early humans and the strategic value that nuclear weapons offered 20th-century superpowers. When it comes to current revolutionary technologies, Kwast says the "the power of space will change world power forever" and that it is up to the United States military to leverage that power:

> "As a historian, reflecting on the fact that throughout the history of mankind technology has always changed world power. But the story of rejecting the new and holding and clinging to the paradigms of the past is why no civilization has ever lasted forever, and values are trumped by other values when another civilization figures out a way of finding a competitive advantage. The nature of power, you either have it and your values rule or you do not have it and you must submit. We see that play out again and again in history and it's playing out now."

As has been common as of late, Lt. Gen Kwast cites rapidly growing Chinese military and technological advances as the reason why the United States must invest heavily in new space-based technologies. "We can say today we are dominant in space, but the trend lines are what you have to look at and they will pass us in the next few years if we do not do something. They will win this race and then they will put roadblocks up to space," Kwast argues, "because once you get the high ground, that strategic high ground, it's curtains for anybody trying to get to that high ground behind them."

Kwast claims China is already building a "Navy in space" complete with the space-based equivalents of "battleships and destroyers" which are "able to maneuver and kill and communicate with dominance, and we [the United States] are not." Kwast's speech centers on the thesis that the United States needs a Space Force in order to counter Chinese advances and win the competition over the economy of the future and, as an extension, which sets the values of the future:

> "Space is the Navy for the 21st century economy, a networked economy that will dominate any linear terrestrial economy in the four engines of growth and dominance that change world power: transportation, information, energy, and manufacturing. [...] Whoever gets to the new market sets the values for that market. And we could either have the market with the values of our Constitution [...] or we could have the values we see manifest in China."

There have been Hints of radical new technologies such as anti-gravity under development by the military and, just as in Kwast's speech, Chinese advances have been cited as the reason why these technologies are needed. China has been rapidly expanding its presence in space in recent years, placing a lander on the far side of the moon in late 2018 in what some say was a push to scout natural resources with which to develop a permanent lunar manufacturing center. China has also been developing "Mothership" aircraft from which to launch space planes and other payloads rapidly and unpredictably into space. The country has also launched several eyebrow-raising satellites in recent years which some analysts claim could be used in anti-satellite warfare. Beyond all this, they have been investing heavily in a traditional space program that includes many facets of manned and unmanned space technologies that rivals, and in some ways, exceeds our own.[11]

SPACE FORCE: 21ST CENTURY WARFARE

Kwast argues that the scientists, engineers, historians, and strategists of today have been pushing the U.S. Congress to more heavily and more rapidly fund the Space Force and associated technologies, but there is still some pushback and confusion as to why these are presently needed. Kwast ultimately makes the case that the United States must be able to bring, non-kinetic power, and informational power to the battlefield cheaper and faster than its adversaries in order to ensure strategic advantage in space.

Around the 12:00 mark in the speech, Kwast makes the somewhat bizarre claim that the U.S. currently possesses revolutionary technologies that could render current aerospace capabilities obsolete:[12]

> "The technology is on the engineering benches today. But most Americans and most members of Congress have not had time to really look deeply at what is going on here. But I've had the benefit of 33 years of studying and becoming friends with these scientists. This technology can be built today with technology that is not developmental to deliver any human being from any place on planet Earth to any other place in less than an hour."

CHAPTER 37: SECRET GOVERNMENT SPACE PROGRAMS

Kwast's comment is only one of several curious comments made by military leadership lately and they do seem to claim that we could be on the precipice of a great leap in transportation technology. We also do not know exactly where he is coming from on all this as it is not necessarily the direct wheelhouse of someone who was running the Air Force's training portfolio, although it does have overlaps. Whether or not the revolutionary aerospace technologies Kwast mentions have actually been developed is one thing, but Kwast's lecture, his recent op-eds, and his supporters make it clear that there are many within the U.S. military and analyst community who have felt that there is a great need to boost investment in American space technologies and the U.S. military's presence in space. That vision is certainly taking root across the Defense Department.

Is all this setting the stage for a new space race that will benefit mankind by furthering scientific and technological development, or is it ushering in the conditions for the first great space war? Only time will tell, but according to Kwast, the technologies needed to win that war may be more science fact than fiction.[11]

PILOTS DIE CHASING UFOS

Timothy Good writes, "The destruction or disappearance of military aircraft during interceptions of UFOs continued apace." As General Benjamin Chidlaw, former commanding general of Air (later Aerospace) Defense Command told Robert C. Gardener (ex USAF) in 1953: "We have stack of reports of flying saucers. We take them seriously, when you consider we have lost many men and planes trying to intercept them." Leonard Stringfield, the former Air Force intelligence officer was told by a reliable source in the 1950s that the "Air Force was losing about a plane a day or 365 a year to the UFOs." Stringfield was reliably informed. [13]

According to US Defense Department figures, from 1952 until the end of October 1956, there were 18,662 major accidents of military aircraft, broken down as follows:

Year	Air Force Losses	Navy Losses
1952	2,274	2,086
1953	2,075	2,325
1954	1,873	1,911
1955	1,664	1,566
1956	1,530	1,358

Of this astonishing total, most involved fast new jets (such as those scrambled in UFO interceptions), of which 56.2 per cent were found to be caused by pilot error; 8.1 per cent by ground-crew or other personnel failure; 23.4 per cent by failure of parts and equipment in the aircraft; 2.8 per cent by various 'unsafe conditions', and −9.5 per cent (1,773) were due to 'unknown factors'. Thanks to Timothy Good' book, "Need to Know" P.172

Ret. USAF Pilot George Filer writes: "When I chased a UFO, we exceeded the red lined aircraft air speed by 20 knots. In the excitement of the chase, it is easy to exceed aircraft capabilities and often some part of the aircraft may fail. Many stories are told of firing missiles and making direct hits on the UFOs that were unharmed only to have them return the fire and destroy the interceptor."[13]

Lt. Chidlaw entered the Air Corps Engineering School at Wright Field, Ohio, and later directed the development of the United States' original jet engine and jet aircraft. He spent most of his career at Wright Field which established Chidlaw as an expert on materiel, especially aircraft. He became chief of the Experimental Engineering Branch, where he monitored the jet engine development. In March 1945, he took command of the Mediterranean Allied Tactical Air Forces and was promoted to major general the next month.

Chidlaw returned to Wright Field in July 1945 as deputy commanding general for operations of what became Air Materiel Command. He was flown to Roswell, New Mexico at the time of the UFO crash announcement in July of 1947. In October 1947, he became deputy commanding general of the command, with rank of lieutenant general, and full commander on September 1, 1949. General Chidlaw was given the task to evaluate UFO sightings and his group agreed the phenomena was real and approved the Research and Development Board. It is likely his team attempted to develop their own flying discs. On July 29, 1951, he received his fourth star and the command of Air Defense Command at Ent Air Force Base, Colorado. Most fighter interceptors were under his command and each Air Force base had a UFO Officer who could launch fighters if UFOs were in the area. [13]

SHOOT THEM DOWN

The US military is estimated at losing 500 aircraft a year attempting to shoot down UFOs. There is some evidence that when missiles or rockets are fired against the UFOS, they can turn them around and use them against the fighter that fired the shot. The History Channel's documentary, *Unidentified*, has documented these stories, and they are widely known amongst Ufologists.

The $21 trillion missing from the U.S. government budget since 1998, as documented by Catherine Fitts and others was spent on the secret space program, Pentagon officials confirm. Much of this technology is now being released to the general public, they say. [14]

Then there is that famous clip that was aired all over the mainstream media news on September 10, 2001, when then Defense Secretary Donald Rumsfeld reported that $2.3 TRILLION was missing from the Pentagon. The day after that, was 911, and the world completely forgot about the missing $2.3 trillion. [15]

Strange Craft is the True Story of an Air Force Intelligence Officer's Life with UFOs U.S. Air Force Major George Filer belongs to the generation of pilots and airmen who first became aware of the strange aircraft showing up in the Earth's atmosphere after World War II. These men – military professionals who flew planes, served as radar operators and air traffic controllers at airfields around the world — began to whisper amongst themselves about encounters with suspected extraterrestrial aircraft.

During secret debriefings at U.S. bases, pilots and air crew told their commanders of seeing UFOs off their plane's wings. Award-winning investigative author John Guerra spent four years interviewing Filer, a decorated intelligence officer and flyer.

From objects in the skies over Cold War Europe to a UFOs over during the Cuban Missile Crisis to UFOs over the DMZ in Vietnam Filer leaves nothing out about his Air Force UFO encounters, Filer's most memorable case — the shooting of an alien at Fort Dix/ McGuire Air Force Base in 1978 — is fully recounted for the first time in this book. As a member of the Disclosure Project, military experts, astronauts, and scientists urge the U.S. government to release all it knows about UFOs to the public. Filer describes his UFO encounters in this incredible book. [16]

SOLAR WARDEN AKA SECRET SPACE PROGRAM

Since approximately 1980, a secret space fleet was code named *Solar Warden* has been in operation yet unknown to the public. Truth or Science Fiction?

Darren Parks researched the secret program through conducting an FOI (freedom of information) request with the DoD (Department of Defense) in 2010, this was his unexpected response by email from them which read:

> "About an hour ago I spoke to a NASA rep who confirmed this was their program and that it was terminated by the President. He also informed me that it was not a joint program with the DoD. The NASA rep informed me that you should be directed to the Johnson Space Center FOIA Manager. I ran your request through one of our space-related directorates and I'm waiting on one other division with the Command to respond back to me. I will contact you once I have a response from the other division. Did NASA refer you to us?"

The program not only operates classified under the US Government but also under the United Nations authority. Well, there are a few people and many others that have tried hard to find out the truth and have succeeded by leaked information or simply asking questions and have government departments slip up and give away information freely, just like what happened when Darren Perks asked the DoD.[17] One notable contributor is Gary McKinnon.

When Gary McKinnon hacked into U.S. Space Command computers several years ago and learned of the existence of "non-terrestrial officers" and "fleet-to-fleet transfers" and a secret program called *Solar Warden*, he was charged by the Bush Justice Department with having committed "the biggest military computer hack of all time," and stood to face prison time of up to 70 years after extradition from UK. But trying earnest McKinnon in open court would involve his testifying to the above classified facts, and his attorney would be able to subpoena government officers to testify under oath about the Navy's Space Fleet. To date the extradition of McKinnon to the U.S. has gone nowhere.

McKinnon also found out about the ships or craft within Solar Warden. It is said that there are approximately eight cigar-shaped motherships (each longer than two football fields end-to-end) and 43 small "scout ships. The Solar Warden Space Fleet operates under the US Naval Network and Space Operations Command (NNSOC) [formerly Naval Space Command]. There are approximately 300 personnel involved at that facility, with the figure rising.

Solar Warden is said to be made up from U.S. aerospace Black Projects contractors, but with some contributions of parts and systems by Canada, United Kingdom, Italy, Austria, Russia, and Australia. It is also said that the program is tested and operated from secret military bases such as Area 51 in Nevada, USA.[17]

Many people around the world are witnessing unknown and exotic spacecraft moving around in the skies and sub space that completely defy gravity. Whether they are part of the Solar Warden secret program, military experimental aircraft or not, people are seeing them, reporting them, and posting videos and photos of them daily all over the internet.

SOLAR WARDEN SPACESHIPS

The secret space program known as *Solar Warden* was first disclosed to the world by the UK Hacker Gary McKinnon who saw a 2002 actual photo of the spaceships while searching for information on UFOs. He accidentally hacked into the US Space Command's website. [18]

The Earth's Interstellar Space Program called SOC began formation back in 1974 under President Gerald Ford when four nations founded the SOC and over the course of the years swelled to twelve nations. SOC consists of 4 Fleets of Ships, 9 Earth Orbiting Space Stations (Cloaked), 69 Space base Research Stations and 47 Planetary Research Stations on 6 different Worlds. The four fleets are run much like individual nations under the Scathe Space Stations and Research Stations are part of each different Fleet. One of the biggest breakthroughs for the EMS and the Propulsion and Power Supplies used in the Ships as well as the ability to fold space and time enabling the ability to cross long distances of Space (Interstellar Space Travel) in very short periods of time.[18]

Solar Warden is said to be made up from U.S. aerospace black projects contractors, with some contributions of parts and systems by Canada, the United Kingdom, Italy, Austria, Russia, and Australia, this is why claims that its somewhat controlled by the United Nations exists. It is also said that the program is tested and operated from secret Air Force bases. We may not know the most recent correct name for Earth's spaceships as classified names are often changed to help keep the information secret.

The newer version of Solar Warden is a research project that would create sunspots and send solar waves to directional nodes in space which in turn aim them at targets like a rogue asteroid or an alien space craft that had flown into our system unannounced, without permission and with hostile intentions. Project Solar Warden has been around for many years and as of yet, has not been successful because of the directional nodes burning up.

Research and experiments of Project Solar Warden have been cut back massively in the last several years because sunspot activity has caused planetary temperatures to increase at an unnatural rate and amount. Also called climate change global warming by scientists unaware of Project Solar Warden.

William Mills Tompkins is one of the most important witnesses to come forward revealing details about the Secret Space Program (SSP) and human interactions with ETs, he died in 2017 after he completed his book and presentation as a whistleblower of the SSP to the Annual MUFON Convention, which was titled: The Secret Space Program. He details the German alliances with Reptilians and Dracos, the infiltration of NASA by these beings as well as the positive contribution by the Nordics to our secret space program over decades since at least the 1920s and perhaps earlier.[19]

Some of these SSP craft were allegedly designed by Tompkins, were based on submarine designs with the ability to overcome gravity. Tompkins had been aboard some of these craft himself. He claimed to have had three friends who were part of Solar Warden since inception who would visit him once every (20) years and they never showed any result of aging. They would visit Earth for a period and then go back into space preferring to live off planet and visit other star systems and interact with other stars and their inhabitants.

The reverse aging process was developed by TRW and Tompkins was aware of this program when he too worked at TRW. A history of that program was discussed by Tompkins on the Jeff Rense Show on November 30, 2016. See, Jeffrense.com

CONTROLLING THE HIGH GROUND

"The Space Force will help us deter aggression and help us control the ultimate high ground," claims President Trump 2018.

Over time, the new Space Force will bring the bulk of Defense Department space programs under one roof. Certain entities that also handle space, like the Missile Defense Agency, National Reconnaissance Office, and National Geospatial Intelligence Agency, will continue as normal. The Space Force is separate from NASA, a civilian agency.

The Dark Fleet worked almost entirely outside the Solar System. They are very martial and offensive. They are extremely classified above the others and were large fleets, they are similar looking Carrier Craft that look like the Star Wars Millennium

Falcon. I have seen photos of these craft, which are posted on my Facebook pages. One really needs to ask the question? Was Star Wars art imitating truth and reality?

McKinnon allegedly discovered information about the craft within Solar Warden. It is said that there are approximately eight cigar-shaped Motherships (each longer than two football fields end-to-end) and 43 small "scout ships". The Solar Warden Space Fleet operates under Air Force Space Command and will be transferred to the new Space Force. McKinnon claims he saw a photo of **a silvery, cigar-shaped object with geodesic spheres on either side. There were no visible seams or riveting. There was no reference to the size of the object and the picture was taken presumably by a satellite looking down on it**. The object did not look manmade or anything like what we have created.[20]

Various Special Access Program SSP's that were small usually had the newer technology. They are very secretive and worked for some of the Secret Earth Governments, Syndicates, and World Military Forces (there could be several independent groups in this category).[21]

It certainly appears that space is the next frontier in the military and U.S. politics. Is the U.S. government finally starting to give the public the disclosure that many have hoped for since the iconic Roswell crash, or is this actually an attempt by the DoD to prime us for the imminent expansion of war beyond the confines of our planetary surface? Or could it be both?[22]

The facts are we have had more DISCLOSURE on UFOs coming out of our government, i.e., Pentagon, Defense Department since President Trump has been in the White House than what we have seen in decades. We may never get full DISCLOSURE from our Government, but we will have more contact, and more controlled disclosure stories.

Notes:
1. Joe Pappalardo, Popular Mechanics: *Declassified: America's Secret Flying Saucer*, https://www.popularmechanics.com/military/a8699/declassified-americas-secret-flying-saucer-15075926/, February 11, 2013
2. *US Air Force Tests 'Flying Saucer'*, https://www.military.com/video/space-technology/spacecraft/us-air-force-tests-flying-saucer/3943489006001
3. Ivana Cardinale, *There Is More Than Just One 'Area 51' & They Exist In Places You Won't Believe Is Possible*, https://www.websitesbb.com/2017/caribflame_dev.com/2016/07/there-is-more-than-just-one-area-51-they-exist-in-places-you-wont-believe-is-possible/, July 1, 2016
4. Ella LeBain, *Who's Who in the Cosmic Zoo?* Book One – Third Edition, See, Chapter: Dulce Underground, pp. 185-202, Skypath Books, 2013.
5. James & Lance Morcan, *Underground Bases, Subterranean Military Facilities and the Cities Beneath Our Feet* (The Underground Knowledge Series Book 7) Sterling Gate Books (November 16, 2015), https://www.goodreads.com/topic/show/1961070-undersea-bases
6. Lloyd A. Duscha, *Underground Facilities for Defense – Experience and Lessons, in Tunneling and Underground Transport: Future Developments in Technology. Economics and Policy*, ed. F.P. Davidson, pp. 109-113, New York: Elsevier Science Publishing Company, Inc., 1987.
7. Richard Sauder, Ph.D., *Hidden In Plain: Sight Underwater & Underground Bases: Surprising Facts the Government Does Not Want You to Know*, Adventures Unlimited Press (May 23, 2014)
8. Gary McKinnon, http://freegary.org.uk/

CHAPTER 37: SECRET GOVERNMENT SPACE PROGRAMS

9. Profile: Gary McKinnon, https://www.bbc.co.uk/news/uk-19946902, December 14, 2012
10. Kerry Cassidy, Project Camelot interviews Gary McKinnon, Nov 20, 2007, https://www.youtube.com/watch?v=_fNsah-0vpY
11. Brett Tingley, *Eyebrow raising claims about advanced space technology made by recently retired US General*, The Drive, https://www.thedrive.com/the-war-zone/31445/recently-retired-usaf-general-makes-eyebrow-raising-claims-about-advanced-space-technology, December 11, 2019
12. Steven Kwast, *The Urgent Need for a U.S. Space Force*, https://youtu.be/KsPLmb6gAdw, December 5, 2019
13. George Filer, Filers File#28, July 9, 2008, *We Have Lost Many Men and Planes Trying To Intercept UFOs*, http://nationalufocenter.com/artman/publish/article_232.php
14. https://missingmoneysolari.com/dod-and-hud-missing-money-supporting-documentation/
15. 9/10/2001 Def. Secy Donald Rumsfeld, $2.3 Trillion Missing from Pentagon, https://www.youtube.com/watch?v=y7ywpfOOn7k
16. John L. Guerra, Strange Craft: The True Story of An Air Force Intelligence Officer's Life with UFOs Paperback, https://amzn.to/2SK3SIQ, December 19, 2018
17. Darren Perks, *Solar Warden – The Secret Space Program*, https://www.huffingtonpost.co.uk/darren-perks/solar-warden-the-secret-space-program_b_1659192.html?guccounter=1, 10.9.2012
18. Joseph Trevithick, THE WAR ZONE, https://nationalufocenter.com/?email_id=418&user_id=18945&urlpassed=aHR0cHM6Ly91Zm8uZmFuZG9tLmNvbS93aWtpL1NvbGFyX1dhcmRlbg&controller=stats&action=analyse&wysija-page=1&wysijap=subscriptions
19. William Tompkins, *Selected by Extraterrestrials, My life in the top-secret world of UFOs., Think-Tanks and Nordic Secretaries*, July 2016, p.417, https://www.amazon.com/Selected-Extraterrestrials-secret-think-tanks-secretaries/dp/1515217469
20. Secret Space Program, http://conspiracy.wikia.com/wiki/Secret_Space_Program
21. Thomas Schrøder Jensen, https://nationalufocenter.com/?email_id=418&user_id=18945&urlpassed=aHR0cDovL2NvbnNwaXJhY3kud2lraWEuY29tL3dpa2kvU2VjcmV0X1NwYWNlX1Byb2dyYW0&controller=stats&action=analyse&wysija-page=1&wysijap=subscriptions
22. Laura Valkovic, Narrated News, August 05, 2018, https://www.libertynation.com/why-are-ufos-making-it-into-the-mainstream-media/?fbclid=IwAR25yA4o9n_ERbwNoLXGeWq3xBr_6tFRlMzewRB98lp2p1BQyWe9yxIRE54,

CHAPTER THIRTY-TWO

REPTILIANS HELPED NAZIS BUILD A SECRET SPACE PROGRAM IN ANTARCTICA

The Truth is Stranger than Fiction!

In COVENANTS – Book Four, I discussed how the Reptilian Draco Aliens helped the Nazis build and start their Secret Space Program in Antarctica, and how by gifting them with spacecraft, they used them to wage war against the God of the Jews, by putting the Nazis under their spell promising them world domination if they would exterminate the people that Yahuah calls His own, the Jewish people. This had very little to do with the Jewish religion, but was an attack on the Jewish race, on Jewish DNA, causing the Nazis to throw everyone into gas chambers who had only one great grandparent or grandparent that was Jewish, regardless of their beliefs. I proved that this attack on the Jews was an attack on the God of the Jews, with whom the Draco Reptilians are in rebellion to and continue to wage war against the Lord of lords, by attacking His people. See, Ephesians 6:12. This war against Yahuah, now extends to all those who are 'grafted in' into the Covenants of Israel, who are Born Again Christians.[1]

WHISTLEBLOWER WHO WORKED FOR EXTRATERRESTRIALS
Who was the late U.S. Navy Intelligence Officer William Tompkins who turned whistleblower in his later years? The following was the resume` of William Tompkins, which accompanies his autobiography, *Selected by Extraterrestrials, Selected by Extraterrestrials: My life in the Top-Secret World of UFOs, Think-Tanks and Nordic Secretaries*, self-published, December 9, 2015. Bill Tompkins was born May 29, 1923, and died unexpectedly on the morning of August 21, 2017, just hours prior to the Total Solar Eclipse over America.

According to Tompkins who told Dr. Michael Salla in his ExoNews TV Interview in April 2016, that leading to the build up to World War II, German secret

CHAPTER 38: REPTILIANS HELPED NAZIS BUILD A SECRET SPACE PROGRAM IN ANTARCTICA

societies and the Nazi SS were guided to three large caverns in Antarctica by Reptilian extraterrestrials. Tompkins described how Reptilians helped the Germans/Nazis build underground bases in remote Antarctic caverns, which were located next to even larger caverns that have been controlled by the Reptilians for thousands of years.[2]

Tompkins claimed that the U.S. Navy learned of the existence of these secret Antarctic bases from U.S. spies implanted within Nazi Germany, who discovered Nazis used these remote bases to launch secret space missions to the Moon and other planets in our solar system, and, most remarkably, to distant interstellar locations.[3]

Tompkins said that the Nazis started moving equipment and supplies to Antarctica in 1913, which coincides historically with the Second German Antarctic Expedition from 1911-1913. This period also coincides with the increasing role of German secret societies in exploring remote global locations for occult knowledge. The movement of equipment accelerated the lead up to World War II. In his interview, Tompkins cites 1934 as the beginning of this acceleration, even though historical records point to 1938 as the launch of the Third German Antarctic Expedition, however, he asserted that the move from Germany to Antarctica was in operation from way before the war started.

In the ExoNews TV Interview, Tompkins revealed that secret agreements had been reached between Hitler's regime and Draconian Reptilians. he discussed one of the major elements of this agreement: "Large portions of equipment were sent down there. But right next to them were three tremendous size caverns which the Reptilians had. Not Grays, but Reptilians. Germany got two more, about a tenth the size of the big Reptilians [cavern]. They were able to ... [go] down, usually by submarine. They built these flat submarines, regular class, so they could ship all this stuff down." [2]

Tompkins remarkable information is consistent with Grand Admiral Karl Dönitz who referred on three occasions to an impregnable fortress being built for Hitler in a remote location using Germany's advanced submarine fleet. In 1943, Donitz is reported to have stated: the German submarine fleet is proud of having built for the Führer, in another part of the world, a Shangri-La on land, an impregnable fortress.

The second incident was in 1944, when he revealed how plans were in place to relocate Hitler so he could launch a new effort for his thousand-year Reich: The German Navy knows all hiding places in the oceans and therefore it was easy to bring the Führer to a safe place should the necessity arise, so he will have the opportunity to carry out his final plans.

Finally, Dönitz's remarks at his Nuremberg war crime trial clearly suggest that it was Antarctica where Germany's most advanced technologies had been secretly relocated by his submarine fleet. At the trial he boasted of "an invulnerable fortress, a paradise-like oasis in the middle of eternal ice." Dönitz's remarks were made plausible in 1966 by cartographer and artist Heinrich C. Berann for the National

Geographic Society. In Berann's depiction of an ice-free Antarctica, he shows underwater passageways that run throughout the Antarctic continent. This provided a plausible way in which submarines could travel under the ice for considerable distances to Nazi Germany's "invulnerable fortress."[4]

Donitz's claims are further supported by documents provided by an alleged German submarine crewman after the war, which described the instructions for U-Boat Captains to reach the Antarctica bases through the hidden passageways. Below is an image of the document with the translated instructions. In the interview, Tompkins describes simultaneous flying saucer programs that had been developed by the Nazis. One was in Germany, while the second was in Antarctica. In his response to a question about where Germany's antigravity craft were being built, he said: "They built the prototypes in Germany. They built pre-protype, something which is ready for production, in Antarctica. They put this stuff in production in the countries all over Germany [Occupied Europe], and they continued to build similar vehicles in Antarctica." [4]

Tompkins shared one of the most astonishing secrets gained by the U.S. Navy spies hidden in Germany. He said that it was only with the help of the Reptilians, that the Nazi's Antarctica Secret Space Program was empowered to successfully launch manned missions to the Moon, planets, and even other star systems. He claimed it was well known that Nazis had multiple space vehicles that flew out of Antarctica into Space and to the Moon and returned to the Antarctica base. On one of their first missions, they ran into some trouble and crashed, and the entire crew died. I wonder if those Google photos of what looks a crashed UFO, that has made tracks in the snow and is buried under the ice, was really the Nazis? At least four years before the war ended, they were moving stuff out of Germany into their Antarctica Base. Tomkins said, he could not verify if it was true or not, but claimed it was stated by his fellow U.S. Navy spies, who claimed that the Germans had gone to other stars and come back.

It is no secret that they were in contact with beings from Aldebaran, as well as from the inner Earth, which I detailed in Book One, of *Who's Who in the Cosmic Zoo?* See, Chapter: Aldebaran. Therefore, Tompkins' claims make sense, in the context of their occult affiliations through trance mediums who channeled to the Nazis via telepathic messages through two entities that claimed to have come from the evil dark star Aldebaran, which were no doubt coming directly from the Reptilians inside the Earth who guided the Nazis in their evil deeds.[4]

Tompkins astonishing claims corroborate the testimony of the MILABs Whistleblower, Corey Goode, who claims that from 1987 to 2007, he had access to smart glass pads, which described the successful Nazi Secret Space Program that operated out of Antarctica. Goode said the Antarctica program was controlled by German secret societies rather than the Nazi SS: "the Nazi remnants that were made up mostly of Secret Societies that created a "Break Away Civilization," kept the most advanced technology secret from even their highest Military and Political leaders, setup enclaves in South America and Antarctica. The locations in Antarctica were

CHAPTER 38: REPTILIANS HELPED NAZIS BUILD A SECRET SPACE PROGRAM IN ANTARCTICA

some ancient civilization ruins that had remained occupied by certain groups in thermal areas that cause areas similar to lava tubes and domes under the glaciers.[5]

Goode goes on to corroborate and confirm Tompkins claim that the Nazi bases were built adjacent to caverns controlled by an ancient-advanced civilization, known as the Draco Reptilians. "There was an underground and under glacier city complex that was already occupied and setup in a couple of locations and the NAZI's renovated an area that was mostly crushed above the surface but had plenty of room under the domed ice, thermal underground energy and caverns (accessible via U-boat under the ice flows and openings that made it ideal for a hidden multipurpose base) that were perfect for them to secretly build out during the entire Second World War."[5]

Tompkins claims are difficult to dismiss due to documents he published in his book, *Selected by Extraterrestrials*, which supports his claim that he participated in a covert Navy Intelligence program that disseminated Nazi Germany's advanced aerospace secrets to selected U.S. aerospace companies, think tanks and universities.[6]

Furthermore, documents, and statements by Admiral Donitz, support Tompkins claims that Nazi Germany had succeeded in locating and building underground facilities under the Antarctic continent. The long route under the Antarctic ice sheets, which the German submarine fleet allegedly took to reach these hidden caverns, gives credence to Tompkins claims that Reptilian extraterrestrials had provided the Nazis with the information necessary to locate the hidden Antarctic caverns, and the under-ice passageways to reach them.[5]

Bill Tompkins was embedded in the world of secrecy as a teenager, when the Navy took his personal ship models out of a Hollywood department store because they showed the classified locations of the radars and gun emplacements. He was personally present at the "Battle of L.A." when a thousand rounds of ammo were fired at UFOs, and one of the Nordic craft may have selected him to be their rep in the evolving aerospace race. This book is a partial autobiography about his life to the beginning of the 1970s including some of his early work for TRW. Selected by the Navy prior to completing high school to be authorized for research work, he regularly visited classified Naval facilities during WWII until he was discharged in 1946.[6]

After working at North American Aviation and Northrop, he was hired by Douglas Aircraft Company in 1950, and when they found out about his involvement in classified work, was given a job as a to create design solutions as a draftsman with a peripheral assignment to work in a "think tank". This work was partly controlled by the Navy personnel who used to work for James Forrestal, who was allegedly assassinated because he was going to publicly reveal what he knew about UFOs. Bill Tompkins was asked to conceive sketches of mile-long Naval interplanetary craft designs. Later, as he became involved in the conventional aspects of the Saturn Program that later became the Apollo launch vehicle, his insight to system engineering resulted in his offering some critical suggestions personally to Dr. Wernher von Braun about ensuring more reliable checkout using the missiles in their vertical position and also some very efficient launch control concepts adopted by both NASA and the Air Force. This story is peppered with very personal interactions with his co-workers and secretaries, some of whom the author believes to be Nordic aliens helping the "good guys" here on Earth. Towards the end of this volume of his autobiography, he sketches what he personally saw on TV when Armstrong was landing on the moon. Born in May 1923, Bill Tompkins is one of the few survivors of the "big war" who is still healthy, married to the same girl Mary, and is willing to tell his story about what he really did during his aerospace life in the 40s, 50s and 60s that relate to aliens, NASA and secrets that now can be told.[7]

Notes:
1. Ella LeBain, COVENANTS – Book Four, Skypath Books, 2018, https://www.amazon.com/COVENANTS-Times-Guide-Aliens-Angels/dp/0692988637/
2. Dr. Michael Salla, William Tompkins Interview Transcript – Reptilian Aliens Helped Germans Establish Space Program in Antarctica, https://www.exopolitics.org/interview-transcript-us-navy-spies-learned-of-nazi-alliance-with-reptilian-extraterrestrials/ April 4, 2016
3. Dr. Michael Salla, William Tompkins Interview Transcript – US Navy Spies Learned of Nazi Alliance with Reptilian Extraterrestrials; https://michaelsalla.com/2016/04/14/interview-transcript-reptilian-aliens-helped-germans-establish-space-program-in-antarctica/, April 14, 2016.
4. Ella LeBain, *Who's Who in the Cosmic Zoo?* Book One – Third Edition, Skypath Books, 2013. pp. 96-97
5. History, Feb. 15, 2020, *Reptilian Aliens Helped Nazi Germany Build Secret Space Program in Antarctica*, https://wokehub.com/history/reptilian-aliens-helped-nazi-germany-build-secret-space-program-in-antarctica/
6. William Mills Tompkins autobiography, *Selected by Extraterrestrials: My life in the Top-Secret World of UFOs, Think-Tanks and Nordic Secretaries*, CreateSpace, Dec.9, 2015.
7. William Tompkins Amazon Bio: https://www.amazon.com/Selected-Extraterrestrials-secret-think-tanks-secretaries/dp/1515217469

CHAPTER THIRTY-THREE

SPACE FORCE

"There will be great earthquakes, famines and pestilences in various places, and fearful events and GREAT SIGNS from HEAVEN."
Luke 21:11

Starfleet of Star Trek could actually be at the point of becoming a reality. Again, what are they hiding from us?! The fact that aliens are real? Are they an enemy? Again, we need to ask ourselves, is Life imitating Art? Or did Art imitate Life?

FROM SPACE COMMAND TO THE NEW SPACE FORCE

President Donald Trump on December 20, 2019, established the US Space Force as America's sixth military service, one of the most significant changes in Air Force history and a milestone in America's exploration and militarization of the cosmos.

The Space Force is the Pentagon's first branch of the military solely dedicated to organizing, training, and equipping personnel to operate and protect military space assets like GPS satellites. It sits within the Air Force and draws on much of the service's existing bureaucracy, while aiming to create its own culture and structure to prioritize and tackle space in new ways.

Trump signed the fiscal 2020 National Defense Authorization Act into law the evening of December 20, at Joint Base Andrews, Md. Under that legislation, Air Force Space Command, the USAF organization that provides space experts and systems to military commanders, is transformed into the Space Force.

"The Space Force will help us deter aggression and help us control the ultimate high ground," Trump said.

Over time, the Space Force will bring the bulk of Defense Department space programs under one roof. Certain entities that also handle space, like the Missile Defense Agency, National Reconnaissance Office, and National Geospatial Intelligence Agency, will continue as normal. The Space Force is separate from NASA, a civilian agency.

Proponents say a Space Force is a needed step to ensure US dominance in the cosmos as the commercial sector sets its sights on orbit and as other countries improve both their space capabilities and their anti-satellite weaponry. Detractors say it is unclear how exactly the new service will improve upon what Air Force Space Command already offered.

The Space Force will be comprised of about 16,000 active-duty military personnel and civilian staffers from Air Force Space Command (AFSPC) at the outset, Air Force Secretary Barbara Barrett told reporters Dec. 20. "They will be assigned to the Space Force effective immediately, though the formal transfer process between services requires people to volunteer for and re-enlist in new jobs over the next few months."

About 3,400 officers and 6,200 enlisted members are eligible to join the Space Force if they so choose. The rest of the service will be comprised of civilians, according to an Air Force official. The remaining 10,000 or so current AFSPC employees, like Numbered Air Force workers and contractors, would stay assigned to the Air Force until further notice.

More people will be added in over time, including space professionals from the Army and Navy, as well as Guardsmen and Reservists, though officials did not pinpoint how large the new service could grow.

"There may come a point where we demonstrate some of our capabilities so that our adversaries understand they cannot deny us the use of space without consequence," Edwin Wilson Deputy Assistant Secretary of Defense said. "That capability needs to be one that's understood by your adversary," he added, they need to know there are certain things we can do, at least at some broad level, and the final element of deterrence is uncertainty. How confident are they that they know everything we can do? Because there's a risk calculation in the mind of an adversary."

Wilson was not the only one to warn that the United States might retaliate forcefully in response to an attack on U.S. military space-based systems. "It's not enough to stand in the ring and take punches. You have to have the will and capability to punch back," U.S. Air Force Chief of Staff David Goldfein also said at the Space Symposium. US. lawmakers from the House and Senate have agreed on a final version of the approximately $716 billion defense spending bill for the 2019 fiscal year, which requires the U.S. military begin work on developing new warning satellites to spot incoming ballistic missiles and weapons to blow them up from space. The draft law requires the Missile Defense Agency to pursue these programs even if it argues against them in an up-coming ballistic missile defense strategy review, which might be setting the Pentagon up for a battle with Congress but might also highlight the opinions of certain senior U.S. military leaders.[1]

An unusual flurry of events and press announcements makes it clear the U.S. government is unveiling its secret space program, multiple sources agree. We are also hearing of a Chinese secret space program, and a Russian secret space program. All these revelations are being accompanied by unusual movements of gold and other financial anomalies. The recent trade agreement between the U.S. and China is also linked to these secret space programs, Chinese and Pentagon sources agree. These are all signs that some sort of mind-boggling planetary event may be coming. One

CHAPTER 39: SPACE FORCE

possibility is an announcement that Earth is being attacked by alien craft and the technologically advanced nations have joined to fight the potential invaders.

The biggest evidence is a series of public announcements by multiple government officials, including the U.S. President. The official launching of a U.S. Space Force by Trump is just the start of a gradual process of disclosure, Pentagon officials say. In a history making win for Trump, the agreement would add a new armed service, dedicated to space, under Title 10 of U.S. Code, which was an action the White House saw as pivotal to solidifying it as a fully independent military branch. The Space Force would be housed within the Air Force and led by the chief of space operations, who would report directly to the Air Force secretary and be a member of the Joint Chiefs.

The media has stories about UFOs and secret U.S. military technology as part of this process. The U.S. Navy has released patents for anti-gravity flying saucers and compact nuclear fusion. Now we have a recently retired Air Force general talking about already existing technology that can take us *"from one part of the planet to any other part within an hour."* US Navy's announcement that the Tic Tac UFOs were indeed *"Unidentified Aerial Phenomena"*. He told this site: *"It took them two years to finally come forward and admit these things in the skies are doing things that they can't explain. Additional Tic Tac tapes are missing, and shortly after the sighting s and radar tracking data the Nimitz Aircraft Carrier was boarded, and the tapes and other data were confiscated.*

We are also getting far more UFO sightings around the world that are being recording on high-quality video. If you get into the habit of looking at the sky as I have, it will not be long before you can see for yourself with your own naked eyes.[2]

The recent trade agreement between the U.S. and China is also linked to these secret space programs, Chinese and Pentagon sources agree. These are all signs that some sort of mind-boggling planetary event may be coming. One possibility is an announcement that Earth is being attacked by alien craft and the technologically advanced nations have joined to fight the potential invaders.

General Jay Raymond, head of AFSPC and US Space Command—the combatant command that carries out daily space operations—is the Space Force first boss. He can serve as chief of space operations for a year without needing Senate confirmation. When the White House nominates Raymond's replacement, that person must go through the Senate vetting process. After one year, the CSO can join the Joint Chiefs of Staff to help set high-level military policy.

The U.S. Space Force". Retired USAF Major George Filer, who is now the MUFON Director for New Jersey and the Eastern Region, spoke to former Congressman, Senator Mike Gravel, during a "Citizen's Hearing on Disclosure" in 2013. Filer's beliefs were confirmed, that "Project Solar Warden" has been in operation for 30-40 years, when Senator Gravel said, in 2013 during the hearing, *"The U.S. had an operational Secret Space Fleet that had been in operation for decades.* In my opinion, the one thing that is at the heart of all this is an extra special technology and treaty that President Eisenhower established through legal agreements, laying the groundwork for the Secret Space program during the 1950's.[3]

The debate over the future of Space Force has now gone mainstream. Academics will be shocked once Space Force roles out the antigravity craft it is inheriting from the USAF SSP: Space Force: What Will the Newest Military Branch Actually Do?4 Protect Earth from Alien Invasion, of course.

What is new is that Space Force will be an open, transparent, and accountable space program, unlike the USAF SSP that has been operating since the 1970s in opaque unaccountable ways. That is a big difference and worth emphasizing as all the sec tech is transferred over, and about to be unveiled as "new".

Mike Burkhalter Simply. Yes. Navy started theirs prior to AF as they recovered a craft from the ocean before the AF with Roswell. Airforce is like the Coast Guard. Navy is more beyond our Solar System. Simply Navy were more accustomed with longer mission as that is what they had been doing for years with ships. Airforce usually out and back in a day.

Hope that helps.

Seriously, are we to believe that both the Navy and Air Force had secret space programs? Did both services try to establish leadership in some sort of turf war over outer space?

A new Space Force Asset? This spacecraft was captured on the ISS Live feed February 21, 2020. UFO? Or Secret Space Program? This went on uninterrupted for 22 minutes. Nobody cut the feed, in fact as you watch this craft ascend into deep space, the ISS camera is focused on its ascent as it appears to go INTERSTELLAR and turns green. This is highly unusual, and NASA has made no comment.

So, what does this really mean? Well, this suggests our understanding of time is outdated, in other words, it is not linear as we have always thought. In fact, everything is always present around us.

"Congress took a seemingly flippant remark and created a rational implementation plan, a Space Force to support Space Command," said Joan Johnson-Freese, a professor of National Security Affairs at the Naval War College in Newport, Rhode Island, referring to a comment President Donald Trump made during a speech in March 2018. (Her views do not necessarily represent those of the Naval War College, the Department of Defense, or the U.S. government.)

"Whether it will evolve into an organization that solves any of the problems that prompted it remains to be seen," Johnson-Freese told Space.com. "On the negative side, it certainly increases the perception that the U.S. is leading the way on the weaponization of space."

Theresa Hitchens is the space and air reporter at the online magazine Breaking Defense and a former senior research associate at the University of Maryland's Center for International and Security Studies. Before that, she spent six years in Geneva as director of the United Nations Institute for Disarmament Research.[5]

Hitchens said there are two big questions at hand regarding the Space Force.

The first is whether and how the Department of Defense (DoD) and the Air Force are going to implement the spirit — and not just the letter — of the congressional mandate in the 2020 National Defense Authorization Act on space acquisition, which created a new position and moved the Space Development Agency to Space Force.

Besides Aliens, Space Force oversees interstellar travel, patrolling the earth around our solar system. That was the purpose behind the Secret Space Program Earth's Solar Warden.

Reminds me of the scripture in Zechariah when the Lord of Heaven and Earth, ordered His horses to patrol the Earth. We all know Horses do not fly. Horses are ancient vernacular for space chariot aka Spaceships 5

> "When the powerful horses came out, they were impatient to go and patrol the earth. And he said, "Go, patrol the earth." So, they patrolled the earth."
> (Zechariah 6:7)

There is another Solar System that is intersecting with ours, Space Force is going to take over what Solar Warden has been doing. Also, the Space technology around the sun. There are multiple objects around the Sun, some are ours, some are not.

I once had a Fundamentalist challenge and unfriend and block me for reporting of the 2 suns. She said "God made the Sun, the moon and stars. There's only one sun." When I pointed out to her that stars are suns and our sun is a star, she unfriended me and then after I explained that most Star systems are binary, some are even trinary. Meaning 3 stars/suns.

Alright, I understand cognitive dissonance. You are taught one thing, but later learn it was a lie, or a bunch of inaccuracies at the very least. Wherever you come from, we need to understand this basic simple point, that we are not alone, not alone in terms of Extraterrestrial life. We are not a singular star System. Our sun has a binary twin, who is an evil twin, a dark star called Nemesis. 5

As we know, the sun is a star. In our galaxy alone, there are estimated to be over 100 billion stars. How many earth-like planets orbit these stars? An estimated 17 billion. "Almost all sun-like stars have a planetary system," said Francois Fressin, an astronomer at the Harvard-Smithsonian Center for Astrophysics. Also, there is estimated to be around 100 billion galaxies in the universe. So, if each galaxy has millions/billions of earthlike planets, and there are hundreds of billions of galaxies, does it make sense to think that we are the only form of intelligent life in the Universe? If this has not blown your mind, this will.6

At the center of each galaxy lies a black hole. Equations show that within the black hole of each galaxy lies ANOTHER whole universe. "Like part of a cosmic Russian doll, our universe may be nested inside a black hole that is itself part of a larger universe. In turn, all the black holes found so far in our universe—from the microscopic to the supermassive—may be doorways into alternate realities." According to the new equations, the matter black holes absorb and seemingly destroy is actually expelled and becomes the building blocks for galaxies, stars, and planets in another reality. Essentially, every black hole contains a smaller alternate universe. And our universe might just exist inside a black hole of a galaxy in a much larger universe. [7]

So, within the 100 billion galaxies in our universe lies 100 billion universes with 100 billion more galaxies each containing another universe with more galaxies, and so on infinitely. What a Creation we live. [6, 7, 8]

> "All of creation groans, that includes all creatures we share this planet with, seen and unseen, known and unknown. The Earth is in distress, For we know that the whole creation groans and travails in pain together until now. And not only they, but ourselves also, which have the first fruits of the Spirit, even we ourselves groan within ourselves, waiting for the adoption, to wit, the redemption of our body."
> (Romans 8:22)

SPACE FORCE LOGO V. STAR TREK'S STARFLEET COMMAND LOGO

According to the press release from Space Force, the logo originated with U S Army in 1942. Historical records of this are from the Army Institute of Heraldry, Star Trek borrowed from it. This "new" logo directly links from the U.S. Air Force Space Command, which has now turned into the US Space Force.

The Air Force Space Command logo you see here — the one that looks like a badge, with the arrow, and the planet, and the stars, has been around and in official use since 2010. There is a version that is even older — slightly darker, but essentially the same, that has been in use since 2004 — the original dates back to 1982. These were variations of the U.S.A.F. Space Command:

CHAPTER 39: SPACE FORCE

Per Ex Astris Scientia, Michael Okuda, graphic designer, and lead designer for Star Trek for many, many years, commented on his design for the Starfleet Command seal – as seen in one incarnation at the head of this article. "The Starfleet Command seal was first seen 'Homefront' (DS9) and later in 'In the Flesh' (VGR), although the agency itself, of course, dates back to the original Star Trek series. The symbol was intended to be somewhat reminiscent of the NASA emblem." More recently, Okuda's also been working with NASA for symbols like the 2017 NASA Flight Operations Emblem.[9]

The official explanation is as follows: "The dark blue disc encircled by a white band with narrow yellow borders inscribed "United States" at top and "Space Command" at bottom in dark blue letters, provides the background and symbolizes the space environment.

"The eagle and shield, a traditional symbol of American strength and vigilance, is positioned above a light blue elliptical globe with light landmasses and dark blue grid lines to represent the expansion of this strength and vigilance into space.

"The globe, as viewed from space, symbolizes the Earth as being the origin and control point for all space vehicles and represents the area of operations of the United States Space Command.

"Encompassing the elliptical globe are two yellow orbital paths crossed diagonally, each bearing a yellow polestar detailed light tan and signifying the worldwide coverage provided in accomplishing the surveillance, navigation and communications missions.

"The gold-brown eagle, detailed in dark brown and highlighted light tan, with a white head and tail; and yellow beak, eye, and talons detailed in light tan, is grasping in his right talon a green olive branch, detailed in dark green, and in his left talon 13 arrows with white arrowheads and feathers, and light tan shafts. The eagle is flanked by an arc of four yellow stars detailed in light tan and symbolizing the fusion of the four-armed services into a unified command." [10]

MEANING BEHIND THE LOGO

First used in 1961, the Delta symbol honors the heritage of the USAF and Space Command.

The silver outer border of the delta signifies defense and protection from all adversaries and threats emanating from the space domain. The black area inside embodies the vast darkness of deep space.

Inside the delta, the two spires represent the action of a rocket launching into the outer atmosphere in support of the central role of the Space Force in defending the space domain.

In the center of the delta is the star Polaris, which symbolizes how the core values guide the Space Force mission. The four beveled elements symbolize the joint armed forces supporting the space mission: Air Force, Army, Navy, and Marines.

The Navy started their support prior to the Air Force, as they recovered a craft from the ocean before the incident at Roswell. Air Force is like the Coast Guard, out and back in a day. The Naval structure is more applicable to operations that occur or will occur beyond our Solar System. The whole structure of command is formulated for longer missions out in the depths, or the abyss of space.

Seriously, are we to believe that both the Navy and Air Force had secret space programs? That the initial purpose of the Secret Space Force programs was to Protect Earth from Alien Invasion. Did both services try to establish leadership in some sort of turf war over outer space? [8]

What this suggests is that time as we know it is irrelevant. In other words, it is not linear as we have always thought. In fact, everything around us is always present.

"Congress took a seemingly flippant remark and created a rational implementation plan, a Space Force to support Space Command," said Joan Johnson-Freese, a professor of National Security Affairs at the Naval War College in Newport, Rhode Island, referring to a comment President Donald Trump made during a speech in March 2018. (Her views do not necessarily represent those of the Naval War College, the Department of Defense, or the U.S. government.)

"Whether it will evolve into an organization that solves any of the problems that prompted it remains to be seen," Johnson-Freese told Space.com. "On the negative side, it certainly increases the perception that the U.S. is leading the way on the weaponization of space."

Theresa Hitchens is the space and air reporter at the online magazine Breaking Defense and a former senior research associate at the University of Maryland's Center for International and Security Studies. Before that, she spent six years in

Geneva as director of the United Nations Institute for Disarmament Research. Hitchens said there are two big questions at hand regarding the Space Force.

The first is whether and how the Department of Defense (DoD) and the Air Force are going to implement the spirit — and not just the letter — of the congressional mandate in the 2020 National Defense Authorization Act on space acquisition, which created a new position and moved the Space Development Agency to Space Force. [10]

HAS SPACE FORCE TAKEN OVER THE SECRET SPACE PROGRAM?

In short, the answer is yes. Again, I cannot stress my overall theme of this book series, *the Truth is Stranger than Fiction!* The very reason we have science fiction such as Star Trek which was all about the 'future' tech, was to prepare the public for what was already being developed, years before it was revealed to the public, which is a space defense system to protect Earth from alien attacks, alien invasions. Since World War II, the US Military has been preparing for a real-life space war with other worldly forces. In fact, the Nazis were already using alien technologies, which became the spoils of war that were taken by both American and Russian forces.

Their competition over these technologies is what spawned the Cold War, which on the surface was about the tension between the US and the USSR over communism vs a free market society, based on American Capitalism. We all know how that turned, when the former Soviet Union disbanded, many who were working on these exotic technologies, as well as tracking UFOs and the Alien Presence on Earth, became whistleblowers, and disclosure of UFOs and Aliens were disseminated around the world to multiple channels within Ufology.

The book, *The Book of Alien Races, Secret Russian KGB Book of Alien Species* was translated from Russian into English by Dante Santori, a former special forces sergeant uncovers the secrets of Russian contact with aliens that was heavily guarded by the KGB until after the dissolution of the former Soviet Union on December 26, 1991, that precipitated after Russian Leader Mikael Gorbachev's decision to allow elections with a multi-party system and create a presidency for the Soviet Union. This began a slow process of democratization that eventually destabilized Russia's Communist control that contributed to the collapse of the Soviet Union. During that slow process, many whistled long held secrets guarded by the KGB, and the disclosure of Aliens and UFOs came out of Russia.

According to the author, Gil Carlson, the USSR, throughout history, has been a treasure trove of information on ETs and their craft. We have long had hints on their abilities to obtain UFOs and make contact with aliens. But this information was well-hidden behind the Iron Curtain. Now, with the release of this book, all that has changed! The original 1946 book was written to inform KGB agents of the various alien races who had visited our planet and also was used as a notebook by secret agents as they constantly made additions and revisions to the original information over the years.

A friend of mine, from the 1990s, who went by the name of John Winston, used to disseminate a chapter each day via his email list, which started as a group of Ufologists on AOL, years before social media, we relied on emails and AOL groups. I was also part of the think tank group over AOL in the early 1990s, known as ISCNI, the Institute for Study of Contact with Non-Human Intelligences, that was led by my old friends, Michael and Deborah Lindemann, where we had group chats with experiencers. It was during this time period, that I was inspired to begin to compile research on my own, which led to writing of my first book, *Who's Who in the Cosmic Zoo? A Spiritual Guide to ETs, Aliens, Gods & Angels*. John Winston, (which was the alias name used for the late John Lennon) fed me chapters from the Russian KGB book on Alien Races. It was through collecting these emails, did I start compiling my own A-Z Compendium on ETs and Aliens, while teaching classes in St. Petersburg, Florida during the 1990s.

The Russian book is full of knowledge of alien species, descriptions, their ships, and place of origin, as well as information on how alien races have contributed to human evolution here on Earth. They also documented negative characteristics of these aliens and speculated on their possible intentions of some alien races. The KGB book also includes information on UFO crashes, Russian contact with aliens, Russian attempts to build their own alien craft, and an amazing section containing communications from aliens! They also documented the history of humans on Earth from the alien's perspectives, along with their contacts and intervention by different alien species. [11]

Needless to say, I was pretty blown away by such disclosures. And I sat on it for about 17 years, until I was ready to publish it, after I had to learn an especially important lesson: Discernment!

My first book teaches spiritual discernment, a theme I continue throughout this book series. As the old adage goes, we teach what we need to learn, and after all the bizarre experiences I had, I really had to learn how to discern these beings from a spiritual perspective, because they have the abilities, technologies to project themselves as benevolent, when in fact they are not, and can be very deceptive. In fact, my abductions and experiences led me back to Jesus, who to date, is the only force I know in the universe that has ultimate power and authority over these alien creatures, and the power to deliver from alien abductions.

I am sharing my testimony and full story in my future book, *CinderElla's Shadow*. But suffice to say, this book series, is not about me, per se, but what I learned, and the research I gathered which I felt compelled to share with the world. Unlike other experiencers, I have chosen to release my story after all my research is published for multiple reasons, which I believe will be evident by the time I releases *CinderElla's Shadow*, which is my story, but is being turned into a fantasy sci-fi novel, because the truth is so bizarre, nobody would believe me. Yet, people easily believe fantasy and science fiction, as if its real, which many of it is, based on reality.

Truth is way stranger than fiction, and harder to swallow. Mark Twain discerned human nature correctly when he said, "It is easier to deceive a man, than to convince that he has been deceived." The cognitive dissonance that I have encountered talking to the public at various science fiction conferences and conventions, when I presented evidence that much of what is deemed science fiction is based on reality, who outright have told me, that they prefer the fantasy over the reality, speaks volumes as the depth of deception being used on humankind for decades, centuries and even millennia.

The Secret Space Program has guarded many secrets, for several reasons, some that were born out of the Cold War days, which was due to competition by other nations vying to become super-powers through the use of alien technologies. However, both the Americans and Russians, may have used their 'cover' that they were enemies on the world stage, when they were really in a working partnership together in one of the spoils of war, the German Antarctica Space Base, where they were able to hide their development of the alien space tech from the rest of the world for 70 years.

It is inevitable that through the creation of the U.S. Space Force, long held secrets guarded by the Secret Space Program will get revealed. These include the Alien presence, and the various types of alien races the U.S. military has been working with, along with their exotic spacecraft, extraterrestrial weapons, interstellar technologies, time-travel technologies, and reverse aging abilities.

President Ronald Reagan forged the path for today's Space Force through establishing what became known as the Star Wars Program, the Strategic Defense Initiative (SDI). While the debate during his tenure ensued on the weaponization of space permeated American Politics and think tanks during the 1980s, meanwhile, the weaponization of space was achieved, another reason for secrecy.

But wait, there's more! Reports of our presence in space along with the achievement of interstellar travel through the galaxy, was coupled with stories that our military and even civilians were forced into slavery by alien forces. This led to the use of military weapons in space! Another reason for the Truth Embargo, UFO Coverup, because many didn't survive, many were taken off planet, never to be returned to Earth, and let's face it, Earth Humans don't need reports of people being abducted off planet, who are never returned, when we have plenty of stories on Earth Humans killing other humans right here on Earth. It would be considered a distraction, as well as the inevitable opening of pandora's box can of worms that even our Military cannot resolve, hence the formation of the Space Force.

We found out, that even our own Moon has been occupied by both human and alien races, and that there are secret human-alien bases throughout our solar system, in particular on the Moon and Mars. According to researcher author Gil Carlson, our government has been involved in activities on other planets and even in other galaxies, if you knew the full extent of where we've been and what we are doing there, it would take you beyond being shocked.[12]

We live in the Age of Knowledge. Much of today's discoveries have become commonplace and widely accepted in today's world, that were once considered imagination and science-fiction fantasies. We can safely say that from what we know and understand today, that nothing is really impossible anymore!

Movies, like *Enders Game* (2017) depict the type of training and space warfare that a Space Force could develop and pursue. If these types of movies are being made in such detail that are pawned off as fantasy sci-fi, they can create a type of smokescreen on the actual technologies that the Secret Space Program has possessed and achieve, that they are now in control of in today's Space Force, technologies that could destroy planets, lifeforms that may threaten our existence, as was the case in the movie, Enders Game, one has to wonder, if these technologies got into the wrong hands, it would be disastrous for our species and our planet. Likewise, put into the right hands, could be used for healings, healthcare, longevity, transportation and much more! [12]

SPACE WARFARE – LEGAL GROUND

Space is a busy place. The Nemesis-Nibiru system intersecting within our solar system, which Wormwood is part of, is dragging asteroids, comets, meteorites, and all kinds of space junk that accompanies it, including alien ships. Self-defense is a universal right. Even Jesus told his disciples in Luke 22:36, "Let the one who has no sword sell his cloak and buy one." This scripture and Words of Christ has been used for millennia to justify self-defense.

In 2019 the Secretary of the Air Force Heather Wilson said the United States may need to a show of force to deter opponents, such as Russia and China, from attacking U.S. military satellites in space. But it was not entirely clear whether this would involve demonstrating the ability to attack an adversary's own space assets only or if it would even show an ability or willingness to attack hostile space-based systems at all. This cuts to a larger issue, that no one really knows for sure what a war in space would actually look like, though the United States is certainly working to answer that question.

Wilson offered her view on the future of potential conflict in space to reporters on the sidelines of the Space Foundation's 35th annual Space Symposium, which began on Apr. 8, 2019. Other Air Force and U.S. officials have also used the event as an opportunity to highlight the very real and growing threats to vital satellites that support critical military capabilities, such as long-distance communications, navigation, weapon guidance, early warning, ballistic missile defense, and intelligence gathering. Defending those assets has been an increasingly hot topic in recent years and President Donald Trump's administration has gone so far as to propose the creation of a sixth military branch, the Space Force, to focus on these issues. In other words, Wilson offered an ominous warning that a show of force was necessary to deter space attacks. But serious questions remain about what an all-out war in space would even look like and what hostile actions would automatically demand a response.[13]

Just like in the 1980s, every time space weapons were debated or discussed, it was always in the context of defending America from hostile earth nations, but what about hostile alien nations that co-exist on Earth?

"There may come a point where we demonstrate some of our capabilities so that our adversaries understand they cannot deny us the use of space without consequence," Wilson said, according to The Daily Beast.

"That capability needs to be one that's understood by your adversary," she added, according to Air Force Magazine. "They need to know there are certain things we can do, at least at some broad level, and the final element of deterrence is uncertainty. How confident are they that they know everything we can do? Because there's a risk calculation in the mind of an adversary."

Wilson was not the only one to warn that the United States might retaliate forcefully in response to an attack on U.S. military space-based systems. "It's not enough to stand in the ring and take punches. You have to have the will and capability to punch back," U.S. Air Force Chief of Staff David Goldfein also said at the Space Symposium.[13]

SPACE RACE OF THE ROARING 20S

The aerospace industry has never been busier. In fact, rocket scientists and space engineers haven't had so much activity since the Cold War days of the space race between the United States and the Soviet Union during the 1960s. At last, humanity is returning to explore the heavens with renewed vigor.

However, it is not just the US and Russia that are dominating this year's space agenda. India, Japan, and China are all planning complex programs and are vying to become space powers in their own rights. Their plans for 2020 include missions to the moon, Mars, and the asteroids. At the same time, the US will inaugurate its Artemis program, which will eventually lead to a series of manned deep-space missions and a space station that will orbit the moon later in the next decade. Europe will be integrally involved in Artemis and will also send its first robot rover to Mars in 2020. For good measure, the United Arab Emirates plans to become a space power in 2020, with its own robot mission to the red planet.[14]

TO THE MOON, MARS, ASTEROIDS & BEYOND

China aims to become the third nation to bring samples of lunar soil back to Earth in the wake of US and Soviet successes decades ago. Its Chang'e 5 robot mission is scheduled to blast off from the Wenchang satellite launch center in Hainan in late 2020. The purpose of the project – named after the Chinese moon goddess, Chang'e – is to collect about 2kg of lunar rocks and return them to Earth. A robot lander will scoop samples into an ascent vehicle, which will be blasted into space to dock automatically with a probe circling the moon. The samples will then be transferred to a capsule and fired back to Earth.

It will be an overly complex business involving several dockings and maneuvers in orbit. By contrast, the last robot lunar sample return – accomplished by the Soviet Union's 1976 Luna 24 mission – did so using a much simpler direct return. Chang'e 5's more adventurous route is considered by many to be evidence that the Chinese are using the mission as a dress rehearsal for manned lunar landings in the near future.

US scientists are also planning a moon mission late next year – but on an even grander scale. The first of the country's Orion capsules is scheduled for launch as part of an unmanned Artemis program test flight. Orion will spend about three weeks in space, including six days orbiting the moon. The craft will have a complete life-support system and crew seats, but no crew. A European-built service module will play a key role in all Artemis missions. It will power Orion capsules after their launch from Nasa's Kennedy Space Center in Florida. Future missions will be manned, however, with the aim being to land "the first woman and the next man" on the moon by 2024. A manned space station in lunar orbit, called Lunar Gateway, is also planned.

In addition, India is to send a new lander mission to the moon in November: Chandrayaan-3. It will attempt what its predecessor failed to achieve. Chandrayaan-2 was India's first attempt at a lunar touchdown, but its main lander craft and robot rover crashed after a communication failure.

Mars, the fourth rock from the sun will become a focus of attention for space engineers this year. In July and early August, Earth and Mars will be in their best positions for craft to be sent to the latter. Nasa will take advantage of this launch window with its Mars 2020 rover, which will seek evidence that Mars was a place where water flowed, and life could have evolved. It will also search for signs of ancient microbial life.

Mars 2020 will also mark the beginning of an extremely ambitious, decade-long Martian exploration program. The robot rover will drill for samples of the most promising rocks. It will place these in metal tubes, seal them and leave them in caches at designated sites on the planet's surface. These caches will be collected by future joint European and US missions and brought back to Earth – by around 2030. About 500g of rock will be returned for scrutiny in laboratories across the world, which should transform our knowledge about past conditions on Mars.

The design of the Mars 2020 rover is based on Nasa's successful Curiosity vehicle but has been upgraded with higher-resolution color navigation cameras, an extra computer "brain" for processing images and making maps, and more sophisticated auto-navigation software.

In addition, Europe will send its own robot rover to Mars this year. In late July or early August, a Russian Proton rocket will blast a relatively small robot vehicle to Mars as part of the European Space Agency's ExoMars project. The rover is British built and has been named Rosalind Franklin after the UK DNA pioneer. Using a drill able to penetrate two meters below the surface, it will retrieve material that has been shielded from the intense radiation that bombards Mars and which may contain evidence of past and possibly even present life on the planet. (Mars 2020 will not be able to drill this deep.)

However, ExoMars has already been delayed by technical problems, and recent failures of its parachute system in trials have caused real concerns for engineers who fear they might have to delay the mission further. Improved chute systems – which will slow the craft down before retrorockets eventually land the probe, gently, on Mars after its nine-month space journey – are being tested. It is too early to know whether they will be ready for this summer's launch, however. "It is going to be very tight getting the probe ready," says ExoMars' manager, Pietro Baglioni. "I think we have only a 50-50 chance that we will be able to go ahead as scheduled." ExoMars' next launch window would then be in 2022.

For good measure, China also plans to send a probe to Mars in 2020. It has tried before to reach the red planet – in partnership with Russia. However, the Russian spaceship that was carrying China's Yinghuo-1 probe crashed in January 2012. After that, China started its own Mars exploration program and has completed a crucial landing test in northern Hebei province. Zhang Kejian, head of the China National Space Administration, said the lander went through a series of tests at a sprawling site littered with small mounds of rocks to simulate Mars's terrain.

The Chinese probe, Huoxing-1, will deploy an orbiter that will circle Mars and a rover that will drop on to the planet's surface. The mission will be launched in July or August with a Long March 5 heavy lift-off rocket. Again, the aim is to find evidence of current and past life on Mars.

Not to be outdone, the United Arab Emirates plans to launch its first deep-space probe, the Hope Mars Mission, which has been built by the Mohammed bin Rashid Space Centre in partnership with the University of Colorado and Arizona State University. The robot spacecraft will study the Martian climate and try to understand why the planet has experienced drastic climate changes.[14]

SCI-FI MOVIES PREP US FOR REALITY

If you thought movies like *Armageddon*,[15] when Bruce Willis and friends at NASA landed on an Asteroid, exploded it, in order to pre-empt it from hitting earth, was just science fiction, think again. The amount of activity in the Space Program to track, land and even disrupt orbiting asteroids from bumping into Earth, is a present reality for today's Space Program.

NASA's Osiris-Rex, which launched in 2016, is currently orbiting Bennu, a small, spheroidal asteroid with a diameter of about 520 meters, made of carbon-rich rock – a material that scientists believe is representative of the cloud of swirling gas and dust from which the solar system formed 4.6bn years ago. This summer, Osiris-Rex will sweep close to Bennu's surface, extend a robot arm, and release a puff of gas that will send pieces of rock and dust flying up from the surface – and into a collection tube. The spacecraft will then start its journey home, releasing a container with samples of asteroid inside as it nears Earth in September 2023, so that the container drops on to the Utah desert.

However, Osiris-Rex is likely to be pipped at the post in its attempt to return pieces of asteroid to Earth by Japan's Hayabusa2 mission. It met with the near-Earth asteroid 162173 Ryugu in June 2018 and began surveying it for a year and a half before taking samples that it is now carrying back to Earth. The probe – and its cargo – are expected to reach Earth in late 2020.[16]

Greenland is another 2020 American disaster film directed by Ric Roman Waugh and written by Chris Sparling. The film stars Gerard Butler (who also co-produced), Morena Baccarin, Roger Dale Floyd, Scott Glenn, David Denman, and Hope Davis. The film follows a family who must fight for survival as a planet-destroying comet races to Earth. Originally scheduled to be theatrically released on June 12, 2020, in the United States, *Greenland* was delayed several times due to the COVID-19 pandemic. The film is scheduled to be released domestically by STX Entertainment through video on demand on December 18, 2020.[17]

The Truth is Stranger than Fiction! The Truth is, NASA has been watching and tracking the Asteroids for decades, especially those with the potential to destroy life on Earth. Currently, NASA is tracking the course of several hundreds of asteroids that could potentially be hazardous to human life on Earth. For this purpose, NASA has deployed multiple Asteroid watching satellites.

In October 2022, a half-mile-wide asteroid called Didymos will approach Earth. The killer asteroid will be accompanied by its 500-foot-wide moon, which will be orbiting it. Given the huge size of Didymos and its moon, ground-based telescopes will be able to detect the asteroid very soon. They will also be able to detect the durational changes in its orbit around the larger asteroid to measure the effects of the impact.

NASA scheduled its DART mission, for a July 2021 launch. The DART mission will test NASA's strategy of slamming a half-ton spacecraft built by the Johns Hopkins Applied Physics Laboratory (APL) into the approaching killer asteroid, just like in the movies! NASA plans to take its asteroid shattering spacecraft, seven million miles from Earth. The refrigerator-sized spacecraft will approach the Asteroid Didymos, but it is not Didymos per se that is NASA's target. NASA's DART mission will be focused on Didymoon, which is big enough to demolish large cities. Before DART smashes into Didymoon at roughly 14,700 miles an hour, the NASA spacecraft will release a shoebox-size camera concocted by the Italian Space Agency. The camera will witness the spacecraft's collision with the Didymoon. It will take pictures of the spray of debris and perhaps even of the resulting crater. The collision could potentially decrease the moon's 12-hour orbit by as much as seven minutes. That will alter the way we measure lunar phases on Earth. [18]

It is only a matter of time, until one of these killer asteroids actually hits the Earth. It's what is predicted to happen in the last days of the End Times in the book of Revelation, which is called Wormwood, a comet like asteroid that hits the Earth that comes out of the Nemesis-Nibiru system and turns one-third of the waters bitter.

Revelation Chapter 8 describes the 3rd of 7 Trumpets that are blown by the Angels of the Lord in the Last Days which depicts the Lord's Judgement on the Earth. The 3rd Trumpet blown by the 3rd Angel is when the Waters are Struck: "Then the third angel sounded: And a great star fell from heaven, burning like a torch, and it fell on a third of the rivers and on the springs of water. The name of the star is *Wormwood*. A third of the waters became wormwood, and many men died from the water, because it was made bitter." (Revelation 8:10-11)

The Asteroid Apophis was discovered in 2004, Apophis is named for the demon serpent who personified evil and chaos in ancient Egyptian mythology. The Apophis discovery caused a wave of fear due to its initial calculations which indicated a small possibility it would impact Earth in 2029. This spawned Christians to immediately think this was the Wormwood of the Bible. However, since then, NASA has updated its calculations and assured the public Apophis will not hit the Earth, but it will come awfully close. As the late comedian George Carlin would say, is it a near miss? Or a near hit?

After searching through some older astronomical images, scientists ruled out the possibility of a 2029 impact. It is now predicted the asteroid will safely pass about 19,800 miles (31,900 kilometers) from our planet's surface. While that is a safe distance, it is close enough that the asteroid will come between Earth and our Moon, which is about 238,855 miles (384,400 kilometers) away. It is also within the distance that some spacecraft orbit Earth. It is rare for an asteroid of this size to pass so close to Earth, although smaller asteroids, in the range of 16 to 33 feet (5 to 10 meters), in size have been observed passing by at similar distances.

During its 2029 flyby, Apophis will first become visible to the naked eye in the night sky over the southern hemisphere and will look like a speck of light moving from east to west over Australia. It will be mid-morning on the U.S. East Coast when Apophis is above Australia. Apophis will then cross above the Indian Ocean, and continuing west, it will cross the equator over Africa. At its closest approach to Earth, just before 6 p.m. EDT, April 13, 2029, Apophis will be over the Atlantic Ocean. It will move so fast that it will cross over the Atlantic in just an hour. By 7 p.m. EDT, the asteroid will have crossed over the United States.

As it passes by Earth, it will get brighter and faster. At one point it will appear to travel more than the width of the full Moon within a minute, and it will get as bright as the stars in the Little Dipper. While NASA is certain it will not hit, it will surely cause a wave of energy that could be disruptive. In fact, it may even spray Earth with a barrage of meteorites and fireballs in its wake that could be trailing it.

Apophis is a 1,120-foot-wide (340-meter-wide) asteroid. That is about the size of three-and-a-half football fields. At its farthest, Apophis can reach a distance of about 2 astronomical units (One astronomical unit, abbreviated as AU, is the distance from the Sun to Earth.) away from Earth. It is expected to safely pass close to Earth — within 19,794 miles (31,860 kilometers) from our planet's surface — on

April 13, 2029. This is the closest approach by an asteroid of this size that scientists have known about in advance. [19]

It is unlikely Apophis will be the Wormwood of Revelation, especially if will not hit the Earth. Also, based on the level of space technologies in use at this time, which is constantly being refined and upgraded, by the end of this decade, NASA will probably be able to thwart incoming asteroids from hitting the Earth, we hope. But there is always that one, that gets away, that is elusive from even the most sophisticated of space technologies. When that day comes, it is the Lord, who will bat last. Till then, keep looking up, for your redemption draws near. (Luke 21:28)

Notes:
1. David Axe, Pentagon Officials Are Preparing for an All-Out Space War, https://www.thedailybeast.com/pentagon-officials-are-preparing-for-an-all-out-space-war, April 11, 2019
2. NASA Astronaut Shockingly Hints at Aliens in Tweet About 'Life Forms', https://sputniknews.com/science/201912141077575518-nasa-astronaut-shockingly-hints-at-aliens-in-tweet-about-life-forms/
3. George Filer, Major USAF ret. Filer's Files, January 1, 2020 New Space Force, https://nationalufocenter.com/blog/2019/12/30/filers-files-1-2020 new-space-force/
4. https://exonews.org/space-force-what-will-the-newest-military-branch-actually-do/
5. Leonard David, Space Force: What will the new military branch actually do? February 09, 2020, https://www.space.com/united-states-space-force-next-steps.html
6. *At Least 17 Billion Earth-like Planets Across the Milky Way Study*, http://news.nationalpost.com/2013/01/07/a-least-17-billion-earth-like-planets-across-the-milky-way-study/, January 7, 2013.
7. Michael Moyer, *Earth-Like Planets Fill the Galaxy*, http://blogs.scientificamerican.com/observations/2013/01/08/earth-like-planets-fill-the-galaxy/, January 8, 2013.
8. Ker Than, National Geographic News, Every Black Hole Contains Another Universe?, http://news.nationalgeographic.com/news/2010/04/100409-black-holes-alternate-universe-multiverse-einstein-wormholes/, April 12, 2010.
9. Chris Burns, Why the Space Force logo looks like Star Trek, and Star Trek looks like NASA, https://www.slashgear.com/why-the-space-force-logo-looks-like-star-trek-and-star-trek-looks-like-nasa-24607686/, Jan 24, 2020
10. U.S. Space Force Next Steps, https://www.space.com/united-states-space-force-next-steps.html
11. Joseph Trevithick, New Space Command's Flag Sure Looks A Lot Like Old Space Command's Flag: Although the idea of an independent Space Force has made headlines for years now, Space Command isn't new. Even its flag hearkens back to the 1980s, https://www.thedrive.com/the-war-zone/29625/the-flag-for-new-space-command-sure-looks-a-lot-like-the-one-for-old-space-command, August 29, 2019.
12. Gil Carlson, The Book of Alien Races, Secret Russian KGB Book of Alien Species https://www.amazon.com.au/Book-Alien-Race-Russian-Species-ebook/dp/B07BTF6WLL/, Blue Planet Project, March 29, 2018.
13. Gil Carlson, Secret Space Program, Blue Planet Project, https://www.amazon.com.au/Air-Force-Secret-Space-Program/dp/1513660276, January 1, 2020.
14. Joseph Trevithick, *USAF Secretary Gives Ominous Warning That Show Of Force Needed To Deter Space Attacks: But serious questions remain about what an all-out war in space would even look like and what hostile actions would automatically demand a response*, https://www.thedrive.com/the-war-zone/27396/usaf-secretary-gives-ominous-warning-that-show-of-force-needed-to-deter-space-attacks, April 11, 2019.

CHAPTER 39: SPACE FORCE

15. Robin McKie, https://www.msn.com/en-us/news/technology/the-moon-mars-and-beyond...-the-space-race-in-2020/ar-BBYCSSC?ocid=spartanntp
16. *Armageddon*, 1998 Film, Director: Michael Bay, Screenplay Writers: Jonathan Hensleigh, J.J. Abrams, https://www.imdb.com/title/tt0120591/
17. Lockheed Martin, Space Exploration, https://www.lockheedmartin.com/en-us/capabilities/space/human-exploration.html?
18. Greenland (2020 movie), https://en.wikipedia.org/wiki/Greenland_(film)
19. Killer Asteroid Hitting Earth In 2022? Here Is What NASA Says, https://www.republicworld.com/technology-news/science/killer-asteroid-approaching-earth-in-2022-asteroid-hitting-earth.html
20. Apophis in Depth, https://solarsystem.nasa.gov/asteroids-comets-and-meteors/asteroids/apophis/in-depth/

CHAPTER THIRTY-FOUR

BIBLE DNA & TODAY'S GENETICISTS

> "And God (Elohim) said,
> Let us make man in our image, after our likeness:
> and let them have dominion over the fish of the sea,
> and over the fowl of the air, and over the cattle, and over all the earth,
> and over every creeping thing that crept upon the earth."
> Genesis 1:26

Have scientists finally caught up to the Bible's explanation of the origin of our DNA? I began this discussion in Book One of *Who's Who in the Cosmic Zoo?* In whose image was the Evadamic race made? Adam was the prototype of modern humans, *homo sapiens*. The original Hebrew uses the word, *Elohim*, where the English word for God is, is actually plural in Hebrew for 'gods.' Hebrew is extremely specific to singular or plural, male or female. I went over all this in Book Two, *Who is God?* This means that it was not just one God, but a group of gods, referred to as Elohim, who I have already established are the Sons of the Most High God, who is El Shaddai, the Almighty. [1,2]

Back in 1990 when I was working as a Counselor on a cruise ship out of Saint Petersburg, Florida, and I was reading *Genesis Revisited*, by Zecharia Sitchin, one of my clients made a comment that I never forgot, that the gods are evolving along with us. They are working out their karma, meaning, they used us as an experiment, and periodically return to Earth to insert another stint in our evolution. I have to say, with all that I have researched, studied, and observed, including the Scriptures, that this appears to be true. They have unfinished business with us, and they are scheduled to return to Earth for a divine appointment and a final showdown over *who* gets to control Earth, and us humans, as they allegedly had a hand in our genetics. [3]

Anyone who has read Book One of this book series, *Who's Who in the Cosmic Zoo?* will remember that within my *Concluding Words*,[4] was my message debunking the popular rhetoric that was birthed out of the 1990s based on Zecharia Sitchin's work showcasing his translations of the Sumerian Cuneiform Tablets published in his 12 book series, the Earth Chronicles, that somehow, people misinterpreted his material to say that reptilian giant aliens created humankind, which has never resonated well with me. Sitchin and I agreed on his interpretation of the word, Elohim in the Hebrew Bible, which clearly is in its plurality meaning, 'gods' with a small 'g.' But these gods were not reptilian, they were human, albeit very tall humans, giant humans, but not reptilians. They created humankind in *their* image, according to *their* likeness. Genesis 1:26 is written in its plurality, they created both male and female and they were equal in their eyes.

Sitchin used to say often in his lectures, "we look like them." Therefore, they cannot be reptilian because we are human. Just because reptilians are humanoid, meaning they are bipedal, with two legs and two arms, does not make them human. Humanoid is the most popular form in the universe. However, reptilians are considered non-human intelligences, which qualifies them as alien. The Annunaki are depicted in stone as human, tall humans, but human, nonetheless. It is important to make these distinctions, for the purpose of discernment. Another point is that the reptilians come from Alpha Draconis, not Nibiru.

Scientific teams have been deciphering the human genome that contains not the anticipated 100,000 – 140,000 genes (the stretches of DNA that direct the production of amino-acids and proteins) but only some 30,000+ — little more than double the 13,601 genes of a fruit fly and barely fifty percent more than the roundworm's 19,098.

It is humbling to find our DNA is not that complicated at least when compared to other life forms. Moreover, there is hardly any uniqueness to the human genes. They are similar to 99 percent of that of the chimpanzees, and 70 percent of the mouse. In fact, cats are just one gene off from human DNA, which I detailed the National Geographic study in Book One. Human genes, with the same functions, were found to be identical to genes of other vertebrates, as well as invertebrates, plants, fungi, even yeast. The findings not only confirmed that there was one source of DNA for all life on Earth. Some organisms are more complex, genetically, than simpler ones, adopting at each stage the genes of a lower life form to create a more complex higher life form – culminating with Homo sapiens. A designer or creator appears to have taken the simple DNA and added more complex DNA forming each life form. Each species, dog, cat, horse, monkey, can be developed by using these building blocks of life.

"Francis Crick, the discoverer of DNA, although an atheist, published a book, *Life Itself: Its Origin and Nature* which subscribed to the theory of intelligent design, that our universe was not simply the result of a series of chemical accidents. He stated, "Life did not evolve first on Earth, a highly advanced civilization became threatened, so they devised a way to pass on their existence. They genetically-modified their DNA and sent it out from their planet on bacteria or meteorites with the hope that it would collide with another planet. It did, and that's why we're here.""[5]

CHAPTER 40: BIBLE DNA & TODAY'S GENETICISTS

Our DNA was encoded with messages from that other civilization. They programmed the molecules so that when we reached a certain level of intelligence, we would be able to access their information, and they could therefore "teach" us about ourselves, and how to progress. And I heard every creature in heaven and on earth and under the earths and on the sea, and all living beings in the universe, and they were singing: "To him who sits on the throne and to the Lamb be praise and honor and glory and might forever and ever!" (Revelation 5:13)

Ancient Sumerian texts as well as the Bible reveals the existence of angels or Anunnaki, translated as *those who from Heaven to Earth came*. Only in recent years has modern science caught up with ancient scientific knowledge. Scientist Dr. Carl Sagan in his book, *Intelligent Life in the Universe* indicated Earth had likely been visited by extraterrestrials. He states: "Sumer was an early–perhaps the first civilization the contemporary sense on the planet Earth. It was founded in the fourth millennium B.C. or earlier. We do not know where the Sumerians came from. I feel that if the Sumerian civilization is depicted by the descendants of the Sumerians themselves to be of non-human origin, the relevant legends should be examined carefully." [p 456] 6

Sagan goes on to ask, "What might an advanced extraterrestrial civilization want from us?" He answered his own question by stating,[6] "One of the primary motivations for the exploration of the New World was to convert the inhabitants to Christianity — peacefully if possible — forcefully if necessary. Can we exclude the possibility of an extraterrestrial evangelism?" [p. 463]

In ancient Mesopotamia, Egypt, and Greece, Judaism, Christianity and in Islam angels were sent as divine messengers to humans to instruct, inform, or command. Many abductees claim healing, love, and miracles from the angels. I met a pastor who stated he was drowning in a lake and was descending deeper with no hope, when an angle lifted him up to safety. He believed he was saved to breach about his miracle.

Erich von Daniken's revealed startling evidence that an alien race helped to create the pyramids of Egypt,[7] a claim he based upon the ruins themselves. And it is these ruins that now provide researchers with a never-ending source of clues, compelling discoveries, revelations, and evidence that Earth was indeed colonized by an alien race: -Research showing that the location and design of the pyramids were uniquely fit for preservation-something the Egyptians couldn't possibly have known. Remains found in the Peru, South America, where mysterious pictures were created 2000 years ago on the plains of the driest desert on Earth. Now these ancient wonders have spawned a modern tourist industry attracting travelers from all over the world. Elsewhere, tourists flock to see monuments that date from the dawn of history, like at Stonehenge in England, and in the surrounding countryside new wonders still appear today, vast, and intricate crop circles that materialize overnight.

SUMERIAN ALIEN CONNECTION

Many abduction cases usually have a similar story to them in that the aliens abducting them will perform medical examination and sometimes experiments having to do with human reproduction. The Sumerian Culture, which dates to 6,000 BC, is the oldest known culture on Earth. Allegedly, 6,000 years ago there were only hunter-gathers and cavemen on Earth. Something happened and a civilization was developed in what is Iraq today. The Sumerians claimed it was the Anunnaki taught them to write.

The time I spent with Archeoastronomer and Biblical Jewish Scholar Zecharia Sitchin, translator of Sumerian cuneiform scriptures, who taught me that a great deal of the Old Testament Book of Genesis was based on the Sumerian documents. As I mentioned in previous books, that during the 70 years when the Jews were held in captivity in Babylon, 605 B.C. to 537 B.C. they were forced to hear the daily and weekly readings of the passages from the Enuma Elish, which led them to write Genesis, which is essentially a synopsis of Enuma Elish, and the early creation stories and records of the great deluge, the origin of Nephilim. This means that the gods of Nibiru are mentioned in today's Bible, just under different names.

ANUNNAKI AND RACISM

Many think the Anunnaki or Igigi were reptilian. Some were tall tan humans. The names Anunnaki and Igigi were just rank names given to people from Nibiru. Their names were dependent on their mission. Similarly, what we call people that go into space astronauts. The Sumerian text states that Anunnaki were the rank name given to the first four hundred heroes from Nibiru that first landed on Earth. Igigi were the rank name of the heroes from Nibiru that manned the earth orbiter and the station on Mars. They were all part of the process to get gold back to Nibiru from Earth.

The text seems to be clear on their appearance. They genetically fashioned humans after their image. So that means we look like them, only they are much taller than we are and seem to be generally light skinned. When people are not close to a sun, they naturally have lesser pigmentation. They also seem to have light hair and blue eyes. As when the Noah in Sumerian story was born, he looked like the gods with white hair and eyes of fire. Blue fire? Genetics seem to trace all people with blue eyes to one place in the Middle East. So, I think we have a good picture of what the Anunnaki look like.

Battles were fought between the Nibiruans over who controls earth and Mars and the rest of this solar system. The battle included the Reptilian Giants, the Serpent Race of humanoids who do like reptilians. This is where the confusion comes from, why many think the Annunaki are reptilians. These ancient battles, in my humble opinion, is at the root cause of racism, xenophobia and antisemitism on Earth.

Meanwhile in Egypt, India, and Babylon, the aliens are depicted as viper like humanoids, green and blue skinned people...contrary to popular belief, there isn't only light and dark when it pertains to skin color. There are statues found in Iraq-Ubaid (City of Ur region) of lizard-looking beings dating back 5000 years. Similar lizard men statues are also found in Peru. However, the Vedic literature tells us of wars between these group, proving there were different color pigmentations, and different types of beings, that battled against each other.

Many cultures around the world recorded in their history and mythologies that there were other beings on the planet that may have had serpent-like qualities. They were called Nagas in India, dragons in Ancient China. These beings have fought over what they believe is their territory, the Earth, which leads people to believe they may have originated here, but they arrived when our Pole Star was Alpha Draconis. They have claimed the Earth as theirs, which is why these battles have been fought throughout millennia and are evidenced in our Judeo-Christian Bible.

These ancient battles ensnared our history with enslavement, racism, and antisemitism on Earth. Each group thinks it is their mission to wipe out the others. There are so many family feuds that have become generational curses within the nations, which have turned into strongholds deeply embedded in our DNA.

ANUNNAKI-NIBIRUANS

Most researchers agree that the Anunnaki were an advanced civilization that came from another planet from within our solar system that orbits and passes Earth approximately every 3,600 years. When the Anunnaki came to Earth, they landed in the Persian Gulf approximately 432,000 years ago. The records they left behind were Cuneiform Tablets, which told their story of creation. However, there were already women on the Earth when they came, which is what I pointed out in Book One of *Who's Who in the Cosmic Zoo?* proving that they couldn't possibly be the creators of the human race, because according to their record they used invitro fertilization to implant the wombs of the females, already on the Earth, which suggests human females were created by another god, before the Anunnaki arrived. Their record tells that they their story of creating man as a slave race to serve them to mine the Earth for gold which they needed to use a shield for their planet or mothership.

It was my conclusion in Book One, that the Anunnaki were *not* the creators of humankind, but that they did indeed genetically manipulate humankind by downgrading human DNA by disabling ten out of the original twelve strands of human DNA, leaving just enough intelligence behind to follow their orders and perform work for them. The promises of the Messiah are clear, that when He returns, humankind will be restored to their glory bodies, which is what they originally were created before the fall of man, which I believe was genetic downgrading of the human beings created in the image and likeness of the Creator.[1]

The Anunnaki left the Earth, leaving what they created behind, including setting up religions to worship them, and empires to carry on their bloodline on the Earth, to keep control of the planet until their promised return at the end of the 3,600 cycles.

DEAD SEA SCROLLS

The Dead Sea Scrolls found in a cave near Jerusalem revealed the Earth was visited with extra-terrestrial visitors. Most religions of the world agree in the concept of Sons of Light and Sons of Darkness, good angels and evil ones, good aliens and bad. My research has indicated there are various groups of aliens with different intentions, their motivations may be vastly different, and they often appear to be fighting one another. "The Zadokite Fragments," also known as the "Damascus Document" discovered in an old synagogue in Cairo in 1896, published in 1910 as Volume I of a series, "Documents of Jewish Sectaries," confirms Genesis and mentions the landing of Spacemen, their giant offspring and immorality. "Because they walked in the stubbornness of their hearts, the Watchers of heaven fell, yea; they were caught thereby because they kept not the Commandments of God. So too their Sons whose height was like the lofty cedars and whose bodies were as mountains, they also fell." [8]

While the Anunnaki visits may have been interpreted in religious terms, the data from the mark they left behind that they were here, is essentially truthful and often the best we have. Biologists search for protoplasm in ancient primordial soup. Archaeologists seek man's origin in the mud. I search with an open heart and shining eyes for our true home in the heavens.

In Book One of *Who's Who in the Cosmic Zoo?* I identified Earth humans as the Evadamic Race of Humans, to distinguish us from Extraterrestrial Humans.[1] As Eve is believed to be the Mother DNA of All Humans, and Adam the father DNA. But after the fall of Adam and Eve, which I proved was actually the genetic downfall by the Anunnaki who downgraded human DNA through genetic manipulation to use humans as their slaves, the descendants of Adam and Eve were wiped out in the floods of Noah. Then it was Noah's descendants who became the mother DNA of today's homo sapiens.

Lee Berger posited in his book, *In the Footsteps of Eve the Mystery of Human Origins* most early homo sapiens developed around Johannesburg, South Africa, interestingly near the best gold mines on Earth. Ancient Sumerian records claim Nibiru and Mars needed gold flakes to save its atmosphere and needed workers in the mines. We can assume that modern man is related to these space visitors who brought wisdom from their civilizations and taught us to read, write and build cities. With an estimated thirty billion earths in our galaxy alone it is reasonable to assume that the Earth has been visited by extra-terrestrial visitors as our ancient texts such as the Bible, Mahabharata, Koran, Torah, and Sumerian and Akkadian clay tablets claim.[9]

Without the wheel we would not have been able to ride in cars, cars, trucks, trains, or ride a bike. Sumerians were also the first to create fortifications, and the first to practice siege warfare. They were also the first to think of composite bows and horseback warriors. The game of checkers was first seen in Sumer and was played by the people around 2500 BC. According to the Sumerian Aliens theory, extra-terrestrial beings inhabited this earth long before man. It is believed that these beings ran out of resources on their own planet and traveled to earth to mine its resources (gold in particular) to bring back to their home planet.[10]

Heather Lynn, PhD., wrote in her recent book, *The Anunnaki Connection: Sumerian Gods, Alien DNA, and the Fate of Humanity (From Eden to Armageddon)*:

> Over 6,000 years ago, the world's first civilization, the Sumerians, were recording stories of strange celestial gods who they believed came from the heavens to create mankind. These gods, known as the Anunnaki, are often neglected by mainstream historians. The Sumerians themselves are so puzzling; scholars have described their origin as "The Sumerian Problem."
>
> With so little taught about the ancient Sumerians in our history books, alternative theories have emerged. This has led many to wonder about the true story behind the Sumerians and their otherworldly gods, the Anunnaki. Lynn traces the evolution of these Mesopotamian gods throughout the Ancient Near East, analyzing the religion, myth, art, and symbolism of the Sumerians, investigating:
>
> The Sumerian civilization, thought to be the oldest on the planet, has left us an enduring portrait of their society, philosophy, religion, and daily life in the countless records they left behind. When the Sumerians wanted to write something down, they used a stylus to press the wedge-shaped marks that composed their alphabet (called cuneiform) into a slab of clay. When the clay was fired in a kiln, the result was a durable document, hundreds of thousands of which have survived to this day. The Sumerian language is well-understood, and for many years, archaeologists have studied the myths told about the Sumerian gods in stories such as the Epic of Gilgamesh.
>
> "The main purpose of these beings, as described in the Sumerian myth *Enki and the World Order*, was to decide the fate of humans. They are also described as residing in the netherworld. Many modern popular accounts depict the Anunnaki as having been worshipped. While this could make some sort of logical sense, there is no hard evidence for their adoration in the archaeological record, with the

exception of only three attestations in administrative texts from the Ur III period, which hint that offerings were made to Anunna (Anunnaki). In the myths of Mesopotamia, their importance does shift. For example, Enki was initially portrayed as less supreme than An and Enlil, and in some cases like the flood myth, Enki intervened. An was the god of the "on-high." He was the first god to rule the universe, the founder of the cosmic order, but in most accounts, he presented as more withdrawn, leaving the power to his son Enlil." [10]

From as early as the 1960s, Archeoastronomer Zecharia Sitchin revolutionized our understanding of these Sumerian myths. Sitchin realized that eleven of the gods described in the Sumerian records were the heavenly bodies of our solar system (the planets plus the sun and moon). These planets were renamed by the Greek Empire, then renamed again by the Roman Empire. Sitchin further realized that the Sumerians had knowledge of the outer planets that, by all rights, they should not have had. Sitchin proved this to the world by submitting to NASA detailed descriptions of the outer planets (as found in the Sumerian records) BEFORE the NASA space probes reached these planets. Images sent back by the probes confirmed that the ancient Sumerians indeed knew what these planets looked like.

Sumerians always described their gods (the planets) starting with Pluto, then Neptune, Uranus with rings, as if they were seeing the planets from a heavenly body (or spacecraft) that was entering our solar system from the outside in. Sitchin's book, *The Twelfth Planet*, his expose` on Nibiru, was named after the Sumerian records, which recorded a twelfth planet, one whose elliptical orbit brought it close to earth for a brief period every 3,600 years. Looked at from this new perspective, the Sumerian religious epics contain a startling story: The twelfth planet, known as Marduk, was inhabited by humanoid beings very much like ourselves, known as Neberu. A problem with their atmosphere sent them on a mission throughout the solar system in search of gold; a metal which they believed could be used to heal their planet. Using rocket ships to shuttle people and supplies between their planet and Earth, during those months when Marduk's elliptical orbit brought it close to Earth, the Neberu established colonies in Mesopotamia (now southern Iraq) hundreds of thousands of years ago. Neberu became Nebiru, which became Nibiru.

They eventually found rich veins of gold in southern Africa, and established mines exploited by the worker element of their society, called the Igigi. Eventually, the Igigi tired of this unpleasant work, revolted, and forced the Anunnaki Nibiru leaders to find another source of labor. Their solution, related in great detail in the Sumerian records, was to create a slave race by splicing their genes with the genes of the most advanced primitive human on the planet at that time (approximately 200,000 years ago). Thus, was born the human race of homo sapiens.

CHAPTER 40: BIBLE DNA & TODAY'S GENETICISTS

Recently scientific discoveries, such as the uncovering of Neolithic gold mines in southern Africa, the tracing of all human DNA back to a single source, called Eve by the geneticists, have tended to confirm Sitchin's interpretation of the Sumerian records. Why science has never found the missing link, that fossils record that would show the evolutionary path from the early hominids to modern man, is also explained by Sitchin's interpretation: the missing link is not a fossil, but rather a genetic experiment performed by beings from another planet in our solar system in an effort to create a slave race for themselves.

The Sumerians claim the visitors brought writing called Cuneiform was written with a wedged-shape stylus on a wet clay tablet and then dried out after. Cuneiform was important because, it made it so Sumerians could record trades, crops, and important events, and it told us how Sumerians lived.

Ancient Sumerian writings claim the Anunnaki brought the wheel first used, which was the potter's wheel. They also developed cities and fortifications, tall buildings, and irrigation. Irrigation was formed in 3,100 BC. It was formed because there was a flood every year from the Tigris and Euphrates rivers. The Sumerians couldn't predict the floods so sometimes their crops would get destroyed so, they made little canals in the ground from the two rivers to the crops and around them so the crops would get enough good water with silt and they built up the banks of the two rivers so the crops wouldn't get destroyed. Since irrigation helped the crops grow, the farmers could harvest a surplus each year and it made life much easier for Sumerians.

The plow was created in 3,000 BC. The plow was an Ox with a digging stick hooked on to it. It was an important invention because if it were not invented it would take way longer to dig holes by digging with just a regular digging stick than an ox with a digging sick attached to it. I think that if the Sumerians didn't invent the plow and it still wasn't invented that we would have a really hard time digging enough holes for the crop's seeds and being able to harvest them in time because of the weather.

The sailboat was invented. The sailboat was invented so the Sumerians would be able to get fish from the middle of deep lakes and rivers. The sailboat was an important invention because if it were not invented then it would be hard for Sumerians to get enough fish. In about 4,000 BC, the sailboat was first just a log with a sail on it. Then about 2,700 years later, the sailboat turned into a big ship with two masts.

The wheel, cuneiform, irrigation, plow, and sailboat are just some of the inventions that us humans could not live without. Only in recent years has modern science caught up with ancient scientific knowledge.[11]

NIBIRU'S POPULATION OF GIANT ALIENS

According to Sitchin, the Anunnaki are from Planet X (Nibiru) and may explain stories of giants which were published in the Bible and apocryphal writings, i.e., the Book of Giants, the Books of Enoch, the Books of Jubilees, etc. Some even say that Planet X has been transformed into a fully inhabitable Dyson sphere. A Dyson sphere is a hypothetical megastructure that completely encompasses a star and hence captures most or all of its power output. It was first described by Freeman Dyson. Dyson speculated that such structures would be the logical consequence of the long-term survival and escalating energy needs of a technological civilization and proposed that searching for evidence of the existence of such structures might lead to the detection of advanced intelligent extraterrestrial life. Different types of Dyson spheres correlate with information on the Kardashev scale.

Since then, other variant designs involving building an artificial structure or series of structures to encompass a star have been proposed in exploratory engineering or described in science fiction under the name "Dyson Sphere". These later proposals have not been limited to solar-power stations. Many involve habitation or industrial elements. Most fictional depictions describe a solid shell of matter enclosing a star, which is considered the least plausible variant of the idea In May 2013, at the Starship Century Symposium in San Diego, Dyson repeated his comments that he wished the concept had not been named after him. [12]

SURVEY NIBIRU – NEW SPACE FORCE MISSION?

When General Jay Raymond was sworn in as the inaugural chief of Space Force Operations at a White House Ceremony led by Vice President Pence, engineers at Tonopah Test Range finalized plans to send a reverse-engineered alien spaceship on a historic mission to the Nibiru system, with the goal of determining why the dark star and its orbiting planets have yet to reach perigee, an event that ought to have seen fruition in 2012, according to a Washington insider speaking under condition of anonymity.

Ever since President Trump announced his intent to create a 6th branch of the United States military—for which he was heavily mocked—civilian, military, and NASA engineers have worked tirelessly to finish construction on an intergalactic ship that allegedly is powered by a gravity generator that warps the fabric of space-time. This non-linear mode of transport allows rapid travel between two points without violating Newtonian physics or Einstein's Special Theory of Relativity.

Discussion on this method of transportation is not new, and was first breached in 1989 by Bob Lazar, a physicist with questionable credentials who claimed to have briefly worked at "S-4," a top-secret subsidiary of Nellis Airforce Base (Area-51) in Nevada. There, Lazar said he worked on a highly compartmentalized project that involved reverse engineering a circular craft's (sports model) propulsion system. His explanation of a craft generating a gravity wave that allows flight and avoids physical

CHAPTER 40: BIBLE DNA & TODAY'S GENETICISTS

detection by bending light around it closely matches a ship the Space Force intends to send to Nibiru. If true, Lazar's claims will be vindicated.

According to a source through retired NY Post Journalist turned rabbit hole researcher, Michael Baxter; President Trump is deeply concerned about Nibiru and has relied on the advice of scientific advisor Kelvin Droegemeier, a meteorologist by trade, to determine what threat, if any, Nibiru poses to Earth. Droegemeier and NASA Administrator Jim Bridenstine told Trump the Nibiru System seemed to have stalled in space. They said optical viewing and radio frequency analysis provided inconclusive answers and that a manned trip to the Nibiru system is needed to determine whether the celestial interloper will achieve perigee and wreak untold havoc on the planet.

Trump told Gen. Raymond that investigating the mysteries of Nibiru would be among Space Force's top priorities. He handed Raymond a structured plan that included names of three astronauts he had handpicked for the mission and said he wanted the ship enroute to the Nibiru system by the 3rd quarter of 2020. The ship is expected to make the 160-million-mile journey in under 16 hours, a milestone for astronomical travel. It will approach to within nine million miles of the brown dwarf, steering clear of Nibiru's fiery twin tails that contain trillions of micrometeorites and impenetrable clouds of red iron oxide dust, and collect imperative astronomical data to either prove or disprove a longstanding scientific belief that Nibiru will get close enough to earth to pelt it with meteorites and to gauge whether its impressive magnetic field is powerful enough to fracture Earth or cause irreparable geomagnetic and geophysical pole shifts that could tear the Earth asunder.

If information is accurate, the U.S. now possesses an extraterrestrial-inspired vehicle that will not only command the domain of space but also give it a decisive advantage over its earthly adversaries, as it seems certain the military will exploit the technology for weapons development platforms. Regardless, there is a zero percent chance the government will share its findings with the public, as it has tried with limited success to conceal existence of the rogue solar system for nearly forty years.[14]

What a time to be alive to see this. According to recent reports from Washington D.C., through the retired NY Post Journalist, Michael Baxter, one of the reasons President Trump wanted to send a reverse-engineered Intergalactic Spaceship to Nibiru was to find out why its apparently stalled? Typically, when planets appear to stall, they are stationing retrograde, so they appear to stop and slow down, as they turn around, then they appear to be going backwards during the entire retrograde cycle from the perspective of the Earth. There is a lot of head scratching as to why Nibiru has not followed the tracked path it was supposed to since 2012. So, according to Baxter, the Space Force will oversee this historic mission.[13]

ANNUNAKI ALIEN THREAT – NEW SPACE FORCE MISSION?

According to retired NY Post Journalist turned Nibiru researcher, Michael Baxter of Twisted Truth reported on Russia's President Vladimir Putin's Address to the Cabinet of Ministers on the Anunnaki in Iran said:

> "Shortly afterward, I contacted the White House and spoke with the American president. I warned him of the imminent attack. Surprisingly, he told me his nation had learned of Iran's intent about the time we had and that he and his officials were weighing a measured response if Iran carried through. I told President Trump, who has recently been receptive to my comments and concerns of Anunnaki intrusions, that attacking Tehran or other Iranian locations would likely provoke Anunnaki in the region to attack his nation."

> "I told President Trump that Anunnaki in the region consider Iranian territory sacred ground. That enraging them might force him into a war his nation is ill-prepared to handle. I told him of our own trials against the Anunnaki and of how we incurred massive losses after engaging these disgusting monsters in Syria. I Informed him that our Special Services are equipped with technology to effectively fight Anunnaki but are not yet ready to move on their coven in Iran. I asked him, in the name of humanity, to not respond to Iran's attack. President Trump told me he would consider my words and take up the situation with his Secretaries of Defense and State. As we now know, President Trump heeded my advice, and may have prevented unstoppable escalation of the Anunnaki conflict."

> "We have repeatedly pointed out that America's presence in Syria and Iraq is unwanted, and there are colleagues of mine who wish war with America because of this. The real threat to our sovereignty comes from space, not overseas, and I emphasized this point to Trump. I also revealed to him what I have already said to all of you: that in coming months I intend to amend the constitution, naming the Anunnaki as the primary threat we must face, not unilaterally, but as part of a global coalition determined to save humanity from an enemy that indiscriminately destroys us."

CHAPTER 40: BIBLE DNA & TODAY'S GENETICISTS

"I have instructed Ministry of Defense of the Russian Federation, the Ministry for Foreign Affairs, and other relevant departments to analyze the level of threat posed by the aforementioned actions of the Anunnaki to our country and take comprehensive measures to prepare asymmetrical response. All assets and finances needed to combat the Anunnaki will be made available. This is not debatable."

"Russia will remain open to an equal and constructive dialogue with the United States of America and other nations that share our goal to eliminate the alien threat and seek to restore confidence and strengthen global security from all hostile extraterrestrials."[14]

Notes:
1. Ella LeBain, *Who's Who in the Cosmic Zoo? A Spiritual Guide to ETs, Aliens, Gods & Angels*, Book One – Third Edition, Concluding Words, pp.415-444, Skypath Books, 2013.
2. Ella LeBain, *Who is God?* Book Two, Chapter: Who is Allah?, p.359, Skypath Books, 2015.
3. Zecharia Sitchin, *Genesis Revisited, Is Modern Science Catching Up with Ancient Knowledge?* (Earth Chronicles) Avon Paperback – October 1, 1990.
4. Ella LeBain, *Who's Who in the Cosmic Zoo? A Spiritual Guide to ETs, Aliens, Gods & Angels*, Book One – Third Edition, Concluding Words, pp.415-444, Skypath Books, 2013.
5. Francis Crick, *Life Itself: Its Origin and Nature*, https://www.amazon.com/Life-Itself-Its-Origin-Nature/dp/0671255622/, January 1, 1981
6. Carl Sagan, I.S. Shklovskii, *Intelligent Life in the Universe*, p.456, 463; Dell Publishing, 1966.
7. Erich Von Daniken, *The Eyes of the Sphinx: The Newest Evidence of Extraterrestrial Contact in Ancient Egypt*, https://www.amazon.com/Eyes-Sphinx-Evidence-Extraterrestial-Contact/dp/0425151301, Berkley, 1st Edition, March 1, 1996.
8. Theodor H. Gaster, *The Dead Sea Scriptures*, New York: Doubleday Anchor Books.; Ex-Monastery Library edition – January 1, 1956
9. Lee R. Berger, *In the Footsteps of Eve: The Mystery of Human Origins*, https://www.amazon.com/Footsteps-Eve-Mystery-Origins-Adventure/dp/0792276825, (Adventure Press) June 1, 2000
10. Heather Lynn PhD, *The Anunnaki Connection: Sumerian Gods, Alien DNA, and the Fate of Humanity (From Eden to Armageddon)* on www.amazon.com/Anunnaki-Connection-Sumerian, New Page Books, March 1, 2020.
11. George Filer, Filer's Files 3, 2020 *Science Catches Up to the Bible*, NUFOC weekly-updates@nationalufocenter.com
12. Gil Carlson, *NIBIRU PLANET X TODAY: Anunnaki Aliens, UFOs, A Blue Planet Project*, Blue Planet Project (April 8, 2018).
13. Michael Baxter, *Space Force to Survey Nibiru*, https://www.twistedtruth.net/featured/space-force-to-survey-nibiru/, January 15, 2020.
14. Michael Baxter, *Putin Addresses Cabinet of Minister on Anunnaki in Iran*, Full Transcript, https://www.twistedtruth.net/featured/putin-addressed-cabinet-of-minister-on-anunnaki-in-iran-full-transcript/, January 15, 2020.

CHAPTER THIRTY-FIVE

CLOUDSHIPS, CHARIOTS & UFOS

"He lays the beams of his chambers in the waters;
He makes the *clouds* his chariot: and walks upon the wings of the wind –
who makes his angels spirits and ministers a flame of fire."
Psalms 104:3-4

"Behold, he shall come up as *clouds*, and his chariots shall be as a whirlwind:
his horses are swifter than eagles."
Jeremiah 4:13

Can Chariots (UFOs) create their own atmosphere? Yes. They can create whirlwind, lightning, fire, storms, but their favorite is clouds. "God clothes Himself with the clouds." Clouds are perfect for cloaking when they come close in our atmosphere. This is accurately depicted in most sci-fi movies about extraterrestrials. This is also happening in our present skies. While meteorologists like to write them off as lenticular clouds, not all are weather clouds, but spaceships hiding within a lenticular cloud. I have been collecting photos for decades, and some even reveal the metallic structure beneath the cloud cover. History merely repeats itself. It has all been done before. There is nothing new under the sun. (Ecclesiastes 1:9)

According to Scripture, Yahuah Himself flies in one of these cloud cloaked vehicles. "And he rode upon a cherub, and did fly, yea, he flew swiftly upon the wings of the wind. He made the dark his secret place; his tent round him was the dark waters and *thick clouds of the skies*," (Psalms 18:10-11). Perhaps this is why Ezekiel referred to the space vehicle as Cherubim?

There are multiple passages of Scripture identifying the space vehicles as Cherubim in Hebrew thinking. I went over this in greater detail in *Who is God?* Book Two in Chapter Two: *Ancient Technology and Biblical Astronauts*[1] and *Who Are the Angels?* Book Three in Chapter Three: *The Celestial Hierarchy of Angels*.[2] Cherubim were a type of robotic angelic servants that guarded spaceships and the garden of Eden. In Hebrew it is pronounced *kerubim*, in its plurality, and *kerub* in its singular.

CHAPTER 41: CLOUDSHIPS, CHARIOTS & UFOs

"When the cherubim moved, the wheels beside them moved; and when the cherubim spread their wings to rise from the ground, the wheels did not leave their side. When the cherubim stood still, the wheels also stood still, and when they ascended, the wheels ascended with them, for the spirit of the living creatures was in them." (Ezekiel 10:16-17) [See also, Ezekiel 10:7; Exodus 37:9; 1 Kings 6:32; 8:7; 2 Chronicles 5:8]

Throughout the Bible we find what is called the *glory of God* manifesting as a cloud. It is sometimes referred to in Hebrew as the Shekinah Glory or the Glory Cloud. Over 3500 years ago Yahuah set up His Tabernacle cloaked with a cloud above the Sinai desert and communicated with Moses and His people for 40 years. "The chariots of God are twenty thousand, even thousands of angels: The Lord is among them, as in Sinai, in the holy place," (Psalms 68:17). Exodus stated that there were lightning, thick clouds, and loud sounds that Moses ascended into to meet with God. On one occasion the Lord revealed Himself to the elders and ate with them in Exodus 24:9-18. In the New Testament a Cloud appears at the baptism, transfiguration, death, and ascension of Jesus. The resurrected Jesus ascended into a cloud; the Angels that were present told His disciples that He will be coming back the same way He ascended *with* the clouds. See, Acts 1:9-11 Matthew 24:30-31.

Myriads of His heavenly hosts have developed and learned to fly the Merkabah. The Merkabah is a space vehicle also known as the clouds of heaven. The Merkabah is mentioned 44 times in the Jewish Bible. It is translated as chariot and is often used with the word fire, that was described as chariots of fire.

The Merkabah is a sacred geometric shape made of three-dimensional triangles that point upwards and spin clockwise, that connect with three dimensional triangles that point downwards that spin counterclockwise. These spinning forms the energy it needs to ascend, it also creates clouds of light from the spinning. If you were to draw it on a two-dimensional piece of paper, you would have the six-pointed Magen David, known as the Star of David, the Seal of Solomon, which is the symbol for the Jewish People and is depicted on the flag of Israel. Its ancient symbolism is about the ancient Merkabah technology. After the ships left the region, people continued to use the symbols in occult practice, in magical rituals, which is how it became associated with the occult. However, the six-pointed star is not about the occult, but about space vehicles that are Extraterrestrial.

What we have learned based on the Bible record, other sacred texts, that what we have called angels and God, all seem to fly on what we call today, UFO's and spacecrafts. The Bible record tells us that some of the angels, who were called fallen angels, interacted sexually with humans. From the Books of Enoch and Genesis in Scripture, these are fallen sons of God who rebelled against their Creator God, who fly counterfeit spacecrafts and obedient son of God (angels) that fly crafts called chariots.

We must be able to discern TRUTH in this day of great deception. The Merkabah ships are spun through the Presence of God and through the spirit of God which comes from love and light bodies. When the fallen sons rebelled, they lost their ability to connect with their Creator, through their rebellion and subsequently being ousted out of their heavenly positions, so they used technology to counterfeit the Merkabah, because they were no longer holy and lost their powers to create and fly in Merkabah chariots. Today their ships may appear to be sophisticated and advanced aerospace technology, but they are clunky, and they lack the connecting presence of love and light that powers the Merkabahs. This is an important discernment in these end times.

Those fallen disobedient ones have allied themselves with the wicked governments of this world and false religion; thus, we see all the wars, corruption, greed, and human suffering. But the good news is: "The kingdoms (governments) of this world are becoming the kingdom of God and His anointed Ones (us)." Of the increase of His government there shall be no end." He is coming with clouds of Heaven in the literal sense, illustrated on this book cover.

For many years we have been prophesying about this time where there would be an increase of UFO (celestial chariots) activity. There is hardly a week that goes by where signs are not appearing in the heavens (sky).[3, 4]

THE 1ST STARGATE MENTIONED IN THE BIBLE

Keep in mind it was common to see spacecraft moving at extremely high speed over 5000 years ago (3000 BC). Besides all the cover for justifying one of the longest wars the American military has been involved in, is not just about occupying Iraq and Afghanistan just for their wealth of resources alone. There is much deeper interest in this ancient region, for their presence, and for the invasion of Iraq, that most would find it hard to believe. As I mentioned in Book One of *Who's Who in the Cosmic Zoo*, that the American military and its Special Forces went to the region to locate and secure multiple Stargates, as well as locate buried ancient technology.[5] The 1994 Hollywood movie, *Stargate*, had more fact than fiction in it. As I mentioned in my previous books, that the first mission the Marines were given during their invasion of the Baghdad Museum in 2003, was to retrieve the ancient Stargate, that was gifted to Sadaam Hussein by Muamar Khadafi from Libya. They succeeded and brought it back to NORAD in Colorado Springs, Colorado. After this, rumors flew about it, so in order to deflect any voracity to this story, they erected a mock Stargate in downtown Colorado Springs, which is now a fountain, for tourists. Interestingly it faces two obelisks, which are situated down the hill from a warehouse that has written on its building with a great big scary alien on its roof, that says, *hiding in plain sight*. Truth is Stranger than Fiction, heh?

The very first mention of a stargate in the Bible is in Genesis 11 which was built by Nimrod. It is called the Tower, later called Babel or Tower of Babel, which was erected in Babylon. It was to serve as a portal from Earth's side to activate an existing Stargate already in place. Babylon in the Akkadian language means, Gateway of the Gods. The Gateway of the Gods was how the Gods entered into and exited Earth's atmosphere-Stargate. It was like a Wormhole or Vortex that allowed "highly advanced extraterrestrial beings" from other worlds, star systems to come to Earth in an extraordinary short time instead of traveling hundreds of light-years. It is important to note, and reiterate from Book One, that not all the highly advanced intelligent beings have humankind's best interest in mind. Technological Advancement does not equate to Spiritual benevolence or Spiritual Evolvement.

Akkadian is an extinct Semitic language spoken in ancient Mesopotamia. The Biblical name for Akkadian or Akkad is Accad- Genesis 10:10. Nimrod was a hybrid, or Nephilim (fallen one), meaning he was mixed with fallen angel/alien and human DNA, due to one of his parents being extraterrestrial.[7]

Nimrod seemed to have possessed the technology and intelligence from his alien ancestry on how to build the Stargate. "Come, let us build a city and tower, whose top may reach unto heaven (stars)." In Genesis 11:4, Nimrod rebels against Yahuah and becomes a hunter, seeking to destroy even the Lord Yahuah. (Genesis 10:8-9) See, the Babylonian "Code of Hammurabi".

Nimrod built a Tower to escape another flood. Yahuah had already declared there would not be another one like that. The purpose for the Tower was for inter-planetary and stellar travel and for him and other Nephilim (hybrids) to connect back with their alien ancestry. The "watchers" of the Grigori Angelic Race otherwise known as the Alien Presence, that left their first estate to mate with humans produced highly intelligent but violent-murderous offspring, as Nimrod was a hunter and not just of animals, but humankind. (Genesis 6:1-5)

> And the angels who kept not their first estate, but left their own habitation, he has reserved in everlasting chains under darkness unto the judgment of the great day. In like manner, Sodom and Gomorrah and the cities around them, who indulged in sexual immorality and pursued strange flesh, are on display as an example of those who sustain the punishment of eternal fire.
>
> (Jude 1:6-7)

Yahuah and his hosts of Heaven saw the ungodly self-serving condition of the human heart and hybrids (Nephilim) and therefore forbid them to reach their goal of inter-planetary or stellar travel through the Stargate. He destroyed it, as evidence of it exists today. "And the LORD (Yahuah) came down to see the city and tower." Genesis 11:5-9 records history as:

> "But the LORD came down to see the city and the tower the people were building. The LORD said, "If as one people speaking the same language, they have begun to do this, then nothing they plan to do will be impossible for them. Come, let us go down and confuse their language so they will not understand each other." So, the LORD scattered them from there over all the earth, and they stopped building the city.

> That is why it was called Babel-because there the LORD confused the language of the whole world. From there the LORD scattered them over the face of the whole earth."

You must realize, this was the "days of Noah," who was alive to see Nimrod his great grandson. Jesus told us that we would see a repeat of what happened over 5000 years ago in Noah's days.

> ". . . because just as it was *in the days of Noah*, so it will be when the Son of Man comes. In those days before the flood, people[a] were eating and drinking, marrying, and giving in marriage right up to the day when Noah went into the ark. They were unaware of what was happening until the flood came and swept all of them away. That's how it will be when the Son of Man appears."
> (Matthew 24:37-39)

Nimrod was also the first antichrist, as he waged a rebellion against the Lord of Hosts, which eventually led to his downfall. He was responsible for the curses on humankind from the Tower of Babel, which confused humankind with different languages. So, when you misunderstand someone from another culture or language, you can blame it on Nimrod. In England, there is a derogatory term used for those who are dumb, stupid, out of sorts, they say, "don't be such a Nimrod," it is considered an insult.

There is a battle within the minds of humans and for the minds of humans by the dark forces, but there is also a battle for planet Earth. This battle will intensify and climax into the final space battle of Armageddon, which will take place over one of the most fought over ancient portals, in Jerusalem, Israel. Har Megiddo is the ancient battlefield outside of the walls of Jerusalem, which in Hebrew translates to, *field of blood*. The end times battle will be fought in space and on the Earth, as the End Times prophecies are clear that the blood of kings will flow like a river and the Lord will call the birds of the air to feast upon it. See, Revelation 16, 19.

The kings of the earth have aligned themselves with the fallen ones in exchange for power, technology, and wealth. Under their demonic leadership they will wage war on the forces of God in the last days of these end times. They know He is coming, but they have been deceived into thinking they could close the Stargates and defeat the King of Glory and the armies of the heavens with alien technology, inter-dimensional beings, and NATO forces.

> "Why do the nations rage and imagine a vain thing? The kings of the earth set themselves, and the rulers take counsel together against the LORD, and against his anointed ones. But he who sits in the heavens laughs at them."
> (Psalms 2:1-2)

Revelation 19 says it will be a blood bath – Jesus, King of kings and the forces of Light will be victorious, who returns and literally ends the final battle with His breath and a Word.

The truth of what is happening around the world is more fascinating than any sci-fi movie that has been produced. It is so far out that it is indeed unbelievable by western programmed minds, thus, the governments of this world have been able to carry on in plain sight their wicked agendas.

Questions to help you to THINK outside the box? Why are there so many prophecies in the Bible about BABYLON to be fulfilled in our time? Here is something to keep in mind, remember, Babylon means, Gateway of the Gods, Stargate. Remember, ancient Babylon is modern day Iraq. Is not the main reason for building the Large Hadron Collider in Cern, Switzerland, to open Stargates, portals into other dimensions?

Could it be that there are many kings of the Earth, i.e., world leaders, that do not want the Gateway of the Gods opened because it would mean the end for their kingdoms of this world system and the beginning of the Millennial Reign of Christ on Earth, which begins a glorious new world? Are there Special Forces trained and deployed around the world to use ancient alien Nephilim technology to prevent certain Stargates from opening and to open others that would assist their satanic agenda? The governments of this world are fully aware and on alert, they know Bible prophecy, but they think they could thwart and prevent these prophesies from fulfilment. [6] Nah, they will not be successful. The LORD of Hosts bats last in this ancient war!

> "Lift up your heads, you ancient gates! (Stargates) Be lifted up, ancient doors, (dimensional portals) so the King of Glory may come in. Who is the King of Glory? The LORD YAHUAH strong and mighty, the LORD, mighty in battle. Lift up your heads, you gates! Be lifted up, ancient doors, so the King of Glory may come in. Who is he, this King of Glory? The LORD of HOSTS of the heavenly armies — He is the King of Glory."
>
> (Psalms 24:7-10)

Stargates around the planet has been opening and are scheduled to be opened in the near future. Most of these openings are taking place during the alignments, conjunctions, eclipses, and movements of various celestial bodies. The unlocking and opening of these Stargates and Vortices have been planned and timed with exact precision thousands of years ago. As they open, more Celestial Chariots (UFOs) will be seen in our atmosphere; this is one reason why there is an increase of UFO sightings worldwide.[7, 8]

BIBLICAL CHARIOTS ARE SPACESHIPS

Chariot is translated as vehicle, cavalry, or millstone, which when it is described as 'flying' is a thick disc-shaped object, i.e., a flying saucer. The number associated with the scripture is 20,000, indicates a fleet of spaceships. The fact that Yahuah has flying chariots that are cloud-like is a direct way to the nature of the clouds associated with the Second Coming of Christ, which include literally thousands of flying vehicles, i.e., spaceships.

> "As they were walking along and talking together, suddenly a chariot of fire and horses of fire appeared and separated the two of them, and Elijah went up to heaven in a whirlwind. Elisha saw this and cried out, "My father! My father! The chariots and horsemen of Israel!" And Elisha saw him no more. Then he took hold of his garment and tore it in two. Elisha then picked up Elijah's cloak that had fallen from him and went back and stood on the bank of the Jordan. He took the cloak that had fallen from Elijah and struck the water with it. "Where now is the Lord, the God of Elijah?" he asked. When he struck the water, it divided to the right and to the left, and he crossed over."
>
> (2 Kings 2:11-14)

While Elijah and Elisha, a bright flying vehicle and cavalry of bright flying objects appear over their heads, and Elijah ascends into the sky by a spinning flying vehicle. The word translated into horses, 'va-soos, aish' means rapid movement of flight in fire. Aish is fire. This cavalry of glowing vehicles is most likely the army of where the angelic hosts appear.

This is proof that the horses of Revelation given to St. John in a vision by Jesus Christ, are not flesh and blood flying horses as he wrote down, but space vehicles, which in the Old Testament uses the same vernacular, because when these scriptures were written down, they didn't have the words to describe them as we do today.

ARMADA OF CHARIOTS OF GOD

> "The chariots of God are twenty thousand, even thousands of angels: The Lord is among them, as in Sinai, in the holy place. Thou hast ascended on high, thou hast led captivity captive: thou hast received gifts for men; yea, for the rebellious also, that the LORD God might dwell among them."
>
> (Psalms 68:17-18)

The vehicle of the Elohim number a myriad of thousands. Twenty = Myriad, see Revelation 5:11.

HORSES AND CHARIOTS OF FIRE

I love to 'horse around' with horses. Horses are amazingly powerful yet highly intuitive and gentle creatures. They are gift from God to humans. Horses can run really fast that it can be euphoric in the sense that it can feel like you are flying. But let's face it, horses DO NOT FLY! The vernacular used in Scripture refers to powerful chariots made of fire, which used the word horses, to denote speed and transportation. Flesh and blood horses are not made of fire. In fact, horses are frightened by fire, and while you can get a herd of horses to gladly get bound to each other and pull a huge chariot, wagon, stagecoach, etc., you will never get horses to pull a chariot made of fire.

Horsepower is a term we use to measure speed. We use in to measure automobile engines, but horses don't fly or pull carts of fire. These scriptures are not describing flesh and blood horses, but vehicles that appeared like horses drawn chariots on fire. This is space technology, and it was described as close to their understanding as possible.

Fire in scripture is also symbolic of anointing. What Elish saw, may have been Extradimensional. The horses and chariots of fire represented the power behind him in the next realm. However, the fact that he was able to smite his enemies with blindness could have had something to do with the fact that they could very well have been blinded by the light of vehicles on fire.

> "And Elisha prayed, and said, LORD, I pray thee, open his eyes, that he may see. And the LORD opened the eyes of the young man; and he saw: and behold, the mountain was full of **HORSES AND CHARIOTS OF FIRE** round about Elisha. 18And when they came down to him, Elisha prayed unto the LORD, and said, smite this people, I pray thee, with blindness. And he smote them with blindness according to the word of Elisha."
>
> (2 Kings 6:17-18)

CLOUDS AS CHARIOTS

Now read Psalm 104:3 again, only with the name of God spelled out, instead of the generic title, as God. He has a name, and His name is Yahuah. "Who lays the *beams of his chambers in the waters*: (USOs) *who makes the clouds his chariot*: who walketh upon the wings of the wind: Yahuah makes *thick clouds his vehicle and travels on the edge of the wind*." (Spacecraft)

THE CHARIOTS OF GOD SPIN AND GLOW:

> "For behold, the Lord will come with 'FIRE' and with His **'CHARIOTS'** like a **'WHIRLWIND'**, to render His anger with fury, and His rebuke with 'Flames of Fire'.
> (Isaiah 66:15)

> "Behold, **he shall come up as CLOUDS, and his CHARIOTS shall be as a WHIRLWIND**: his horses are swifter than eagles. Woe unto us! for we are spoiled."
> (Jeremiah 4:13)

Zechariah looks up and sees four flying vehicles coming from in between two copper-colored mountains.

> "And I turned, and lifted up mine eyes, and looked, and behold, **there came four CHARIOTS out from between two mountains; and the mountains were MOUNTAINS OF BRASS**."
> (Zechariah 6:1)

ANCIENT STAR WARS CULMINATES IN DIVINE END TIMES APPOINTMENT

As imagined on the front cover of this book, Christ leading an army of thousands of cloud chariots with thousands of angels, extraterrestrial messengers. Imagine Him arriving in Jerusalem surrounded by this heavenly army aka heavenly hosts, as He takes up His throne to rule over all the earth as King of Kings. (Zechariah 6:12-13 and 14:9,16) This will be the biggest event ever to have happened to Planet Earth.

Isaiah 66:15 is another scripture that relates to the Lord's chariots. Christ shall come with Fire and with His Chariots like a Whirlwind, to pour forth His anger and rebuke3 upon the unrepentant humankind (Revelation 9:20-21; 16:9, 11). Remember His Redeemed are NOT appointed to His Wrath, See, 1 Thessalonians 5:9, "For God has not appointed us to wrath, but to obtain salvation by our Lord Jesus Christ." Therefore, before this event, of His Rebuke and His Wrath, His redeemed are promised to be removed from the Earth, in another End Time event known as the Rapture or Mass Ascension.

That with His Flaming Fury, God's Justice will be applied with Holiness and Righteousness. The Coming of the Lord draws near as both Judge and Jury. He stands before the door, (James 5:8-9) God's Word declares that His Day of Wrath shall come. (Revelation 6:17) Who shall be able to stand?

Christ is full of pity and mercy (James 5:11), but make no mistake, He continually calls humankind to repent. 2 Peter 3:9 says "The Lord ... is longsuffering toward

CHAPTER 41: CLOUDSHIPS, CHARIOTS & UFOs

us, not willing that any should perish, but that all should come to repentance (Revelation 2, 5, 16, 21, 22 and 3:3; 19).

If you doubt that God can be a God of a Wrath as well as a God of Love, then examine these scriptures:
- Wrath of the Lamb – Revelation 6:16; 11:18; 16:1; 19:15
- Day of the Lord comes with Wrath – Isaiah 13:9
- Break them with a Rod of Iron – Psalm 2:9
- Day of Wrath – Job 21:30
- Proverbs 11:4
- Zephaniah 1:15
- Romans 2:5

Will you raise your arms in praise, fall on your knees in awe and worship when Christ returns and stand among the Chariots of God and angels, or will you be in the camp of those who will run away in shame to find a place to hide from His Glory?

If you are hiding now from His Presence and His Spirit, it is because of the conviction and guilt of your mistakes. Maybe you are angry at God for not giving you loving parents? Or maybe you are angry at God for the bad behavior of so called, religious people who have disappointed you? It does not matter what your story is, as me being orphaned and raised a Jew have a similar story. We are all broken, cracked pots if you will in the potter's hand. Yet, when we are broken, is when the light of Christ can reach us, for He understands all these feelings, He too was betrayed and disappointed by religious people. The Pharisees were so puffed up with pride and full of themselves and their legalism, that they were blinded by it, and subsequently missed recognizing their long-awaited Messiah. Be that as it may, He is returning and all those people including you, will get another chance to ask for forgiveness, and be put right with God. Invest in your spirit now, walk in the assurance of the complete salvation of Jesus Christ, who promises not just to forgive all sin, but to break the power of sin if you repent to Him. Do not be the one hiding when He returns, because the window of Grace, that we enjoy now, will be closed, just like a stargate.

> "…when the Lord Jesus shall be revealed from Heaven with His mighty angels," "In 'Flaming Fire' taking vengeance on them that know not God, and that obey not the gospel (the good news) of our Lord Jesus Christ."
> (2 Thessalonians 1:7,8)

> The mountains quake at Him, and the hills melt, and the earth is burned at His presence, yea, the world, and all that dwell therein. "Who can stand before His indignation? And who can abide in the fierceness of His anger? His fury is

poured out like 'Fire', and the rocks are thrown down by him."
(Nahum 1:5:6)

For the great day of His wrath is come, and who shall be able to stand?
(Revelation 6:17)

But the day of the Lord will come as a thief in the night; in the which the heavens shall pass away with a great noise, and the elements shall melt with fervent heat, the earth also and the works that are therein shall be burned up.
(2 Peter 3:10)

Looking for and hasting unto the Coming of the Day of God, wherein the heavens being on 'Fire' shall be dissolved, and the elements shall melt with fervent heat?" Nevertheless we, according to His promise, look for new heavens and a new earth, wherein dwelleth righteousness." Wherefor, beloved, seeing that ye look for such things, be diligent that ye may be found of him in peace, without spot, and blameless.
(2 Peter 3:12-14)

CHRIST RETURNS WITH BLAZING FIRE

Enoch, the seventh from Adam, also prophesied about them: "Behold, the Lord is coming with myriads of His holy ones."
(Jude 1:14)

"...This will take place when the Lord Jesus shall be revealed from heaven with His mighty angels," In **'FLAMING FIRE'** taking vengeance on them that know not God, and that obey not the gospel of our Lord Jesus Christ.
(2 Thessalonians 1:7-8)

There the Angel of the LORD appeared to him in a blazing fire from within a bush. Moses saw the bush ablaze with fire, but it was not consumed!
(Exodus 3:2)

CHAPTER 41: CLOUDSHIPS, CHARIOTS & UFOs

Mount Sinai was completely enveloped in smoke because the LORD had **DESCENDED ON IT IN FIRE**. And smoke rose like the smoke of a furnace, and the whole mountain quaked violently.
(Exodus 19:18)

As I continued to watch, thrones were set in place, and the Ancient of Days took His seat. His clothing was white as snow, and the hair of His head like pure wool. His throne was flaming with fire, and its wheels were all ablaze.
(Daniel 7:9)

For the Son of Man will come in His Father's glory with His angels, and then He will repay each one according to what he has done.
(Matthew 16:27)

Truly I tell you, some who are standing here will not taste death until they see the Son of Man coming in His kingdom."
(Matthew 16:28)

When the Son of Man comes in His glory, and all the angels with Him, He will sit on His glorious throne.
(Matthew 25:31)

Then He will say to those on His left, 'Depart from Me, you who are cursed, into the eternal fire prepared for the devil and his angels.
(Matthew 25:41)

It will be just like that on the day the Son of Man is revealed.
(Luke 17:30)

Repent, then, and turn back, so that your sins may be wiped away,
(Acts 3:19)

For our God is a consuming fire.
(Hebrews 12:29)

And by that same word, the present heavens and earth are reserved for fire, kept for the day of judgment and destruction of ungodly men.
(2 Peter 3:7)

Notes:
1. Ella LeBain, *Who is God?* Book Two in Chapter Two: *Ancient Technology and Biblical Astronauts*, pp.21-46, Skypath Books, 2015.
2. Ella LeBain, *Who Are the Angels?* Book Three in Chapter Three: *The Celestial Hierarchy of Angels*, pp.48-60, Skypath Books, 2016
3. CLOUDSHIPS – UFO CLOUDS -Federation of Light - UFO Cloud Ships?, htttps://youtu.be/jwpsTZIkEiA, September 16, 2008
4. UFO cloud mystery worldwide sightings - Secret experiments Alien spaceships weather, https://youtu.be/i8c87harZCw, January 27, 2010
5. Ella LeBain, *Who's Who in the Cosmic Zoo?* Book One – Third Edition, Skypath Books, 2013.
6. Order of Melchizedek, UFO, ALIENS & THE BIBLE, http://www.atam.org/UFO.html, January 22, 2011
7. Order Of Melchizedek, *Celestial Chariots of God: Aliens & UFO's* http://atam.org/celestial-chariots-of-god/, April 11, 2011.
8. UFO cloud mystery worldwide sightings - Secret experiments Alien spaceships weatherhttps://youtu.be/i8c87harZCw, January 27, 2010

CHAPTER THIRTY-SIX

THE CASE AGAINST FLAT EARTHERS

*"It ain't what you don't know that gets you into trouble;
It's what you know for sure that just ain't so."
~Mark Twain*

Let's start with the obvious, the Earth is not Flat! How do I know for sure? Well let's start with actual flesh and blood living, breathing witnesses. To date, approximately 600 Astronauts, (and counting) between 11 Space Agencies across the globe, have been to Space. Some of which have been to the Moon, all of whom have returned safely to Earth, and not a single one of them reported back to us that the Earth was Flat! I am a retired Paralegal/Legal Assistant, and one thing is certain, eyewitnesses are considered legal evidence in a Court of Law. In fact, the more eyewitnesses who share similar testimonies, the more credibility their evidence has with the Court, which often leads to successfully proving their case. Eyewitnesses are Experiencers. They all have a story to tell, and people need to listen and connect their stories to others who share similar experiences.

So, whenever I bring up this sticking point with Flat Earthers, they actually think they are all lying, and are part of some vast 'global' conspiracy. Astronauts who cross 11 different Space from 11 different nations of the world, are not all part of some conspiratorial deception. If the Earth were Flat, at least one of the 600 Astronauts would have spilled the beans by now, just as many have leaked other more sensitive information, such as aliens on the moon and the fact that we are not alone in this universe. Yet to date, not a single Astronaut who has returned to Earth alive, reported a flat earth, in fact, just the opposite and they returned with photographic and videographic evidence to support their eyewitness of a spherical shaped round global Earth. This is the best type of Witness in a Court of Law, those who have vids and photos to prove what they saw.

For the record, I personally met several of the Apollo Astronauts who did indeed fly to the Moon and were witness to the Lunar landings. One thing every

Astronaut says, is without a shadow of a doubt, they witnessed a round global spherical Earth from space and from the Moon.

In August of 2019, I had the privilege of working closely for 2-1/2 days with Apollo 15 Commander Astronaut Al Worden, as well as Apollo 11 Commander Astronaut Michael Collins, and Apollo 13 Astronaut-Pilot Fred Haise. These are all men of immense integrity who all share hero status in my book. If you know anything about what happened on Apollo 13, it was nothing short of a miracle from God, that they survived and returned back to Earth. When I met Fred Hayes, I discussed with him why he thought there was now this pervasive issue plaguing the internet and all discussion about Space, about Flat Earth? He told me, with grace and humility, that it's because there are people who simply don't want to believe that they achieved what they set out to do, and that the Apollo missions were successful and as a matter of fact they really did go to the Moon several times. I invite anyone reading this book, in disbelief to visit my Facebook page, and go to my Album on Spacefest and you will see my photos with these amazing American heroes.

So why are we even having this conversation? Why have flat earth cults been trolling the internet? The real truth behind this end times cult may shock you, which is why I am giving it a chapter in *The Heavens* – Book Five.

Being a paralegal, as far as I am concerned, our eyewitnesses have all spoken, unanimously, that the Earth is not flat, so normally I would just leave it there – CASE CLOSED!

But this delusion has become a dangerous religious end times Hebrew Roots cult and Knowledge Empowers – Ignorance Endangers, and that is what this book series is about, discerning truth from lies, clarity from delusion.

FLAT EARTH THEORY FABRICATED TO DISTRACT FROM PLANET X

Before we get into the debunking, I also want to add, that "Flat Earth" is a distraction from several issues, that the establishment wishes to keep secret. Nothing to see here, now run along, type attitude, so they started this End Times Flat Earth Hebrew Roots Antisemitic Cult, who troll the internet and try to disrupt just about any conversation about our current Space Age, the Space Force, Planet X, Nibiru, UFOs and get this, the Inner Earth. I am convinced from all that I have researched and am presenting here, that this is what they're covering up and trying to distract you from. Because if you believe the Earth is flat, then you will never believe there's underground bases, whether they be by human or of an alien presence.

I could not resist including herein this excellent submission from the Society of Flat Earth Debunkers. They have exposed a new and dangerous dimension to this ongoing and intensifying Psychological Operative (PSYOP) known as the Flat Earth Theory as the Ultimate PSYOP to Wage War Against The Truth Movement.[1]

The Flat Earth Theory is not a theory at all; rather, it is a meticulously engineered hoax that is zealously advanced on the Internet by a rogue government black operation known as the Flat Earth Society. Their central goals and purposes are

CHAPTER 43: THE CASE AGAINST FLAT EARTHERS

numerous and by no means mutually exclusive. However, there is one goal in particular that reigns supreme on their "agenda of disruption": To discredit every other truth movement by infiltrating them with their utter babble. Flat Earthers seek to bring down the real truthers by the mere association with their fallacious and preposterous drivel.

Flat Earthers Are Being Sent Into Every Real Truth Movement To Disrupt And Destroy Them, Especially The Planet X Groups! If ever there was a PSYOP designed to distract, divert, and misdirect, the thoroughly insane Flat Earth Theory (FET) is the one. Never in the history of black operations have the co-conspirators been so fierce and fanatical in their mission to deceive and dupe.[1]

Just what is that covert mission? To effectively blow up every internet chat room, forum and blog that has meaningful discussions about Planet X, formerly known as the 10th Planet. Planet X is the real megillah that everyone has been trying to hide for decades. Simply put, its profound ramifications are so far-reaching and consequential that TPTB cannot risk the people catching on. Hence, along comes the Flat Earth Theory to change the conversation into absolute and utter gibberish.

Anyone who knows how COINTELPRO really works will tell you that these Flat Earthers have been carefully cultivated in some serious mind-control programming experiments. Whenever they encounter any resistance to their absurd FET hallucinations, they literally go ballistic. However, they do so with great purpose and calculation. Their studied responses are quite manipulative, just as their triggered reactions are practically inhuman.

Just go to one of the big chat rooms and hang out in one that has made the Flat Earth Theory the topic of discussion. Watch the intensity on the side of the Flat Earthers, especially the way that they deceptively advance their daft arguments. It is almost as though these folks (probably droids) have been fabricated in some laboratory somewhere in Andromeda. Their thought process is not human, and the practiced sophistry that they routinely employ in downright otherworldly as in AI gone awry in an evil Grey lab in Arca 51.[1]

Flat Earthers have proven themselves to be incapable of exercising human reason. Likewise, they completely lack any common sense and ability to utilize basic logic. This is because they harbor demons. The Flat Earthers were somehow brainwashed in such a way that they can spontaneously suspend the use of their intellect. Most of them lack the capacity to even discern between a simple right and wrong answer to a question which is obviously black and white. For real, these dunces cannot even answer "Yes" or "No" when asked if the sky — RIGHT NOW — is blue or green, when it is a perfectly deep blue sky. Yes, they have completely left the reservation of rationality, and are either permanently Out To Lunch or AWOL. We are talking the complete absence of even an iota of intellectual integrity. Do you now get the picture?

If you really knew that this whole Flat Earth PSYOP was a full-blown DARPA-conceived, CIA-directed, NSA-monitored, DIA-driven black operation, would you continue to give them any energy whatsoever? Well, now you know. Perhaps some folks enjoy the journey into the realm of the extremely deranged; however, be aware that these Flat Earthers know exactly what they're doing and that you may find yourself trapped in a fictitious space-time continuum of puerile phantasmagoria and mentally challenged pseudo-entertainment. BEWARE, and be aware, of where they might be taking you.[1]

The Society of Flat Earth Debunkers (SFED)[2] perfectly understand that NASA does not tell the truth about anything, unless they must, for National Security purposes. The SFED also knows that there are many scientific laws, theories, and hypotheses, which are falling apart by the day. We put no stock in the many falsehoods propagated by the scientific establishment. However, that does not mean that everything that comes out of NASA is bunk. After all, they did know enough to launch a space shuttle every now and then and yes, they did make it to the Moon! More importantly, the Russians would have given up the International Space Station years ago if all NASA science and technology were falsely concocted. Would you send up your cosmonauts if the math were all wrong?[22]

LEARN TO DISCERN THE SPIRITS FOR GOD'S SAKE!

> It is easier to fool people, than to convince them that they have been fooled.
> ~ Mark Twain

Flat Earthers have been duped! They are under the influence of an alien implant. A Spiritual limitation device that is programmed to confuse, distort, and pervert the Truth and Reality. Spend an hour with me and the Christ in me will kick those demons out of you and disable and dissolve your implants, and you'll wake up from a dark spell and come back to your senses, you may at first be overcome by a sense of deep regret, shame and even humiliation, at first, but after you wake up you'll wonder how on god's big green global earth did you ever believe such utter Rubbish?

The Flat Earth theory, the modern form, the one that mysteriously showed up just a few years ago, is right out of an alphabet agency dark project, to see if they could actually co-opt a good number of gullible minds to distract them from what's really going on, the return of the Nemesis-Nibiru System. Flat Earthers are being mind-controlled by the very entities they claim to combat. The Flat Earth theory, the modern form, dates back to 1849, plus its rooted in antisemitic propaganda.

> Jesus said, "Beware of false prophets, who come to you in sheep's clothing, but inwardly they are ravenous wolves. <u>You will know them by their fruits.</u>"
> (Matthew 7:15-16)

CHAPTER 43: THE CASE AGAINST FLAT EARTHERS

I have my carpets cleaned roughly twice a year. In the fall of 2018, I purchased a Groupon for carpet cleaning. I come downstairs after my husband let the technician in, only to walk right into a very bizarre confrontation in my own home. He was asking my husband if I was geodesic after noticing a picture hanging on my wall of the Earth taken from Space by the Apollo Astronauts. My husband didn't exactly know what that meant? As soon as I heard that, I knew he was referring to flat earth. Well, that's when it all started. I said, no, I'm not a flat earther, which I said, was a lie. Then he started to argue their usual points, but first he was asking about my faith in Jesus. I told him yes, of course I believe in Jesus. Then he went on that the Bible says the earth is flat. I told him, no, it does not say that at all. He gave me a long spiel and he then he started spouting all kinds of classic antisemitic propaganda to me, which I found offensive in my own home. I told him, he was misled and shared that I am a Messianic Jew, and I know the Hebrew Bible, and nowhere does it say that, just the opposite. Tactfully, I explained that the Hebrew was mistranslated by the Church of Rome, who literally wrote the doctrine of Antisemitism, which has created all kinds of confusion and false beliefs, and flat earth was one of them.

I shared with him, what I'm sharing in this chapter, and he looked at me with his jaw hanging low and his eyes wide open. He knew deep down he had been lied to. He shared with me that he got saved at a Flat Earth Conference in Fort Collins, Colorado about 6 months prior. I was pretty shocked to learn there was such a thing in my own state! Anyway, I shared a lot of information that was already published in my book series, which he seemed intrigued by. So, instead of the $88 he was supposed to collect from me, in addition to the Groupon to scrub my stairs and hallway, he walked away with 4 of my books, as a trade, because at the time, I was still putting this book together, and this chapter is dedicated to him and all those like him, who have been sorely misled.

These are Christians for goodness' sake who actually believe that the Holy Spirit revealed to them that the Earth is flat. The HOLY SPIRIT is the spirit of TRUTH! IT CANNOT LIE. However, the Father of lies is Satan, the god of this world, the counterfeit spirit mimicked by Gray aliens, who he uses to collect data, memory of humans so they're easily implanted and manipulated to run their counterfeit false programming that creates strife, divisions, and disunity in the body. Remember, Satan masquerades as an angel of light! (2 Corinthians 11:14)

WAKE UP! These Christians are under a spell. They ignorantly believe that the Jewish-Hebrew Bible says the Earth is a Pizza Pie, spinning like a top hat in space. The Hebrew Scriptures certainly does NOT say THAT! My discernment of all their so-called theological arguments is they are blinded by good old-fashioned Antisemitism! Call it a generational curse or an ancestral sin, but it's a proclivity in these types who glom onto these false teachings, because somehow it justifies their malignment of Jewish people and Israel.

Between their ignorance and anti-Semitism which steals Jewish identity, as is almost always the case with these Hebrew Roots Cults, is a case of stolen valor from

Jewish scholars, who pervert and distort Hebrew words. These people are so deep in ERROR, that they think good is evil and evil is good because they believe the lie as truth. This is evidence of the fruits of a Counterfeit Spirit and Fallen Angel influence (a Gray Alien Demon) who masquerades as god, and Spirit, more precisely, a Religious Spirit.

There is an ancient curse on those who distort the truth and reality, which is written in Isaiah 5:20 that says: "Woe to those who call evil good, and good evil; who put darkness for light, and light for darkness; who put bitter for sweet, and sweet for bitter!"

I am not going to mention his name, because this is not about public shaming of one individual, but a whole cult. But there is a well-known man on Facebook who identifies as a Hebrew Roots Flat Earther. He comes across as a know it all. His fruits are antisemitic, and he is always sick, in some type of trouble, financially and legally. He can be caustic and bitter, especially towards Israel and anyone who challenges him.

However, the fruits Jesus was referring to in Matthew 7:16, were His fruits, the fruits of His Spirit. Galatians 5:22-23 says, "But the fruit of the Spirit is love, joy, peace, forbearance, kindness, goodness, faithfulness, gentleness and self-control. Against such things there is no law." This spiritual discernment can be applied to all false prophets, who think they know better than the Lord's inspired ancient scribes. Those who are truth seekers, are not antisemitic, they just want God's truth, even though it may be packaged in Hebrew within a Jewish Bible or Jewish Apocryphal writings. Truth is truth, without prejudice.

DEBUNKING FLAT EARTHERS

It still literally blows my mind that people in this day and age actually believe in a flat earth and fall for such nonsense. If you are one of them, then you have been the victim of a mind control-PSYOPS Project and are being used by the enemy to spread lies. This chapter is being dedicated to you with hopes of saving you from this implantation and deception! I encourage everyone of you to get deprogrammed and delivered ASAP!

Flat Earth Society Internet Social Media trolls are irritating and annoying, especially so-called Christians, which are bound here by a spirit of unbelief, which networks with the deaf and dumb spirit, that according to Jesus, is rooted in pride that creates blindness coupled with a dose of ignorance and a few drops of stupidity. When it comes to stupidity, it only takes one drop to spoil the tea. True Christians who are born again into the Kingdom of God, are given new hearts, they are guided by the Christ Spirit, which has a teachable spirit, it is humble, understanding, forgiving, flexible and kind, they are also given a new mind.

The opposite of pride is rooted in unbelief, or unbelief is rooted in pride, suffice it to say, spiritually speaking they network together as spiritual strongholds. Many, including myself, have befriended other believers on Facebook, only to learn they

CHAPTER 43: THE CASE AGAINST FLAT EARTHERS

are in bondage to this Flat Earth mind control, which is a mental disorder. Actually, all mind control can be considered a mental disorder. In fact, if you have read my Book Four: *Covenants*, you will know that all mind control is witchcraft.[3] Believing that the earth is flat is a very dark spell. Time to snap out of it!

So, next you need to ask yourself, as I have, why Christians? Well, if you are a believer, you should already know why, it is a form of satanic attack on the mind, to distract the believer off of what Christ wants to do or guide them into. They are temporarily derailed.

Science, and scientific thought, requires analytical thinking as well as spiritual discernment. Let's discern.

These Christians believe that the Bible teaches that the earth is flat or a flat disk, kind of like a pizza pie. Both assumptions are incorrect. There are numerous Scriptures that indicate a spherical earth. These scriptures were penned centuries before a spherical earth was confirmed.

Isaiah 40:22, Job 22:14 and Proverbs 8:27-28, all use the same Hebrew word, *khuug*, when describing the s*phere* of the earth.

Job 22:14 says that God "walks on the vault (*khuug*) of heaven," again suggesting something solid.

Proverbs 8:27-28: "He drew a circle (*khuug*) on the face of the deep...and made firm the skies above."

Isaiah 40:22 states that, "God sits above the circle of the earth." The Hebrew word used here that is translated into English as circle, is *Khuug*, which is translated as *sphere, globe, ball*, and being that *a ball is a circle*, and a *compass*, somehow the word *circle* stuck in English bibles.

A ball is a three-dimensional circle which is a sphere. Biblical Hebrew is an ancient 'functional' language that represents the practicalities of physics. The language of Hebrew is in and of itself a language based on mathematics and logic. All words come from three root letters, and those root letters build a tree, with many words that are derived from the root. The root and the tree branches illustrate its functionality and logical application.

Khuug was mistranslated into the Strong's Concordance, and misunderstood by those who carry this delusion, not only of the science but more dangerously their stubbornness to understand and accept the original Hebrew meaning which I discern as antisemitism. The Strong's Concordance is not the Bible, it is an index of Biblical words, with an attempt at translations of both Hebrew and Greek, put together by well-intentioned Gentile Christians, but were in error on the correct translations and understandings of several Hebrew words. Their misunderstandings have spawned two very dangerously false end times Hebrew Roots Cults who hang their hats on these mistranslations and misunderstandings which are not only antisemitic but antichrist. The false name 'Yahusha' is another one.[7] See *Who is God?* Book Two, Chapter 13: *The Name Above All Names*, subheading: *What's in a Name?* pp. 213-228

Unlike English, where there are exceptions to every rule, Hebrew is pretty much a logical language, made up of only twenty-two letters. Biblical Hebrew was only letters, no vowels, until 750AD when the Masoretic Jewish Scribes from Babylon decided to rewrite the Jewish Bible and add vowels. Those of you who have read Book Two: *Who Is God?* are familiar with this history, and how they deliberately transposed the vowels for the word Adonai and Elohim over the sacred name of the Father, which is known as the Tetragrammaton, thereby creating the false name Jehovah in order to confuse the Gentiles and prevent them from knowing the Father's true name who is Yahuah. [7]

The Masoretic Jews certainly succeeded in confusing the Gentiles. And so are all those who are part of this False Christian Antisemitic Hebrew Roots Cult who in this day in the Age of Knowledge, Space Travel, and Sophisticated Space Telescopes, believe the earth is flat, based on their misunderstanding of the original Hebrew and their antichrist demons of antisemitism. And God forbid you should point it out to them, they argue it thinking they have some secret esoteric knowledge, and the rest of the world is deceived by NASA that the earth is flat. They are deluded, just as the Lord Himself promises to cause delusion and delusional thinking to all who reject Truth.

> "…and with every wicked deception directed against those who are perishing **because they refused the love of the truth** that would have saved them. **For this reason, God will send them a powerful delusion so that they will believe the lie,** in order that judgment will come upon all who have disbelieved the truth and delighted in wickedness."
> (2 Thessalonians 1:10-12)

There are real conspiracies, and yes, paranoids do have real enemies! However, it is important to note, that NASA is considered a military arm of the US Military Industrialized Complex. NASA protects the national security interests of the United States of America. NASA is not the only space agency on Earth. NASA is one of eleven space agencies that are based in 11 different nations. As I mentioned before, there have been well over 600 Astronauts who have successfully been to outers space, that includes landing on the moon and have returned to Earth. Not a single one of them, have returned to tell us the earth is flat. Flat Earthers are obsessed with NASA's lies, but it is not just NASA that is lying, they think it must be a worldwide conspiracy.

NASA has been known to cover up UFOs, as I have proved in Book One of *Who's Who in The Cosmic Zoo*, and in this book on my chapters exposing the UFO Coverup. It is partly true that NASA would not let their Astronauts go public with that they witnessed with their own eyes in space and during the Lunar Landing Missions on the Moon. However, as I presented herein, multiple Astronauts became whistleblowers and told their tale of seeing spaceships and alien bases on the Moon.

CHAPTER 43: THE CASE AGAINST FLAT EARTHERS

NASA is also a public front for the Secret Space Program that has been going on behind the scenes as long as NASA has been established by German Rocket Scientists who came over after WW2. They have been working on alien technologies, which became the spoils of WW2 to the allied forces of Americans and Russians, which was the underlying cause for the Cold War, besides the cover reason of fighting to dissolve Communism. The Soviet Union disbanded, they moved away from Communism, and our new enemy is the Chinese Communist Party who too have been competing with the US in a Secret Space Program.

So, PSYOPS projects like flat earth serve the CIA deep state like an Octopus, designed to distract from multiple secret projects, and truths about the inner earth and what is really going on underground. The PSYOPS is a Machiavellian divide and conquer type program designed to keep people arguing and divided, and to kick up just enough dust so people cannot see what they're being distracted from. From the denial of the Lunar Mission, which is a big core belief of all Flat Earthers, to UFOs, to underground bases, inner Earth civilization, the SSP, which are all happening under our very noses.

Flat Earthers deny that we went to the Moon. They say nobody can get out of Earth's atmosphere, because they think there is a big dome covering us, and that our Earth is a flat disk being held up by a turtle. They actually believe that nobody gets out of here, that it's impossible to get into outer space, let alone travel to another planet in our solar system, because like the *Truman Show*, they believe we live inside a dome, like a snow globe. When you explain to them it is called the ionosphere, magnetosphere, all different layers of the Earth's atmosphere and ozone, is not solid but gaseous to absorb radiation from space and sun, to protect the Earth.

We know from multiple whistleblowers, that NASA has been known to photoshop photos of the Moon released to the public, in order to cover up the fact that the Moon is inhabited by alien colonies, giant structures, spires, and underground bases. NASA is also known to photoshop photos of Mars which essentially reveals more of the same, underground facilities, ancient ruins, and yes, aliens. However, NASA has never covered up the fact that the earth is a sphere, just like all the other planets are spheres.

Flat Earthers actually believe that NASA is covering this truth up by photoshopping photos from space of the spherical global three-dimensional round earth. And just to put the cherry on the cake, they believe that NASA are deceivers because they think the Hebrew word that means 'to deceive' just happens to be *nasa* pronounced, nuh-sa. Actually, the Hebrew word, is nuh-sha. The Shin is a 'sh' sound, not an 's' sound. But again, these Hebrew Roots people are lacking understanding of the specifics of the Hebrew language. The verb נָשָׂא (/nasa/) actually means lift or carry (or marry). However, the archaic verb נָשָׁא (/nasha/) does mean to deceive, though it is not attested in that form in the Bible, only in derived forms such as נִשָּׁא /nisha/ and הִשִּׁיא /hishi/. A little knowledge is a dangerous thing.

Linguistically speaking, this train of logic is as credible as saying that the English word, *fuck*, is no longer considered offensive or foul because it means 'melons' in

457

the Thai language. In Thailand, they make *fuk soup*, which is melon soup. They also use the word, *fuku* meaning good luck. Truth be told, I once ate in a Thai Restaurant in Cape Canaveral called, *Thai Fuku*, when we asked them what *Fuku* meant, they told us, good fortune. The food and service were excellent, so I suppose that was our good fortune.

Ok, you get where I am going with this, just because one word means one thing in one language, does not mean that same meaning gets projected onto the same sounding word in completely different language. The facts are, that NASA is an acronym for National Aeronautics and Space Administration. The original English word, *fuck* began as an acronym decreed by King Henry the Eighth, which stood for, Fornication Under the Consent of the King. This was because under the Catholic Church it was illegal to divorce, therefore, they were forced to go to King Henry VIII to receive a decree to leave their wives in order to be cleared to have sexual relations with another.

Words have meaning, but it is all in how you use them, and appropriate for each language. However, when you start to conflate one word in one language with a similar sounding word in another language, you are back to the Tower of Babel's confusion, deception, and misunderstanding. Language is used to communicate thoughts, ideas and tell stories. There are many words that cross over from the Latin languages into the English language, this is due to our history. However, Hebrew is a specific pictorial language of practicality and functionality. In fact, for those of my readers who read Book One in the *Who's Who in The Cosmic Zoo?* series, you will remember that the twenty-two letters of the Hebrew language all represent pictures of constellations.[5] Hebrew is truly a cosmic script, a space language, the language of the Angels in Heaven, otherwise, why would the Lord have taken Enoch up multiple times, to teach him the Hebrew language, (ancient Aramaic-Hebrew) in order to pen his books? He did not teach Enoch Spanish, French, or Latin? It was Hebrew. So, while I encourage all students of the Bible to learn Hebrew for greater understanding and clarification, please do not butcher the language due to ignorance and misunderstanding.

Now, back to debunking Flat Earthers, NASA is not the only space agency sending pictures of the heavens back to earth. Yes, NASA has been covering up their knowledge of UFOs, but the big Truth Embargo that they are holding on to, is the exact location, velocity and speed of Planet X aka Nibiru and the rest of the Nemesis exoplanets, and most importantly, *when* Nibiru is scheduled to pass the earth, not once but twice. But they are most certainly not covering up the fact that the earth is a sphere just like all the other planets in our solar system.

This PSYOPS mind control implant of Flat Earth is designed to keep people distracted from researching what's really going on with Planet Earth, and that is the Pole Shift due to the presence of multiple exoplanets and the approaching Planet of Crossing, which is said to be four-five times the size of Earth. That is big news, especially for the 7+ billion surface dwellers. So, the idea behind this PSYOPS is to keep people busy arguing over whether the earth is flat or round, which is as regressive as medieval times aka dark ages when they believed in such obliviousness.

Unfortunately, cults are built based on ignorance, brainwashing through lies, and alien implantations. In the case of Flat Earthers, they're victims of implantation from MK Ultra Mind Control Projects, which was also done to use ignorant believers to cover up the fact that the earth is not only a sphere, but is more accurately shaped like a Torus, with a bigger bulge around the equator, due to the oceans and the placement of the North and South poles with two huge holes in each of its poles, causing it to be shaped more like a squashed sphere, which is one of the its unique characteristics that causes it to wobble. As we go through pole shift, this will shift, and coastal areas will inevitably get flooded, as the new magnetic north and new magnetic south poles will have shifted on the spherical globe. They essentially want to keep as many people as possible in the dark about what's going on, so they blame humans for global warming, guilt tripping surface humans into the false belief that the earth is warming and it's all our fault, to sell curly lightbulbs that takes a Hazmat Suit to dispose of when they break. [6]

The other truth being obscured by MK Ultra to these unsuspecting ignorant Christians, is that if they got them to believe the lie that the earth was flat, then they would never believe the truth that the earth is hollowed out like a honeycomb, which is where all the inner earth alien and ancient human beings live in vast caverns *inside* the earth. There are space bases *inside* the earth, that they know about and are protecting.[7] So, throwing out the implant that the earth is flat, and the Bible says so, is nothing other than a bad spell, on undiscerning, ignorant, and anti-Semitic Christians.

Most Flat Earthers are focused on NASA's deceptions and cover-ups, yet totally oblivious to the fact that nine nations of the earth have all successfully travelled into space, gone to the Moon, Mars and frequently travel back and forth to the International Space Station. (ISS). Flat Earthers believe that NASA is covering up the flat earth in some grand conspiracy. And, like I always say to Flat Earthers, what about the Astronauts? Multiple astronauts from America, Russia, China, Japan, the UK, and Germany have all witness the Earth as a globe from space. They actually believe that ALL of them are part of a vast conspiracy by NASA to cover up the flat earth. This is beyond insanity!

Why haven't any of the international Astronauts from Russia, China, Japan, France, England, Israel, Iran, and North Korea, aren't coming clean with the flat earth? Are they all in on the conspiracy as well? In the nine nations that possess the technology to orbit around the Earth, some of whom are actual real enemies on the Earth, why wouldn't they out the others if the Americans were covering up a flat earth?

The truth is, that this flat earth heresy is nothing but mental disorders, caused by a bad spell of mind control, brainwashing and the implantation of lies to create division and confusion. Sadly, this disease seems to affect only Christians, who are all in bondage to religious spirit demons otherwise known as Gray aliens.

Another misunderstanding is the word, *firmament*, and the language translated from the original Hebrew, that the Lord secured the earth in its place in space. They think this means it cannot be moved. What the scriptures are actually saying, is its orbit can't be moved, meaning another extraterrestrial group with the technology to tractor beam it out of its orbit and relocate it into another solar system or galaxy, can't be done, because the Lord of Heaven and Earth has secured the orbit of the Earth. However, let's not forget, the same Lord promises to recreate the Heavens and Earth after this Age is over, which I go into great detail why, in my chapter on *The Signs in the Heavens - All Things Nibiru*, because every 3,600 years this rogue planet intersects in its elliptical corkscrew counter-clockwise orbit with our solar system and always seems to wreak havoc, creating earth changes and even crashing into other planets in the ancient past.

Because God is omnipresent, He looks down upon the earth from every direction. Therefore, from God's heavenly perspective, looking down upon the earth from every location, the earth would appear round from every perspective ONLY if it were a sphere. If the earth were a flat circle for instance, then from most angles the earth would appear as an oval or even a straight-line, if perpendicular to it. Therefore, Isaiah 40:22 indicates a spherical earth.

Furthermore, Jesus said, "For as Jonah was three days and three nights in the belly of the great fish, so will the Son of Man be three days and three nights in the heart (belly) of the earth." (Matthew 12:40) Of course a person cannot be in the heart or belly of a flat circle any more than he can be in the belly of a flat fish. And Ephesian 4:9 corroborates when it explains that Jesus "first descended into the lower parts of the earth."

Luke 17:34-36 also implies a spherical earth. Jesus said that at His return some would be asleep at night while others would be working at daytime activities in the field. This is a clear indication of a revolving earth, with day and night occurring simultaneously.

In addition, Job 26:7 explains that the earth is suspended in space, the obvious comparison being with the spherical Sun and moon. See also Job 26:10 and Proverbs 8:27. The Bible is the Creator's Word, so it is not surprising that recent scientific discoveries are confirming what the Creator has said millennia. "For the word of the LORD is right, and all His work is done in truth." (Psalm 33:4)

Here is what most Gentiles do not understand about the Hebrew language: Hebrew words describes functions, not labels. The three root letters that create all Hebrew words, are based on science, physics, functionality. Unlike English, which has labels for everything, Hebrew words represent practical function. The Hebrew word *Khuug*, used in Isaiah 40, is more about function, than labels. Today, the same word is used to describe ball and sphere, so yes, the Hebrew scriptures does indeed describe a spherical, if you understand the Hebrew language. Looking at it superficially, and depending on mistranslations, mistransliterations, creates misunderstandings, which is the case with the flat earth heresy that uses the Bible to validate its false scientific claims.

The flat earth delusion is no different than those who suffer from atheism, or occultism, or confusion, all of which are under all kinds of witchcraft, which is mind control.

Flat earthers contribute absolutely nothing to today's discussion, and oddly enough, they are all Christians. It is a false cult. All cults are ruled by Religious Spirits which are alien-demonic forces that war against the Spirit of Christ, which is the Spirit of Truth, Knowledge, Wisdom and Love, also known as the Holy Spirit. The Holy Spirit never taught anyone that the Earth was flat. Mind controllers do that. Christians should be ATTRACTING people to the Gospel of Salvation, not turning people off because of Funda-MENTALISM, which is a type of mental illness. To not have a teachable spirit, is not the fruit of Christ. To be so stuck on what you think you know, that just ain't so (as Mark Twain aptly said) is what gets you into trouble. It is a type of spiritual and mental blindness, that no amount of arguing will change, but only God can deliver from all illness. (Isaiah 53:5)

For the record, the reason the earth is not a perfectly round sphere, but a little squashed like the shape of a Torus, with a bigger bulge around the equator, is due to the oceans and the placement of the North and South poles. As we go through pole shift, this all will shift and coastal areas will inevitably get flooded, as the new north and new south will have shifted on the spherical globe.

It may seem round when viewed from space, but our planet is actually a bumpy spheroid.

I am attaching something from Sir Isaac Newton on the actual shape of the Earth, which the ancients knew well, going all the way back to Pythagoras, Aristotle, and the Original Hebrew Scriptures. The Books of Enoch do not teach flat earth, but that Earth is a globe.

According to Diogenes Laertius, "Pythagoras was the first Greek who called the Earth round; though Theophrastus attributes this to Parmenides, and Zeno to Hesiod." Early Greek philosophers alluded to a spherical Earth, though with some ambiguity.

Isaac Newton first proposed that Earth was not perfectly round. Instead, he suggested it was an oblate spheroid—a sphere that is squashed at its poles and swollen at the equator. He was correct and, because of this bulge, the distance from Earth's center to sea level is roughly 21 kilometers (13 miles) greater at the equator than at the poles.[6]

It has been known that the Earth was round since the time of the ancient Greeks. I believe that it was Pythagoras who first proposed that the Earth was round sometime around 500 B.C. He based his idea on the fact that he showed the Moon must be round by observing the shape of the terminator (the line between the part of the Moon in light and the part of the Moon in the dark) as it moved through its orbital cycle. Pythagoras reasoned that if the Moon was round, then the Earth must be round as well. During the time between 500 B.C. and 430 B.C., Anaxagoras determined the true cause of solar and lunar eclipses – projecting the shape of the Earth's shadow onto the Moon during a lunar eclipse was also used as evidence that the Earth was round.

As countless photos from space prove, Earth is a round global "Blue Marble," as astronauts have affectionately dubbed it. Appearances, however, can be deceiving. Planet Earth in fact, is not perfectly round, this is not to say Earth is flat. Well before Columbus sailed the ocean blue, Aristotle and other ancient Greek scholars proposed that Earth was round. This was based on a number of observations, such as the fact that departing ships not only appeared smaller as they sailed away but also seemed to sink into the horizon.

Isaac Newton first proposed that Earth was not perfectly round. Instead, he suggested it was an oblate spheroid—a sphere that is squashed at its poles and swollen at the equator. He was correct and, because of this bulge, the distance from Earth's center to sea level is roughly 21 kilometers (13 miles) greater at the equator than at the poles.

Instead of Earth being like a spinning top made of steel, explains geologist Vic Baker at the University of Arizona in Tucson it has "a bit of plasticity that allows the shape to deform very slightly. The effect would be similar to spinning a bit of Silly Putty, though Earth's plasticity is much, much less than that of the silicone plastic clay so familiar to children."

Our globe, however, is not even a perfect oblate spheroid, because mass is distributed unevenly within the planet. The greater a concentration of mass is, the stronger its gravitational pull, "creating bumps around the globe," says geologist Joe Meert at the University of Florida in Gainesville.

Earth's shape also changes over time due to a menagerie of other dynamic factors. Mass shifts around inside the planet, altering those gravitational anomalies. Mountains and valleys emerge and disappear due to plate tectonics. Occasionally meteors crater the surface. And the gravitational pull of the moon and Sun not only cause ocean and atmospheric tides but earth tides as well.

In addition, the changing weight of the oceans and atmosphere can cause deformations of the crust "on the order of a centimeter or so," notes geophysicist Richard Gross at the Jet Propulsion Laboratory in Pasadena, Calif. "There's also postglacial rebound, with the crust and mantle that were depressed by the huge ice sheets that sat on the surface during the last ice age now rebounding upward on the order of a centimeter a year."

Moreover, to even out Earth's imbalanced distribution of mass and stabilize its spin, "the entire surface of the Earth will rotate and try to redistribute mass along the equator, a process called true polar wander," Meert says. [6]

To keep track of Earth's shape, scientists now position thousands of Global Positioning System receivers on the ground that can detect changes in their elevation of a few millimeters, Gross says. Another method, dubbed satellite laser ranging, fires visible-wavelength lasers from a few dozen ground stations at satellites. Any changes detected in their orbits correspond to gravitational anomalies and thus mass distributions inside the planet. Still another technique, very long baseline interferometry, has radio telescopes on the ground listen to extragalactic radio waves to detect changes in the positions of the ground stations. It may not take much technology to understand that Earth is not perfectly round, but it takes quite a bit of effort and equipment to determine its true shape. [6]

When the student is ready, the teacher is there. One must have a teachable spirit, which comes from the Holy spirit to be able to understand Biblical truths, whether you understand the original languages or not. It is not about knowing Hebrew or Greek; it is about knowing the Spirit of the Lord.

This is why there are millions of people around the world who are saved through the agency of the Holy Spirit alone, who do not have any knowledge of the Bible, which is just as well, because as they say, a little knowledge is a dangerous thing, meaning it leads to arrogance, as displayed by this so-called Christian's who displayed absolutely no fruits of Christ. By their fruits, you will know them.

> "Beware of false prophets, who come to you in sheep's clothing, but inwardly they are ravenous wolves. You will know them by their fruits. Do men gather grapes from thorn bushes or figs from thistles? Even so, every good tree bears good fruit, but a bad tree bears bad fruit. A good tree cannot bear bad fruit, nor can a bad tree bear good fruit. Every tree that does not bear good fruit is cut down and thrown into the fire. Therefore, by their fruits you will know them.
> (Matthew 7:15-20)

My books expose mistranslations, mistransliterations, misinterpretations from Hebrew-English, which all lead to misunderstandings of what the original Hebrew scribes wrote down. I also include many of the Great Rejected Texts that were rejected, deleted, and edited by the Church of Rome along with its history, politics, and linguistics. I graduated, matriculated from Sde Boker with Biblical Hebrew, I do not know everything, but I do know that so much of today's Christians are deeply and sadly deceived and in bondage to many lies. Flat earth is one of them.

I include in my books, proof that Enoch, Elijah, and Ezekiel were all 'frequent flyers', meaning they were all taken up from the earth, ascended into the heavens. The Hebrew word used for heavens, which is *Shamayim*, is the same word used for sky. Today's Modern Hebrew came from Biblical Hebrew.

I agree with Erich Von Daniken's Ancient Astronaut Theory and have been since I was thirteen years old. My book series proves his theory correct through Hebrew Linguistics that the Scriptures do indeed point to people ascending from the earth and being returned, through space craft. All describe seeing the sphere and circle of the earth from space. I literally spend chapters on this, too much to repost here. See, Book Two; *Who Is God?* which is packed with all these corroborating scriptures. See also, Book Three: *Who Are the Angels?* where I expose and reveal the

Ten Raptures (Ascensions) listed in Scriptures and the Three End Times Raptures (Ascensions/Abductions) from the earth. [7,8]

WHAT IS THE FIRMAMENT?

The word "firmament" comes from the Latin *firmamentum*, meaning "sky" or "expanse" or "heavens" is mentioned 17 times in the King James Version of the Bible and refers to the expanse of the heavens above the earth.

Nine of the occurrences of firmament are in the first chapter of the Bible as part of the creation account. Genesis 1:6-8 says, "And God said, Let there be a *firmament* in the midst of the waters, and let it divide the waters from the waters. And God made the *firmament* and divided the waters which were under the firmament from the waters which were above the *firmament*: and it was so. And God called the firmament Heaven. And the evening and the morning were the second day."

The "firmament" is called "heaven," i.e., it is what people see when they stand outside and look up. It is the space which includes the earth's atmosphere and the celestial realm. It can also be called, "outer space," which is above Earth's atmosphere. In the *firmament*, we see the Sun, moon, and stars; and in modern Bible translations the word, *firmament* is often called the "expanse" or the "sky."

Genesis says that the firmament "separated the water under the expanse from the water above it" (Genesis 1:7). Originally, God created the earth with water "under" the sky (terrestrial and subterranean water) and water "above" the sky—possibly a "water canopy" which enwrapped the earth in a protective layer. Or the waters above the firmament could simply be a reference to clouds.

During the first floods, known as Lucifer's Floods, also referred to as the Floods that Sunk the Ancient Civilization of Atlantis, there were waters that flooded the Earth from both above and below the Earth. [9]

Firmament is used again in Psalm 19:1: "The heavens declare the glory of God; and the firmament shows his handywork." And again, in Psalm 150:1, "Praise ye the LORD. . ... Praise him in the *firmament* of his power."

Firmament is used again in five times in the Book of Ezekiel and once in the Book of Daniel. In Ezekiel, each occurrence takes place within a vision. For example,

> "Then I looked, and behold, in the *firmament* that was above the head of the cherubim there appeared over them as it were a sapphire stone, as the appearance of the likeness of a throne."
> (Ezekiel 10:1)

> "And they that be wise shall shine as the brightness of the firmament; and they that turn many to righteousness as the stars for ever and ever."
> (Daniel 12:3)

CHAPTER 43: THE CASE AGAINST FLAT EARTHERS

British Dictionary definitions for firmament:
1. Firmament /ˈfɜːməmənt/ noun1. the expanse of the sky; heavens
2. fixed above the earth
3. to spread out, but in Syriac meaning to make firm or solid,
4. Separate

The word first appeared in the King James Bible and is found in its offshoot translations that copy King James, only without the thee's and thou's. Because of its Latin root, it was translated from the original Hebrew word, which is *raqia*, meaning to be stretch out or expand. Yet, due to a combination of Antisemitism, out-right Jewish persecution and oppression of all things Jewish, which began through Constantine's Creed, the original doctrine of Antisemitism and Replacement Theology, these mistranslations and misunderstandings can all be traced back to Babylon, where all the false religions are born. Let's discern:

> "And God said, "Let there be an expanse [raqia] between the waters to separate water from water." So, God made the expanse and separated the water under the expanse from the water above it. And it was so. God called the expanse 'sky.' [shemayim] And there was evening, and there was morning — the second day."
> (Genesis 1:6-8, NIV)

> "Let there be a firmament in the midst of the waters, and **let it divide the waters from the waters.**" And God made the firmament and divided the waters which were *under* the firmament from the waters which were *above* the firmament; and it was so. And God called the firmament sky heaven."
> (Genesis 1:6-8)

In verse 8, the same word for sky *shamayim* is also the same word for heavens. It is clear from the splitting of the waters that is under the firmament land of earth, and the other waters are above the firmament land, and above that is the sky or the heavens. In this verse, the word heavens refer to outer space, or the skies, not the state of heaven, which is a spiritual state and spiritual dimension.

So, here is how lies fester, through mistranslations and misunderstandings. The word 'firmament' in Genesis 1:6 in the *Logos Bible Software*, is used to justify this flat earth nonsense. These flat earth Hebrew roots Cult people claim this is from Hebrew, yet some Jewish Rabbis would have been all over this 'great' disclosure a long time ago, because they want to be regarded as the brightest and most clever in the universe. The work, *raqia* does not in any way suggest flatness, in fact the word means, *expanse*. The word expanse means it is stretched in its vastness, breadth and its spread over a large area or region. The land expands and stretches all around the Earth to separate the waters under the earth and on top of the earth.

The term *raqia*, here is translated as "expanse," implying something that has been spread out or stretched out; it is a related to the verb *raqa*, meaning, "to spread out or stretch out." It is important to point out, that there is no 'material' substance inherent in the term *raqia*, it is an action verb, that is immaterial. Therefore, what is being spread out must be discerned from its context. The context of *raqia* in the Genesis narrative does not imply any sort of solid structure. In fact, it is describing the *shemayim*, which is the heavens, or the sky. Genesis 1:8 states that God called the *raqia* – shemayim, thereby equating the *raqia* with the "sky" or "the heavens." The term raqia of the shemayim, or "expanse of the sky" or "expanse of the heavens," occurs four times in the creation narrative: Genesis 1:14-15,17, 20. Birds are said to fly "in the open expanse of the sky." (Genesis 1:20)

Obviously, the sky is not a solid structure. *Raqia* (firmament) is just the sky, then, how on Earth did anyone ever get the idea that the *raqia* was a solid structure? Such as a vault, a dome, or an inverted metal bowl?

Here again, we have another classic example of words being lost in translation, misunderstood, and even manipulated to fit into some human idea or agenda. I have asserted and proved in *Who is God?* Book Two of this book series, that English Bibles are mistranslated from multiple Hebrew words and Names of God,[7] and this word, *raqia* is most certainly one of them. Due to the fact, that most English-speakers have been influenced by the King James Version's translation of *raqia*, to mean "firmament," is why they mistakenly think it suggests the idea of something firm and solid, when the original Hebrew actually conveys just the opposite, which is the sky, the heavens aka outer space, that is intangible, and most will agree is a non-solid and may even compare outer space to moving through oceans. [9]

In the beginning all was void, and all became water. "Spread out the earth above the waters," means there where land rising out of the waters, like how volcanic islands is formed even today (nothing new under the Sun), whole continents formed the same way.

> "It is He who sits above the circle of the earth, and its inhabitants are like grasshoppers, who stretches out the heavens like a curtain and spreads them out like a tent to dwell in."
>
> (Isaiah 40:22)

The LORD God sits OUTSIDE this round ('circumference [circle] of the earth') expanding universe of ours. The 'water above the waters' in Genesis? Easy:

> "He binds up the WATERS in his thick clouds; and the cloud is not." (Job 26:8)

CHAPTER 43: THE CASE AGAINST FLAT EARTHERS

"Praise him, you heavens of heavens, and you waters [RAIN] that be above the heavens." (Psalm 148:4)

"If the clouds be full of rain, they empty themselves on the earth:" (Ecclesiastes 11:3)

Josephus calls it 'Crystallinity' (NOT firmament). This is the atmosphere. Crystallinity refers to the degree of structural order in a solid. In a crystal, the atoms or molecules are arranged in a regular, periodic manner. [10]

"He has fixed the earth firm, immovable." (1 Chronicles 16:30)

"Thou hast fixed the earth immovable and firm …" (Psalm 93:1)

"He has fixed the earth firm, immovable." (1 Chronicles 16:30)

"Thou hast fixed the earth immovable and firm ..." (Psalm 93:1)

"He has fixed the earth firm, immovable ..." (Psalm 96:10)

"Thou didst fix the earth on its foundation so that it never can be shaken." (Psalm 104:5)

"...who made the earth and fashioned it, and himself fixed it fast..." (Isaiah 45:18)

The fact that the Earth is fixed in the position, means just that. That it is in a fixed position relative to other objects. It is not confirming the Earth is flat but that its position in the solar system is fixed, it's orbital path around the Sun is fixed, it turns on its axis daily, its position in all these activities is fixed and immovable, meaning, it can-not just fly away any time soon, or be tractor beamed into some other solar system.

We know outer space is made up of particles that contain water vapors, perhaps this was from early creation remnants, but suffice it to say, space is fluid, not solid. And yes, it is rocket science, so I will leave it here for those who really need to get into the physics of outer space, who are rocket scientists and astronauts.

Interestingly, the origins of the word "firmament" goes way back to around 300 BC, and the mistranslation took root and like the game of telephone, the wrong message was passed on and remained uncorrected. The Septuagint was the Greek translation of the Hebrew Scriptures, that was published around 250 BC that was a collaborative effort by 70 Jewish scholars in Alexandria, Egypt, at the order of Ptolemy Philadelphus, who was the Hellenistic ruler of Egypt. He wanted the Septuagint to be included in the famous vast library of Alexandria. It is apparent that these Greek translators were influenced by then-popular cosmological notions that included the idea that the sky was a stone vault. They translated *raqia* into Greek as *stereoma*, which implies a "solid structure."

Then over 600 years later, in around 350AD when Jerome was translating the Hebrew Scriptures for the first time into the Latin Language Bible known as the Latin Vulgate Bible, he was influenced by the Septuagint, and translated *raqia* into the Latin word *firmamentum*, meaning a strong or steadfast support. Finally, some 1200 years later, when English scholars were translating the Scriptures into what would become the most influential English Bible—the King James Version---Jerome's Latin term *firmamentum* was simply transliterated into English as "firmament." Its original cosmic meaning was completely lost in translation.

While the English word, firmament originates from the Latin firmamentum signifying firmness or strengthening, the Hebrew word, raqia, has no such meaning, but conveyed the "expanse," of that which was stretched out. Of course, the sky was not regarded as a hard vault in which the heavenly orbs i.e., planets were fixed upon, like you would stick Styrofoam balls to Velcro. This is ridiculous idea has no basis in reality, science or astrophysics, and because Biblical Hebrew was always and originally a language of physics, with great specificity in participles, and how each word is a root that branches into many words, as well as representing numerical values, indicating mathematics, which is the cousin of physics, therefore there is absolutely nothing in the original language that suggests that the Jewish writers were influenced or even misled by the ideas false theologies of heathen pagan nations. [11]

A key point in history during the mid-19th century, Sir Austen Henry Layard found a mound near Mosul, Iraq, which turned out to be the site of biblical Nineveh, a treasure trove of clay tablets with cuneiform inscriptions. Layard stumbled onto the ruins of a royal library amassed by the ancient Assyrian king Ashurbanipal. On some of these tablets were found the Babylonian creation story known as the *Enuma Elish*, believed to have been written around 1,100 BC in Babylon. However, around 1890, German Assyriologist Peter Jensen translated the Babylonian word appearing on tablet IV, line 145, as Himmelswölbung ("heavenly vault"). At about this same time, a school of German critics of Scripture began promoting a theory known as "pan-Babylonianism," which held that most of the Old Testament was written during the Babylonian captivity, and the Jewish writers of Scripture were heavily influenced by Babylonian cosmology. The idea that the Babylonians believed in a vault of heaven, combined with the idea that the Bible writers were influenced by Babylonian cosmology, led to the idea that *raqia* meant a solid vault amongst Gentiles. Soon, Hebrew lexicons and Bible commentaries began to reflect this idea that the raqia was a solid vault or dome, likely composed of metal. [9]

Pan-Babylonianism is a false theology based on misunderstanding of both Jewish identity and the original Hebrew language. Are we are expected to believe that Hebrew scribes, who were considered the 'guardians' of the sacred Hebrew Scriptures that was the foundation of Jewish-Hebrew identity, would willingly adopt the Babylonian (Pagan) worldview? Despite the fact that Babylon was an arch enemy of Ancient Israel – aka the Jewish-Hebrew nation. Remember his was after the Babylonians conquered the Hebrews, and destroyed the most sacred and treasured Jewish building, Solomon's Temple, did the Jews who wrote the Bible, and especially the Masoretic Babylonian Jews, were the first to pen the Hebrew Scriptures with vowels, during Babylonian captivity, is why they deliberately concealed the sacred name of Yahuah, by using the vowels of Adonai and Elohim, so the Gentiles wouldn't invoke the sacred of name or use it in vain. Again, this notion, is rooted in antisemitism, and is entertained by those with little to no understanding of Jewish Roots, despite having some knowledge of Hebrew Roots. Anyone can learn a language, but not everyone can understand or appreciate what holds Jews together, which has come to be shared persecution over the centuries by enemy gentiles.

The original meaning is rooted in the understanding of astrophysics, which was suppressed and misunderstood on Earth particularly by the Pan-Babylonians, who established this misunderstanding thereby believing and perpetuating its lie, that has followed us all the way into present times, laying the foundation for the End Times Antisemitic Hebrew Roots Cult of Flat Earth.

Furthermore, conservative Christians and Jews have long held the belief it was Moses who wrote the Book of Genesis around 1,500 BC which was before the Enuma Elish was written. This is why the Torah is also called, 'The Five Books of Moses' beginning with Genesis. Therefore, if any ancient pagan cosmology could be expected to be reflected in the Genesis narrative, it would be that of ancient Egypt, where Moses was raised and educated, not that of the Babylonian civilization, which came into play much later on the timeline.

There is little support among scholars that raqia signifies a solid vault or dome, despite the errors of Pan-Babylonianism. In 1975, Assyriologist W. G. Lambert re-examined this issue, he found there was no evidence for the idea that the Babylonians conceived the sky as a solid vault. The only "evidence" was Jensen's apparently unjustified mistranslation of the term in Enuma Elish as "heavenly vault." Lambert thought there was some support for the notion that the ancient Babylonians viewed the cosmos as a series of flat, superimposed layers of the same size separated by space, held together by a rope. But there was no dome or vault in Babylonian cosmology. [11] It is obviously an error that was misunderstood, due to mistranslations and the false imaginings of ignorant humans.

However, the larger and more fundamental problem with the *raqia*-as-solid-dome theory is that it assumes that the Bible reflects only the human wisdom and understanding of its human writers. Whatever the cosmology was of ancient Israel, that cosmology must be reflected in the Hebrew Scriptures. This idea ignores the Bible's own claim that all Scripture is inspired by God (literally "God-breathed") (2 Timothy 3:16). If "holy men of old spoke as they were moved by the Holy Spirit" (2 Peter 1:21), then Scripture will reflect more than human wisdom. We err if we assume that the Hebrew Scriptures reflects only ancient Near Eastern cosmology. The Hebrew language itself is extraterrestrial, reflecting a language of physics, mathematics which follows a logical pattern, sequence and meaning.

Just like the Roman Catholic Church which went to great lengths to conceal the truth about ancient aliens in the Bible Scriptures, by rejecting dozens of ancient Jewish Scriptures because of their adherence to Constantine's Creed, this knowledge was lost to the masses, coupled with multiple gross mistranslations that have created misunderstandings, false beliefs that established false theologies, that appears to be foundational to today's End Time's Antisemitic Hebrew Roots Flat Earth Cult, which is riddled with all kinds of errors, both scripturally and scientifically, making it delusional as its not rooted in reality but the false imaginings and misunderstandings of humans.

The attitude one brings to the Jewish Scriptures ultimately determines where we stand on this issue. If you believe the Torah reflects only human ideas, then you can argue that *raqia* indicates a solid vault or dome, despite it having no basis in the original language. However, if you believe that the original Hebrew Scriptures were inspired by the Lord and God of Heaven and Earth, revealing more than mere human wisdom, then the *raqia* of Genesis is simply the "expanse of the sky."

As a student of science raised in the 1960s and 1970s, I was taught to apply *Occam's Razor*, which states, that the simplest solution is usually the correct one. In this case, *raqia* means the expansion on the heavens aka outer space aka the endless sky. End of story!

THE END TIMES ANTISEMITIC FLAT EARTH CULT

So, where does the Flat Earth Hebrew Roots End Times Cult even get this ridiculous notion that the Bible suggests a flat earth? Again, it comes down to their misunderstanding, mistranslation of Hebrew and Jewish Roots, caused by their blindness of their proclivity towards antisemitism.

The first line of Isaiah 40:22 reads, "It is he [i.e., God] who sits above the circle of the earth." Some have argued from this that Scripture teaches the earth to be a flat disc, rather than a globe. However, even if the original Hebrew is correctly understood to refer to a circle, this does not necessarily indicate something flat; a sphere appears as a circle when seen from above—and indeed from whatever direction it is viewed. Moreover, there is good reason to believe that the word translated 'circle' might be better translated 'sphere.' [12]

But what most Hebrew Roots Flat Earth cults lack is understanding of Biblical Hebrew and how it is used as poetry. While inserting the word, "sphere' in place of circle in verse 22, it would not have the same poetic value. The reason he wrote that was because of how it looked from his perspective, suggesting that Isaiah, had to have been a frequent flyer, like Elijah and Ezekiel, and was taken up in the cloudships, and witnessed the Earth from the space above the Earth, which depending on how far out he was, could have easily looked like a circle. Thereby describing the Lord who sits above it all, the circle of the sphere of the planet earth. It is absolutely ludicrous to suggest or even think, that all the other planets are spheres, including the moon and our Sun, but earth sits like a flat pancake in its order of our solar system.

The Hebrew word in question is khûg (חוג) which is also found in Proverbs 8:27 where, in many Bible versions, it is translated 'vault'. For example, the New American Standard Bible reads, "Clouds are a hiding place for Him, so that He cannot see; and He walks on the vault of heaven." Clearly 'vault' carries the sense of something three-dimensional and is given as the primary meaning of khûg in the well-known Brown-Driver-Briggs Hebrew and English Lexicon.[4] In modern Hebrew, a sphere is denoted by khûg, along with kaddur, galgal, and mazzal. In Arabic (another Semitic language), kura means ball and is the word used in the Van Dyck-Boustani Arabic Bible (1865) to translate khûg in Isaiah 40:22.

A case can also be made from modern European terms denoting sphericity. Philologists have discovered a number of Indo-European words that appear to be related to Semitic words, whether of shared origin or having been borrowed in the distant past.[3] While there is no specific evidence confirming a link in the case of the Hebrew word khûg, it may be significant that, in Indo-European languages, there are similar-sounding words that definitely refer to a spherical object, examples being kugel (Middle High German), kula (Polish), kugla (Serbo-Croatian) and gugā (their Proto-Indo-European root).[22]

Hebrew-Latin *polygot* Bible edited by Benedictus Arias Montanus and first printed in 1528. This uses the Latin word *globus* to translate the Hebrew word *khûg* in Isaiah 40:22. Various sixteenth century Latin Bibles indicate that medieval scholars understood *khûg* in Isaiah 40:22 to refer to the *sphericity* of the earth. For example, Santes Pagnino translated this sphaera, and Benedictus Arias Montanus and François Vatable *globus*. The seventeenth century Giovanni Diodati Bible also used globus and the eighteenth-century Dutch Hebraist Campeius Vitringa used orbis.[7] More recently, the Spanish Jerusalem Bible used 'orb' and the Italian Riveduta Bible '*globo*'. [13]

Conclusion

While most modern Bible versions translate khûg as 'circle', a good case can be made that 'sphere' was the sense intended by the original Hebrew. Historically, scholars have often taken this view, preferring the Latin words *sphaera, globus* and *orbis*. The recent preference for 'circle' may have arisen from the belief that people living in Isaiah's time were too primitive to realize the true nature of the earth. This would seem unlikely, however, as Job 26:7, probably written several centuries before, states that God "hangs the earth on nothing," indicating that the ancient Hebrews had quite a sophisticated understanding of cosmology.

Everyone agrees that khûg carries the sense of roundness, and common usage makes clear that this can refer to either a two or three-dimensional geometry. Hence, it cannot be argued that Isaiah 40:22 clearly teaches the earth to be a disc. Moreover, even if khûg does refer to a circle here, this does not necessarily indicate flatness as a globe appears as a circle from whatever direction it is viewed.

Another theory put forth is Flat Earthers make the Jews look stupid and not Christianity, this comes from the kingdom of Antichrist. [14]

Flat Earth theory misleads seekers deep down a rabbit hole, to make both Jews and Christians look stupid. You will not find a single Jew promoting it publicly, because they do not want to be regarded as stupid. Wise of them, but of course, they know it is psyop as they most probably created the PSYOP.

Beware Christians are under a Psyop which is an implant powered by demons from the antichrist kingdom set out to make Christianity and Christians look stupid. Nobody can find any Jews promoting this Flat Earth nonsense, not now nor in the ancient past. Jews do not want to make fools out of themselves. They prefer Christians do it. The Psyops agenda is to align the Christian worldview with the sages of the Jewish Talmud, so as to bring Christianity into Babylonian Talmudism that feeds right into the anti-Semitic, anti-Zionist and antichrist kingdoms. Jews will not purport it, so give the PSYOPS to the Gentile Christians who are susceptible to religious spirits and the kind of funda-*mentalism* that breeds cult like behaviors.

Even Christian scholars, like Mike Heiser was fooled into the misunderstanding of the nature of the earth, that the sages of the Torah believed the world is flat. That is NOT what the Jewish scriptures suggested and certainly not what the Hebrew words suggest. While I agree with Heiser that the earth is not flat, but his position that the sages of the Talmud say it is flat, is disturbing and adds to the Jew hating that already exists over this and other misunderstood Jewish scriptures. Even one of his followers Sylvain Durand wrote, "To get that out of the way, I don't believe the earth is flat. You are however convincing me the Bible says it is. This is a problem for me."

Mike Heiser wrote, "My position is straightforward. The biblical writers do indeed describe a flat round earth (with other features the flat earthers skip; see below). They wrote about the world this way because they lived at a time before knowledge of the natural world was sufficient to demonstrate otherwise. But I don't believe the earth is really flat "because the Bible tells me so." The knowledge the biblical writers had of their physical surroundings isn't a truth proposition for biblical theology. Anyone who uses my work to prop up this idea without providing a disclaimer that I reject modern flat earth thinking is unprincipled and deliberately dishonest." [15]

In his zeal to prove flat earthers wrong, he is completely misunderstanding the biblical writers, who in fact did not describe a flat earth. This is a misunderstanding and mistranslation of the original Hebrew. The truths that many of these Christians fail to see, is that many of the prophets in the Jewish Bible, (known as the Old Testament to Christians) were in fact frequent flyers, as I proved extensively in Book Two: *Who Is God?* Enoch, Elijah, Ezekiel, and even Isaiah all described a spherical round earth, because they saw it from space. They did not write the earth was flat, therefore, Mike Heiser is misinterpreting the Jewish scriptures which suggests that Jews were foolish and stupid.

This comes from antisemitism based on jealousy of Jews, their birthrights, and the ancient positions of Jews in the Bible. This is how Replacement Theology was formed, which comes from ignorance and the arrogance that Christians replaced Israel and the Jews, because of their unbelief that Jesus was the Messiah. Truth alert, all the books of the Bible, both Old and New Testaments were written by Jews. Jesus was Jewish, and Christians would all be pagan Gentiles if it were not for books written by Jews and a Savior who was Jewish.

There will come a time when it will no longer be cool or fashionable to malign Jews, distort their history and steal Jewish identity, which is exactly what many Christians try to do by pawning it off as some Christian cult that they claim is rooted in Judaism or Hebrew Roots. Mike Heiser's Midwestern University of Wisconsin degree in Biblical Hebrew does not make him Jewish, and while one certainly does not need to be Jewish to learn or understand Hebrew, however, a degree in Hebrew, unless it comes from an Israeli University, will not offer the understanding that most Jews inherently have, who grew up with Biblical Hebrew, Ancient Jewish concepts, traditions, and Christian persecution. No Jew would be caught dead promoting a flat earth, let alone saying it is from the Torah!

Heiser questions in his article, "If we are to take Israelite cosmology as literal scientific reality, where is the dome over the earth?" This issue of the dome over the earth is grossly misunderstood by Flat Earthers and Mike Heiser as well, who mistakenly thinks that ancient Israelite cosmology actually communicated a flat earth with a dome around it, as flat earthers suggest.

For the record, I do like and respect Mike Heiser, we're friends on Facebook, and I'm not trying to besmirch him, I am simply pointing out an error, that unfortunately has spread like wildfire, which has created even more errors, and quite frankly I'm tired of seeing these misinformed believers quote him on the internet, because he has a PhD. He is also wrong about Zecharia Sitchin, a well-respected, brilliant Jewish Scholar who was proficient in 5 languages, who certainly never once said, that the Jewish Bible or even the Ancient Sumerians who he wrote 12 books about, suggested a flat earth. Just the opposite.

Unfortunately, Heiser spent a lot of time using his PhD in trying to debunk Sitchin too. Every time Sitchin's name comes up in Christian circles, people post Heiser's spiel on 'Sitchin Was Wrong.' No, I respectfully disagree, Sitchin was NOT wrong! Sitchin pioneered Sumerian research. These same people abuse Heiser's words by trying to prove flat earth, and this is an unfortunate pattern of errors that Heiser began, who himself, doesn't even believe the Earth is flat, but he mistakenly thinks Ancient Jews did. They did not, they do not now, and it is both a linguistically and scientifically false notion, based on the *astrophysics* aspect of Hebrew – something they don't teach people at the University of Wisconsin. If he ever comes across this piece, I sure hope he would do the right thing and detract or correct his article. It is simply inaccurate.

I am certainly not alone in my discernment of the Modern Flat-Earth Movement and Anti-Semitism Connection. While most flat-earthers don't *appear* to be anti-Semitic, at first, but what really changed my mind is that this is classic age-old antisemitic propaganda being regurgitated and circulating again, yet now in another antisemitic Christian cult. I read excerpts from Edward Hendrie's book, *The Greatest Lie on Earth: Proof That Our World Is Not a Moving Globe*, first published in 2016, but with nine editions since. The first publication date coincides with the year I became aware of the flat-earth movement, so this book came out relatively early in the movement's history. I was shocked by both the extent and the character of anti-Semitism in this book.

Hendrie is not the only anti-Semitic flat-earther. Eric Dubay, who is probably more responsible for reintroducing a flat earth to the 21st century, is also anti-Semitic. Like attracts like. For instance, his book *200 Proofs Earth Is Not a Spinning Ball* features anti-Semitic imagery. Dubay wrote the following antisemitic propaganda and blatant lies on his Atlantean Conspiracy website: "Adolf Hitler was actually a vegetarian, animal-lover, an author, an artist, a political activist, economic reformer and nominated for a Nobel Peace prize. He enacted the world's first anti-animal cruelty, anti-pollution, and anti-smoking laws. Unlike the demonic portrait that the 'allies' painted of him, Hitler was beloved by his people, he wanted nothing but peace, and never ordered the extermination of a single Jew. The largely Jewish-controlled mainstream media has ever since painted an evil picture of Hitler and the Jew World Order has even enacted laws in 16 European countries prohibiting free-speech on the issues of Judaism, Hitler and the Holocaust."

You just need one gulp to taste the ocean!

This is the guy who started the antisemitic racist phrase that trolls the internet, 'Jew World Order.' I guess he doesn't know that most of the Illuminati aren't all Jewish. They mistakenly think that Jews rule the world. I say, if Jews ruled the world, then why is there rampant antisemitism, Jew hatred and antizionism? If Jews really ruled the world, then these forms of hatred would be ruled out, illegal and obliterated. In fact, one day, Yeshua HaMashiach (Jesus the Christ) will return to Earth and establish His Kingdom on this planet, and He will rule over all the nations. That is when a Jew will rule the world, and in His Kingdom, there will be no racism or antisemitism.

But for now, this is what Jesus has to say to those who hate Jews: "The Gospel of the Kingdom is first for the Jew, and then for the Gentile. (Romans 1:16) "There will be tribulation and distress for every human being who does evil, the Jew first and also the Gentile, but glory and honor and peace for everyone who does good, the Jew first and also the Gentile." (Romans 2:9-10)

Without writing a whole essay on the Spiritual Legal Ground based on the Word of God against those who are antisemitic, suffice to say, there are blessings and curses attached to the ancient covenant from Yahuah, on how the nations treat Jews and the descendants of Abraham. The LORD Yahuah decreed in covenant to

Abraham: "I will bless those who bless you, and whoever curses you I will curse; and all peoples on earth will be blessed through you." (Genesis 12:3)

Christians have the right to disagree and question, but they do not have the right to be mean-spirited to Jews within the Body of Christ. I have had horrible things said and done to me by so called Christians, after they find out I am Jew for Jesus and a Woman. So, like it or not, it is my calling to expose them as frauds and false Christians. I cannot be false, I'm a Jew for Jesus, something Christians have been trying to get Jews to believe in for centuries. Now all of a sudden, someone like me is considered an outcast.

Antisemitism has no place in the Body of Christ nor is it condoned in the Kingdom of Heaven. Romans 11 specifically lays out the spiritual legal ground for the invitation to Gentile believers to be grafted into the covenants of Israel and Abraham. That means, if they don't bear fruit, they too will be pruned and burned off the tree. Those branches who are grafted in and curse the roots, will too be cut off.

> "If some of the branches have been broken off, and you, though a wild olive shoot, have been grafted in among the others and now share in the nourishing sap from the olive root, do not consider yourself to be superior to those other branches. If you do, consider this: You do not support the root, but the root supports you. You will say then, "Branches were broken off so that I could be grafted in." Granted. But they were broken off because of unbelief, and you stand by faith. <u>Do not be arrogant</u>, but tremble. For if God did not spare the natural branches, he will not spare you either.
>
> Consider therefore the kindness and sternness of God: sternness to those who fell, but kindness to you, provided that you continue in his kindness. Otherwise, <u>you also will be cut off</u>. And if they do not persist in unbelief, they will be grafted in, for God is able to graft them in again. After all, if you were cut out of an olive tree that is wild by nature, and contrary to nature were grafted into a cultivated olive tree, how much more readily will these, the natural branches, be grafted into their own olive tree! I do not want you to be ignorant of this mystery, brothers and sisters, so that you may not be conceited: Israel has experienced a hardening in part until the full number of the Gentiles has come in, and in this way **all of Israel will be saved**. As it is written:
> "The deliverer will come from Zion;
> he will turn godlessness away from Jacob (Israel).
> And this is my <u>covenant</u> with them
> when I take away their sins." (Romans 11:17-27)

EARTH'S ANCIENT DOME OF WATER

The issue of the dome, or vault as many flat earthers misinterpret the Hebrew word, *raqia*, to mean, with a flat earth below it, suggesting we're living in some type of snow globe, which is flat on the bottom with a dome over it that makes snow, is as preposterous as thinking that we are alone in the universe.

There is evidence that the atmosphere encircling the early earth was vastly different than it is today. Prediluvian earth enjoyed a warm tropical environment and there was enhanced oxygen in the atmosphere. Organisms grew larger and lived longer as a result.[16]

Here is the truth about the firmament and the so called 'dome' around the earth. Firstly, during the first floods of Atlantis, which Genesis 1:1 refers to the Spirit hovering above the waters of the earth, which was due to the first major flood, also known as *Lucifer's Flood*, the ancient earth was more like a round cell, with the earth in the middle surrounded by a bubble of water, kind of like a giant round green house, which created fauna and green life to thrive. This bubble of water or firmament canopy also served to protect the earth from the gamma rays of space. It was also believed that earth used to have two moons, one of them was destroyed by a passing of Nibiru, which caused the bubble of water around the planet to burst and flood the earth. After the first flood, the atmosphere was held in place, and a quarantine was placed around the earth, which meant that the fallen angels who were imprisoned *inside* the earth could not escape nor could those coming from outer space have easy access to earth. I wrote about this quarantine in detail in Book One of *Who's Who in The Cosmic Zoo?* (See, pp. 23,50-62) [5]

Many creationists have attributed this to a water vapor canopy that was created by God on the second day, the *waters above the firmament* (Genesis 1:7). This theory holds that a "vast blanket of invisible water vapor, translucent to the light of the stars but productive of a marvelous greenhouse effect which maintained mild temperatures from pole to pole, thus preventing air-mass circulation and the resultant rainfall (Genesis 2:5). It would certainly have had the further effect of efficiently filtering harmful radiation from space, markedly reducing the rate of somatic mutations in living cells, and consequently, drastically decreasing the rate of aging and death." (Scientific Creationism, p. 211.) [17]

Somehow, this Flat Earth heresy has been used to create even more division amongst Christians, along with turning off non-believers who are searching for Christ. This Flat Earth Christian Cult is just another reason for non-believers, secularist, spiritual seekers to stay as far away from Christians as possible. It's truly a black eye in the body of Christ, it serves to make a mockery of the Jewish bible, and it's using today's Christians to do it, while they make fools of themselves in front of the rest of level-headed Christians. As if Christians don't have enough division arguing over End Times Prophesies, and creating cliques based on whether you are pre-trib, mid-trib or post trib with respect to the Rapture, and then of course, there's the evils of Replacement Theology, all based on arrogant haughty religious demons. Since when did becoming a Christian give one the license to judge, and act superior

CHAPTER 43: THE CASE AGAINST FLAT EARTHERS

to others? Jesus taught the opposite, and history records how he contended with Religious Spirit Demons through the Pharisees and Sanhedrin, who conspired to have him crucified. How much more should Christians, flush out this enemy from within?

Therefore, the New World Order One World Religion gets a foothold around the world, people are simply sick and tired of people being judged and divided over religious viewpoints and belief systems. Yes, there is only one Savior and Messiah, but there are many counterfeits, and herein lies the rub. If loving one another does not take center stage in your heart for Christ, then perhaps seeking deliverance of religious spirits would be in order for the Church.

Another reason why, Jesus promised to return to judge the Church in the Great Sheep and Goat Judgement. Do not be a goat.

It is astounding to me, that in this day and Age of Knowledge and Space Travel, I am here writing a chapter on debunking Flat Earthers. But this is my work to expose the Religious Spirit Demon, and it's been at work overtime with this stubborn bunch of Christians, who no matter if the evidence is staring them right in the face, as it often is through countless photos and videos taken from space, they still insist it's all fake. Now, there are real conspiracies going on, but this is just not one of them. This is mental illness and the bondage to Psyops' demons piggy backing on the spiritual legal ground of the Religious Spirit. This cult believes they have some special type of secret knowledge into the past, when in reality they are missing the entire point of the Jewish scribes, and this implant blinds them to the real conspiracies, which is the alien presence *inside* the spherical torus shaped hollowed earth, and the spaceships in the Bible, which reveals the real *Star Wars* we are all a part of as terrestrial dwellers.

Let me reiterate the litmus paper for spiritual discernment, come from the very words of Jesus Christ Himself, 'By their fruits, you will know who my disciples are.' (Matthew 7:20) That they show love for one another, (John 13:35) that they feed and look after the poor, so your light shines like the noon day Sun, (Isaiah 58:10) take care of the orphans and widows, for this is the *true religion* (ways) of the Lord. (James 1:27).

When so-called Christians are puffing themselves up with spiritual pride, rejecting other believers because they do not believe in their cultish flat earth nonsense, their lights have gone dark, and they are in bondage and being used by the enemy of the saints. Let's face it, in warfare, whether its physical or spiritual, when the enemy infiltrates the camp, it takes over from within. Great invasions happen in secret, in this case, between one's ears.

Here is what Jesus taught about guarding one's mind and heart from infiltration from the enemy camp:

> "Above all else, guard your heart, for everything you do flows from it."
>
> (Proverbs 4:23)

> Be ye transformed by the renewing of your mind daily, that ye may prove what is that good, and acceptable, and perfect, will of God. (Romans 12:2) "Be made new in the attitude of your minds;"
>
> (Ephesians 4:23)

> Take every thought captive: "We demolish arguments and every pretension that sets itself up against the knowledge of God, and *we take captive every thought* to make it obedient to Christ."
>
> (2 Corinthians 10:5)

Currently, does it really matter as long as the human hearts are demon possessed be it on a flat or round Earth? What does evangelism for the Kingdom of Heaven and the Saving Grace through Jesus Christ have to do with evangelism of a Flat Earth? Does being a flat earther save anybody from purgatory or hell? How could it, when there is a special place in hell reserved for those who mislead and misguide others through false teachings, false prophesies, and deceptions?

> "Why do the heathen rage, and the people imagine a vain thing?"
>
> (Psalm 2:1)

> "The natural man does not accept the things that come from the Spirit of God. For they are foolishness to him, and he cannot understand them, because they are spiritually discerned."
>
> (1 Corinthians 2:14)

As many whom I have communicated with on Social Media have expressed their opinions on the Flat Earthers is essentially leading truth seekers deep down a rabbit hole which is essentially a black hole, that make Christians look stupid. This idiocrasy is promoted only by some pretending to be 'Christians'. You will not find a Jew or Muslim, nor any Humanists, Secularist, Scientists, nor a Buddhist promoting this foolishness. Sadly only 'Christian pastors' who have a proclivity for antisemitism and antizionism. Doesn't that tell you something?

> Test the Spirits: "Beloved, do not believe every spirit, but test the spirits to see whether they are from God, for many false prophets have gone out into the world."
>
> (1 John 4:1)

CHAPTER 43: THE CASE AGAINST FLAT EARTHERS

> "but test everything; hold fast what is good. Abstain from every form of evil."
>
> (1 Thessalonians 5:21–22)

> "For God hath not given us the spirit of fear; but of power, and of love, and of a *sound mind*."
>
> (2 Timothy 1:7)

If you come across any 'Christians' promoting this PSYOP, then ask them: Can you name me a few Jewish people supporting this Flat Earth thing publicly? You will not find a single Jew promoting it, because they do not want to be regarded as stupid. Jews are regarded by many as the most intelligent minds on Earth. Not only do Jews dismiss this nonsense, they also have never believed that their ancestors who wrote the Bible scriptures suggested this as well. Therefore, even insinuating that they do, comes from anti-Semitic leanings, which I have identified many times as a deep-rooted jealousy for Jews that they were 'chosen' for something, which is layered with deep-rooted resentments for Jews, leading to identify theft, then perverting it and regurgitating it as something else is, by very definition antisemitism and the evils of replacement theology. Jewish scholars of Tanakh who understand the Hebrew words in the Torah in the correct context they are written in, who grow up studying Tanakh in Yeshivas, have never suggested such idiocies that the Jewish Bible suggests a flat earth.

Jews have suffered for centuries to protect and preserve knowledge, traditions and most importantly their identity. Most Jews reject Christ because of the persecutory behaviors of Christians. This is the area the Church needs reformation and maturity, which it is sorely lacking. There first needs to be awareness of the problem, before they can even begin to take steps to correct themselves before it is too late, when the Lord return and judges them as goats.

Christians do not make a fool out of yourself and Christianity, by pretending to be Christian Flat Earth. This PSYOP is one of multiple *internet religions*, that spreads like wildfire through social media. People have become so entrenched in their phones and Facebook apps, that they actually believe they are *living* their lives vicariously through it, and are judged or accepted based on their memes, posts, and status updates. Jesus warned us of all of this, he taught that what matters most to Him is what you do in private what you do secretly in your prayer closet. Your good deeds, your alms, your tithes, and your prayer and fasting, are not to be broadcast to the world, but for your Heavenly Father to reward you, as He alone sees what you do in secret.

> "Be careful not to practice your righteousness in front of others to be seen by them. If you do, you will have no reward from your Father in heaven.

> "So, when you give to the needy, do not announce it with trumpets, as the hypocrites do in the synagogues and on the streets, to be honored by others. Truly I tell you, they have received their reward in full. But when you give to the needy, do not let your left hand know what your right hand is doing, so that your giving may be in secret. Then your Father, who sees what is done in secret, will reward you.
>
> "And when you pray, do not be like the hypocrites, for they love to pray standing in the synagogues and on the street corners to be seen by others. Truly I tell you, they have received their reward in full. But when you pray, go into your room, close the door, and pray to your Father, who is unseen. Then your Father, who sees what is done in secret, will reward you. And when you pray, do not keep on babbling like pagans, for they think they will be heard because of their many words. Do not be like them, for your Father knows what you need before you ask him."
>
> (Matthew 6:1-8)

Social Media has turned into a religion for people, and its created mental disorders and mental illnesses cross the board. I have never heard of people who actually believed the earth is flat in modern times especially based on their misinterpretation of the Bible until I first saw this Psyops on Facebook. Because it is so pervasive, I felt it imperative to add a chapter discerning and debunking it in this book on The Heavens – Book Five of the *Who's Who in The Cosmic Zoo?* book series, with the hopes that it will save just one Flat Earther out there and lead them to the Truth.

While I generally make it a point to accept and love all people unconditionally, until they give me a reason not to, posting pictures of Earth from space has attracted its fair share of flat earth trolls, who have become to me like swatting flies and killing wasps, as I am forced to delete and even block some on Facebook due to their irrational, outrageous behavior, and stupid comments. I have taken the time to prove them wrong, based on science, evidence from Astronauts and of course the correct translations from Hebrew to English, and due to this deep implant penetrated deep within their psyches, they have proven to be the most stubborn and irritating bunch of so-called Christians, I've come across. And people wonder why there is an apostasy, and why so many Christians run to the New Age for solace, light, and acceptance. The church is divided and half of them have gone raving mad. This is due to implantation. For more details, see, my chapter on *Alien Implants* pp. 471-521, in COVENANTS: Book Four. [18]

> Be good, whatever your conclusion.
> It is not the mountain we conquer but ourselves.
> Edmund Hillary

CHAPTER 43: THE CASE AGAINST FLAT EARTHERS

Notes:
1. Society of Flat Earth Debunkers, Flat Earth Theory Fabricated By TPTB To Distract From Planet X, http://themillenniumreport.com/2016/01/flat-earth-theory-fabricated-by-tptb-to-distract-from-planet-x/, January 20, 2006
2. Society of Flat Earth Debunkers, Submitted: 1/20/06, http://m.beforeitsnews.com/alternative/2016/01/busted-flat-earth-theory-used-to-wage-war-against-the-truth-movement-3284388.html
3. Ella LeBain, COVENANTS - Book Four, Skypath Books, 2018.
4. Brown, F. et al., Brown-Driver-Briggs Hebrew and English Lexicon: With an Appendix Containing the Biblical Aramaic, Hendrikson Publishers, USA, p. 295, reprinted January 1999 from the 1906 edition; biblehub.com/hebrew/2329.htm.
5. Ella LeBain, Book One of *Who's Who in The Cosmic Zoo?* (See, pp. 23,50-62) Skypath Books, 2012.
6. Charles Q. Choi, Strange but True: Earth Is Not Round: It may seem round when viewed from space, but our planet is actually a bumpy spheroid, https://www.scientificamerican.com/article/earth-is-not-round/, April 12, 2007.
7. Ella LeBain, *Who is God?* Book Two, Skypath Books, 2015.
8. Ella LeBain, *Who Are the Angels?* Book Three, Skypath Books, 2016.
9. David Read, Raqia: 'Expanse or Vault?' http://advindicate.com/articles/1494, June 26, 2012
10. Crystallinity in the Atmosphere, https://en.wikipedia.org/wiki/Crystallinity
11. Vine's Expository Dictionary of Old and New Testament Words, 1981, p. 67.
12. John Gill's Exposition of the Bible, footnote to Isaiah 40:22; biblestudytools.com.
13. Dominic Statham, *Isaiah 40:22 and the shape of the Earth*, http://creation.mobi/isaiah-40-22-circle-sphere, August 11, 2016.
14. J. Johansen, Flat Earthers Make Jews and Christians Look Stupid, March 18, 2016.
15. Mike Heiser, *Who Believes the Earth is Really Flat? Does it get any dumber than this?* http://drmsh.com/christians-who-believe-the-earth-is-really-flat-does-it-get-any-dumber-than-this/
16. The Pre-flood Atmosphere, http://www.genesispark.com/exhibits/early-earth/atmosphere/
17. Henry M. Morris, Scientific Creationism: Study Real Evidence of Origins, Discover Scientific Flaws in Evolution, p. 211, Master Books (August 1, 1985), https://www.amazon.com/dp/B006E8ZNJA/
18. Ella LeBain, COVENANTS - Book Four, Chapter 31: Implants and Spiritual Limitation Devices, pp.479-521, Skypath Books, 2018.
19. Ben-Yehuda, E. and Ben-Yehuda, D, Hebrew Dictionary, Pocket Books (Simon & Schuster), USA, p. 252, 1961.
20. Levin, S., Semitic and Indo-European: The Principal Etymologies, vol. 1, John Benjamins, USA, 1995.
21. Buck, C.D, A Dictionary of Selected Synonyms in the Principal Indo-European Languages, University of Chicago Press, Chicago, pp. 907–8, 1949.
22. López-Menchero, F., Proto-Indo-European Etymological Dictionary, indo-european.info/indo-european-lexicon.pdf, 2012. Asociación Cultural Dnghu; dnghu.org.

CHAPTER THIRTY-SEVEN

THE REAL STAR WARS

Our relationship with God is the only thing that will last all of eternity. So, make sure it is your number one priority on Earth.

My Panel Talks at the annual StarFest Conventions in Denver's Tech Center, are titled, *The Truth is Stranger than Fiction*, which I began in 2013, sharing slideshows of Exoplanets around the Sun, as well as what appears to be giant spacecraft that resembles the shape of the Star Ship Enterprise. I also include Nibiru and Star Wars in the Zohar and Jewish Scriptures in my panel discussions which have always been popular despite the cognitive dissonance. So many people have been conditioned to believe and associate Nibiru with the ancient Sumerians, but very few see its connection to the Jews, the God of Israel Yahuah and Yeshua aka Jesus.

Rabbi Yuval Ovadia, an Israeli filmmaker who has made several award-winning films on the subject of Nibiru and Planet X is one of a few Jews to connect the dots between NASA's 2017 discovery of 7 Earth-sized planets and the description in the ancient Jewish Zohar, written 2,000 years ago. The Zohar is the foundational work of Jewish mysticism on the Torah, which predicted the appearance of a red star with seven 'stars' orbiting it as the sign in the heavens before Messiah comes.

> "I see him, but not now; I behold him, but not nigh; there shall step forth a star out of Yaakov, and a scepter shall rise out of Yisrael, and shall smite through the corners of Moab, and break down all the sons of Seth."
> (Numbers 24:17- The Israel Bible™)

In February of 2017, NASA announced the discovery of a relatively close star system, but according to one opinion, the discovery comes 2,000 years after it was first described in detail by a classic Jewish text as a necessary element preceding the Messiah.[1]

NASA Discovery of 7-Planet System Conforms Exactly to Zohar's Description of Pre-Messiah "Nibiru." Israeli filmmaker and lecturer Yuval Ovadia revealed the connection claims the new astronomical discovery goes by many names but in his

CHAPTER 44: THE REAL STAR WARS

video on the subject, he calls it by its most popular: Nibiru, which was given to us by Archeoastronomer Zecharia Sitchin.

NASA research revealed seven planets orbiting a dwarf star, called Trappist-1, roughly eight percent the size of the Sun and located 39 light years away from Earth. Four of its seven planets are similar in size and mass to the Earth, and three are in what scientists believe is a habitable zone. The recent discovery required NASA's most advanced telescope, the Spitzer space telescope, to verify its results, but Jewish sources have been discussing the appearance of a seven-planet star system preceding the Messiah for thousands of years.

This is the claim made by Yuval Ovadia, an Israeli filmmaker who has made several award-winning films about Nibiru and Planet X. Ovadia spoke to Breaking Israel News about his last video, "Amazing Revelations." Ovadia emphasized the similarity between NASA's recent discovery and the description in the ancient Jewish sources.

2,000 years ago, the Zohar, the foundational work of Jewish mysticism, predicted the appearance of a star with seven 'stars' orbiting it. After forty days, when the pillar rises from earth to heaven in the eyes of the whole world and the Messiah has appeared, a star will rise up on the east, blazing in all colors, and seven other stars will surround that star. And they will wage war on it.

"Jewish sources say that this astrological phenomenon is a necessary part of the geula (redemption)," Ovadia said. He explained that if the star does not appear, then any claim that the Messiah has arrived will be rejected by Judaism. He pointed to an example: 400 years ago, Rabbi Yaakov Sasportes used the absence of a new star as an argument against Shabbetai Tzvi, a Jew who falsely claimed to be the Messiah. [1]

The belief in a planet-sized object that will catastrophically collide with or closely pass by Earth in the near future has resurfaced in recent years, with the astral body referred to as Nibiru or Planet X. Many rabbis have acknowledged the possibility of the object hitting Earth. "This is not a foreign idea that was incorporated into Judaism," Ovadia, an observant Jew, told Breaking Israel News. He noted that the Bible contains a prophecy of a star signaling the arrival of the Jewish Messiah.

I see him, but not now; I behold him, but not nigh; there shall step forth a star out of Yaakov, and a scepter shall rise out of Yisrael, and shall smite through the corners of Moab, and break down all the sons of Seth. (Numbers 24:17)

Ovadia lectures on the subject and has been following it closely for many years. He believes that data is being suppressed. "There is not enough knowledge on this subject because the authorities, especially NASA, obscure the facts," Ovadia claimed. "One day they will say it is impossible, then the next day they will say it is far away. It can't be both."

The appearance of this pre-Messianic star is predicted to be catastrophic, accompanied by extreme natural disasters. The Vilna Gaon, an 18th century Jewish scholar, stated that the power of this star system could destroy the world. Ovadia explained that this may not be caused by the star colliding with the Earth but rather by God exerting his influence on global weather events via the star.

"We are already seeing its influence. In the last ten years we have seen a sharp rise in the number and intensity of natural catastrophes," Ovadia told Breaking Israel News. "Authorities admit that the number of meteors striking the earth has gone up enormously in recent times."

The NASA discovery seems to confirm his claims, but for Ovadia, the conclusion is more important. "The fact that the star exists is incontrovertible," Ovadia states emphatically. "It exists. But we knew this all along. It is part of our tradition. The rabbis who wrote about it tell us that Jews have a major part to play. Our actions, our relationship with God, can influence the outcome." [2]

CONNECTING THE DOTS FROM JEDIISM TO JUDAISM: STAR WARS AS JEWISH ALLEGORY

I share this in memes and pictures during my StarFest Panels, titled, "The Truth is Stranger than Fiction!" Connecting the dots to some of the Jewish elements – whether it be coincidental or otherwise – of Star Wars.

A long time ago in a place far, far away...

It is a period of civil war. A new government has declared the practice of the old faith a crime punishable by death, disbanding an ancient order of sages and sending many into exile. Rebel fighters, striking from a hidden base, have won their first major victory against the evil Empire, stirring a spirit of defiance among the populace. Out-armed and vastly outnumbered, the ragtag band of rebels – aided by an all-powerful, all-permeating Force that binds together all life in the universe – remain the only hope for restoring peace and freedom to their people. [3]

HANUKKAH AND STAR WARS – THE UNCANNY PARALLEL

It is one of the greatest epics known to mankind was not Star Wars, the story of Hanukkah, the eight-day Jewish festival that commemorates the miraculous victory of Israelite insurgents against the tyrannical Seleucid Empire roughly 2,200 years ago. Daniel Perez of Aish.com shared his thoughts after watching *Star Wars Episode VII – The Force Awakens*; on the strange parallels to the history of the ancient Israelites who have become known as the Jews.

I agreed with him, as I too thought about how certain aspects of the Star Wars universe are eerily like the history, beliefs, and teachings of the Jews. The Empire is no more. In its place, The First Order has risen from the ashes of Darth Vader's legacy and aims to consolidate its power across the galaxy. It reminded me of the fall of Ancient Israel, which started the Diaspora, and the land laid into a barren wasteland for a little over 1800 years. Now, I am not suggesting that Star Wars creator, George Lucas set out with the intention to create a fantasy universe full of Jewish references, but call it coincidental or synchronicity, the connections are there, nevertheless. There is a funny meme that goes around during Hanukkah, "put the "Han" back in Hanukkah" with Harrison Ford, who by the way, is half Jewish, as was Princess Leia, Carrie Fisher. Just saying. This is inside joke amongst Jews to continue to fight for our freedoms. To take up the gauntlet as Han Solo did against tyrannical forces that seek to destroy us.

One of the main psychological reasons, the Star Wars universe has resonated so deeply and profoundly with people from all over the world, is, it's essentially because of its underlying story of father issues within a dysfunctional family, which let's face it, who doesn't have father issues? But if we take the mythologies of Star Wars, we can apply them to Greek Mythologies as well, and the galactic dramas as written in the Books of Enoch about the fallen rebel angels. It's all relative.

The Star Wars universe teaches that what goes around comes around. When individuals try to enslave others, their attempts become the tools of their own destruction. Leia kills Java the Hut with the very chain which he used to imprison her. Palpatine died in the hands of the man he corrupted and used to take over the world.[3]

> This is an interesting lesson in discernment about the Hanukkah story: After weeks of recitation about the Book of Maccabees, as the Hanukkah story to kindergarteners, one of the parents asked their son, "Why did the Maccabees fight against the Hellenized Jews?" The son answers: "Because they accepted Greek things, like idols and all that."
>
> "No. The problem with the Hellenized Jews wasn't that they adopted Greek values. It was that they adopted the *wrong Greek values.*"
>
> They could have adopted the Socratic paradox: One must know that one doesn't know. They could have adopted Athenian standards and methods of learning, as did the Jewish sages of later generations. They could have adopted democratic values, as we do today.[4]
>
> Instead, they chose to adopt hedonism, idol-worshiping, and the cult of the physical body. When Matityahu and his sons refused to worship an idol, they chose the Jewish strand of their identity over the various values and pursuits of Hellenist culture. When they found one sealed flask of oil, they chose to use it and reject bountiful vessels of impure oil.[4]

> The Maccabees won both the spiritual and political battle of religious freedom. This is why we celebrate, because they chose the right options, instead of indulging on everything without discernment.
>
> In this world of information overload, endless entertainment, and social media, we need this message more than ever.[4]

IN A GALAXY OF HEBREW NAMES

The Book of Maccabees details the Hanukkah story, when the evil King Antiochus tried to obliterate all the Jews. Had he succeeded, well, for one, we wouldn't be having this discussion, and two there would be no Yeshua, who was born a Jew, who by the way celebrated Hanukkah, the miracle of light. See, John 10:22–23.

In parallel, the heroes of the Star Wars series are members of a "rebel alliance," who are basically Maccabees in outer space. It is right there in the name: Jedi. The Hebrew letter yud is transliterated in English as a "J," and syllables occasionally get truncated in translation. Hence, a Biblical name like "Yehoshua" makes its way into English as "Joshua." It is not much of a stretch to see how "Jedi" can be derived from the original Hebrew word for Jew, "Yehudi."

Luke Skywalker's Jedi Rebbe, Grand Master Yoda? Other Jews have pointed out the peculiar syntax of Yoda is uncannily similar of someone whose first language is Yiddish ("Yodish")? More to the point, his name sounds a lot like the singular masculine conjugation: "yoda, yodea" and, "yada," the Hebrew words meaning "I know" in both masculine and feminine participles. [3]

And how about those Skywalkers? Luke Skywalker might sound like a gentile name, but that name was clearly chosen to alliterate with his twin sister Leia (Leah). Also keep in mind that their parents were an interfaith couple. The father, Anakin Skywalker, played by the unmistakably un-Jewish Hayden Christensen, tried to convert to Jediism, but as we know he ultimately turned to the Dark Side instead. Their mother was Queen Amidala, portrayed by the beautiful and talented Israeli-born actress Natalie Portman. Suffice it to say their marriage did not end well, and it was not until much later in life that their children discovered their Jedi-ish identity.

Aish author Daniel Perez, put these interesting parallels together about the Jedi Knight Learning Academy and Yeshiva which is Torah school: "When an aspiring Jedi Knight goes to the Academy, he or she must complete what is essentially an apprenticeship with one more learned in Jediism than they are. Similarly, a future rabbi's yeshiva experience will consist largely of chavruta learning, studying with a partner – lit. "friendship". Fun fact: The name for a young, unmarried yeshiva student, "bochur," means "chosen" as in "The Chosen People". The idea of a foretold "Chosen One" who would "restore balance to the Force" was a theme running throughout the Star Wars films, wherein Anakin Skywalker was recognized for his extraordinary potential as a Jedi. As mentioned above, he went "off the derech" (off the pathway) and became the villainous Darth Vader. In Return of the Jedi, however, Vader/Skywalker fulfills the "prophecy" when he does teshuva, our term for repentance, which literally means "return." Whoa. Return of the Jedi, thwarting Emperor Palpatine to save his son's life, and ultimately, the galaxy.[3]

Of course, if you tell a young rabbi-in-training that he is the "Chosen One," it sounds cool and dramatic and is technically true, but then, the same can be said of all his classmates. While the Star Wars films do not feature Jedi trainees delving into sacred texts, because let's face it, doesn't make for the most exciting movie montage, some of the greatest rabbinic books of ethics and Jewish philosophy would be right at home in any Jedi library. "Duties of the Heart," "The Path of the Just" .... tell me these do not sound like the reading list for a hero of the Light Side.[3]

THE FORCE

While Jediism isn't a theistic religion per se, its practitioners do teach of a Force that, in the words of Reb Obi-Wan Kenobi "...is what gives a Jedi his power. It is an energy field created by all living things. It surrounds us and penetrates us; it binds the galaxy together." That almost sounds like some sort of Chasidic teaching – just replace "energy field" with "entity" or "consciousness," and "created by," with "that creates," and what you have starts to come across less like new age hippie talk and more like an introduction to Kabbalah, Jewish mysticism.

One idea that devout Jews of all stripes share, is that God, the creative "Force" that sustains all, is the source of a Jew's power. "Ein od milvado," there is none besides Him. The Jew expresses his or her connection to the universe by striving for an even closer relationship with its Creator.

Another aspect of Jedi belief is the notion of balance, the idea that the Light Side and the Dark Side are both aspects of the same Force seeking equilibrium. The religions that branched off from Judaism tend to show the Creator and Satan, or "The Devil," in an adversarial relationship, almost a sort of de facto dualistic theology with a God and an anti-God, if you will. Judaism maintains that the Satan (lit. "Accuser") is the angel associated with temptation, and prosecution in the Heavenly Court. He is basically Slugworth to God's Willy Wonka. He has got a dirty job to do, but in the end, we are both serving the same Boss.[3]

Judaism also teaches that the source of Light and Darkness are One and the same, as it says in the prayer book: "Blessed art Thou, Lord our God, King of the Universe, Who forms light and creates darkness, Who makes peace and creates all things." The source for this line of liturgy can be found in the Hebrew Bible, Isaiah 45:7: "Who forms light and creates darkness, Who makes peace and creates evil; I am the Lord, Who makes all these."

Incidentally, one of the traditional names for God – invoked particularly by the Jewish mystics – is HaMakom, literally "The Place." The deeper idea conveyed by this name is that the Creator does not exist within the universe; the universe exists within Him. It sounds a lot like The Force. The key conceptual difference between the fictitious all-uniting Force of Star Wars and the Shechinah or "Divine Presence" is that the former is impersonal and passive, the latter is an omnipotent consciousness that actively intervenes in human history, speaking with Prophets and working miracles until this very day.[3]

Those who are considered the Jedi in Star Wars, have uncanny parallels to the characters of the Hanukkah story historically recorded in the Book of Maccabees. Maccabaeus, Judah, and his brother, the Je(hu)di rebels of their generation, do battle with Seleucid Imperial troops.[3]

*"Yesterday is history,
tomorrow is a mystery
but today is a gift
that's why it is called the present"*

Aish.com author Keren Keet said, "In the Empire Strikes Back, Luke trains under Jedi Master Yoda. His knowledge of The Force becomes more profound as Yoda explains and demonstrates its power. During Luke's training, he practices moving small rocks around with The Force. Yoda then insists that he raise his Sunken spaceship from the swamp in a similar way. Luke tries but is defeated, 'It's too big!' he gasps as he struggles to catch his breath. Yoda admonishes him for attributing such importance to physical size, which is irrelevant in the face of the all-powerful Force. But Luke is unconvinced and scoffs, 'You want the impossible!'[5]

Then to his amazement, and a rousing musical surge from John Williams, Yoda raises the ship clean out of the water and back onto land with just his two little outstretched yellow claws and a concentrated expression.

Luke is dumbfounded and exclaims, 'I can't believe it!' and Yoda replies, 'That is why you fail'.

Keet makes the comparison to Jewish philosophy is astounding. Rabbi Dessler explains that Nature is simply a series of miracles we have become used to. There are many levels in which humans can relate to Nature, but those on the highest levels see Nature not as something independent from God, running along its own principles and laws, but as a function of God's ongoing and continuous intervention. They feel God in every single detail of the world around them (feel The Force!). Furthermore, since natural processes are merely manifestations with God's will, these people can perform 'miracles.[5]

Rabbi Dessler brings the famous example of Rabbi Chanina Ben Dosa. His daughter poured vinegar into the Shabbos lights instead of oil, but Rabbi Chanina was unphased, saying 'Let the One Who says that oil should burn say that vinegar should burn,' and it did! For Rabbi Chanina, oil only burns because Hashem says it should, so if Hashem wants vinegar to burn, so be it. Rabbi Akiva Tatz points out that a sign of someone on this level is that he will register no surprise at all if vinegar will burn. Everything is just a function of Hashem's will, so what's the difference between oil burning or vinegar burning? [Tatz, Worldmask, Targum Press, 1995, p34]

Ok, so lighting Shabbat candles is hardly as dramatic as raising a spaceship out of swamp, but the principle is the same, just substitute 'God' with 'The Force'. Luke's inability to believe in the omnipotence of The Force means that he cannot perform 'miracles', he is too bound by the 'natural' reality – and that is why he fails. Luke needs to work on his Emunah (faith)![5]

And what about the requirement for the Jedi to control his emotions? This is Jewish character development, *tikkun hamiddot*, to its core! Yoda and Obi Wan repeatedly warn Luke against anger, fear, aggression and hatred, all characteristics that lead to the Dark Side. These emotions are as abhorrent in Judaism as in Jediism, and Luke is continually battling to control himself. As a Jew is required to conquer his negative characteristics, so is a Jedi.⁵

Star Wars is by no means a Jewish text, and George Lukas isn't promoting Judaism, nevertheless some parts of Star Wars do resonate with the philosophical beliefs of Torah and ancient Judaism. May The LORD be with you!

THE LAST JEDI AND THE FORCE OF FAMILY

Star Wars resonated with so many people because of its dysfunctional family storyline that resonated with so many in our pop culture. One of the greatest moments in cinema history is when Luke discovers his true parentage. It was this revelation that took the original movies to a new level of intensity in terms of the struggles of the hero, forcing him to wrestle with the Light and Dark in himself – in his bloodline – passed down from his fallen father. Talk about Generational Curses and Ancestral Sins! Ultimately, Luke resists the temptations of the evil Emperor, and manages to save Darth Vader/Prince Anakin.

For this reason, I thought the Prequels, which focused on the life of Anakin, before he transformed into Darth Vader, were so fascinating to watch, and took the Stones song, "Sympathy for the Devil" to a new level.⁶ The incredible heart and soul losses Anakin suffered, first losing his mother, then his betrothed and eventually his offspring, left him with permanent psychic wounds, resonated with me, after being orphaned at age 6, losing my mother, and then losing my father at age 15 through estrangement.

I identified with his pain, and his reason for turning to the dark side which evoked empathy and compassion in me. My teenage daughter fell in love with Anakin, which became a running joke in our family, affectionately, that my daughter is in love with Darth Vader. Truly, it was who he was before he became Vader, that intrigued us so much. He had it all, the handsomeness of a true prince, the power of a Jedi, the adoration of his mother and the love of his true love, Senator Padme Amidala, and Queen of Naboo. Their secret romance and forbidden marriage for the sake of their love and attraction unfortunately proved to have dire consequences for the galaxy, making Anakin the Disappointing Jedi. But in the end, was he really? Anakin was born to defeat the Sith, and as Darth Vader, he fulfilled his destiny, nevertheless.

Star War helps us to understand psychologically how human nature gets sucked into the dark side, through the psychic woundings and heartbreak of the soul.

Episode 7 continued to explore these dynamics, but now the good side is represented by the parent generation, Han, Leia, and Luke, the uncle, while the child, Kylo Ren, has chosen the Dark Side.⁶ Ren was immediately established in *The Force Awakens* as an exceedingly wicked and heartless villain, murdering innocents within the first few minutes, but his most heinous crime was his cold-blooded murder of his own loving father Han, a complete reversal of Luke's actions. In *The Rise of*

Starwalker, we watch Kylo Ren repent, as he joins Leia who together join forces and end up saving Rey in the final installment, after the revelation that Rey is Palpatine's granddaughter. Together they destroy the Sith Lord, and after Leia passes, she sends all of her force into Kylo Ren to save Rey. This was one of the most profound scenes of the movie.

The centrality of the parent-child connection cannot be lost on the Jewish viewer. Judaism is passed down from mother to child, and it is the parents' responsibility to educate the child in the ways of the Torah, hoping and praying that the child will choose to follow and embrace it of their own accord. The relationship between a parent and child is so vital and powerful that it forms the 5th of the Ten Commandments, and is grouped amongst the laws between Man and God, and not between Man and Man. The reason is that through your relationship with your parents you understand your relationship with God, and they are your connection to Him.[6]

This is not to neglect another major theme in Star Wars that resonates with the Jewish viewer – that of the teacher and pupil. Ability, talent and gifts within the force is passed down through bloodlines, but Jedi tradition is taught by a mentor. Again, the relationship between teacher and pupil is central in Star Wars, as it is in Judaism. Pirkei Avot (Ethics of the Fathers) exhorts the individual to 'Make for yourself a teacher' because everyone needs someone who they can learn from, in addition to parents, just like a Jedi needs to learn from a Master. [6]

Proverbs 11:14 says, "Where there is no counsel, the people fall; but in the multitude of counselors there is safety." Proverbs 15:22 reiterates, "He who is wise has many counselors, for without counsel, plans go awry, but in the multitude of counselors they are established.

Keren Keet shared, this theme of family and the transmission of traditions are inherently the most Jewish themes in Star Wars.[6] This really resonated with me while watching *The Force Awakens* now as a mother, as my Jewish-Messianic daughter has chosen the 'Light' of Messiah within the Torah.

The anguish of Han and Leia about the loss of their son struck a chord with me in a way it never could have with the child I was when Star Wars made its debut. Perhaps this was a conscious decision on the part of the filmmakers who know their original audience has become parents. But it's not all doom and gloom, as Leia says about her son, 'There is light in him', and though Ren is a very extreme example of a child rejecting his parents and mentors, the light is always there, and ultimately, there's always an opportunity for redemption. [6]

That is a message that resonates with us here, and with those in a galaxy far far away…

CHAPTER 44: THE REAL STAR WARS

MOVIES AND SUBLIMINAL MESSAGES

Stars Wars director JJ Abrams says it is a "spiritual journey." Aish.com has also mentioned Star Trek and other mega movie blockbuster superimposing Jewish values within the story. Even as far back as Popeye saying "I am what I am and that's all that I am" sounds eerily similar to "I am that I am" or *Ayer Asher Ayer*, which spells the name of the God of Israel. That was God's answer to Moses as to 'who' Moses was encountering.

The list goes on and on giving credulous to the notion of Jews clout in the media world. I do not think it is a bad thing nor should Jews take offense to the allegation. I think Jews are more comfortable with defending injustices of antisemitism and injustice than being conceived as media, political or financial power houses.

It is no secret that Jewish people have related to many of the Star Wars themes, as many I meet in StarFest Conventions, Cosplaying Jedis and Star Wars characters, who could easily be cosplaying Ancient Israelites following Judaism. I had a couple in one of my StarFest panels, who identified as Orthodox Jews, yet believe cosplaying Jedis is synonymous with their beliefs in the Torah and Judaism. Certainly, there is no coincidence there. We can only make comparisons, as the jury is out, on whether or not George Lukas used ancient Jewish texts to inspire Star Wars. Whether it was done purposely or unconsciously, the *Truth is Stranger than Fiction!*

STAR TREK AND GOD

While Sci-fi films often ignores the tough questions about human existence, and who God is, after working closely with Star Trek Stars during Star Fest Conventions and learning that most if not all of the original cast members were atheists, I found this article by Jonathan S. Tobin about Star Trek intriguing and full of irony. After all, Gene Roddenberry was an atheist, secularist and absolutely hated and condemned all forms of religion, but worshipped science instead. Well, I have already proved in Book One of *Who's Who in the Cosmic Zoo?* that science is the mind of God, and understanding scientific concepts leads one to understanding the high mind of God. In fact, one of the world's leading physicists, Michio Kaku has shared those sentiments, in the discovery of the God particle.

"Star Trek" is an all-encompassing vision of the future that has given generations of its fans an alternative universe to ponder. The original series and its generally well-made successors have reflected the culture and controversies of our own time rather than being an educated guess about what will be bothering humanity in the 24th century. In that sense the original Trek did a fine job exploring the tough issues of the 60's, such as racism and the balance of terror between competing superpowers.[7]

In the original "Trek" as well as "The Next Generation," "Deep Space Nine," "Voyager" and "Enterprise," faith in God, or indeed any sort of religious belief, was largely confined to the less advanced populations of planets that were not likely to be joining the United Federation of Planets. The humans of the future were, by and large, clearly atheist even if they traveled among a great many races that held a variety of peculiar beliefs. The only thing in which Star Fleet officers genuinely believe is the "Prime Directive." That is a law that requires space travelers not to interfere in other cultures though, as fans of the show know, it is a rule that is often observed in the breach by officers like Kirk who are often as judgmental about other civilizations as the explorers of an earlier age.[7]

For example, when Captain Picard, the "Enterprise's" leader in "The Next Generation" comes into contact with the inhabitants of a primitive planet that was being observed by Federation sociologists, they naturally assume he is a god because of the amazing tricks he can perform with space age technology. The generally likeable Picard regards these natives with the sort of condescension reserved for slow children. The frame of reference of the "Enterprise" crew is not that the residents of this planet had wrongly mistaken the captain for the true Master of the Universe but that they are children for believing in any sort of deity.

It should be noted that one famous exception to this rule came in a throwaway line concluding an episode during the original "Trek's" second season. At the end of a show in which the intrepid Kirk and Spock had interacted with a planet where something resembling the Roman Empire had endured until the equivalent of the 20th century replete with televised gladiator fights, we were informed that the slaves fighting for freedom were not worshippers of the Sun, but the son of God. This was the only evidence that I can recall that the "Enterprise" crew was even aware of the tenets of any faith that its audience might observe.[7]

One religious breakthrough took place thirty years later in the "Star Trek: Voyager" show which was first shown in the mid-1990s. In that show, the "Voyager's" first officer was a Native American who still practiced the faith of his ancestors making him the first person from Earth to actively profess a religion. [7]

But specifically, Jewish content remains absent from "Trek." Though both of its two main original stars — William Shatner and Leonard Nimoy — are members of the tribe, as is Walter Koenig, who played Cherkov, in the original cast. However, there has yet to be any character in the show who wore a kippah or needed to be absent from his station on the bridge because of Yom Kippur. Much was and continues to be made of Nimoy's "invention" of the Vulcan hand salute ("live long and prosper") that he derived from the priestly benediction extended to the congregation by Cohanim Priests that is used during the Aaronic Blessing in Numbers 6:23-27. "The Lord bless you and keep you; the Lord make his face shine on you and be gracious to you; the Lord turn his face toward you and give you peace."

By the way, I just can't help but make this connection, in this book about Nibiru, that Vulcan was another ancient name to describe Nibiru. It could have been another one of its 7 planets in the Nemesis-Nibiru system. There was some buzz about it, a few decades back, and it is considered an asteroid in today's astrology. So just saying, that for all we know, there may have been some truth within Spok's home planet of Vulcan.

Though viewers have experienced every sort of alien ritual over the years on the show, the observance of Judaism, or even Christianity, is something that appears to be extinct in a "Star Trek" universe where appreciation of other relics of our time such as baseball still exist.

In no small measure this disdain for faith stems from the mindset of its creators who envisioned a future that was based on belief in humanity's steady and inexorable advance from primitive beliefs to rational scientific thought. Doubts about faith in technology have crept into the show over time. The computers and devices that would define the "Trek" world could as easily be used for evil as for good. But it was for the most part a world in which the Supreme Being as understood by the Judeo-Christian worldview (as opposed to the understanding of the fictional Klingons, Romulans, Cardassians and Bajorans) is absent. Gene Roddenberry who was not Jewish despite the persistent rumors to that effect, was atheistic, and a secularist, who saw the future in a basically optimistic way and for him that implied the absence of traditional Western religion.

As such, for all of the enjoyment that "Trek" has given its fans (and I will stipulate that I have been one from the show's earliest days though I have never attended a "convention" or sported a "Star Fleet Academy" bumper sticker on my car.), it has always punted on some of the toughest questions that intelligent beings on this planet or any other are bound to ask about the universe we inhabit. The "Trek" world is one that is without any true God but a belief in humanity and science, a faith that, as we learned to our collective sorrow in the 20th century, is a poor guide for any civilization, no matter how superficially advanced. [7]

BATTLESTAR GALACTICA

An interesting contrast to the atheism of "Trek" is the way religious faith was depicted on the recently concluded "Battlestar Galactica" revival. The new version of the show was a brilliant and dark exploration of the story told from the perspective of the post 9/11 world. Though its ending left most of its fans unsatisfied, in its pilot and first seasons the show's chilling depiction of the genocide of humanity by a race of robots called Cylons made one ponder what it would actually be like to survive such an event or how to deal best with terror and a war against an enemy bent on annihilation.

The most chilling aspect of "Galactica" was its treatment of religion. But as unsettling as all of that might be, the most chilling aspect of "Galactica" was its treatment of religion. The humans on "Galactica" seem to be the product of an advanced civilization but they are also polytheists who always referred to deity in the plural, even when cursing. Their "gods" were those of the ancient Greek myths. But their foes, the implacable Cylons, rejected the pagan religion of the humans and instead worshipped a single God whom they believed was the only true Supreme Being.

How can we account for this curious paradox? Were these monotheistic mass murdering robots a reflection of the fundamentalist Islam that brought down the 9/11 cataclysm on the United States? Perhaps. But as was the case with many other Americans who found the notion of a clash of civilizations between Islamists and the West too painful to contemplate, the show eventually changed course and humanized the Cylons to the point that they were just as sympathetic, if not more so, than the show's humans.

As for faith, it wasn't entirely clear what the survivors of "Galactica's" cataclysmic war between people and machines believed in by the time the show ended but it looked as if something akin to the monotheism of the Cylons seemed to prevail over the pagan myths that drove the humans.

If space operas, like the Westerns that came before them, are the ultimate morality plays, then maybe it is not too much to ask that they take faith seriously. Unlike "Trek's" breezy and optimistic atheism, "Galactica's" dark and confusing attempt to think seriously about faith marked an advance for the genre. A world without God, even a fantasy world, is one that will inevitably be as much at a loss for ultimate answers as the one we currently inhabit.[7]

KNIGHTS TEMPLAR AND STAR WARS

There is also a mysterious connection between the history of the Knights Templars and Star Wars. At the time of writing various episodes of the Star Wars saga, George Lucas used the most terrestrial story to give life to a galaxy far, far away.

From my research it seems to be the opinion of several scholars, that the original trilogy and its Jedi Knights bear similarities to the legend of King Arthur and the history of the Templars. In fact, numerous authors have drawn connections in how the original trilogy has many peculiar similarities to Arthurian legends and how Jedi Knights were modeled based on medieval warriors.

While the Jedi Knights hold noble values, wisdom, and mysticism, so does the Knights of the Order of the Temple. In 1979, Marilyn Sherman said that there is a parallel between the two stories. "These knights are custodians of peace and justice in this galactical civilization, and they are armed with appropriate weapons. Luke Skywalker's Excalibur is a lightsaber, not a clumsy stormtrooper blaster that kills at random, but a clean clear ray that dispatches its deserving victim with finality." Marilyn Sherman wrote two years after the release of the first Star Wars film.

The character of Arthurian Merlin could well be translated as that of Obi-Wan Kenobi. Angela Jane Weis wrote in her book *The Persistence of Medievalism*; "while George Lucas likes to claim that Star Wars is a myth for modern times, it is striking that among the variety of mythic narratives he suggests, his strongest inspiration is clearly the medieval Arthurian romance."

And when you start to think about it, you clearly find several similarities that become evidence immediately. Don't you think it is fascinating that both Luke Skywalker and King Arthur, grow up without knowing their fathers?

Both, however, know a spiritual mentor: Merlin, in the case of Arthur, and Obi-Wan Kenobi, or later Yoda, in the case of Luke. In addition, both are prepared to finally claim an almighty saber, such as Excalibur and the lightsaber.

The love triangle between Luke, Leia, and Han Solo, also has a correlation in the Arthurian narrative, through Arturo, Guinevere, and Lancelot. The difference between both is that in Star Wars the love triangle, breaks when Luke and Leia discover that they are brother and sister which clears the way for Han and Leia to become a romantic couple. Whereas Guinevere and Lancelot's affair, leads to tragedy that contributes to the fall of Camelot.

Here is another uncanny parallel, in the first Star Wars scripts, the Jedi Knights are called Jedi Templars. This order, like the Order of the Templar, was based on votes of austerity, devotion, and moral purity. They also practiced poverty, chastity, and were revered for their honesty, wisdom, and courage.

If you look at Star Wars and the purpose of its characters in the film and compare it to the Knights Templars, you can see that while Star Wars' Jedis were guardians of peace for the galaxy, the Templars had a similar goal in history, they were assembled to protect the conquered Holy Land and its Christian pilgrims from Muslim armies and brigands.

It is also noteworthy to mention how the political institutions of "Star Wars"—such as the Senate, Republic, and Empire—and the pseudo-Latin names of characters such as chancellors Valorum and Palpatine echo those of ancient Rome.

As Tony Keen notes in "Star Wars and History," the architecture on the planet Naboo resembles that of imperial Rome, and the pod race in "The Phantom Menace" rivals that of the Roman chariot race seen on screen in "Ben-Hur."[8]

It is a recurring discussion among Templar enthusiasts that the Order established its headquarters deliberately at the Temple of the Mount (see earlier posts) in order to basically dig down and unearth some secrets held beneath the ancient Temple. One theory is that the Templars were seeking the secrets of the Kabbalah. This is a series of teachings that are completely outside the usual Jewish texts and describe a rather mystical/magical way of getting to know God. [8]

Let's not forget the famous scene on Tatooine with Luke at the backdrop of two setting suns. This was reiterated in The Force Awakens. Could Tatooine be shadowing the myths of the Nemesis-Nibiru system intersecting within our solar system, causing us to see twin sunsets?

It is no coincidence that in 2011 SETI Astronomers named an exoplanet Tatooine in a solar system with two Suns. [9]

THE TREE OF LIFE IN KABBALIST THOUGHT

I do not want to repeat descriptions of the Kabbalah given elsewhere so I can recommend what seems to be a fairly accurate Wiki entry here. In one account of how the Kabbalah was transmitted to Jewish rabbis, it came via the fallen angel Raziel. This angel, the keeper of secrets, had special access to God and would write down everything He said in a big book. This book was handed by Raziel to Adam and Eve – forbidden knowledge of course. Other angels got it back and chucked the offending book in the sea.

The Kabbalah, a gift to mankind from Raziel, is a magical text that allows mortals who embrace it to communicate with the 'other' world of the spiritual. If you are detecting a dualist view of the universe as expounded by Plato, then you are on the right track. Plato didn't just influence Christians – he also exerted a huge influence on Jewish thought. While Kabbalists have always insisted, they believe in the unity of a single God, it is a stream of thinking that is always sounded suspiciously dualist to mainstream Judaism.

Modern Kabbalists also like to claim that the small text anticipates modern atomic theory, the spherical shape of the earth the existence of other solar systems, etc. I have heard the same claims made for the Koran and Nostradamus. But is it so unusual in the ancient world for people to have grappled with the concept of atoms and other universes? Simple answer – no. The Greeks went much further without recourse to magical mumbo-jumbo even calculating the distance between the earth and the moon and the earth and the Sun with remarkable accuracy – by use of pure mathematics and a good grasp of rudimentary astronomy. [10]

Today's Kabbalah is a mix of science and religion, astronomy and astrology, mainstream Judaism with dualist ideas. It is written in code – of course – and has a multi-layered meaning. To get to grips with it is to lift the curtain that separates our physical world to another realm beyond – the true realm. The true Kabbalah was split after its occultic abuse. However, the original Kabbalah, which is Hebrew for the "the receiving" or "the gift," is based on the Torah's philosophies and spiritual beliefs.

A Zodiac from the floor of a 6th century synagogue – magic and religion mixing. So – how do the Templars fit in to all of this? Well, according to some – they weren't set up to protect pilgrims. Forget all that nonsense – that was a cover for their true purpose. They grabbed a spot on the Temple Mount in Jerusalem as their headquarters because, like Indiana Jones, they were after a very big prize. And off they went digging for the Kabbalah underneath their own feet. [10]

Evidence that they discovered the Kabbalah was their effective rejection of Christianity and adoption of magical and mystical symbols and beliefs. Look at the weird imagery you find in Templar churches like pentagrams for example. Having got the secret, they set about destroying all those through crusade who might also have knowledge of the Kabbalah. The Templars wanted to be the only force that truly understood this magic. [10]

CHAPTER 44: THE REAL STAR WARS

So, did the Templars really find the Kabbalah? There is no doubt Kabbalism does pop up with vigor in Europe in the Middle Ages, mainly among Jews living in Provence in modern France and more importantly, Muslim ruled Spain and Portugal. Karen Armstrong, in her book 'The Battle for God', argues that Jews in Spain turned to Kabbalism as their situation grew steadily worse at the hands of invading Christian forces that eventually took the entire Iberian peninsula and drove many Jews in to Morocco. In effect, Jews turned to magic as their real earthly power and influence diminished – it was a belief system based on utter despair. [11]

> "What I see for them is not yet, What I behold will not be soon: A star rises from Yaakov, A scepter comes forth from Yisrael; It smashes the brow of Moab, The foundation of all children of Set."
>
> (Numbers 24:17)

Most notably, a section of the Zohar, the foundational work of Jewish mysticism, describes in detail an astronomical occurrence that must necessarily precede the Messiah.

> "Then one awful star the color of crimson will appear in the middle of the heavens, burning and sparkling in the eyes of the entire world. And a fire will appear as one flame in the northern part of the sky, for forty days, shocking the entire world. At the end of forty days, the fire and the star will battle...At that time, the Messiah will be awakened from a place called the 'bird's nest', and he will be revealed in the Galilee."

Many in Christian circles, who are rooted in fear of all things Jewish, think Jewish mysticism is occult, and quite frankly have little to no understanding of how to extract wisdom from these ancient writings, hold the belief that because the above statement comes from Jewish Mysticism, it can't be true, and must be a prophecy about the Antichrist.

Ovadia and some rabbis believe this describes the arrival of the Planet Nibiru, though this belief is not universally accepted.

> "Pythagoras said that the universal Creator had formed two things in His own image: The first was the cosmic system with its myriads of Suns, moons, and planets; the second was man, in whose nature the entire universe existed in miniature." ~ Manly P. Hall

Notes:
1. Adam Eliyahu Berkowitz, *NASA Discovery of 7-Planet System Conforms Exactly to Zohar's Description of Pre-Messiah "Nibiru"* February 28, 2017, https://www.breakingisraelnews.com/wp-content/uploads/2017/02/TRAPPIST-1-planets
2. Yuval Ovadia, *NASA Discovery 7 Planet System Conforms Exactly Zohar's Description Pre-Messiah Nibiru*, https://www.breakingisraelnews.com/84337/nasa-discovery-7-planet-system-conforms-exactly-zohars-description-pre-messiah-nibiru/#UOLRLaZxISKXoowb.99
3. Daniel Perez, *From Jediism to Judaism: Star Wars as Jewish Allegory*, https://www.aish.com/j/as/From-Jediism-to-Judaism-Star-Wars-as-Jewish-Allegory.html, December 5, 2015.
4. Rachel Sharansky Danziger, *Star Wars and the Hanukkah Story Have More in Common Than You Think*, https://www.kveller.com/star-wars-and-the-hanukkah-story-have-more-in-common-than-you-think/, December 10, 2015.
5. Keren Keet, *Confessions of a Jewish Jedi*, https://www.aish.com/j/as/Confessions-of-a-Jewish-Jedi.html, December 12, 2015, Tatz, Worldmask, Targum Press, 1995, p34, *used with permission*.
6. Keren Keet, *The Last Jedi and the Force of the Family*, https://www.aish.com/j/as/The-Last-Jedi-and-the-Force-of-Family.html, December 10, 2017, *used with permission*.
7. Jonathan S. Tobin, *Star Trek and God, Sci-fi fun often ignores the tough questions about human existence*, https://www.aish.com/j/as/48970691.html, May 11, 2009.
8. Tony McMahon, *The Templar Knight, Mysteries of the Knights Templar* https://business.facebook.com/QuestForTheTrueCross/
9. Brian Vastag, From 'Star Wars' to reality: Astronomers discover Tatooine world with two Suns, https://www.washingtonpost.com/national/health-science/from-star-wars-to-reality-astronomers-discover-tatooine-world-with-two-suns/2011/09/15/gIQAChBzUK_story.html, September 15, 2011
10. Tony McMahon, *The Templars and the Kabbalah*, May 29, 2011, Facebook Post
11. Tony McMahon, *The Knights Templar today - who and where are they? In "Middle Ages"* 2011 Facebook Post.

CHAPTER THIRTY-EIGHT

SIGNS AND WONDERS IN THE HEAVENS

Jesus said, there would be:
"signs in the Heavens, in the Sun, Moon and stars." (Luke 21:25)

"And I will show wonders in the sky above and signs on the Earth below, blood and fire, and vapor of smoke." (Acts 2:19)

"And there will be great Earthquakes, and in various places plagues and famines; and there will be terrors and great signs from Heaven." (Luke 21:11)

In the Torah, the Lord Yahuah said through Jeremiah, "Thus says the LORD, do not learn the way of the nations, and do not be terrified by the *signs of the Heavens*, although the nations (goyim; gentiles) are terrified by them." (Jeremiah 10:2)

I believe we have been seeing these signs in the Heavens that Jesus warned us about, over 2,000 years ago, that would happen prior to His Return. I believe these signs are the exoplanets in the Nemesis-Nibiru system that is now intersecting within our own. I also believe that this is the cause of the exponential increase in UFO sightings, based on the monthly and annual reports provided by MUFON (Mutual UFO Network)[1] proves that, by nation, the United States of America has 600% more reported sightings than another other country in the world. Within the United States, California consistently comes out on top of all the other states as having the most reported UFO sightings in the nation, with Texas coming in second.

On my Public Facebook Book Page[2] for *Who's Who in the Cosmic Zoo?*, I shared that the reason the data always has the USA as having 600% more UFO sightings than any other country has more to do with the Secret Space Program, that has been going on covertly for 70 years, which I detailed in Book One of *Who's Who in the Cosmic Zoo?* as a joint effort and coalition with factions of the Alien Presence inside the Earth, known as the Extraterrestrial Military Industrialized Complex.[3]

Together with all sorts of exotic technologies that are being used by our *Star Wars Program*, the SDI (Strategic Defense Initiative) was implemented by President

Ronald Reagan in the 1980s. The Secret Space Program of today, known as Solar Warden, which not only monitors the Sun but all the kinds of crafts around the Sun, would qualify as the Sign and Wonders in the Heavens that Jesus predicted would happen prior to His return.

The Sun Simulator, one of the exotic technologies, a part of Solar Warden, is followed by some type of spacecraft jet that communicates with the Artificial Prism Sun in preparation when one of the large Nemesis planets is getting ready to eclipse our Sun. Not only is this a sign in the Heavens that the Planet X system is already here, but it is also an indication that one of its giant planets is actually inside our solar system, eclipsing our Sun's light from us on Earth. That means the planet is somewhere between Earth and the Sun, especially since it covers up our Sun. This has been going on for years. During the summer of 2017, when the Solar Eclipse of the Century was occurring on August 21, 2017, there was more than the Moon eclipsing the Sun during those days.

As of July 2018, the Nibiru system is directly behind the Sun. However, we have been seeing its sister planets, Helion, Arboda, Ferrada and their Moons.

INDICATIONS PLANET X IS HERE

Chemtrails have been amped up in the past decade, to befuddle and confuse the population from discerning the Nemesis-Nibiru System and its objects, planets, dwarf stars and Moons at times on a daily basis. The chemtrails serve to obfuscate the ability to discern clearly what is going on around the Sun daily. Planes are ordered to spray the skies daily, creating an opaque cloud covering that can blur and befog our vision together with the brightness of the Flashlight Artificial Sun Simulator, which serves to conceal the incoming Nemesis-Nibiru System. Chemtrailing seems to be especially amped up in the afternoons to create cloud cover to obscure Sunsets, which depending on geographical location and timing, often reveals what appears to be two Suns setting.

The planet Venus is often used as a debunking tactic, not just for Nibiru but for UFOs. It is a fact that the planet Venus is reported as a UFO more than any other object in our night sky. This is because Venus is so bright that as it twinkles it creates oscillations and can easily be misconstrued as moving.

Then on a biochemical level, humanity is being bombarded with chemical warfare, via our foods, water, and air. GMO foods, pesticides, vaccinations, prescription drugs and breathing in chemical aerosols that are used to whitewash our skies.

Chemtrails create passivity and apathy, like fluoridation does, only on another level. One thing I found interesting in my research was that buoys at sea are not compromised by chemtrails because it is an unpopulated area, and the chemtrails are used to conceal the objects in the skies from the human population.

Many people who look at a UFO may say, "I can't believe my eyes!" This is a form of cognitive dissonance, and when the time comes, it is going to be impossible

CHAPTER 45: SIGNS AND WONDERS IN THE HEAVENS

to conceal Nibiru with chemtrails. The whole world will see it, perhaps all of a sudden, because this phase is being used to confound the mind and create a false sense of security as a distraction from those people who actually know about it and have taken steps for preparation. Others however will inevitably be caught off guard and when they see Nibiru coming, the Bible says, their hearts will fail them. Despite the chemtrails and Sun Simulator, their fears will not necessarily be of what they see, but as Marshall Masters rightly pointed out, be due to the realization that they're unprepared.[4] Being prepared is crucial, both mentally, spiritually and physically.

Let's be real, the goal of all life is death. So, preparing our souls, spiritually, emotionally, and mentally, through making right with God through Christ, will insure the best outcome for our individual souls and our collective souls, our soul group, our tribe.

Preparing physically may not work out as one prepares for all kinds of scenarios, with food, water, medicine, supplies, but if your base of operations gets destroyed, then what good will that help you in your preparedness?

> "For what does it profit a man to gain the whole world but forfeit his soul?"
>
> (Mark 8:36)

Death is the stripping away of all that is not you. Eckhart Tolle said that the secret of life is to "die before you die" --- and this is how you find that there is no death. Death of our little egos frees the soul to fulfill its destiny in the Spirit. However, on a celestial and cosmic level, the hard truth is that our Sun is dying. There are multiple exoplanets and solar cores draining it, like vampires.

> "The Sun will be turned to darkness and the Moon to blood before the coming of the great and dreadful day of the LORD."
>
> (Joel 2:31)

Nibiru is getting closer and will cause it to go dark. This is why the Prophesies say that the Lord and Creator of the Cosmos (and us humans) will return to recreate the Heavens and Earth. The Prophesies also predict that Christ will return, and His Light will replace the Sun. In the Age to come, known as the Millennial Reign, 5th Dimension, Earth will receive its light from the Son of God. He becomes the Sun/Son. The End Time Prophecy in Malachi 4:2 says, "The Sun of Righteousness which can also be translated as Son of Vindication, refers to the Terrible Day of the Lord. But it is only terrible if you are in rebellion to Him. To those who are of His Redeemed who respect His name and the power of His name, the promise and covenant is, that you will receive Healing from His Wings, also translated as Rays, and that you will go forth and grow up and leap for joy as calves released from their stall of bondage.

OUR CITIZENSHIP IN HEAVEN

"For our *citizenship is in Heaven*..." (Philippians 3:20)

Citizenship implies a government, and the Scriptures describe the exopolitical hierarchy of the Kingdom of Heaven, governed by the Kingdom of God. When Yeshua-Jesus walked the Earth, He talked more about the Kingdom of God and Kingdom of Heaven than any other teaching. When I put all of Christ's teachings together, what I glean is this: The Kingdom of Heaven is a place that is coming to Earth, which to date, has not yet arrived. However, the Kingdom of God is already here, within us. This implies, one is an actual place, or Grand Mothership as Revelations 22 describes as the Holy City, the New Jerusalem, coming out of the Heavens, and literally landing on a scorched Earth, where the Lord of lords, King of kings, who comes with many crowns, sits on His Throne in the middle of this Phantasmagorical City, and rules over all the nations of the Earth, from the New Jerusalem.

However, Jesus told His disciples that the Kingdom of God is within us. This implies government. When the Lord promised to put His Spirit into His followers and all those who put their trust in Him, and in the last days, that He promised to write His Laws on our hearts, this is what the Kingdom of God within us means. We are born again of His Spirit, and intuitively follow His precepts. Jesus made a clear distinction when trying to teach the Pharisees, that it was more important in the Kingdom of God to follow the spirit of the law, than it was to follow the letter of the law. He also called lawyers a brood of vipers, as did His forerunner, John the Baptist.

Legalism networks with arrogance, which creates a critical accusing spirit, also known as the Religious Spirit. Knowledge networks with wisdom, these are incomplete without each other. When you do not have the wisdom to apply knowledge, it is arrogance and pride. Just because you may have great knowledge, does not mean you have great wisdom. The beginning of wisdom is the fear (respect) of the Lord. It is humility. Humility defeats arrogance and pride in the spiritual battle over our souls. When we abide in humility, we can be assured that the Holy Spirit lives with us.

Spiritually speaking, this is the difference between the legalistic Religious Spirit, which implies the personality of satan's accusing spirit, versus the Holy Spirit of Grace, which promises that where error (sin) abounds, Grace abounds even more. This is the hope and promises of God, that it is not about us being perfect in our own self-righteousness, but about us being made perfect through Him. "He will perfect that which concerns me." (Psalm 138:8) "He is the author and finisher of our faith." (Hebrews 12:2)

CHAPTER 45: SIGNS AND WONDERS IN THE HEAVENS

"The LORD said to my Lord: "Sit at *My right hand* until I make Your enemies a footstool for Your feet."
(Psalm 110:1)

"After the Lord Jesus had spoken to them, He was taken up into *Heaven* and sat down at the *right hand of God*."
(Mark 16:19)

The Right Hand of God is the Seat of Power. All Power and Authority has been given to the Son. Not many come to the Father except through the Son. Yeshua/Jesus is the Gatekeeper to Heaven.

"For it became him, for whom are all things, and by whom are all things, in bringing many sons unto glory, to make *the captain of their salvation* perfect through sufferings."
(Hebrews 2:10)

We are familiar with Captain America, Captain Marvel, and the Captain of Ships both sea and air, but to think of the Lord as the *Captain of Salvation* is a whole new paradigm shift. But truth be told, this is exactly what the Original Hebrew Bible has been saying all along yet it has been lost in translation. The words, *Lord of Hosts*, were translated from the Hebrew, *Adonai Tzebayoth*, which literally translates to the Lord of Celestial Armies. He is the Captain of fleets of spaceships. The top Admiral, the top Gun, the Captain of all the extraterrestrial legions of angels.

"And being made perfect, he became **the author of eternal salvation** unto *all of them that obey him*; Thou maddest him a little lower than the angels; thou crowned him with glory and honor, and didst set him over the works of thy hands:"
(Hebrews 2:5, 7-9)

He became the *author of eternal salvation* to *All of them that obey him*. That includes extraterrestrials, aliens, angels, animals, birds, fish, and humans. Why we as humans tend to think it is all about us. This thing we're in, on Earth, the opportunity to be redeemed from the curse of the Lucifer, is an extraordinary path to becoming actual *citizens* of Heaven's Kingdom through the *author and finisher of our faith*, which is the Lord Yeshua aka Jesus. He is the Way, the Truth, and the Life, He is the only Gatekeeper you need to worry about, to ensure your place in His Kingdom.

"Wherefore when he cometh into the world, he saith, Sacrifice and offering thou would not, but a body hast thou prepared me:"
(Hebrews 10:5-12)

We are to offer up our bodies as living sacrifices, so He may dwell within us, as us. This is the Body Temple; this is the Body we are called to Prepare for Him to house the indwelling of His Holy Spirit. "I beseech you therefore, brethren, by the mercies of God, to present your bodies a living sacrifice, holy, acceptable to God, which is your spiritual service." (Romans 12:1)

Christ is the Way through His ultimate and final sacrifice. What He has done for us, spiritually and legally, is open the doors and gates of Heaven through His sacrifice that was sufficient to take all our sins away, break the curses of sin and death over humankind. No animal sacrifice can match or outdo His because He was the Son of God. This was for *all Creation*. Not just humankind.

> "For by a single offering he has perfected for all time those who are being sanctified. And the Holy Spirit also bears witness to us; for after saying, "This is the covenant that I will make with them after those days, declares the Lord: I will put my laws on their hearts, and write them on their minds," then he adds, "I will remember their sins and their lawless deeds no more." Where there is forgiveness of these, there is no longer any offering for sin."
>
> (Hebrews 10:14-18)

> "All creation waits in eager expectation for the revelation of the sons of God."
>
> (Romans 8:19)

All Creation!

> "**All creation** is eagerly waiting for God to reveal **who** his children are. Creation was subjected to frustration but not by its own choice. The one who subjected it to frustration did so in the hope with the hope that the same creatures shall be delivered from the slavery of corruption into the glorious liberty of the sons of God. We know that **all creation** has been groaning with the pains of childbirth up to the present time. And not only they, but ourselves also who have the first fruits of the Spirit, even we ourselves groan within ourselves, waiting for the adoption, that is to say, the redemption of our body."
>
> (Romans 8:19-23)

> "**And there shall be no more curse**: but the throne of God and of the Lamb shall be in it; and his servants shall serve him:"
>
> (Revelation 22:3-5)

EXOPOLITICAL EVANGELISM

Astronomer Dr. Carl Sagan posed an interesting question in his 1966 book, *Intelligent Life in the Universe*, about not excluding the possibilities that Extraterrestrials may come here to evangelize us. Sagan indicated Earth had likely been visited by extraterrestrials numerous times. He stated, "Sumer was an early–perhaps the first civilization in the contemporary sense on the planet Earth. It was founded in the fourth millennium B.C. or earlier. We do not know where the Sumerians came from. I feel that if the Sumerian civilization is depicted by the descendants of the Sumerians themselves to be of non-human origin, the relevant legends should be examined carefully." [5] (p. 456)

He goes on to ask, "What might an advanced extraterrestrial civilization want from us?" He answered his own question by stating, "One of the primary motivations for the exploration of the New World was to convert the inhabitants to Christianity — peacefully if possible — forcefully if necessary. Can we exclude the possibility of an extraterrestrial evangelism?" [5] (p. 463)

What if, it is the other way around? What if we, who have been given the opportunity to transform ourselves out of the Kingdom of Darkness into the Kingdom of Heaven through the work of Christ, are here as Ambassadors of that Kingdom not only to witness to other Earth humans, but unbeknownst to ourselves, to the Extraterrestrials who watch humans? Perhaps Carl Sagan was onto something in his postulation, but because he was a non-believer, he could not see it from the opposite perspective.

The simple protocol of our roles verses God's role in dealing with the world of the unredeemed is, firstly, our job as Ambassadors of the Kingdom to shine our light into the darkness. It is our role to be "loving" and "kind" to the unbeliever because we represent the Grace of God through Christ. When we reach out positively, then the job of the Holy Spirit is to convict their conscience and draw them to repentance. For the promise of the Word says, "it's the Goodness of God, which leads one to repentance." Repentance is a turnaround from going in the wrong direction and from error. Repentance does not imply punishment, but the promise of God's Grace being poured out through Forgiveness. It is the very essence of the Gospel (*Gospel* means *Good News* in Greek).

Judgement belongs to the Lord. It is His job to judge them. Our job is to be vessels of His Grace and Love. Just as we were saved behind enemy lines, so is it our duty to go back to enemy lines of where we used to be and show others the way out.

> "Judgment will come because the ruler of this world has already been judged."
>
> (John 16:11)

Often, it takes only a crack of light to illuminate a dark room. Likewise, it can take words spoken in the right way, at the right moment, to ignite someone's God Spark within them. God can turn anyone around. We are to be His Catalyst.

It is well established that the Lord is Extraterrestrial (John 18:36). He will have the last Word. He will be the Judge – not us. He will judge rightly, especially those who behave like Lucifer/satan.

> "We know that we are children of God, and that the whole world is under the control of the evil one."
> (1 John 5:19)

Satan is the one who said he wanted to be exalted above the Most High God. (Isaiah 14:14) So, when Christians who claim to be saved, turn on the unsaved by judging them, or others, then in their self-righteousness they are serving the Accuser of the Brethren, who is satan, who wants to usurp, steal, and block the power of God. This is the influence of the Counterfeit Religious Spirit, which wars against the Holy Spirit of Christ.

What and where is the power of God when it comes to the unbeliever? It is in His Grace. The same Grace that saved you and I out of the pits of despair and out of the clutch of the enemy. When people who do get saved, and God begins to use them to draw others to Him through them, then comes the finger pointing of how false they are for fraternizing with sinners or speaking to them in their own language. Yet, Christ did exactly the same thing. He ate with the tax collectors, the gluttons, the drunkards, this is how He fulfilled His calling, to save the Lost House of Israel. Surely, we who are of the called, and of the redeemed, are expected to do the same.

The Funda-mentalism of Judgmentalism is none other than the work of satan, who wants to thwart the power of God through us, and more importantly wants to block those who want to reach others for the Kingdom.

The field of Exopolitics is about policy and protocol for dealing with the Extraterrestrial presence. This is not limited to the Alien Presence on Earth but is inclusive of our relationship to the Lord through His Spirit on Earth.

HEAVEN IS INSIDE THE EARTH & ABOVE US

Henry David Thoreau profoundly said, "Heaven is under our feet as well as over our heads." This means there is a Heaven below as well as above.

The quintessential End Times Prophesy is focused on the final events when the Lord Himself returns to Earth to collect His own people. While Christians can argue till the Lord Returns, (notice I didn't say, till the cows come home), on whether this event known as the Rapture-Ascension, will happen before the Tribulation, halfway through the Tribulation or at the end of the Tribulation. Many Christians identify themselves as Pre-Trib; Mid-Trib; and/or Post Trib, depending on how they perceive these scriptures. But without getting into the weeds about timing, as I have

CHAPTER 45: SIGNS AND WONDERS IN THE HEAVENS

already gone over all these positions along with all their supporting scriptures in the three respective positions, that this event described in 1 Thessalonians 4, is further explained as to "when" it will take place, in 1 Corinthians 15:52 "in an instant, in the twinkling of an eye, **at the last trumpet**. For the trumpet will sound, the dead will be raised imperishable, and we will be changed."

However, what I want to focus on within 1 Thessalonians 4, are the words, *the dead in Christ will be the first to rise*. This clearly implies that those who die in Christ, go to the Paradise, aka Abraham's Bosom, which is located *inside* one of Earth's many dimensions.

The Return of the Lord

"Brothers, we do not want you to be uninformed about those who sleep in death, so that you will not grieve like the rest, who are without hope. For since we believe that Jesus died and rose again, we also believe that God will bring with Jesus those who have fallen asleep in Him.

By the word of the Lord, we declare to you that we who are alive and remain until the coming of the Lord will, by no means, precede those who have fallen asleep. For the Lord Himself will descend from Heaven with a loud command, with the voice of an archangel, and with the trumpet of God, and *the dead in Christ will be the first to rise*. After that, we who are alive and remain will be caught up together with them in the clouds to meet the Lord in the air. And so, we will always be with the Lord.

"Therefore encourage one another with these words."

(1 Thessalonians 4:13-18)

Where, O Death, is Your Victory?

(Isaiah 57:1-2; Hosea 13:14)

"Now I declare to you, brothers, that flesh and blood cannot inherit the kingdom of God, nor does the perishable inherit the imperishable.

Listen, I tell you a mystery: We will not all sleep, but we will all be changed—in an instant, in the twinkling of an eye, at the last trumpet. For the trumpet will sound, the dead will be raised imperishable, and we will be changed. For the perishable must be clothed with the imperishable, and the mortal with immortality.

When the perishable has been clothed with the imperishable and the mortal with immortality, then the saying that is written will come to pass:

"Death has been swallowed up in victory."
"Where, O death, is your victory?
Where, O death, is your sting?"

(1 Corinthians 15:50-55)

Notes:
1. MUFON, Mutual UFO Network, https://www.mufon.com/
2. *Who's Who in the Cosmic Zoo?* Facebook Fan Page, http://www.Facebook.com/whoswhointhecosmiczoo
3. Ella LeBain, *Who's Who in the Cosmic Zoo?* Book One – Third Edition, Skypath Books, 2012.
4. Marshall Masters, *Surviving the Planet X Tribulation: A Faith-Based Leadership Guide* 2nd ed. Edition, Your Own Worlds Books, 2016.
5. Carl Sagan, I.S. Shklovskii, *Intelligent Life in the Universe*, p.456, 463; Dell Publishing, 1966.

CHAPTER THIRTY-NINE

SUN OR SON?

"The Sun shall from the grave arise,
And tread again the summer skies."

WHO IS THE GREATER LIGHT?

Solar Mythology and the Jesus Story: The Solar Mythology, the birth of Jesus Christ, is actually a metaphor for what is the oldest and most important story humans observed, recorded, and wrote down--that of the annual passage of the seasons, based on the changing positions of the Sun against the celestial sphere.

Christians Verses Christmas

Every year, during the holidays, a certain group of modern-day Pharisees, bearing the Religious Spirits, and believing they are holy, holier than we are, who claim to be real Christians, engage in a lot of argument on why Christians should ignore Christmas due to its Pagan Roots. An old saying, "a little knowledge is a dangerous thing," applies to them.

> "The beginning of wisdom is this: Get wisdom. Though it cost all you have, get understanding."
> (Proverbs 4:7)

This means having a little bit of knowledge about the pagan root of Christmas does not negate the celebration of it by spirit-filled, faithful Christian believers. I am writing this to impart this spiritual discernment and understanding to all those Christians who judge others for celebrating Christmas, and why they should too.

I included a rather lengthy chapter in Book Two: *Who is God?*[1] that lays out the historical timeline of the 16 gods who were venerated on December 25th since deep antiquity. As I pointed out, most people alive today never even heard of these 16 gods, except for Jesus Christ.[2]

> You cannot cut Christmas out of the calendar,
> Nor out of the heart of the world.
> Christmas is not a date. It is a state of mind.

"Christ is all and is in all." (Colossians 3:11)

The point is, it really doesn't matter who was once venerated on December 25th, nor does it matter that Jesus wasn't actually born in December. Conceived perhaps, but most Bible scholars, including me, agree that Yeshua, Emmanuel, Jesus Christ, was born during the Fall Feasts, his date change did in fact occur under the auspices of the Church of Rome by the Emperor Constantine. The decision by Constantine was to vacate their god Mithras and replace him with Jesus for the Winter Solstice Celebration of the Return of the Sun God on Christmas Day, with the Son of God, who is Jesus.

This is why this is important to every faithful, spirit-filled Christian alive, and that is, the ultimate and final End Times Bible Prophecy, found in Malachi.

> "But for you who revere my name, the Sun of righteousness will rise with healing in its wings. And you will go out and leap like calves released from the stall."
> (Malachi 4:2)

This blessing is promised to those who fear the Lord and are ready for His return. "Sun of righteousness" can also be translated "son of vindication." The context concerns the Day of the Lord, the time when God vindicates His people and judges sin. This vindication will be clear to all, like the bright light of the Sunrise.

The One described as the "Sun of Righteousness" can be no other than Jesus Christ Himself. The Lord is called "the LORD your righteousness" in Jeremiah 33:16. And the coming of the Messiah is pictured as a Sunrise in several passages. "Arise, shine, for your light has come, and the glory of the LORD has risen upon you" (Isaiah 60:1).

The fact that the Sun of righteousness rises with "healing in its wings" invokes the picture of the wings of a bird stretched across the sky, offering healing to those below. A healing effect will infuse the Earth during this time, removing the negative impact of past sins (Isaiah 30:26; 53:5). When Christ returns, God's righteousness and peace will flood the Earth (Isaiah 11:9; Habakkuk 2:14).

Christmas has always been about the path and return of the Sun's light. On the Winter Solstice, the Earth is furthest away from the Sun, causing it to receive the least amount of Sunlight, making it the longest night of the year, and the shortest day. Astronomically speaking, it takes three days for the Earth to turn around on its elliptical orbit around the Sun, where it appears to be standing still for three days, until it actually moves back towards the Sun from turning an orbital corner, so to speak. This happens on Christmas Day, which is the return of the Sunlight. Christmas Day the days begin to increase in Sunlight daily, until it is equal day and equal night at the Vernal Equinox leading up to the longest day of the year and the shortest night, on the Summer Solstice.

CHAPTER 46: SUN OR SON?

Christmas celebrations were traditionally all about the return of the light. This is why the Church of Rome adopted this day to celebrate the birth of Christ, mainly because they switched out the Sun god Mithras for Jesus Christ. Christians who reject Christmas because of its pagan roots, need to look in the mirror and examine the grace of their salvation from pagan roots. But here's why it's so important to see this both spiritually and prophetically: Those who believe in Jesus as the Light of the World need to understand that the quintessential End Times Prophesy is the Return of Christ to Earth, after Earth is nearly destroyed and the Sun and Moon no longer give their light.

> "For the stars of Heaven and the constellations thereof shall not give their light: the Sun shall be darkened in his going forth, and the Moon shall not cause her light to shine."
> (Isaiah 13:10)

> "Immediately after the distress of those days 'the Sun will be darkened, and the Moon will not give its light; the stars will fall from the sky, and the Heavenly bodies will be shaken.'"
> (Matthew 24:29)

The Bible Prophesy says that when Christ returns, He becomes the Light of the World, and we will no longer depend on the Sun for light but will receive light from the Glory of His Presence, because He is Light. Here is the rub, our Sun is dying. This is understood by many astrophysicists and astronomers today who study the cycles of the Sun. It is Bible Prophecy that the Sun will go dark. But the Redeemed for the Kingdom of Heaven inherit the Earth, and the Light of the World is Jesus, and He becomes both the Sun and the Son of Righteousness, who returns with Healing in His Wings (Rays).

This is why during this darkening time of the year, it is so important to lean on Him as your Light, and forget the Pagan traditions, but honor and celebrate Jesus returning as the Sun of Righteousness with Healing in His Wings.

And finally, another point I want to make, which I have articulated in *Who is God?* Chapter 27: *Who Are the Christmas Gods?*, is that when someone becomes a Christian, by accepting Jesus as your Savior and Lord of your Life, it's a process of transformation, through repentance and receiving His Grace, to make you, His Vessel. Think about all the stuff you let go of, or are presently letting go of, that you are surrendering to the lordship of Christ in your life. This is known as sanctification and deliverance, so that you may be cleaned out from all the dark stuff Christ saved you from and become a vessel for His Light and His Love to be expressed in this world. You do this through repentance and being filled with His Holy Spirit. We are all saved behind enemy lines.

Each time you let go of something negative or ungodly and invite His Spirit to take more territory in your mind, your heart, your relationships, your body, you become a brighter vessel to accomplish the Divine calling on your life by Christ, so that His Divine Will may be expressed and accomplished through you. Anyone who has been through this deliverance and sanctification process will relate, that your life has more and more of Christ in it, and less and less of your old self.

How much more can a date on the calendar be vacated by the forces of darkness and taken over as "territory" for Christ and His Kingdom of Glory? If you think the Lord does not want anything to do with something that used to be Pagan, think again. That is the very reason He came to Earth in the first place, to shine through as the God of gods, King of kings and Lord of lords. That includes your body, and that includes every day on the Calendar, not just holy days (holidays) and appointed times. In this way, Christmas is an appointed time, because one day, Malachi 4:2 will be fulfilled, and instead of us waiting for the return of Sunlight every year on December 25, we will have the return of the Son's Light, as the Sun of Righteousness with Healing in His Wings.

Those who judge Christians or Jews, or anyone for that matter, for celebrating Christmas, belong to the Antichrist Religious Spirit which wars against the Holy Spirit. How petty does one need to be, to see that celebrating someone's birth into this world, on what has become a unanimous date of the year for souls worldwide, does not have to be on that person's actual birth date? Many of us, whose birthdays fall in the middle of the week, end up postponing our celebration for a date that is mutually agreeable for all family and friends to get together.

Birthdays are not legalistic. If three billion people choose to celebrate Christ's birthday on Christmas Day, and you claim to follow Christ and refuse to do so, then perhaps you need to examine which spirit you follow in the Church. The Religious Spirit, aka modern-day Pharisee, hung up on legalism and pettiness? Or the Holy Spirit, the Spirit of Peace, Love and Joy, that is the Spirit of FREEDOM? If you choose the Religious Accusing Spirit, then I say to you, stop being such a Spiritual Grinch, by aligning yourselves with atheists, and every other Antichrist spirit in the world, by teaming with the demons who try to steal the Christmas Spirit from those who get it.

Be careful, that you don't end up in the camps of Spiritual Goats when Jesus returns to Earth to Judge the Church in His Sheep vs. Goat Judgment. **Matthew 25:31-46** has to be the scariest scripture to Christians, let alone in the entire Bible. For it is not a Judgement on the unbeliever, but a Judgment on the False Church, the Church who follows the Religious Spirit, instead of the Holy Spirit of Christ.

THE SHEEP AND GOAT JUDGMENT

"When the Son of Man comes in his glory, and all the angels with him, he will sit on his glorious throne. All the nations will be gathered before him, and he will separate the people one from another as a shepherd separates the sheep from the goats. He will put the sheep on his right and the goats on his left.

"Then the King will say to those on his right, 'Come, you who are blessed by my Father; take your inheritance, the kingdom prepared for you since the creation of the world. For I was hungry, and you gave me something to eat, I was thirsty, and you gave me something to drink, I was a stranger and you invited me in, I needed clothes and you clothed me, I was sick, and you looked after me, I was in prison and you came to visit me.'

"Then the righteous will answer him, 'Lord, when did we see you hungry and feed you, or thirsty and give you something to drink? When did we see you a stranger and invite you in, or needing clothes and clothe you? When did we see you sick or in prison and go to visit you?'

"The King will reply, 'Truly I tell you, whatever you did for one of the least of these brothers and sisters of mine, you did for me.'

"Then he will say to those on his left, 'Depart from me, you who are cursed, into the eternal fire prepared for the devil and his angels. For I was hungry, and you gave me nothing to eat, I was thirsty, and you gave me nothing to drink, I was a stranger, and you did not invite me in, I needed clothes and you did not clothe me, I was sick and in prison and you did not look after me.'

"They also will answer, 'Lord, when did we see you hungry or thirsty or a stranger or needing clothes or sick or in prison, and did not help you?'

"He will reply, 'Truly I tell you, whatever you did not do for one of the least of these, you did not do for me.'
"Then they will go away to eternal punishment, but the righteous to eternal life."

(Matthew 25:31-46)

We will all be judged by our words and deeds in the end by Christ, not Religious Spirits. Let me remind all Christians of this fact, Christ was Crucified by Religious Spirits. Those who carry the Religious Spirit, instead of the Holy Spirit, keep crucifying Him over and over again by persecuting those who are vessels for Christ's Spirit of Freedom, Peace, Love and Joy. I also want to remind these so-called Christians, that once a person is given salvation through Jesus Christ, they are called to the Great Commission to share the good news of the Gospel of Grace to all creation. Yes, that includes aliens too! Jesus said, "I tell you the truth, anyone who believes in me will do the same works I have done, and even greater works, because I am going to be with the Father." (John 14:12)

It is a fact that Christmas and Easter are the two biggest evangelical opportunities of the year. Thousands of people come to Church on Christmas Eve who may never step foot into a church setting all year, and either give or rededicate their lives to Christ. So, I ask you, who are you to judge them? If you are one of those who thinks Christmas is pagan, then how come thousands of people are getting saved on this day? Why do churches have the biggest attendance in altar calls during Christmas services than any other time of the year, except for Easter? Me thinks, Christians need not only understand the historical battles here, between the Church of Rome and the Jews, but also need to understand the Will of God, which is save all of creation, through the Grace of God through Jesus Christ.

Rabbi and Jewish scholar Arthur O. Waskow concluded in his book, *Seasons of Our Joy*, about the Jewish holidays of the year, can relate to all the seasonal holidays:

> "In a world in which any moments of rest, celebration, meditation — the festivals of any religious tradition or spiritual orientation — are often seen as 'a waste of time' because they detract from productive work, it is crucial to remind ourselves that work is not the only valuable behavior. Doing, making producing, must be part of a rhythm in which being also has its place. This is the ultimate message of the seasons of our joy: that there is joy in the seasons themselves, in our very decision to join in noting them, in celebrating them, in walking the spiritual path — the spiral — that they make."[3]

The following was a Word of Knowledge given to a group of believers. It is believed to have come from the Holy Spirit. No other name is attributed to this Word:

This time I came I came as a lamb,
I humbled myself and I walked as a man

But this time I am coming like a mighty Lion,
And I will cause my church to have a backbone like iron

I will move by my Spirit and cause you to pray,
You shall pull down the strongholds that stand in the way

For I am raising up a people never seen before
Ones who are fearless and ready for war

This battle will not be fought with ammunition,
but with songs of praise, you will take your position

Your ears will be tuned to the captain's command,
You will take over cities, you will march through the lands

Men Women and children will take their stands,
All colors, all races will walk hand in hand

This army will be filled with the Holy Ghost,
And their commander and chief is the Lord of Hosts

Everyone marching in one accord,
All in this army sold out to the Lord

All with their eyes on the crucified one,
And everyone moving in unison.

ANCIENT HUMAN STARGAZING

In ancient times humans lived by the Sun and the stars. When it was night, they had nothing better to do then stargaze. They used the stars to navigate through endless terrains and vast deserts. They did not overcome the darkness with the flip of a switch. When Sunset came, the day ended. This is why in the Jewish Calendar, Sunset began, marking the end of a day and the beginning of the next. They observed how stars moved across the sky during the night, and how different constellations are visible at various times of the year. This is how they created calendars.

Some would have you believe that it was our ancestors who made up stories as they connected the dots to create the star pictures. Yet one of the earliest Jewish scriptures, the Book of Enoch, tells us that it was a group of 200 fallen sons of Heaven, that have become known as fallen angels, who were the ones who taught men astrology. Now if these sons of Heaven who rebelled against their Creator were the ones who gave the original meanings of the Star Pictures to humankind, then it would only be logical to deduce that they knew God's original astrology, which as I lay out in Book Six, *Heaven's Gospel*, was the first Gospel in the Stars.

Later on, the star picture stories took on a life of their own, as they were changed according to the ruling empire and elite, who changed the original star pictures to reflect and immortalize their own pagan gods. It is a known fact that the Greek gods were merely renamed from those of the Roman pantheon. Some constellations continue to hold a measure of the original truth in them, some have become completely lost in the annals of time as the stories they attached were passed down from generation to generation, while the original knowledge of what the stories represented was lost.

Therefore, it comes as no surprise that in our age of knowledge and technology that we've lost the original astronomical meaning of the original Divine Plan of Salvation as it was first written into the stars, along with digital clocks, wall and computer calendars, that results in a generation of people who never look to the stars for direction nor do they pay any attention to the true position of the Sun in the sky.

However, a growing group of Sky Watchers and researchers are noticing that the skies have changed, that today's position of the Sun is not where it used to be in years past. In Colorado, we measure west by the boundary of the Rocky Mountains. We always know when we see the Front Range Mountains and Foothills, we are looking due west. In the winter months, the Sun naturally sets in the southwest. However, in recent years, the Sun is setting way further south down the Front Range than ever before. We know this, because every mountain, every Fourteener is a landmark. We can see in our neighborhoods that the position of the Sun setting across the Front Range has shifted miles south, so far that some locations cannot see it, as they used to. The same goes for the summer months, when the Sun sets furthest behind the northern mountains, and has shifted markedly further south. This indicates a coming shift.

OUR STORY BEGINS AND ENDS IN HEAVEN

> "But our citizenship and conversation is in Heaven. where the Lord Jesus Christ lives. And we are eagerly waiting for him to return as our Savior."
> (Philippians 3:20)

When you go outside to look up at the night sky, you are literally looking at the Heavens. You see all the stars that make up all the constellations. The origin of the Jesus Story takes place not on Earth, but in the Heavens, the starry night sky above us.

Comparing Jesus's ministry to the journey of the Sun through all 12 zodiac constellations may be a conundrum to some, but it is the secret to understanding the Divine Plan of Salvation. For this reason, I have devoted my next book, *Heaven's Gospel* -Book Six and the final book of my book series to revealing these mysteries.

Right after Sunset is a good time to observe the Sun's position in relation to the stars in the Heavens, as the stars become visible just after Sunset. You know the Sun is just below the horizon, so you can get a good idea of where the Sun is located within the celestial sphere of stars, the Milky Way.

CHAPTER 46: SUN OR SON?

THE ECLIPTIC CROSS

The cross Jesus is crucified on represents the division of the four seasons of the year: Winter, Spring, Summer, Autumn. The four posts of the Cross divide the year into the four seasons. The circle you often see in the center of the Cross, the symbol of Christianity, represents the ecliptic, the circular path of the Sun. A planisphere is a map of the starry night sky. A planisphere also doubles as a time of year calendar. On the planisphere along with the stars is drawn the ecliptic — the apparent path the Sun takes through the stars during the year. On the planisphere the ecliptic is close to a circle. You can divide the planisphere into the four seasons of the year by drawing a cross on it. Thus, you end up with a cross with a circle in the middle. The Cross on top of Goleta Presbyterian Church has the traditional circle in the middle of it.

THE SOLSTICE CROSS

The Bible is mostly astrological allegories. The reason the book of Revelation doesn't make any sense is because it's an astrological allegory, and includes prophecies that are astronomical events, like comets hitting the Earth, causing the waters to turn to blood, the Sun to go dark and the Moon to turn red. Revelation's symbolism and metaphor relates to Astrological Constellations. September 23, 2017, was believed to be the fulfillment of Revelation Chapter 12, believed to have been the fulfillment of the first two verses. To many of us, it makes sense when you start to interpret it in that manner.

Revelation 12:1-3 describes "a woman clothed with the Sun, with the Moon under her feet and a crown of twelve stars on her head. She was pregnant and cried out in pain as she was about to give birth." The woman is believed to be the Virgin in the Constellation of Virgo, and September 23, the Sun and the Moon were located under her feet. The nine stars of the zodiac constellation Leo, plus three planets (Mercury, Venus, and Mars), will be at the head of Virgo — "on her head a crown of 12 stars." I found in the past 1,000 years, this same arrangement in the sky has happened at least four times, in 1827, 1483, 1293, and 1056.

Revelation 12: 3-4 describes: "Then another sign appeared in Heaven: a huge red dragon with seven heads, ten horns, and seven royal crowns on his heads. His tail swept a third of the stars from the sky, tossing them to the Earth. And the dragon stood before the woman as she was about to give birth, ready to devour her child as soon as He was born."

Her pregnancy was that of the transit of the planet Jupiter which passed through her belly, however, there was one distinction: the claim that transiting with Jupiter was the Red Dragon, Nibiru, that passed through her belly on that day, heralding the Apocalyptic End Time Rapture of the Church, which did not happen. The fact that this astronomical lineup has happened before means it will happen again. However, according to Nibiru researchers, the red comet-like planet did enter into our solar system at that time and is traveling toward the Sun. Nibiru has long been called the Dragon, and it has two tails of comets in its path, which is also why it has been called The Destroyer. This could be a future event that will take place in Virgo, when Nibiru does sweep through our solar system as it has done before, approximately every 3,650 years.

Old Testament stories coincide with astrological allegory. The story of Adam and Eve is the story of Virgo and Bootes, the constellation next to Virgo. The story of Joseph and his coat of many colors, the youngest and most favorite of 12 brothers, represent Pisces, the sign that blends all the rest into one. Joseph was one of 12 sons of Jacob, who began the 12 tribes of Israel. The number 12 again indicative of an astrological allegory of the 12 months of the year and the 12 signs of the Zodiac. Later, there were 12 Jewish Disciples who followed Jesus and began Messianic Christianity. No coincidence there.

THE THRONE OF DAVID

> "The LORD has sworn to David, A truth from which He will not turn back; 'Of the *fruit of your body* I will set upon your *throne*."
>
> (Psalm 132:11)

Psalm 132 is prophetic in nature, referring to the Lord's promise made to David through the prophet Nathan years before. The promise had involved the establishment of a permanent king who would sit down on the throne of David and rule forever. This King would be a descendant of David. Where do we look for the fulfillment of this promise? Has it been fulfilled as yet, or is it something to be fulfilled in the future?

> "And in the days of these kings shall the God of heaven set up a kingdom, which shall never be destroyed: and the kingdom shall not be left to other people, but it shall break in pieces and consume all these kingdoms, and it shall stand forever."
>
> (Daniel 2:44)

NATHAN'S PROPHECY

The prophesy of the succession of the Throne of King David is found in 2 Samuel 7:12-16 spoken through the Prophet Nathan. This is what he spoke from the LORD YAHUAH, to King David:

> "When your days are complete and you lie down with your fathers, I will raise up your descendant after you, who shall come forth from you, and I will establish his kingdom. He shall build a house for My name, and I will establish the throne of his kingdom forever. I will be a father to him, and he will be a son to Me; when he commits iniquity, I will correct him with the rod of men and the strokes of the sons of men, but My lovingkindness will not depart from him, as I took it away from Saul, whom I removed before you. And your house and your kingdom shall endure before Me forever; your throne shall be established forever."
>
> (2 Samuel 7:12-16)

Nathan's prophecy is for David's bloodline. His son Solomon was the first to manifest fulfillment of this prophecy by building the Lord's First Temple and establishing it in Jerusalem. But it did not end there, the prophecy extends down to his descendants. A future King and future Kingdom through Yeshua Ben Joseph, Ben Miriam, both were descendants of King David.

What many of today's scholars forget, was that during Old Testament times, your bloodline was identified by your Mother. This was for practical reasons, as men were sent to out to hunt and to do battle, they did not always return. A child's identity was attached to his mother's bloodline. Yes, Old Testament people were linked to their father's bloodline, but when the father was absent, they were dependent on their mother. This is why this rule was later changed and incorporated into Rabbinical Judaism, who today measures a Jew based on their maternal lineage, even though the Torah says otherwise. Be that as it may, this is somewhat silly, because today we have DNA tests to tell *who is who* in one's family.

However, this detail about carrying on the bloodline, was why Miriam was chosen to be the mother of Yeshua, because she was a descendant of King David. As a matter of fact, Joseph was as well. So, Yeshua's ancestry could not be questioned, at least in God's eyes. We all know how that story ended. And what really matters in the end, about all these controversies, is what God knows about us. He is the ultimate Judge and Jury of our lives. He knows 'who' His Son is, and 'who' the true Messiah is.

The prophetic about building the Lord's house in verse 13, is in one sense, Solomon who built the Lord's house in his glorious temple in Jerusalem. But Yeshua/Jesus also built a house for God; made of living stones; a spiritual temple; the body of Christ:

> "Do you not know that you are the temple of God and that the Spirit of God dwells in you? If anyone destroys God's temple, God will destroy him. For God's temple is holy, and you are that temple."
>
> (1 Corinthians 3:16-17)

> "Now, therefore, you are no longer strangers and foreigners, but fellow citizens with the saints and members of the household of God, having been built on the foundation of the apostles and prophets, Jesus Christ Himself being the chief cornerstone, in whom the whole building, being fitted together, grows into a holy temple in the Lord, in whom you also are being built together for a dwelling place of God in the Spirit."
>
> (Ephesians 2:19-22)

> "Coming to Him as to a living stone, rejected indeed by men, but chosen by God and precious, you also, as living stones, are being built up a spiritual house, a holy priesthood, to offer up spiritual sacrifices acceptable to God through Jesus Christ."
>
> (1 Peter 2:4-5)

In 2 Samuel 7 verse 12 of Nathan's prophecy, foretells of the establishment of Kind David's descendant's kingdom, through which the Lord established through Solomon's kingdom. Under Solomon, the Kingdom of Israel prospered and reached its zenith. However, the physical kingdom of Israel was not forever, as it does not exist today, but it did give birth to God's spiritual kingdom of God on Earth which was the job of Yeshua/Jesus to create this spiritual kingdom that is eternal which "cannot be shaken."

> "And Jesus was saying to them, "Truly I say to you, there are some of those who are standing here who will not taste death until they see the kingdom of God after it has come with power."
>
> (Mark 9:1)

> Jesus answered, "My kingdom is not of this world. If My kingdom were of this world, then My servants would be fighting so that I would not be handed over to the Jews; but as it is, My kingdom is not of this realm."
>
> (John 18:36)

> "He has delivered us from the domain of darkness and transferred us to the kingdom of his beloved Son,"
> (Colossians 1:13)

> "But you have come to Mount Zion and to the city of the living God, the heavenly Jerusalem, to an innumerable company of angels, to the general assembly and church of the firstborn who are registered in heaven, to God the Judge of all, to the spirits of just men made perfect, to Jesus the Mediator of the *new covenant*, and to the blood of sprinkling that speaks better things than that of Abel.
>
> Hearing from Him who speaks from the Heavens: "See that you do not refuse Him who speaks. For if they did not escape who refused Him who spoke on earth, much more shall we not escape if we turn away from Him who speaks from heaven, whose voice then shook the earth; but now He has promised, saying, *"Yet once more I shake not only the earth, but also heaven."* Now this, "Yet once more," indicates the removal of those things that are being shaken, as of things that are made, *that the things which cannot be shaken may remain.*
>
> *Therefore, since we are receiving a kingdom which cannot be shaken*, let us have grace, by which we may serve God acceptably with reverence and godly fear. For our God is a consuming fire."
> (Hebrews 12:22-29)

In 2 Samuel 7 verse 14 of Nathan's Prophecy, it speaks about correction when he sins or falls away. King Solomon had it all, flying carpets, and as many wives as he wanted, along with children, but he did fall away for a time when he followed after one of his wives' pagan gods, which nearly cost him his kingdom. However, the Lord was faithful, and corrected him, and he was reconciled back to God. Some of the most thought-provoking writings of the wise man Solomon are made about his mistakes and what he learned from them.

However, Jesus bore no sin Himself (Hebrews 4:15), but he did suffer correction "with the rods of men and strokes of the sons of men" for our sins. He bore our sins through his scourging and his death on the cross paving the way for our redemption.

> "For we know that if the tent that is our earthly home is destroyed, we have a building from God, a house not made with hands, eternal in the heavens."
> (2 Corinthians 5:21)

In either case, whether Solomon or Jesus, we find that Nathan's prophecy has indeed been fulfilled. This is not a prophecy about a kingdom yet to be established in our future when Jesus comes again. King Jesus is already reigning over His kingdom. (Revelation 1:5,9)

> "Brethren, I may confidently say to you regarding the patriarch David that He both died and was buried, and His tomb is with us to this day. And so, because he was a prophet, and knew that God had sworn to him with an oath to seat one of his descendants upon his throne, he looked ahead and spoke of the resurrection of Christ, that He was neither abandoned to Hades, nor did His flesh suffer decay. This Jesus God raised up again, to which we are all witnesses. Therefore, having been exalted to the right hand of God, He has poured forth this which you both see and hear."
> (Acts 2:29-33)

According to the Holy Spirit, through Peter and the apostles, the promise to "seat one of (David's) descendants upon his throne" was fulfilled by the resurrection and exaltation of the Lord Jesus to the right hand of God. "Therefore, let all the house of Israel know for certain that God has made Him both Lord and Christ-this Jesus whom you crucified." (Acts 2:36)

The Hebrew writer refers to Nathan's prophecy as being fulfilled as well; "For to which of the angels did He say, 'Thou art My Son, Today I have begotten Thee'? and again, 'I will be a Father to Him, and He shall be a Son to Me'?" and "But of the Son He says, 'Thy throne, o God, is forever and ever, and the righteous scepter is the scepter of His kingdom." (Hebrews 1:5,8)

> "Truly I say to you, there are some of those who are standing here who shall not taste death until they see the Son of Man coming in His kingdom."
> (Matthew 16:28)

Jesus said that those living in the first century would "see the Son of Man coming in His kingdom." In Book Two, *Who is God?* I proved this can only happen through reincarnation. That the prophesy in Revelation 1:17, Revelation 1:7 – Behold, He is coming with the clouds, and every eye will see Him, even those who pierced Him; and all the tribes of the earth will mourn over Him, could only happen through those very souls who were part of his *piercing* aka crucifixion, would have to reincarnate at the same time on Earth, when He returns to the Earth, in order for them to see Him and mourn over him. [6] See Chapter 28: *What Happens When You Die*, [pp. 455-479].

Jesus spoke more about the Kingdom of Heaven and Kingdom of God interchangeably than any other topic. The Kingdom of Heaven has not arrived on Earth yet, which is why He instructed His followers to pray it in, with the Lord's Prayer, by praying, *Thy Kingdom Come, they Will be done, on Earth, as it is in Heaven.* (Matthew 6:9-13). This clearly articulates a future kingdom, which is not of this world, just as Jesus said on more than one occasion, that His Kingdom was not of this world, and that He was not of this world, clearly defining Him as Extraterrestrial. As I pointed out in Book One of this book series, discerning the differences between an alien and an extraterrestrial. An alien is non-human intelligence. Not all aliens are extraterrestrials as we have an alien presence on Earth. And likewise, not all Extraterrestrials are alien, as many are human.

Jesus lived, died, and was resurrected as Human. We are made in His image and likeness. He was called Son of Man as well as Son of God, because of his unique hybrid status along with His divinity, which He needed in order to fulfill the prophecies of being the suffering servant in Isaiah 53, and fulfilling the spiritual legal ground that was necessary for a perfect human to pay the price for all humans who were born into the sin of the genetic errors, genetic manipulations, and genetic downgrading from the Evadamic Race.[7] See, *Who's Who in the Cosmic Zoo?* Book One, Chapter Two: *Exopolitics & Divine Jurisprudence, Spiritual Legal Ground*. [pp.43-69]

However, Jesus said, that the Kingdom of God was already here. That the Kingdom of God already lives inside of us in Luke 17:21-22. The Kingdom of Heaven is *governed* by the Kingdom of God. But the Kingdom of Heaven is located at the moment, in the Heavens, and is scheduled to come to Earth when the Lord returns, at which time, He will establish His Throne on Earth and rule over all the nations.

The prophet Zechariah declared: "Behold, the day of the Lord is coming ... For I will gather all the nations to battle against Jerusalem ... Then the Lord will go forth and fight against those nations, as He fights in the day of battle. And in that day His feet will stand on the Mount of Olives, which faces Jerusalem on the east ... Thus, the Lord my God will come, and all the saints with You ... And the Lord shall be King over all the earth" (Zechariah 14:1-5; Zechariah 14:9).

Jesus explained that the kingdom He established in the first century was spiritual in nature, and that all Jews and Gentiles, both men and women who put their faith in Him, His Grace and accept Him as 'Lord' and 'Savior' by being servants of His spiritual kingdom, would most certainly inherit and rule in His Heavenly Kingdom when it comes to Earth. This is the age-old metaphysical truth, that the spirit creates that which manifests as physical. Our earthly lives are but tests and training ground for a much more glorious experience in the Heavenly Kingdoms, which as I understand it, is limitless, as are *The Heavens*.

His return will not happen in secret, for everyone alive will see Him (Matthew 24:30; Revelation 1:7). Great supernatural sounds will accompany that event: "For the Lord Himself will descend from heaven with a shout, with the voice of an archangel, and with the trumpet of God" (1 Thessalonians 4:16). The Kings of the earth will attempt to make war against Him, but He will swiftly overcome them (Revelation 17:14) bringing the victory of peace. [11]

Jesus sat down on David's spiritual throne who rules over spiritual Zion, a kingdom which cannot be shaken.

> "And that He may SEND JESUS CHRIST, who was preached to you before, whom heaven must receive UNTIL THE TIMES OF RESTORATION OF ALL THINGS which God has spoken by the mouth of all His holy prophets since the world began."
> (Acts 3:20-21)

> "After this I will return and will REBUILD THE TABERNACLE OF DAVID which has fallen down; I will rebuild its ruins, and will set it up; so that the rest of mankind may seek the LORD, even the Gentiles who are called by My name, says the LORD who does all these things."
> (Acts 15:16-17)

THE PROPHECY OF THE MESSIAH

Chapter 53 of the book of Isaiah in the Torah, is considered today by most Jews and Rabbis to be the Forbidden Chapter. This is because it caused so many arguments and confusion amongst Jews, that the Rabbis decided to take it out of the Haftorah readings. One thing, I can say, based on my education in Hebrew Linguistics, is the original Hebrew most certainly points to a man, one being, as the entire chapter is written in the singular participle. I am asserting these facts to those Jews, who mistakenly think it relates to the Jewish people or the land of Israel. One thing I learned that I always remember from my Hebrew language classes, was how specific the Hebrew language is when it comes to singular, plural, masculine or feminine. The wording, and the vernacular is about a singular man, who suffers greatly for all the people of Israel and takes all the sins of the world upon himself.

The Prophet Isaiah's very name in Hebrew, *Yesha'ayahu*, translates to, *God is Salvation*. So, the prophet's divine mission was to point to the Savior, the Salvation of the Lord Yahuah. His very name is close to the Messiah's full name, *Yahushuah*, truncated to, *Yeshua*, which means, the Lord is our Salvation/Savior/Messiah.

He prophesies that the Messiah would be rejected by his people, suffer, and die in agony and that God would see his suffering and death as an atonement for the sins of humanity. Isaiah lived and prophesied about 700 BCE. According to his prophecy in chapter 53 the leaders of Israel would recognize they had made a mistake at the *end of days* when they rejected the Messiah, so Isaiah put the prophecy in past tense and because he saw himself as part of the people of Israel, he used third person plural (we). [8]

Isaiah begins to introduce the "Servant of the Lord" at the end of chapter 52 Isaiah writes an introduction to chapter 53: "Behold, my servant shall prosper…"

The term "servant" is supposed to connect back to sections earlier in the book that speak of "the Servant of the Lord" (for example, in chapters 42, 49 and 50, where the Messiah is described as a servant that suffers).[8]

> "He will be high and lifted up and greatly exalted."

This is to emphasize the eminence of the Messiah who would in fact rise from the dead and ascend to the heavens and sit next to the Father. His actions would give him a higher status than every human king or ruler.

> "Just as many were appalled at You—His appearance was disfigured more than any man, His form more than the sons of men."

Before the Messiah is exalted, he would suffer and be humiliated. His body would be abused and tortured so badly that he would be completely disfigured and unrecognizable.

> "So, He will sprinkle many nations. Kings will shut their mouths because of Him, for what had not been told them they will see, and what they had not heard they will perceive."

Despite the horrific suffering the day would come when even kings would come to look to him with reverence. Chapter 53 begins with:

> "Who has believed our report?" [4]

Isaiah describes the lack of faith by discerning the spirit of unbelief among the people of Israel who do not believe what they have heard.

> "To whom is the arm of Adonai revealed?" [9]

Isaiah calls the Messiah the "Arm of the Lord". Earlier, in chapter 40 Isaiah declares that the "Arm of the Lord" would rule for him. In chapter 51 the gentiles put their hope in the "Arm of the Lord", and the "Arm of the Lord" would redeem. In chapter 52 the "Arm of the Lord" brings salvation. Now, in 53, Isaiah reveals to us that the "Arm of the Lord" is in fact the Messiah. The Messiah is very much part of God himself. [8]

> For He grew up before Him like a tender shoot,
> like a root out of dry ground.
> He had no form or majesty that we should look at Him,
> nor beauty that we should desire Him. [4]
>
> He was a tender shoot planted in spiritually dry ground as there had been no word from God for 400 years. "He had no beauty that we should desire Him". [9]

He was not appealing to us. We did not want him. His appearance was not particularly glorious or impressive, and the way he showed up did not cause people to desire him. In contrast to what rabbinic Halacha teaches today, according to this prophecy, the Messiah would not be born to a prestigious rabbinic family or grow up in the grand residences of wealthy rabbis. We can say with near certainty that the external appearance of the Messiah was nothing extraordinary at all. [8]

> He was despised and rejected by men,
> a man of sorrows, acquainted with grief,
> One from whom people hide their faces.
> He was despised, and we did not esteem Him.[9]

The life of the Messiah was characterized by pain, rejection, and suffering. He did not get the honor due to the Messiah but was despised and rejected by the leaders of his people. We considered him some kind of social misfit – someone we might hide our faces from when we pass someone on the street that we are embarrassed to see.

We did not think he was the Messiah. We did not even register it could be him.

> Surely, He has borne our griefs,
> and carried our pains.
> Yet we esteemed Him stricken,
> struck by God and afflicted. [9]

The Messiah suffered in our place – he carried our sicknesses, our suffering, our pain… and the sins we committed, while our people – while we – thought he was being punished, and that his suffering was God's punishment for sins that he himself had committed. We did not understand that it was for OUR sin. [8]

> But He was pierced because of our transgressions,
> crushed because of our iniquities.
> The chastisement for our shalom was upon Him,
> and by His stripes we are healed. [9]

The Hebrew says wounded, pierced. He died. Like someone who has fallen wounded, or someone perforated with bullets – not for any fault of his own, but it was our wrongdoing. He was crushed because of our inequities, our sins – the punishment and discipline we deserved went to him. The "stripes" are hard blows that leave marks, and by his scars we are healed. In exactly this way, hundreds of years later, the prophecy was fulfilled. Yeshua went to the cross in order to take the death we deserved. [8]

> We all like sheep have gone astray.
> Each of us turned to his own way.
> So, Adonai has laid on Him the iniquity of us all. [9]

The Hebrew talks of going astray like sheep wander off and get lost. We all, people of Israel, ignored him and went on our way, but despite this, God put all our sin and iniquity on him – on the Messiah. [3]

> He was oppressed and He was afflicted,
> yet He did not open His mouth.
> Like a lamb led to the slaughter,
> like a sheep before its shearers is silent,
> so, He did not open His mouth. [9]

The Hebrew says he was exploited, abused... his dignity and right to a fair trial were taken from him. The Hebrew says he was afflicted – tortured – but he did not open his mouth. This shows that he did not resist his unjust sentence. He didn't try to rebel or escape, and he didn't take legal representation in spite of the fact he was facing a death sentence, but he was led like a sheep to the slaughter, or to be sheared without resisting the injustices being done to him. [8]

> Because of oppression and judgment, He was taken away.
> As for His generation, who considered?
> For He was cut off from the land of the living,
> for the transgression of my people—
> the stroke was theirs. [9]

They arrested him and took his to trial. As a result of the trial, he was "cut off from the land of the living". A death sentence. Not for his own crimes, but those of his people. In the Scriptures, "My people" always means the people of Israel. The Messiah would die not for his own sin but for the sin of his people – the people who should be taking the punishment for their own sins – but the Messiah took it upon himself. He is the one who died.

His generation would not care to bring him up in conversation but would rather sweep his existence under the carpet. So, for the last 2000 years, Yeshua the Messiah has been the best kept secret in Judaism, and this is precisely why he was labelled "Yeshu" in Judaism, which stands for "May his name and memory be blotted out."[3]

> His grave was given with the wicked,
> and by a rich man in His death,
> though He had done no violence,
> nor was there any deceit in His mouth. [9]

Even though he was taken out to be executed like a criminal, even though he did nothing wrong, and never lied, in his death he was to be buried in the fancy tomb of a rich man. Yeshua really was killed on the cross and was buried in the grave of a rich man a member of the Sanhedrin, Joseph of Arimathea. It is a clear symbol of the ironic situation in which the Messiah receives honor for the noblest deed of them all – taking the death sentence we deserve on himself. [3]

> Yet it pleased Adonai to bruise Him.
> He caused Him to suffer.
> If He makes His soul a guilt offering,
> He will see His offspring, He will prolong His days,
> and the will of Adonai will succeed by His hand. [9]

So, who is responsible for the death of the Messiah? "The Jews"? As so many Catholics have accused us of in the past. Was it the Romans? As they were the ones who actually performed his crucifixion? No. You and I, we were responsible. He died for all of us.

"God was pleased to bruise him." God is the only one able to forgive and bring salvation to the world and he turned himself into a sacrifice. What kind of sacrifice? A guilt offering. The death of the Messiah was no accident – God used his own stiff-necked people as priests in order to bring about the forgiveness of sins not only for his people Israel, but for the whole of humanity. In contrast to the Yom Kippur sacrifice which was only valid until the following year and just 'covered over' sin, the atonement of the Messiah took away the power of sin once and for all! None of us as human beings are perfect – we are not able to be that perfect sacrifice. Only God himself could do that. [8]

After that comes an interesting statement:

> "He will see His offspring, He will prolong His days," [4]

In spite of the fact, he would be killed, he would also prolong his days. He would rise again from the dead and would see the "fruit of his seed," planted in his resurrection.

> As a result of the anguish of His soul
> He will see it and be satisfied by His knowledge.
> The Righteous One, My Servant will make many righteous
> and He will bear their iniquities. 4

The Messiah would see and be satisfied by his labor, because many would be made righteous by the suffering he endured, as a righteous man when he took on himself the sins and iniquities of many. All who recognize him as the Messiah will be his "seed" in a spiritual sense.

> Therefore, I will give Him a portion with the great,
> and He will divide the spoil with the mighty—
> because He poured out His soul to death,
> and was counted with transgressors.
> For He bore the sin of many,
> and interceded for the transgressors. 9

The Messiah was the one interceding for us an advocate for us as sinners before a holy God as Melchizedek, the King of Righteousness. The Messiah took on his shoulders the sin of all who believe in him. It is an encouraging prophecy of hope and a future. God is not just interested in forgiveness expressed in words but also demonstrated in actions. That is why he took on the appearance of a servant and took upon Himself, the punishment that we deserve.

The Jewish Sages thought Isaiah 53 was about the Messiah. It is important to understand we are not just talking about a Christian interpretation here – the Jewish Sages of ancient times always interpreted Isaiah 53 to be about the Messiah. In fact, the well-known term "Messiah ben Yosef" originates from this very text.

In the ancient Jewish translation of Yonatan ben Uzziel (Targum Jonathan) from the first century opened the section with the words "The Anointed Servant" that is to say Ben Uzziel connected the chapter to the Messiah, the Anointed One. Rabbi Yitzhak Abravanel who lived centuries ago admitted that "Yonatan ben Uzziel's interpretation that it was about the coming Messiah was also the opinion of the Sages (of blessed memory) as can be seen in much of their commentary." 8

The Zohar recognizes the principle of substitution that the suffering of the Messiah would come to take upon the suffering that others deserved for their sins. Isaiah 53:4 says, "Surely He has borne our griefs," the Zohar says, "There is in the Garden of Eden a palace named the Palace of the Sons of Sickness. This palace the Messiah enters, and He summons every pain and every chastisement of Israel: All of these come and rest upon Him. And were it not that he had thus lightened them off Israel and taken them upon himself, there had been no man able to bear Israel's chastisements for the transgression of the law."

Midrash Konen in discussing Isaiah 53 puts the following words in the mouth of Elijah the prophet: "Thus says the Messiah: Endure the sufferings and the sentence your Master who makes you suffer because of the sin of Yisroel. Thus, it is written, "He was wounded because of our transgressions, he was crushed because of our iniquities", until the time the end comes." (Isaiah 53:5)

Tractate Sanhedrin in the Babylonian Talmud (98b), writes about the name of the Messiah: "His name is 'the leper scholar,' as it is written, "Surely he hath borne our griefs, and carried our sorrows yet we did esteem him a leper, smitten of God, and afflicted".[9]

In Midrash Tanhuma it says, "Rabbi Nachman says, it speaks of no one but the Messiah, the Son of David of whom it is said, here a man called "the plant", and Jonathan translated it to mean the Messiah and it is rightly said, "man of sorrows, acquainted with grief".

Midrash Shumel says this about Isaiah 53: "The suffering was divided into three parts: One for the generation of the Patriarchs, one for the generation of Shmad, and one for the King Messiah".

The prayers for Yom Kippur, the ones we all know also relates Isaiah 53 to the Messiah. The prayer added for Yom Kippur by Rabbi Eliezer around the time of the seventh century: "Our righteous Messiah has turned away from us we have acted foolishly and there is no one to justify us. Our iniquities and the yoke of our transgressions he bears, and he is pierced for our transgressions. He carries our sins on his shoulder, to find forgiveness for our iniquities. By his wounds we are healed."

The deeper we go into this prayer for Yom Kippur the more significant it gets. The prayer brings the sense that the Messiah left his people. "The righteous Messiah turned [away]". That is to say, the Messiah has already come and left. Also, the Messiah suffered in the place of the people, and the sins of people were put on him then after the Messiah suffered, he left them that was the reason for their concern and so the people are praying for his return. A large part of this prayer is taken straight out of Isaiah 53, so from this we can prove that up to the 7th century the Jewish perception – also among the rabbis – was still that Isaiah 53 was about the Messiah. [8]

King David revealed in his writings, that not only did he know who the Lord and Messiah was spiritually, but he knew that He was sitting at the right hand of the God of Heaven.

Psalm 110:1, David says, "The LORD says to my Lord: 'Sit at my right hand, until I make your enemies your footstool'" (ESV). In Matthew 22:44, Jesus quotes this verse in a discussion with the Pharisees in order to prove that the Messiah is more than David's son; He is David's Lord.

The clause the LORD says to my Lord contains two different Hebrew words for "lord" in the original. The first word is Yahweh, the Hebrew covenant name for God. The second is adonai, meaning "lord" or "master." So, in Psalm 110:1, David writes this: "Yahuah says to my Adonai. . .." To better understand Jesus' use of Psalm 110:1, we can discern the identity of each "Lord" separately.

The first "Lord" in "the LORD says to my Lord" is the eternal God of the universe, the Great I AM who revealed Himself to Moses in Exodus 3. The second "Lord" in "the LORD says to my Lord" is the Messiah, or the Christ. Psalm 110 describes this second "Lord" as follows:

- He sits at God's right hand (v. 1)
- He will triumph over all His enemies and rule over them (v. 1–2)
- He will lead a glorious procession of troops (v. 3)
- He will be "a priest forever, in the order of Melchizedek" (v. 4)
- He will have divine power to crush kings, judge nations, and slay the wicked (v. 5–6)
- He will find refreshment and be exalted (v. 7)

In Matthew 22:44, Jesus unmistakably identifies the second "Lord" of Psalm 110:1 as the Messiah, and the Pharisees all agree that, yes, David was speaking of the Messiah. When David wrote, "The LORD says to my Lord," he distinctly said that the Messiah (or the Christ) was his lord and master—his Adonai.

"Son of David," was a common title for the Messiah back in Jesus' day and was based on the fact that the Messiah would be the descendant of David who would inherit the throne and fulfill the Davidic Covenant (see 2 Samuel 7). Jesus capitalizes on the Jewish use of the title "Son of David" to drive His point home to the Pharisees in Matthew 22. "While the Pharisees were gathered together, Jesus asked them, 'What do you think about the Messiah? Whose son, is he?' 'The son of David,' they replied. He said to them, 'How is it then that David, speaking by the Spirit, calls him "Lord"? For he says, '"The Lord said to my Lord: "Sit at my right hand until I put your enemies under your feet." If then David calls him "Lord," how can he be his son?'" (Matthew 22:41–45)[10]

Jesus' reasoning is this: "Son of David" is your title for the Messiah, yet David himself calls Him "Lord." The Messiah, then, must be much more than just a son—a physical descendant—of David. According to Psalm 110:1, this "Son of David" was alive during David's time and was greater than David. This is called a 'Christophany' when Christ appears in the Old Testament both in the flesh and in spiritual visions. See, *Who Is God?* Book Two for all those scriptures and instances. [6]

Suffice it to say, that all of this information is contained in the statement that "the LORD says to my Lord." Yeshua/Jesus is David's Lord; He is the Christ, the Jewish Messiah, and Psalm 110 is the promise of Yeshua/Jesus' victory at His second coming.[10]

Notes:
1. Ella LeBain, *The Christmas Gods*, Chapter 27, *Who is God?* Book Two, Skypath Books, 2015. http://www.whoswhointhecosmiczoo.com
2. Ella LeBain, *Should Christians Celebrate Christmas?*, https://spirituallydiscerning.com/f/should-christians-celebrate-christmas
3. Rabbi Arthur O. Waskow, *Seasons of Our Joy - A Modern Guide to the Jewish Holidays*, The Jewish Publication Society, 1985.
4. Solar Mythology and the Jesus Story, http://www.solarmythology.com/lessons/christ2002.htm
5. Solar Mythology and the Jesus Story Explained, http://www.solarmythology.com/david16.htm
6. Ella LeBain, *Who is God?* Book Two - Chapter 28: *What Happens When You Die: Jesus Taught Reincarnation*, pp. 455-479, Skypath Books, 2015.
7. Ella LeBain, *Who's Who in the Cosmic Zoo?* Book One, Chapter Two: *Exopolitics & Divine Jurisprudence, Spiritual Legal Ground.* [pp.43-69]
8. Dr. Eitan Bar, *Isaiah 53 – The Forbidden Chapter*, https://www.oneforisrael.org/bible-based-teaching-from-israel/inescapable-truth-isaiah-53/, Article later added to his books, *Refuting Rabbinic Objections to Christianity & Messianic Prophecies*, ONE FOR ISRAEL Ministry (January 5, 2019).
9. Isaiah Chapter 53 verses 1-12, King James Version translation from the Hebrew Masoretic Text.
10. Got Questions Ministries: *What does it mean that "the Lord said to my Lord"?* https://www.gotquestions.org/Lord-said-to-my-Lord.html, January 2, 2020.
11. Jon W. Quinn, *The Nature of God, From Expository Files*, 3.7; July 1996

Concluding Words

There is a time for everything,
and a season for every activity under the Heavens:
a time to be born and a time to die,
a time to plant and a time to uproot,
a time to kill and a time to heal,
a time to tear down and a time to build,
a time to weep and a time to laugh,
a time to mourn and a time to dance,
a time to scatter stones and a time to gather them,
a time to embrace and a time to refrain from embracing,
a time to search and a time to give up,
a time to keep and a time to throw away,
a time to tear and a time to mend,
a time to be silent and a time to speak,
a time to love and a time to hate,
a time for war and a time for peace.
Ecclesiastes 3:1-8 NIV

THE METAPHOR OF MOVIES

The Truth is Stranger than Fiction!

"Life's riddles are answered in the movies."
~ Steve Martin, *Grand Canyon*, 1991

Star Wars, Marvel and DC comics movies, all depict truths from both our past, our future, and the interdimensional multiverse in which we co-exist with a variety of creatures, aliens, extradimensional and extraterrestrial beings. In all the films there seems to be strong correlations to the disclosure of alien creatures, ancient aliens in our history, and what is to come in our future, which is why these films strike such a cord within our psyches.

Discernment is powerful tool, but when Christians begin to connect the dots yet run on fear, they are not empowering the collective with the Gospel of Christ. By seeing all these films as 'evil agendas, demonic, and sinful', they are doing a great disservice to others, who resonate for all the right reasons. All these films have one thing in common, Ephesians 6:12 which says:

> "For we do not wrestle against flesh and blood, but against the rulers, against the authorities, against the cosmic powers over this present darkness, against the spiritual forces of evil in the Heavenly places."

The Battles are always against the different forms of evil, the good guys win. The reason these films are so popular is because they are inspirational to viewers who live vicariously through them as they play out their own personal battle with invisible enemies through the fantasy books and films.

Nobody wants to see the evil demons win. For Christians to suggest it's a sin to read or watch these movies is taking the funda-mental disorder too far, by fear-mongering that Christians will sin if they patronize these science fiction/fantasy films is absolutely wrong on many levels. 1. Becoming Modern Day Pharisees, failing to see the forest from the trees; 2. Religious Spirits, exhibiting spiritual arrogance, pride; 3. Fear mongering; 4. Missing the point of the Great Commission, which is to preach the Gospel to ALL Creation, and to ALL the nations of the world. That includes Nephilim, miscreants, half breeds, hybrids, hu-brids, and fallen angel offspring on the Earth, even to Satan's seed. (See, Genesis 3:15; John 8:37-59) Discerning Abraham's Seed vs. Satan's Seed.

If the Bible says that Satan's seed is on the Earth, which I have already discerned in more detail in this Book Series (See, *Who Are the Angels?* Book 3) then that means they have the opportunity to repent and hear the Gospel of Salvation. "Faith comes by hearing, and hearing by the Word of God" (Romans 10:17). If the Great Commission is to <u>all creation</u>, then that includes those born with Satan's seed, Nephilim, and hybrids. This is the Great Commission:

> "And Jesus said to them, 'Go into all the world, and <u>preach the gospel to every creature</u>."
> (Mark 16:15)

Why are Christians oblivious to this? Jesus said, 'by their fruits you will know them'. Those fruits are the will of God. That is what this book series is about, this is the thesis I began in my Conclusion of Book One of *Who's Who in the Cosmic Zoo?*[1] That as human lives are miraculously transformed through the grace of God through Christ, as the *Watchers* watch humans, they too are being given the opportunity towards repentance. "Or do you presume on the riches of his kindness and forbearance and patience, not knowing that God's kindness is meant to lead you to repentance?" (Romans 2:4).

In the end times, people will faint from fear of what is coming upon the Earth, which includes *seeing* aliens. If people are not prepared, and not equipped spiritually, and not grounded in their authority in Christ, then how can they discern rightly? What about what they are here to do? If Christians are spreading fear and bigotry towards other worldly creatures, aliens, hybrids, mythological creatures, whom God did not destroy, then that means they are denying God's will. The Others too have an opportunity to be led to repentance for salvation.

> "Men's hearts failing them for fear, and for looking after those things which are coming on the Earth: for the powers of Heaven shall be shaken."
> (Luke 21:26)

The ancient way human beings could obtain power over the elements, powers of the mind and over other's minds, is by contracting and fellowshipping with supernatural beings. God's angels are not permitted to go there, but fallen angels lend their powers to humans by means of soul contracts and transactions that ultimately imperil the human soul. This is the ancient fascination with magic, magical beings, mythological beings, fantasy, and fairytales. The reason these types of Sci-Fi/Fantasy films have grown to be so popular within our culture is because people are vicariously working out their own struggle with these dark forces, through generational curses on their own bloodlines, breaking their ancient soul ties with rebel ETs, aliens, fallen angels, through these films. See, Ephesians 6:12.

END TIMES DISCLOSURE

Some will rejoice, others will die in fear, a handful will know the truth, and many will be deceived. Cognitive dissonance will kick in when those who believe aliens have been advancing humankind find out that aliens have been manipulating us for thousands of years. No, it is not all about us. The so-called gods, the ET aliens, are evolving right along with us. Book Four-COVENANTS focuses on these ancient contracts and agreements made between the so-called gods, the God of this world and the overall permission to allow them to play out this Grand Experiment by the God of gods, whose ultimate Divine Plan will be played out despite the upset.

Advanced technology does not necessarily equate to evolution or spiritual maturity. Don't believe me? Just take a look at how evolved and spiritually mature some humans are today with some pretty sophisticated technology. As above, so below. Know this, when full disclosure happens, many will be deceived, and others will be in shock, which makes humans vulnerable to manipulation, mind control and implants. Learn the truth. Learn how to DISCERN.[2]

Just remember, Disclosure Day, December 18, 2017, happened under President Trump. If he did not want the Pentagon to release those UFO files, it would not have happened. Presidents do not control the UFO Cover-up. Every President that promised Disclosure is held back. Just look at what happened to JFK over his policy to Disclose. For the time being we are in drip-drip Disclosure. We will get a story here and there, piecemeal, and it is up to the researchers to piece the puzzles pieces together. There are many issues, on both sides, of why full Disclosure will never happen from government sources, as expected, despite being lobbied by the Disclosure Movement. Besides National Security issues, which is a given, there are real enemies to the United States in this world, but more realistically, the extraterrestrial and alien presence is complicated to say the least. It goes back further than we could imagine. It would confuse everyone, create major cognitive dissonance, and would disrupt and even destroy the status quo of our cultures and religions. These are just a few of the reasons full Disclosure has not happened yet. Besides of course, the Alien Presence, who have made it clear, they do not want to be revealed yet.

I went into the Disclosure scenario in COVENANTS – Book Four,[2] and how it is being used by the Vatican to manipulate what the Bible says is an evil agenda, a drive to create a One World Government, the New World Order, that according to prophecy is run by one man who is the Man of lawlessness, the Antichrist, who is an alien, a composite human host and Satan incarnate. World Religion plays a big role in why Full Disclosure will never happen. [2]

And lastly, let's face it, we earth humans are just not going to get our heads around some of the exotic technologies along with understanding the cosmic dramas that have been going on for millennia between various extraterrestrial and alien races. We can get a glimpse of it though, but to really understand, *who* is *who* in the Cosmic Zoo, we must first drop all our preconceived notions, religious brainwashing, such as all aliens are demons, which is a form of cosmic bigotry, something we on earth tend to practice continually.

I am reminded of the scripture where the Lord says, "For my thoughts are not your thoughts, neither are your ways my ways," declares the LORD. "As the heavens are higher than the earth, so are my ways higher than your ways and my thoughts than your thoughts. ... This will be for the LORD's renown, for an everlasting sign, which will not be destroyed." (Isaiah 55:8-9)

Finally, in my humble opinion, as a UFO researcher, Experiencer and Witness of the reality of Jesus, Angels, Demons and Aliens, Full and Total Disclosure will happen at the time when Christ returns to Earth and brings with Him legions of celestial armies, and the New Jerusalem Mothership, Holy City, Kingdom of Heaven, when it lands over the scorched Earth, and He sets up His Kingdom on Earth and Rules over all the Nations for a Millennium. See, Revelation 21 and 22.

Then we will know the TRUTH about *who* is *who* in the Cosmic Zoo. Who are the good ETs that serve the Kingdom of Heaven, and who are the evil aliens that Christ will defeat and wipe off the face of the earth, once and for all! This is the quintessential End Times Prophecy, which cannot be denied, which will fulfill all prophecies, and end all controversies, wars, and mysteries. The truth of Full Disclosure will definitively clarify our true place in this universe.

This is another reason why there could never be full disclosure because of patents and aerospace technologies that should stay secret, for reasons of national security. However, on the other hand, I feel our government has a lot of explaining to do, hiding interstellar and anti-gravity technology from the public for 50-70 years. Another reason why it is wishful thinking to believe that they will come clean about it all is because they would certainly open up a can of worms. They would spend the rest of their days back peddling instead of doing what they have been doing, interfacing, and protecting us from negative aliens. Shhh!

The big announcement is coming soon. The reason Jesus warned us in Matthew 24:24 about a deception coming in the last days, when even *the elect* could be deceived, is for the church to be ready. Sadly, most Christians are afraid of what is going on. Fear paralyzes anyone from research or study. Then out of their fear and cognitive dissonance there are countless superstitions, lacking in discernment, which are fear based, particularly about what the ancient scriptures actually did say. This is the reason I wrote my books, to hopefully educate and teach discernment in connecting the dots from Bible Scriptures to the available evidence.

KNOWLEDGE EMPOWERS. Ignorance endangers.
(Hosea 4:6)

Cognitive dissonance is a psychological term that describes a mental state reacting to new information that does not fit into old models. The alien presence is such a phenomenon that provokes cognitive dissonance. People have literally made up their minds before even learning of the evidence.

Christians tend to think in absolutes -- for example, that all aliens are demonic fallen angels, whereas New Agers tend to think all aliens are benevolent space brothers here to help us. They cannot both be right. Ufologists take a more scientific approach and tend to be skeptical unless there is hard evidence, photos, videos, physical evidence that they can conclude fits reality. But even ufologists know that we are dealing with a phenomenon that simply does not fit into our reality.

How do you know *who* is *who* in the Cosmic Zoo? Well, you can start by reading my books. Book One is an A-Z Compendium of ALL the beings experienced by contactees, government whistleblowers and experiencers (abductees). I have been an Experiencer, since the age of 2. My personal testimony is shared piecemeal throughout my books, with the entire testimony in my book, *CinderElla's Shadow*, which will be released after all my research material and its conclusions have been successfully published.

Yes, there is and has already been a mass deception going on for nearly 70 years covering up the fact of our government's involvements with the alien presence on earth. I wrote my books from the perspective of dealing with the reality of the alien presence, not necessarily getting caught up with having to prove that this is real, but by taking the next step, and asking *who* is *who*? For those who are still stuck in skepticism, there is a plethora of information available in Ufology that already proves this reality. I, however, start from the premise that this reality is a given, and here is where we need to go next.

It's a mixed bag to learn the result of the past five decades of deliberate movie scripts being released to the public for the purpose of conditioning us to accept the reality of aliens, ETs, UFOs, has had on the American psyche. While people are fascinated by Star Trek, Star Wars, sci-fi movies, people have also become jaded in believing the real thing even if it smacks them right between the eyes! We live in an age where skepticism has turned into a spiritual stronghold of unbelief. Jesus said that the spirit of unbelief blinds, deafens, and mutes the spirit (Mark 9:24,25; Matthew 12:22; Luke 11:14; Zechariah 3:2; Isaiah 35:5,6).

The spirit of unbelief is not limited to a belief in God or Jesus but is the very spirit that operates in cover-ups of all kinds, which includes the spiritual veil cast over countless Christians, Jews, Muslims, to think that satan is just some made up being that represents humanity's fears and paranoia. Not true! This is exactly satan's modus operandi, to operate in the darkness, and confuse you, so you do not see his influence behind your own personal problems, or the problems of this world. If the boogie man is not real, then people can all just relax and stop imagining things. But the truth is, he is real, and people who have been told that satan is false, are operating under the influence of demons.

The spirit of unbelief causes spiritual blindness, spiritual deafness, and spiritual muteness. The physical correspondences to this spiritual condition are ADD, ADHD, tinnitus, deafness, blindness, and muteness. When these demon spirits were rebuked and cast out of people, people literally woke up, the deaf hear, the blind see, and the mute speak. The spirit of unbelief, the Bible says, is rooted in haughtiness and pride.

Yes, it is true, there are many aliens that are actually demonic entities that use humanity, like humans use cattle. But the Bible also reveals that the aliens are outnumbered 2:1, that the Lord of Hosts commands legions upon legions of extraterrestrial mighty warrior angels. These are not fluffy light beings, but warrior beings that serve the Kingdom of Heaven. This is the DISCERNMENT that I am teaching in my book series.

As a Messianic Jew, I cannot deny the knowledge passed down to me from my ancestors, the body of knowledge that was dissected, altered, and rejected by the Church of Rome who were the OPPRESSORS of Jews and Israel. For millennia, for this very reason, Jews rejected that Jesus was their Messiah. I have seen the light, both spiritually and scripturally, that Yeshua/Jesus was and is indeed the Jewish Messiah as prophesied by the Old Testaments, came in the flesh, and was resurrected. The spirit of unbelief has blinded the eyes of those who read scripture from seeing this, in addition to the heaviness of persecution by the Church of Rome on Jews, is it any wonder why Jews reject Jesus because of the bad behavior of Christians? Only the Holy Spirit can reveal the truth behind scriptures. I share these truths in my book series.

With that said, many Jews are waking up to the fact that Yeshua/Jesus is their Messiah, which is the fulfillment of end times Bible prophesies. I am a remnant and a witness to that fact. Just as many Gentiles are completing the times of the Gentiles,

CONCLUDING WORDS

as the Lord promised, in Luke 21:24. However, these are the End Times, and the mass deception that Matthew 24:24 predicted is about to come upon us all, and sadly the Church is not prepared, because it lacks knowledge and discernment.

The Religious Spirit is a demon, that blinds and creates discord and strife inside the church. The religious spirit is a false spirit, which comes out of the Church Thyatira and often points fingers of accusation to those who are actually speaking and teaching Truth, just as Jezebel did to Elijah in 1Kings, which is a tactic from satan to discredit the truth and prevent his works from being exposed. History tells us, that this spirit lost to Elijah and to King Jehu who defeated it. Jezebel represents false prophets, which the world is full of now, because of confusion and a rejection of truth, facts, and knowledge.

I am at a point in my walk where I couldn't care less what people think, I know the Truth, and the truth has set me free, and I'm as real and down to earth as they come, just ask anyone who knows me personally. I have been on this path of digging and investigating this subject for over 41 years, when I first got saved in South Africa and joined the Assemblies of God and served as their missionary for two years around apartheid South Africa from 1979-1981. They told me, knowing that I had books on this subject, "to have nothing to do with the UFO phenomenon, because when the Rapture happens, it's a plot from the devil to lie to the rest of the world, that believers were abducted by aliens in UFOs." That has got to be one of the most memorable and life changing words in my life, which put me on a path of intense research.

There is some truth to that statement, however, half-truths are still lies. The truth of the matter is that the Church of Rome has covered up this reality since they edited and canonized the Bible in 325AD. One really needs to question, 'why?'

Claiming all aliens are demons is NOT discernment. That is ignorance. Looking at the evidence and comparing it to Scripture develops discernment. You cannot discern the counterfeit from the truth without first knowing the original. Unfortunately, the Church has been put under a spell by the Constantine and Nicean Creeds, the original doctrines of antisemitism, via the Church of Rome who later became the Roman Catholic Church, who are now in control of this knowledge, and will use it to purport their own end time agenda, to be the final authority for 'god' on earth. However, the question that should be on everyone's mind is, "which god?". In *Who is God?* Book Two, I spent over 560 pp discerning between the gods and the God of gods in the Bible, and how to tell them apart from what religions have taught. [3]

Yes, this is a multi-layered event, multi-layered subjects, and it requires not just intellectual and spiritual discernment but spiritual guts and the courage to face the truth and accept it when it shows up.

> Jesus said, "I have not given you a spirit of fear, but of power, love and sound mind."
>
> (2Timothy 1:7)

Remember God is the Creator, satan is the great Counterfeiter. Whatever he has copied, he distorted from the original. [4]

> "In the last days,' The Lord God says, 'I will pour out my Spirit upon ALL people. Your sons and daughters will prophesy. Your young men will see visions, and your old men will dream dreams."
> (Acts 2:17)

Remember the words, ALL people, not just men, but women too. In *Who Is God?* Book Two of my book series, I prove through scripture that sexism was NOT created by the LORD of Hosts, but is an agenda inserted into the Bible by the Church of Rome. I prove it through comparison to multiple scriptures in the Torah, the Jewish Bible, aka the Old Testament. This Word in the book of Acts proves just that, that the movement of the Spirit of God is for ALL people, men and women are equal.

Is the Vatican part of the coming Alien Deception? Fr. Guiseppe, Tanzella-Nitti, Professor of the Vatican University said, "Very soon, we will not have to deny our Christian faith, but there is information coming from another world, and once it is confirmed, it is going to require a re-reading of the Gospel as we know it."

> "A large part of the available UFO literature is intricately linked with mysticism and the metaphysical. It deals with subjects like mental telepathy, automatic writing, and invisible entities as well as phenomena like poltergeist [ghost] manifestation and 'possession.' Many of the UFO reports now being published in the popular press recount alleged incidents that are strikingly similar to demonic possession and psychic phenomena." ~ Lynn E. Catoe, UFOs and Related Subjects: USGPO, 1969; prepared under AFOSR Project Order 67-0002 and 68-0003

The reason I put my Book Series together is to teach both spiritual, scientific, and scriptural discernment. Because I encountered two diametrically opposed viewpoints in my 41 years of soul searching and investigating this phenomenon, between Ufologists, the Spiritual New Age community and Christians and realized they both can't be right at the same time, so as the ancients have said, the truth is usually in the middle ground, which is what I have sought out through discernment. In conclusion, while most Christians believe all aliens are demons, here is the truth, yes, all demons are aliens but not all aliens are demons.

CONCLUDING WORDS

Yes, fallen angels became demonic but not all angels are aliens. Just do the math: Only 1/3 of Heavens' angels rebelled and fell, while 2/3 remains faithful to the Creator, these are the good ETs, aka good aliens, also called angels which simply means messenger. God created them. God also created many other types of creatures, also mentioned in Scriptures that are simply out of this world and classified as 'alien'. Not all are evil. Book One is written as an A-Z Compendium which is being republished as an updated revised 4th Edition this year. I identify the evil demonic aliens, verses those who are simply alien creatures mentioned in the Bible that serve the Kingdom of Heaven and the Lord Himself.

After being educated in Israel and learning the Bible in Hebrew, Jesus appeared to me in 1979 and told me he was the Messiah and said, 'follow me,' so I did. He taught me where the good aliens are and where the evil aliens are in the original Hebrew Scriptures which are shared in my book series. Learn to discern. The Truth is Stranger than Fiction!

While I go into great lengths to compare the 4 living creatures of Ezekiel to that of Revelations which I analyze from original Hebrew that these living creatures are not angels but alien beings that serve the thrones. The thrones in Ezekiel are spaceships. The thrones in Revelation are in Heaven. They both serve the Lord. They are not demons but alien creatures.

In Revelation 4:6–8, four living beings are seen in John's vision. These appear as a lion, an ox, a man, and an eagle, similar to Ezekiel but in a different order. They have six wings, whereas Ezekiel's four living creatures are described as only having four wings.

The Earth plane is for is testing. We are being tempted, tried, and tested here. This is the ground we must overcome. God does not tempt. Satan is the tempter. He is the god of this world according to scripture. Satan does not live in the Kingdom of Heaven, therefore there will not be any tempting or testing there as it is all in the presence of the Lord. Testing and tempting do not come from the Lord God who is the Creator and Savior made of love and light. Testing and tempting comes from the satans.

> "When tempted, no one should say, "God is tempting me."
> For God cannot be tempted by evil, nor does he tempt anyone;"
>
> (James 1:13)

I go over this very issue packed with corroborating scripture in Book Two, *Who Is God?* which discerns *who* is the Lord God vs *who* is Satan and how Bible translators got it confused because of what's revealed in original Hebrew. The Angels that were tempted were persuaded by Satan to follow him in his rebellion. God did not tempt the angels. He created us and the angels with free will. 1/3 of Heavens angels fell into temptation, 2/3 of Heavens angels remained faithful.

Likewise, New Agers, who tend to glorify aliens, also need to learn discernment, as they give their power to the fallen angels who masquerade as light beings, or magical wizards to say the least. And I know many New Agers cringe at this discernment, and that is, the litmus test must be, *who* are these beings in relation to the Lord of Spirits *who* is the Lord of lords? Are they in rebellion? Or are they serving the Lord? Have they lost their way through galactic warfare, and are now serving the dark lord who is leading the rebellion against the Kingdom of Heaven? Any New Ager, which gets deep into the research has to acknowledge there is a classic clash of the Titans, when it comes to *who* these aliens and extraterrestrials are.

In Hebrew, we need to remember that the same word for Heaven is the same word used for skies which is "shamayim." This is relevant because many verses actually relate to beings that come out of the skies, most important example would be the return of Jesus Christ who is coming out of the Heavens (the skies) with the clouds of Heaven which I have proved in great detail both here and in Book Two, *Who Is God?* are spaceships.

Secondly, when you understand that the word "angels" means messengers, then the history of the fallen angels makes sense. The fallen angels were extraterrestrials that rebelled and followed one of Heavens sons, Lucifer the adversary or Satan, who according to Isaiah 14 challenged the Almighty for power over the Universe and the Earth.

God created angels and humans with free will, this is clear throughout ALL the Ancient scriptures, including the great rejected Jewish texts like the Books of Enoch, Jasper, Jubilees, Giants, War Scrolls, all corroborate this fact.

One of the reasons I was led to put this Book Series together was to clarify discernment between *who* is God the Creator of all life and *who* is Satan – the god of this world *who* is the great counterfeiter and deceiver of humankind.

The answers lie in understanding *who is who* in the Bible scriptures' cast of characters and how spiritual legal ground works in the Courts of Heaven (an actual place, not the skies) and the spiritual realms. This is understood by focusing on the laws of the Creator God, which is written in His Word.

The Bible scriptures contain the Word of God, but they also contain the words of Satan, evil kings, evil Queens and fallen men. I make this distinction because Christian fundamentals think the Bible is God's Word in totality. It contains God's Word, and historical records of other's words as well which simply are not God's Word, but relevant to His-story.

> "Study to shew thyself approved unto God, a workman that need not to be ashamed, rightly dividing the word of truth."
> (2 Timothy 2:15 (KJV)

My contribution to Bible believers are the correct translations of the Hebrew Scriptures that were lost in English translations, specifically discerning and exposing *who* the cast of characters are in the Bible, and why they got to do what they did, based on the history and/or myths of fallen beings from the Heavens, i.e., Cosmos.

The Jewish texts are clear that there are evil stars and death stars and that the Heavens are under a curse. This is why Yeshua/Jesus Christ promised in both Old and New Testaments to recreate the Heavens and the Earth in the Age to come.

> "See, I will create new Heavens and a new Earth. The former things will not be remembered, nor will they come to mind."
> (Isaiah 65:17)

> "Then I saw "a new Heaven and a new Earth," for the first Heaven and the first Earth had passed away, and there was no longer any sea."
> (Revelation 21:1)

Because we are in the last days of these End Times, we are essentially "in Disclosure" which is what the word, Revelation/Apocalypse means, when knowledge is increased. The Bible Prophesies have been 100% accurate thus far. The History of the Bible is also accurate, albeit missing pieces. The difference in doctrines, not so much.

The editing of the Old and New Testaments, the many discrepancies and versions, the Catholic version has 73 books, Eastern Orthodox 80, and an Ethiopian Orthodox version contains 81 books. All of these versions were written mostly during the "400 years of silence." Yet, there were numerous books, deleted from the original Jewish Scriptures, in both Old and New Testaments, due to multiple controversies over authority is problematic to say the least. However, in my opinion, the biggest problems are the gross mistranslations from Hebrew to English, which have created huge misunderstandings due to mistranslations and mistransliterations, that have led to false theologies and counterfeit religions. I exposed some of the biggest ones in my 2nd book *Who is God?* [3]

We are living through unprecedented psychological and spiritual warfare, where lies are pushes as truth and truth is discredited as lies. Isaiah 5:20 is an ancient curse on all those who twist truth into lies and lies into truth. "Woe to those who call evil good, and good evil; who substitute darkness for light and light for darkness; who substitute bitter for sweet and sweet for bitter! Woe to those who are wise in their own eyes and clever their own sight!"

MAKING HEAVEN MORE CROWDED

Making Heaven more crowded is the goal of everyone called to the Great Commission, to spread the good news of the gospel of Grace to the whole world. It's not the job of Christians to become like some exclusive club that if you don't adopt their particular brand of theology then you're outcast. Just the opposite, we are called to salvage souls from the kingdom of darkness. Essentially, this means, because everyone on Earth is saved behind enemy lines, that we are called to reclaim victory from the jaws of defeat, when it comes to leading souls back to the Light of the World, who is Christ.

> "For we know that if the tent that is our Earthly home is destroyed, we have a building from God, a house not made with hands, eternal in the Heavens. For in this tent we groan, longing to put on our Heavenly dwelling, if indeed by putting it on we may not be found naked. For while we are still in this tent, we groan, being burdened—not that we would be unclothed, but that we would be further clothed, so that what is mortal may be swallowed up by life. He who has prepared us for this very thing is God, who has given us the Spirit as a guarantee."
>
> (2 Corinthians 5:1-5 ESV)

Heaven is not just an Afterlife belief; it is a vast Kingdom governed by the Kingdom of God. Jesus talked more about His Kingdom than anything else. His words go back and forth from a place called the Kingdom of Heaven to what he said was already with us, the Kingdom of God, which is inside everyone who accepts his Grace of salvation and is baptized in His Spirit. Therein lies the Kingdom of God. Therefore, it is the job of every believer to make Heaven more crowded.

The Kingdom of Heaven is a real place located in different dimensions from Earth, nevertheless, happening as an overlay simultaneously.

> Jesus said, "in my father's house there's many Mansions."
> (John 14:12)

As I have written in Book Two, *Who is God?* the word *Mansion*, means abode, but also means Stars in Sanskrit. To the Ancients, Mansions in the sky, were spaceships. To not connect these dots, is to ignore the vernacular of the Ancients.3

The word *mansions* is actually a celestial term, for celestial houses, (i.e., star systems). What I found remarkably interesting was that in the Vedic literature, the word *mansion* is used as a description for spaceships. These are otherwise known as Vimana and also called "flying cities," which are, otherwise, known today as "mother ships." Those who live inside spaceships, makes these flying cities, their homes, therefore we can also call them, flying celestial houses.

CONCLUDING WORDS

The Vedic epic, the Ramayana, refers to the Vimana as "the Aerial Mansion of Ravana." The following is a translation by Swami Tapasyananda, from his book Sunkarakandam of Srimad Valmiki Ramayan [1]:

> "That heroic son of the Wind-god Sawa in the middle of that residential quarter the great aerial mansion-vehicle called Puspaka-Vimana, decorated with pearls and diamonds, and featured with artistic windows made of refined gold.
>
> Constructed as it was by Visvakarma himself, none could gauge its power nor affect its destruction. It was built with the intention that it should be superior to all similar constructions. It was poised in the atmosphere without support. It had the capacity to go anywhere. It stood in the sky like a milestone in the path of the Sun...
>
> It was the final result of the great prowess gained by austerities. It could fly in any direction that one wanted. It had chambers of remarkable beauty. Everything about it was symmetrical and unique. Knowing the intentions of the master, it could go anywhere at high speed unobstructed by anyone, including the wind itself...
>
> It had towers of high artistic work. It had spires and domes like the peaks of mountains. It was immaculate like the autumnal Moon. It was occupied by sky-ranging Rakshasas of huge proportions with faces brightened by their shining ear-pendants. It was delightful to look at the spring season and the bunches of flowers then in bloom. It had also for protecting it numerous elementals with round and deep eyes and capable of very speedy movements.
>
> Hanuman, the son of the Wind-god, saw in the middle of the aerial edifice a very spacious construction. That building, half a yojana in width and one yojana in length, and having several floors, was the residence of the king of the Rakshasas....
>
> Visvakarma constructed in the Heavenly region this Puspaka Vimana or aerial mansion-vehicle of attractive form, which could go everywhere, and which augmented the desire nature of its occupants. Kuvera by the power of his austerities obtained from Brahma that aerial mansion which was decorated entirely with gems, and which received the homage of the resident of all the three worlds. It was by

> overcoming Kuvera that Ravana, the king of the Rakshasas, took possession of it."

In Vedic Astrology, the mansions of the Moon are connected to fixed stars. When Jesus spoke of His Father's House, The Heavens, having many mansions, I believe He was referring to the vastness of the Kingdom of Heaven, which is made up of many stars, many mansions, many levels, and many places of abode, including huge mother ships or flying cities where literally hundreds, maybe thousands of people can dwell safely. Jesus knew the hidden meaning of the word *mansion* when He said, "in my Father's house there are many mansions." He knew what He meant; only the true meaning has been obscured to us for centuries because many did not have the reference point to understand it until now. No one can fully comprehend the word of God without the revelation known as *Rhema* (Greek) which can only come from the Holy Spirit. This is why the true meaning of the logos (Greek for *Word*) has been hidden too so many for so long.

The description of the Aerial Mansion of Ravana, speaks of a mother ship which is eight miles by eight miles in size. No wonder the scribes of the Vedas described them as flying cities. And how about the reference to the beings whose job it was to protect the Vimana? "Elementals with round and deep eyes" sounds like today's description of a Gray alien.

In the Bible, particularly in the book of Ezekiel, he describes the Lord's spaceships as divine chariots, also called the Throne with Wheels. Ezekiel describes them as having wheels within wheels that were powered and protected by Cherubim who each had four faces and four wings, had the face of a lion, an ox, an eagle, and a man. Completely different than the Vimana spacecraft, but obvious similarities, in that they could both fly and both had beings protecting them and powering them.[2]

PREPPING YOUR SOUL

> "He shakes the Earth from its place,
> so that its foundations tremble."
> Job 9:6

Today there are those who are focused on survival. The Elite and the wealthy have been building underground communities and bunkers to wait out the Tribulation and catastrophe coming upon the Earth. Several Planet X Researchers believe humanity will survive as it has past Fly-Bys of Nibiru and that rebuilding will follow, as will the opportunity to go to the stars. [6] This is why there is a push for Space Exploration in recent decades, which suggests that humans have already established bases on the Moon and Mars. This has been known by conspiracy theorists as Alternative 3, otherwise known as the Secret Space Program, to terraform Mars, so humanity may move to there while Earth is in upheaval. But what Alternative 3 does not take into consideration is that the cataclysms coming to

CONCLUDING WORDS

Earth are not limited to this one planet but to all the planets in our solar neighborhood. Mars has been hit before, so who is to say it will not happen again?

There are those who believe they will survive on Earth by taking refuge *inside* the Earth, yet the Bible Prophesy says otherwise about the Day of Reckoning:

> "Then the kings of the Earth and the great men and the commanders and the rich and the strong and every slave and free man hid themselves in the caves and among the rocks of the mountains;"
>
> (Revelations 6:15)

> "To go into the clefts of the rocks, and into the tops of the ragged rocks, for fear of the LORD, and for the glory of his majesty, when he rises to shake terribly the Earth. Go enter into the rock, and hide thee in the dust, for fear of the LORD, and for the glory of his majesty…"
>
> (Isaiah 2:10,19)

> "In that day men will cast away their idols of silver and gold—the idols they made to worship—away to the moles and bats. *They will flee to caverns in the rocks and crevices in the cliffs*, away from the terror of the LORD and from the splendor of His majesty, when He rises to shake the Earth. Put no more trust in man, who has only the breath in his nostrils. Of what account is he?"
>
> (Isaiah 2:20-22)

> "For this is what the LORD of Hosts says: "Once more, in a little while, I will shake the Heavens and the Earth, the sea and the dry land. I will shake all the nations, and they will come with all their treasures, and I will fill this house with glory, says the LORD of Hosts."
>
> (Haggai 2:6-7)

> "But for now, I will send for many fishermen, declares the LORD, and they will catch them. After that I will send for many hunters, and they will hunt them down on every mountain and hill, even from the clefts of the rocks."
>
> (Jeremiah 16:16)

> "And it shall come to pass, while my glory passes by, that I will put thee in a cleft of the rock, and will cover thee with my hand while I pass by:"
>
> (Exodus 33:22)

"To dwell in the clefts of the valleys, in caves of the Earth, and in the rocks."
(Job 30:6)

"At that time 'they will say to the mountains, "Fall on us!" and to the hills, "Cover us!"
(Luke 23:20)

"They will suffer the penalty of eternal destruction, separated from the presence of the Lord and the glory of His might,"
(2 Thessalonians 1:9)

"At that time, His voice shook the Earth, but now He has promised, "Once more I will shake not only the Earth, but Heaven as well."
(Hebrews 12:26)

"And they will all come and settle in the steep ravines and clefts of the rocks, in all the thorn bushes and watering holes."
(Isaiah 7:19)

"Therefore, I will make the Heavens tremble, and the Earth will be shaken from its place at the wrath of the LORD of Hosts on the day of His burning anger."
(Isaiah 13:13)

"Behold, the LORD lays waste to the Earth and leaves it in ruins. He will twist its surface and scatter its inhabitants. . . . Whoever flees the sound of terror will fall into a pit, and whoever escapes from the pit will be caught in a trap. For the windows of Heaven are open, and the foundations of the Earth are shaken. The Earth is utterly broken apart, the Earth is split open, the Earth is shaken violently."
(Isaiah 24:1; 18-19)

"Every city flees at the sound of the horseman and archer. They enter the thickets and climb among the rocks. Every city is abandoned; no inhabitant is left."
(Jeremiah 4:29)

"Abandon the towns and dwell among the rocks, O residents of Moab! Be like a dove that nests at the mouth of a cave."
(Jeremiah 48:28)

> "Now I will arise," says the LORD. "Now I will lift Myself up. Now I will be exalted."
> (Isaiah 33:10)

> "Tell them that this is what the Lord GOD says: 'As surely as I live, those in the ruins will fall by the sword, those in the open field I will give to be devoured by wild animals, and those in the strongholds and caves will die by plague.'"
> (Ezekiel 33:27)

> "The high places of Aven, the sin of Israel, will be destroyed; thorns and thistles will grow over their altars. Then they will say to the mountains, "Cover us!" and to the hills, "Fall on us!"
> (Hosea 10:8)

> "You too will become drunk; *You will go into hiding and seek refuge from the enemy.*"
> (Nahum 3:11)

FAITH IN GOD'S FAITHFULNESS

The Final Prophecy for the Body of Christ. The "Body" of Christ are all about human vessels that carry the Holy Spirit and serve Christ's Kingdom of Heaven on Earth. He is faithful to save us on our day of trouble.

> "Your steadfast love, O LORD, extends to the heavens, your faithfulness to the clouds."
> (Psalms 36:5)

> "O LORD God of hosts, who is mighty as you are, O LORD, with your faithfulness all around you?"
> (Psalms 89:8)

The Bible clearly shows that faith is not just a one-time decision to follow Christ (which is certainly important and has eternal consequences) but we are called to increase in faith. Faith has been described like a muscle. A muscle in our body will atrophy if it is not used. If you have ever had a cast on an arm or leg for a few weeks, you know what I mean. Just from not using the muscle it withers away and is almost useless. Our faith is the same way. We need to use our faith and to grow like Christ and increase our faith. Pray to God for him to increase your faith in Him today. Praise Him!

"Seek the LORD and his strength; seek his presence continually!"
(1 Chronicles 16:11)

"But the Lord is faithful. He will establish you and guard you against the evil one."
(2 Thessalonians 3:3)

"Let us hold fast the confession of our hope without wavering, for he who promised is faithful.
(Hebrews 10:23)

"He will cover you with his pinions, and under his wings you will find refuge; his faithfulness is a shield and buckler."
(Psalms 91:4)

"Now faith is the assurance of things hoped for, the conviction of things not seen."
(Hebrews 11:1)

"And without faith it is impossible to please him, for whoever would draw near to God must believe that he exists and that he rewards those who seek him."
(Hebrews 11:6)

"I have not hidden your deliverance within my heart; I have spoken of your faithfulness and your salvation; I have not concealed your steadfast love and your faithfulness from the great congregation."
(Psalms 40:10)

"His master said to him, 'Well done, good and faithful servant. You have been faithful over a little; I will set you over much. Enter into the joy of your master.'"
(Matthew 25:21)

"if we are faithless, he remains faithful—for he cannot deny himself."
(2 Timothy 2:13)

THE KINGDOM OF HEAVEN

> "But our citizenship is in Heaven. And we eagerly await a Savior from there, the Lord Jesus Christ."
> (Philippians 3:20)

The Kingdom of Heaven is coming to Earth soon. This Kingdom will rule overall. The very presence of the Lord, on the Earth, will inevitably raise the vibration of the Earth, which is what the ancients refer to as the 5th world, or the 5th Dimension. He is and will be the Light of the World. He will replace the Sun as the Son of Heaven. All things will be restored unto Him, for Him and by Him. This is the hope that sits at the foundation of many faiths, which will eventually converge into the Kingdom of the Redeemed, on Earth. The battles are over who gets to rule and own the Earth? After this age is completed, the prophesies are clear that it is the meek, the poor in spirit, the faithful, the long suffering, who will inherit the Earth and be given the Kingdom. We have glimpses into this Kingdom on Earth, but no man can fully imagine what God has planned for those who put their faith and hope in Him. It will be phantasmagorical to the say the least.

> "The LORD has established his Throne in the Heavens, and His Kingdom rules overall."
> (Psalm 103:19)

> "Set your minds on things above, not on Earthly things."
> (Colossians 3:2)

The time is near, for this the end of this Age which will converge into the Kingdom of Heaven on the Earth. This is good news! There is no bitterness, anger, resentments, grudges, jealousy, racism, anti-Semitism in the Kingdom of Heaven. As Aldous Huxley rightly discerned, "Maybe this world is another planet's hell?" Yes, indeed it is, and not just for humans, but aliens as well, who live and move amongst us. They can masquerade as human, animals, or vapor but they are somehow bound to this Earth, through past deeds and curses. This is why Earth's history is wrought with so much drama. All the powers and principalities that have held humans in bondage on Earth for millennia have been conquered through the power of Christ, who shares that power with His people whom He empowers with His Freedom and His Spirit, through His Presence.

"To rest in God's power when your own weaknesses seem to be screaming at you? That is grace! To be confident in who God is for you when you feel overwhelmed by odds against you? That is peace! To stand alone against massive intimidation. That is trust! To know beyond any shadow of a doubt that God is bigger, and therefore you cannot lose. That's the faith that moves mountains!"

Despite what is going on around you on Earth, it is possible to create your own heaven or hell experience on Earth. You do have a choice. Therefore, live each day as if it is your last day on Earth and one day, you are sure to be right! One minute you are here and the next you are not. Choose this day whom you will serve. Will you serve the Kingdom of Heaven that is governed by the Kingdom of God on Earth? Or will you serve the hounds of hell?

> Be kind for everyone you meet is fighting a hard battle. In today's world, kindness shines like a light in a darkening world. All the more reason, to have courage and be kind!
> ~*Cinder-Ella*~

It is difficult to hold on sometimes - but someday beyond our tears and all the world's wrongs, beyond the clouds and all we can see and touch...there will be love, compassion and justice, and we shall all understand.

~ Finis ~

Notes:
1. Ella LeBain, *Who's Who in the Cosmic Zoo?* Book One – Third Edition, Concluding Words, pp. 415-444. Skypath Books, 2013.
2. Ella LeBain, *COVENANTS* - Book Four - *An End Times Guide to ETs Aliens gods and Angels*, Skypath Books, 2018. https://www.amazon.com/gp/aw/d/0692988637/
3. Ella LeBain, *Who is God?* Book Two, Chapter Four: *Motherships of the Lord*, p.71-85, Skypath Books, 2015.
4. Ella LeBain, *Who Are the Angels?* Book Three, Skypath Books, 2016.
5. John Pipe, *Battling the Unbelief of a Haughty Spirit*, http://www.desiringgod.org/messages/battling-the-unbelief-of-a-haughty-spirit, December 18, 1988
6. Swami Tapasyananda, Sunkarakandam of Srimad Valmiki Ramayan, The Ramayana, Published by The President, Sri Ramakrishna Math Printing Press, Mylapori, Chennai, India, 2006. pp.46-48

NOTES AND BIBLIOGRAPHY

INTRODUCTION
1. Ella LeBain, *Who Are the Angels?* Chapter Sixteen: The Second Coming and Nibiru pp.338-363, Skypath Books, 2015
2. Sarah Lewin, *New Model: Nearby Exoplanet TRAPPIST-1e May Be Just Right for Life*, https://www.space.com/36349-trappist-1e-just-right-for-life.html?utm_source=sp-newsletter&utm_medium=email&utm_campaign=20170405-sdc, April 05, 2017.
3. Universe Today, *Do the TRAPPIST-1 planets have atmospheres?*, https://www.universetoday.com/146958/do-the-trappist-1-planets-have-atmospheres/#more-146958
4. Anna Merlan, "Here's Why Gallup Polled Americans About UFOs for the First Time in Decades, if they're asking about aliens, they must be real." (February 25, 2020) https://www.vice.com/en_us/article/n7j4bb/why-gallup-polls-americans-about-ufo-aliens
5. Lydia Saad, Americans Skeptical of UFOs, but Say Government Knows More, https://news.gallup.com/poll/266441/americans-skeptical-ufos-say-government-knows.aspx, September 6, 2019.
6. Ella LeBain, *Who's Who in the Cosmic Zoo?* Book One, Trafford, 2012.
7. Leonard David, Space Force: What Will the Newest Military Branch Actually Do? https://exonews.org/space-force-what-will-the-newest-military-branch-actually-do/, February 9, 2020
8. Third Phase of the Moon, Space Force New Asset? An Investigative report in regards to the ISS and a UFO close encounter, February 22, 2020, https://youtu.be/xhCNgVg7enY

CHAPTER ONE – SIGNS IN THE HEAVENS
1. Ella LeBain, *Who Are the Angels?* Book Three of *Who's Who in the Cosmic Zoo? A Guide to ETs, Aliens, Gods & Angels*, Chapter: The Second Coming and Nibiru, pp. 338-363. Skypath Books, CO. 2016
2. Jason M. Breshears, *Annunaki Homeworld, Orbital History and 2046AD Return of Planet Nibiru*, The Book Tree, San Diego, CA 2011.
3. Immanuel Velikovsky, *Worlds in Collision*, Paradigma Ltd; Later Printing edition (October 1, 2009)
4. Ella LeBain, *The Hype and Anticlimax of Blood Moons*, http://spirituallydiscerning.com/?cat=1, 2014.
5. Sheldan Nidle, *You Are Becoming a Galactic Human, Spiritual Education Endeavors*, 1st edition (April 1, 1994).
6. Elizabeth Tenety (January 3, 2011). "May 21, 2011: Harold Camping says the end is near". Washington Post. Kimberly Winston (March 23, 2011). "Judgment Day: May 21, 2011". Washington Post.
7. Nelson, Chris (June 18, 2002). "A Brief History of the Apocalypse; 1971 – 1997: Millennial Madness". Retrieved June 23, 2007.

8. "Harold Camping Says End did come May 21, spiritually; Predicts New Date: October 21". International Business Times. Retrieved May 23, 2011.
9. "Did Harold Camping Ever Teach the End Was Coming In 1994?" on YouTube. July 30, 2009. Retrieved December 16, 2012.
10. "Harold Camping False Prophet: Ministry Probably Doomed". International Business Times. October 21, 2011.
11. Ella LeBain, *Who is God? Book Two of Who's Who in the Cosmic Zoo? A Guide to ETs, Aliens, Gods & Angels*, Skypath Books, CO. 2015

CHAPTER TWO – LIFE ON OTHER PLANETS?
1. Biblehub.com Commentary on 1 Corinthians 15:22 and 45
2. Gary Bates, Did God create life on other planets? Otherwise, why is the universe so big? https://creation.com/did-god-create-life-on-other-planets, November 2009
3. Ella LeBain, *Who's Who in the Cosmic Zoo?* Book One – Third Edition, Chapter Three: From Adam's Failure to the Second Adam's Victory, pp 71-74 – Tate. 2013
4. Isaiah 53:6,10; Matthew 20:28; 1 John 2:2, 4:10
5. Ibid, Concluding Words, pp. 415-446.
6. Ella LeBain, *Who Are the Angels?* Book Three, Chapter Fourteen: The Harvest of Angels, pp. 289-302. Skypath Books, 2016.
7. Ella LeBain, *COVENANTS* – Book Four, Chapter Two: How Disclosure of ETs Impacts Religion, pp. 5-13; Skypath Books, April 2018.
https://www.amazon.com/COVENANTS-Times-Guide-Aliens-Angels/dp/0692988637/
8. Ella LeBain, *Who's Who in the Cosmic Zoo?* Book One – Third Edition, Tate, 2013. https://www.whoswhointhecosmiczoo.com
9. *The Lost Book of the Bible*, https://www.gotquestions.org/lost-books-Bible.html
10. Zecharia Sitchin, *The 12th Planet*, Book One of the Earth Chronicles, Avon, NY, 1991
11. Ella LeBain, *COVENANTS* – Book Four, Skypath Books, 2018.
12. Ella LeBain, *Who is God?* Book Two, Skypath Books, 2015.
13. Ella LeBain, *Who Are the Angels?* Book Three, Skypath Books, 2016.
14. Ella LeBain, *Who's Who in the Cosmic Zoo?* Book One – Third Edition, https://www.amazon.com/Whos-Who-Cosmic-Zoo-Third/dp/1629942065/, 2013.

CHAPTER THREE – CONTACT WITH ETs FROM PLANET X
1. Ella LeBain, *Who's Who in the Cosmic Zoo?* Book One, pp. 110-119, 339.
2. Ella LeBain, *Who's Who in the Cosmic Zoo?* Book One – Third Edition, https://www.amazon.com/Whos-Who-Cosmic-Zoo-Third/dp/1629942065/, 2013.
3. Ella LeBain, *Who Are the Angels?* Book Three, Skypath Books, 2016.
4. Ella LeBain, *Who's Who in the Cosmic Zoo?* Book One – Third Edition, https://www.amazon.com/Whos-Who-Cosmic-Zoo-Third/dp/1629942065/, 2013.
5. Ella LeBain, *Who's Who in the Cosmic Zoo?* Book One – Third Edition, https://www.amazon.com/Whos-Who-Cosmic-Zoo-Third/dp/1629942065/, 2013.
6. Ella LeBain, *Who's Who in the Cosmic Zoo?* Book One – Third Edition, pp. 90-91, The Cosmic Drama/Galactic Warfare/Class Between Two Kingdoms: The Human-Reptilian Wars, i.e., "Draco Wars," Chapter Four: How to Tell Who is Who?
7. Ella LeBain, *Who's Who in the Cosmic Zoo?* Book One – Third Edition, https://www.amazon.com/Whos-Who-Cosmic-Zoo-Third/dp/1629942065/, 2013.
8. Ella LeBain, *COVENANTS* – Book Four, Skypath Books, 2018.

9. *Peter Townsend Won't Get Fooled Again*, The Who, Album: Who's Next, 1985.
10. Ella LeBain, *Who is God?* Book Two, Chapter Twenty-Eight: What Happens When You Die? pp. 455-483.

CHAPTER FOUR – CONNECTING THE DOTS TO NIBIRU

1. Skywatch Media News, YouTube Channel, Nibiru-Planet X System & It's impact on our Solar System.
2. R. S. Harrington, *The Location of Planet X*, The Astronomical Journal, Volume 96, Number 4, U.S. Naval Observatory, Washington, DC, October 1988.
3. *When Cosmic Cultures Meet*, An International Forum presented by the Human Potential Foundation, held in Washington, D.C., May 27-29, 1995, The Proceedings, Zecharia Sitchin, p. 163-166, The Past Holds the Key to the Future, The Human Potential Foundation, 1996.
4. Dr. Michael Layne, Renowned Physicist Says, 'There Is A God.', https://godtv.com/renowned-physicist-says-there-is-a-god/, February 28, 2019.
5. Ella LeBain, *Who is God?* Book Two of Who's Who in the Cosmic Zoo? A Guide to ETs, Aliens, Gods and Angels, Skypath Books 2015.
6. Ella LeBain, *Who's Who in the Cosmic Zoo?* Book One, Trafford, 2012.
7. Sebastian Kettley, Planet Nine SHOCK: Telescopes in Hawaii are hunting down mystery planet larger than Earth, https://www.express.co.uk/news/science/1109962/Planet-nine-proof-subaru-telescope-hawaii-planet-9-discovery-planet-x-michael-brown/amp, April 8, 2019.

CHAPTER FIVE – THE NEW ASTRONOMY

1. Immanuel Velikovsky, *Worlds in Collision*, Macmillan Publishers, April 3, 1950.
2. Ella LeBain, *Who Are The Angels?* – Book Three of Who's Who in the Cosmic Zoo? A Guide to ETs, Aliens, Gods or Angels, Skypath Books, 2016. https://www.whoswhointhecosmiczoo.com
3. Zecharia Sitchin, *The 12th Planet* (Book One of the Earth Chronicles) Bear & Company, NM, January 1, 1977
4. Ella LeBain, *Who is God?* Book Two, Skypath Books, p.7, 2015. https://www.amazon.com/dp/0692911529

CHAPTER SIX – ALL THINGS NIBIRU

1. Ella LeBain, *Who is God?* Book Two, Chapter: Names of God, What's in a Name, Skypath Books, 2015.
2. Lee Billings, *The Book That Predicted Proxima b*, https://www.scientificamerican.com/article/the-book-that-predicted-proxima-b-excerpt/, September 8, 2016.
3. Zecharia Sitchin's Book *The Lost Book of Enki: Memoirs and Prophecies of an Extraterrestrial God*, Bear & Company; First Edition (November 15, 2001)
4. Zecharia Sitchin, *The 12th Planet*, Book One of the Earth Chronicles, Avon Books, NY, 1976
5. Dr. Danny R. Faulkner, https://answersingenesis.org/astronomy/solar-system/Nibiru/, March 24, 2017
6. Sarah Lewin, *New Model: Nearby Exoplanet TRAPPIST-1e May Be Just Right for Life*, https://www.space.com/36349-trappist-1e-just-right-for-life.html?utm_source=sp-newsletter&utm_medium=email&utm_campaign=20170405-sdc, April, 2017.

7. Ella LeBain, *Who Are the Angels?* – Book Three, Chapter 16: on The Second Coming and Nibiru, pp. 338-359, Skypath Books, 2016.
8. *World's in Collision* by Immanuel Velikovsky, Doubleday; Second Edition, 1951.
9. http://webcamsdemexico.com/webcam-punta-cancun-poniente
10. Ella LeBain, *Who's Who in the Cosmic Zoo?* Book One – Third Edition, pp. 248-289, Tate, 2013.
11. https://answersingenesis.org/astronomy/solar-system/Nibiru/
12. Zecharia Sitchin, Book One of The Earth Chronicles, *The 12th Planet*, Avon Books, 1976.
13. https://www.drclaudiaalbers.com/
14. https://www.youtube.com/user/claalb1

CHAPTER SEVEN – NIBIRU: MYTH OR SCIENCE?

1. Ella LeBain, *Who's Who in the Cosmic Zoo? A Spiritual Guide to ETs, Aliens, Gods & Angels*, Book One, Chapter Five: Annunaki (Nibiruans), pp. 110, Third Edition, 2013.
2. Mountain Wolf Blog: *Evidence for Nibiru or "Planet X"*, http://mtnwolf63.wordpress.com
3. Leslie Mullen, *Sun's Nemesis Pelted Earth with Comets, Study Suggests*, https://www.space.com/8028-sun-nemesis-pelted-earth-comets-study-suggests.html, March 11, 2010
4. *Chilean astronomer CARLOS MUÑOZ FERRADA Predicts Hercolubus aka Planet X*, Excerpts Reprinted from the JOURNAL OF TODAY, of El Salvador, https://www.rawgist.com/chilean-astronomer-carlos-munoz-ferrada-predicts-hercobulus-aka-planet-x/, May 19, 2015.
5. *Carlos Ferrada Predicts Planet X*, Aug 20, 2013, https://www.youtube.com/watch?v=N9f-Bhub0Lg
6. The *Who's Who in the Cosmic Zoo* Facebook Page, https://www.Facebook.com/Whoswhointhecosmiczoo
7. *Russ Report Finding 10th Planet* (AP) February 12, 1960, New York, NY, Reprinted in The Milwaukee Sentinel – February 13, 1960.
8. *Solar System Warming–and its Implications*, http://prof77.wordpress.com/2010/08/07/why-all-planets-in-our-solar-system-are-warming-and-its-implications, August 7, 2010.
9. Zecharia Sitchin, *The 12th Planet*, Book One of the Earth Chronicles – Astonishing Documentary Evidence of Earth's Celestial Ancestors, Avon Books, New York, 1976.
10. *Planetary Defense Conference* Exercise – 2019 – NASA, https://cneos.jpl.nasa.gov/pd/cs/pdc19/
11. Zecharia Sitchin, *The Lost Book of Enki – Memoirs of an Extraterrestrial God*, Bear & Company, Vermont, 2002.
12. Ella LeBain, *Who is God?* Book Two, Skypath Books, 2015
13. Marshall Masters, *Being in it For the Species*, CreateSpace, 2014
14. Immanuel Velikovsky, *World's in Collision*, Doubleday; Second Edition, 1951.
15. Immanuel Velikovsky, *Ages in Chaos*, (1948), Doubleday Reprint Edition, February 1, 1952.
16. Immanuel Velikovsky, *Earth in Upheaval*, (1952), Dell; Mass Paperback Edition, 1972.
17. Elisheva Velikovsky, *Stargazers & Gravediggers: Memoirs to Worlds in Collision*, 1983.

18. Erich Von Daniken, *Chariots of the Gods*, Bantam Book, Putnam's Edition, February 1970.

CHAPTER EIGHT – OUR SOLAR SYSTEM IS PERTERBED
1. Burak Eldem, 2012: *Rendezvous with Marduk, Book One of the Trilogy: The Hidden History* (Inkilap (2003), Turkey.
2. Joakim RS Nilsson, *Nibiru*, Stockholm, Sweden, used with permission, https://astromantra.se/english/nibiru2.html
3. World News Australia, *Scientists discover Solar System's 'Planet X'*, February 28, 2008 http://news.sbs.com.au/worldnewsaustralia/scientists_discover_solar_system39s_39 planet_x39_541620
4. Immanuel Velikovsky, *Worlds in Collision*, 1950.
5. Immanuel Velikovsky, *Stargazers and Gravediggers: Memoirs to Worlds in Collision*, William Morrow & Co; 1st edition (March 1, 1983)
6. NASA, Jet Propulsion Laboratory, California Institute of Technology, *When The Moon Rang Like a Bell*, Podcast, Episode 2: *Music of the Sphere*, https://www.jpl.nasa.gov/podcast/content, November 5, 2018.
7. Amy Shira Teitel, *Does the Moon Sound Like a Bell?*, May 27, 2016 https://www.popsci.com/does-Moon-sound-like-bell
8. Ella LeBain, *Who's Who in the Cosmic Zoo?* Book One, Trafford, 2012.
9. Ibid, Conclusion of *Who's Who in the Cosmic Zoo?* Book One, 2012
10. Ella LeBain, *Who is God?* Book Two, Chapter Twenty-Nine: Who Created Sexism, Skypath Books, 2015.
11. Rutherford H. Platt, *The First and Second Books of Adam and Eve*, Kessinger Publishing, LLC (September 10, 2010)
12. Zecharia Sitchin, *Are We Alone in the Universe?* Amazon Prime Video, 1978
13. Erich Von Daniken, *Chariots of the Gods*, Bantam Books; Unabridged edition (1972)
14. Carl Sagan, Ann Druyan, *Contact*, Robert Zemekis, Director, 1997 Film.
15. Norman Kagan (2003). *"Contact". The Cinema of Robert Zemeckis*. Lanham, Maryland: Taylor Trade Publishing. pp. 159–181. ISBN 0-87833-293-6.
16. Robert Zemeckis, Steve Starkey, DVD audio commentary, 1997, Warner Home Video
17. Norman Kagan (2003). *"Contact". The Cinema of Robert Zemeckis*. Lanham, Maryland: Taylor Trade Publishing. pp. 159–181. ISBN 0-87833-293-6.
18. *Contact* (1997 American film)

CHAPTER NINE – REPEAT AFTER ME, PLUTO IS A PLANET
1. Zechariah Sitchin, *The 12th Planet*, Avon Books, NYC, NY, 1976.
2. Pluto Facts, https://space-facts.com/pluto/
3. Elizabeth Howell, Neil deGrasse Tyson Rejects Pluto Planethood Proposal on 'Colbert', https://www.space.com/36126-pluto-planethood-slammed-by-neil-degrasse-tyson.html, March 20, 2017.
4. Tim Prudente, The Baltimore Sun Writer, Scientist leads effort to restore Pluto's planetary stature, https://www.seattletimes.com/nation-world/scientist-leads-effort-to-restore-plutos-planetary-stature/, March 17, 2017.
5. Elizabeth Howell, What Is a Planet?, https://www.space.com/25986-planet-definition.html, April 7, 2018.

6. Mike Wehner, Pluto might become a planet again because astronomers can't make up their mind, https://bgr.com/2018/09/07/pluto-planet-argument-definition/, September 7, 2018.
7. Zechariah Sitchin, *The 12th Planet*, Avon Books, NYC, NY, 1976
8. John Dinardo, The Planet 9 bait-and-switch involving NASA, Caltech and Planet X / Nibiru, https://planetxnews.com/2016/01/27/the-planet-9-bait-and-switch-involving-nasa-caltech-and-planet-x-Nibiru/
9. Thomas O'Toole, Possibly as Large as Jupiter, Mystery Heavenly Body Discovered, The Washington Post, https://www.washingtonpost.com/archive/politics/1983/12/30/possibly-as-large-as-jupiter/1075b265-120a-4d40-9493-a8c523b76927/, December 30, 1983.
10. Nemesis Star Theory: The Sun's 'Death Star' Companion, https://www.space.com/22538-nemesis-star.html, July 21, 2017.
11. Christian deBlanc, How the Media separated Nibiru from Planet X from Nemesis, Planet X, Writers, August 10, 2017, https://planetxnews.com/2017/08/10/media-separated-Nibiru-planet-x-nemesis/
12. John DiNardo, Planet X, Washington Post flip-flops on Planet X / Nibiru, January 8, 2016, https://planetxnews.com/2016/01/08/washington-post-flip-flops-on-planet-x-Nibiru/
13. Darwin Malicdem, Planet Nine: Scientists Find 'Solid' Evidence Of Super-Earth Planet Beyond Neptune, Mar 8, 2019, https://www.medicaldaily.com/planet-nine-scientists-find-solid-evidence-super-Earth-planet-beyond-neptune-430846
14. The Planet X / Nibiru system and its potential impacts on our Solar System. July 12, 2016 ·Planet X·, https://planetxnews.com/2016/07/12/planet-x-Nibiru-system-potential-impacts-solar-system-2/
15. Stephen Luntz, Astronomers Find Evidence Of Two Undiscovered Planets In Our Solar System, https://www.iflscience.com/space/signs-planet-x-and-maybe-y/, November 16, 2014.
16. Mike Wall, Mysterious Planet X May Really Lurk Undiscovered in Our Solar System, https://www.space.com/28284-planet-x-worlds-beyond-pluto.html, January 16, 2015.
17. Joachim Hagopian, Global Research, The Nibiru Planet X System and Its Potential Impacts on Our Solar System, A Scientific Case for Nibiru/Planet X System https://www.globalresearch.ca/the-niburu-planet-x-system-and-its-potential-impacts-on-our-solar-system/5459788, September 6, 2015.
18. Amber William, Nasa Warning – Clear Signs That Planet X is Affecting Earth in a Bad Way, https://www.mydailyinformer.com/nasa-warning-clear-signs-that-planet-x-is-affecting-Earth-in-a-bad-way/, June 13, 2018.

CHAPTER TEN – TRACKING AND CLOAKING
1. Ella LeBain, *Who is God?* Book Two, Chapter Twenty-Three: Babylon History: Where it All Began, pp.367-386, Skypath Books, 2015.
2. Ella LeBain, *Who is God?* Book Two, Chapter Two: Ancient Technology & Biblical Astronauts, pp. 21-46, Skypath Books, 2015.
3. Walter Manuel Montano, *Behind the Purple Curtain* Hardcover, Literary Licensing, LLC, July 27, 2013.
4. Ella LeBain, *COVENANTS* – Book Four, Skypath Books, 2016.
5. Mark Passio, Lecture, Free Your Mind 2 Conference, April 27, 2013, New Age bullshit and the suppression of the sacred masculine. http://youtu.be/Q511_E8Tlp0, July 2, 2013.

6. Origins: 11 Unreal Places and their Artifacts Explained, http://hidden-truth.net/2017/07/origins-11-unreal-place-artifacts-explained/, July, 2017.
7. Solar Terrestrial Relations Observatory, STEREO, https://www.nasa.gov/mission_pages/stereo/main/index.html
8. Solar Terrestrial Relations Observatory (STEREO), https://www.spaceweatherlive.com/en/solar-activity/solar-images/stereo.html
9. Lucifer in the Vatican – Father Malachi Martin, https://thewildvoice.org/lucifer-vatican-malachi-martin/
10. Gerald Clark, YouTube, Abrahamic Religions, 2017, https://youtu.be/-ygSDL7qdvA?list=UUFUH_0JRBPG9g5k5M3AgJ4g, http://www.geraldclark77.com/uploads/4/6/0/7/46076627/7th_planet_broadcast_outline_episode_003.pdf
11. Zecharia Sitchin, *When Time Began – The First New Age*, Book Five of the Earth Chronicles, Avon Books, NY, 1993.
12. David Flynn, *Cydonia: The Secret Chronicles of Mars*, (used with permission) 2002. http://www.mt.net/~watcher/stones.html
13. *The Apocryphal First and Second Books of Adam and Eve*
14. Zecharia Sitchin, *War of Gods & Men: Evidence of Extraterrestrial Warlords who destroyed Ancient Civilization*, Book Three of the Earth Chronicles, Avon Books, NY,1985.
15. Ella LeBain, *Who is God?* Book Two, Skypath Books, 2015.

CHAPTER ELEVEN – THE MESSIANIC STAR

1. Zecharia Sitchin *Genesis Revisited* p30, 324-328, 34 Avon Books, NY, 1990.
2. Kapiel Raaj, *Planet Nibiru Facts & Secrets Revealed: Can you feel its Presence?* http://www.krschannel.com/Nibiru-.htm
3. Stephen Wagner, *Is Nibiru Approaching?* May 24, 2019 http://paranormal.about.com/library/weekly/aa021102a.htm
4. Andy Lloyd, *Dark Star*, 2005, Nibiru the Messianic Star, http://www.darkstar1.co.uk/ds6.htm
5. E.C. Krupp *In Search of Ancient Astronomies* pp215-219, Penguin, NY, 1984.
6. Zecharia Sitchin *The Lost Realms* p 268, Avon Books, NY, 1990.
7. Laurence Gardner, *Genesis of the Grail Kings*, Bantam, NY, 1999.
8. Nancy Lieder, Zetatalk, http://www.zetatalk.com
9. Rabbi Yuval Ovadia, Planet X-Nibiru, February 8, 2018, https://www.youtube.com/watch?v=xS2FknRh3t8
10. Ella LeBain, *Who Are the Angels?* Book Three, Chapter Sixteen: The Second Coming and Nibiru, p. 338-347, Skypath Books, 2016.
11. http://www.abideinchrist.com/messages/mat2745.html
12. Ella LeBain, *Who is God?* Book Two, Skypath Books, 2015.
13. Robert Bauval and Adrian Gilbert *The Orion Mystery*, p.202 Mandarin, 1994.
14. Brad Aaronson, *When Was the Exodus?* Jerusalem Institute of Ancient History, (JIAH), http://ou.org
15. Zecharia Sitchin, *The 12th Planet*, The First Book of the Earth Chronicles, Mass Market Paperback, 1976
16. Ella LeBain, *COVENANTS* – Book Four, Chapter Twenty-Nine: Messianic Prophecies and the Lost Tribes of Israel, p. 510, Skypath Books, 2018.
17. *The Amplified Bible*, Zondervan (February 1, 2001)

18. Rev Willem J.J. Glashouwer, Jerusalem, the UN and the times of the Gentiles, http://www.whyisrael.org/2017/01/12/jerusalem-the-un-and-the-times-of-the-gentiles/, January 2017.

CHAPTER TWELVE – NIBIRU IN BIBLE PROPHECY

1. Anderson Reed, *Shouting at the Wolf, A Guide to Identifying and Warding Off Evil in Everyday Life*, Library of the Mystic Arts, Citadel; First Edition, August 31, 1998.
2. Ella LeBain, *Who is God?* Book One, Chapter Four: The Motherships of the Lord, The Cloudships and the Clouds of Heaven, pp.71-88, Skypath Books, 2015.

CHAPTER THIRTEEN – WHO ARE THE ANUNNAKI?

1. Zecharia Sitchin, *The Twelfth Planet*: Book I of the Earth Chronicles (The Earth Chronicles) Harper, reprint edition, 2007.
2. Ella LeBain, *Who is God?* Book Two, Skypath Books, 2015.
3. Gregg Prescott, M.S., *What the Church Isn't Telling You About Nibiru And The Annunaki*, August 24, 2016 | In5D.com
4. Ella LeBain, *Who is God?* Book Two, Chapter: Who Are the Annunaki? P.264, Skypath Books, 2015.
5. Ella LeBain, *Who's Who in the Cosmic Zoo?* Book One – Third Edition, Concluding Words, pp. 415-444, Skypath Books, 2013
6. Zecharia Sitchin, *Genesis Revisited, Is Modern Science Catching Up with Ancient Knowledge?* (Earth Chronicles) Avon Paperback – October 1, 1990.
7. Ella LeBain, *Who is God?* Book Two, Chapter: Who is Allah?, p.359, Skypath Books, 2015.
8. Gregg Prescott, M.S., *What the Church Isn't Telling You About Nibiru And the Annunaki*, August 24, 2016 | In5D.com
9. Ella LeBain, *Covenants* – Book Four, Skypath Books, 2018.
10. Ella LeBain, *Who Are the Angels?* Book Three, Skypath Books, 2016
11. Marshall Masters, *Two Suns in the Sky, Who Lives, Who Dies?* Audio Transcript. CreateSpace, June 2017

CHAPTER FOURTEEN – SAVING EXTRATERRESTRIALS

1. Guy Consolmagno, Paul Mueller, *Would You Baptize an Extraterrestrial? . . . and Other Questions from the Astronomers' In-box* at the Vatican Observatory, in 2014.
2. YouTube: http://youtube/ua26wcJ8Z80
3. Ella LeBain, *Covenants* – Book Four, Skypath Books, 2018.
4. Thomas Horn, Cris Putnam, *EXO-VATICANA, Petrus Romanus, Project L.U.C.I.F.E.R. And the Vatican's Astonishing Plan for the Arrival of an Alien Savior*, Defender Press, 2013.
5. https://archive.org/details/ArtBellAndMalachiMartin
6. Ella LeBain, *Who's Who in the Cosmic Zoo?* Book One, Annunaki p.110, 2012.
7. Ibid, p.415
8. Gregg Prescott, M.S., *What the Church Isn't Telling You About Nibiru And the Annunaki*, August 24, 2016 | In5D.com
9. Ibid
10. Ella LeBain, *Covenants* – Book Four, Skypath Books, 2018.
11. https://newspunch.com/exposed-the-church-isnt-telling-you-about-nibiru-and-the-anunnaki/

12. *Are We Alone in the Universe* – Zecharia Sitchin, UFOTV, 1992, 2003.

CHAPTER FIFTEEN – SALVATION FOR ALL CREATION
1. Ella LeBain, *Who's Who in the Cosmic Zoo?* Book One – Third Edition, https://www.amazon.com/Whos-Who-Cosmic-Zoo-Third/dp/1629942065/, 2013.
2. Ella LeBain, *Who's Who in the Cosmic Zoo? A Spiritual Guide to ETs, Aliens, Gods & Angels*, Trafford, 2012.
3. *Are We Alone? The Search for Extraterrestrial Life: How Does SETI Search for Alien Civilizations*, https://www.etupdates.com/2019/02/23/how-does-seti-search-for-alien-civilizations/
4. George Hunt Williamson, *Secret Places of the Lion*, pp. 184-185; Warner Destiny Books, NYC, 1958.
5. Ella LeBain, *Who's Who in the Cosmic Zoo?* Book One, Chapter Two: Exopolitics and Divine Jurisprudence, Spiritual Legal Ground, p. 45-55; Trafford, 2012.
6. Brian Nelson, https://www.mnn.com/Earth-matters/climate-weather/stories/magnetic-north-shifting-by-40-miles-a-year-might-signal-pole-r, February 5, 2019.

CHAPTER SIXTEEN – NIBIRU AND MENTAL DISORDERS
1. Ivan Petricevic, The Kolbrin: A history-changing 3,600-year-old 'lost' Bible, https://educateinspirechange.org/alternative-news/the-kolbrin-a-history-changing-3600-year-old-lost-Bible/, October 28, 20-17.
2. Greg Jenner, Nibiru and the Kolbrin Bible – Correlations, darkstar1.co.uk http://humansarefree.com/2011/09/nibiru-and-kolbrin-Bible-correlations.html
3. Jeremiah Jameson, Mondo Frazier, Mental Health Problems in the End Times, https://endtimesprophecyreport.com/2016/10/30/mental-health-problems-in-the-end-times/, October 30, 2016.
4. By Elizabeth Howell, *What Is a Planet?* https://www.space.com/25986-planet-definition.html, April 7, 2018.
5. Mike Wehner, Pluto might become a planet again because Astronomers can't make up their mind, https://bgr.com/2018/09/07/pluto-planet-argument-definition/, September 7, 2018.
6. Elizabeth Howell, Neil deGrasse Tyson Rejects Pluto Planethood Proposal on 'Colbert', https://www.space.com/36126-pluto-planethood-slammed-by-neil-degrasse-tyson.html, March 20, 2017
7. Tim Prudente, The Baltimore Sun (TNS), Scientist leads effort to restore Pluto's planetary stature, https://www.seattletimes.com/nation-world/scientist-leads-effort-to-restore-plutos-planetary-stature/, March 17, 2017
8. "Robin Williams," by James Kaplan, US Weekly, January 1999, p. 53.

CHAPTER SEVENTEEN – REVIVED ROMAN EMPIRE
1. *The Revived Roman Empire – What is it?*, https://www.compellingtruth.org/Revived-Roman-Empire.html
2. Ella LeBain, *Who is God?* Book Two, Chapter Twenty: The God of this World, pp. 311-318, Skypath Books, 2015.
3. Paul Carter, https://ca.thegospelcoalition.org/columns/ad-fontes/w11.ho-is-the-whore-of-babylon-and-why-does-it-matter/, June 14, 2018.

4. Kevin Cullen, More than 80 percent of Victims since 1950 were Male, Report says, https://archive.boston.com/globe/spotlight/abuse/stories5/022804_victims.htm, February 28, 20014.
5. The Associated Press contributed to this report. https://www.nbcnews.com/news/world/u-n-report-vatican-policies-allowed-priests-rape-children-n22531
6. Dr. C. Peter Wagner, *Freedom from the Religious Spirit*, Chosen Books (May 10, 2005).
7. *The Book of the Watchers*, Chapters 1-36, in the Books of Enoch, translated by Dr. Richard Laurence in 1821.
8. *What The Church Isn't Telling You About Nibiru And The Anunnaki*, In5D.com https://newspunch.com/exposed-the-church-isnt-telling-you-about-nibiru-and-the-anunnaki/
9. Ella LeBain, *Who is God?* Book Two, Chapter Two: Ancient Technology & Biblical Astronauts, pp. 21-47, Skypath Books, 2015.
10. Ella LeBain, *Who Are the Angels?* Book Three, see, Chapter Two: The Angelic Government, subsection Who Are the Living Creatures? Pp34-37, Skypath Books, 2016.

CHAPTER EIGHTEEN – NEMESIS & THE DEATH OF OUR SUN
1. Hebrews 12:26-29-Berean Study Bible, Biblehub.com
2. *Plato, Allegory of the Cave*—Book VII of the Republic, Life Lessons on How to Think for Yourself, https://www.mayooshin.com/plato-allegory-of-the-cave/, January 2018

CHAPTER NINETEEN – SOLAR SIMULATOR
1. United States Patent 3,325,238 SOLAR SIMULATOR, https://patents.google.com/patent/US3325238A/en
2. Astronomer Terral Croft, http://terral03.com
3. https://watchers.news/2015/03/08/inuit-elders-tell-nasa-Earth-axis-shifted/
4. New documentary recounts bizarre climate changes seen by Inuit elders. Originally published, October 19, 2010, Updated, May 2, 2018, https://www.theglobeandmail.com/arts/film/new-documentary-recounts-bizarre-climate-changes-seen-by-inuit-elders/article1215305/
5. Jeff P – Nibiru Watcher YouTube Channel, https://www.youtube.com/channel/Uci62JvN-lUn7hVL3ffofADA, The Very Best Of Jeff P Proof of Planet X and Sun Simulator, https://youtu.be/AK-EjmP-8cQ
6. https://www.usatoday.com/story/tech/2015/09/21/breakthrough-cloaking-technology-grabs-militarys-attention/72544510/
7. What force in the universe is the strongest or the most powerful?, www.physlink.com/education/askexperts/ae268.cfm
8. Blue Beam Project by NASA, http://politicalavengernews.com/index.php/2018/03/31/blue-beam-project-nasa/
9. Serge Monast, 1994, http://educate-yourself.org/cn/projectbluebeam25jul05.shtml
10. https://www.groundzeromedia.org/9-13-18-hammer-smacc-weaponizing-natures-fury-w-christopher-fontenot/
11. Massive Pole Shifts are Cyclic According to Declassified CIA document, https://www.Exoplanets.org/massive-pole-shifts-are-cyclic-according-to-declassified-cia-document/

CHAPTER TWENTY – INTERLOPERS IN OUR SOLAR SYSTEM
1. Kojima's Blog, EMP Caused by Planet X, http://poleshift.ning.com/profiles/blogs/emp-caused-by-planet-x, March 18, 2017
2. Nancy Leider, ZetaTalk: Magnetic Clash, http://www.zetatalk.com/index/zeta300.htm, July 1, 2006.
3. Michael Salla, PhD., The Exopolitics Institute, https://www.exopolitics.org

CHAPTER TWENTY-ONE – POLE SHIFT
1. Jonathan Erdman, It Just Hit 100 Degrees Fahrenheit in Siberia, the Hottest Temperature on Record So Far North in the Arctic, https://weather.com/news/climate/news/2020-06-21-siberia-russia-100-degrees-heat-record-arctic, June 21, 2020.
2. Ella LeBain, *Who's Who in the Cosmic Zoo?* Book One – Third Edition, Chapter: Life on Mars, pp. 310-313, Skypath Books, 2013.
3. Seth Borenstein / AP Updated: Earth's North Pole Shifts About 34 Miles Per Year, Report Finds http://time.com/5520537/Earth-north-pole-shifts/, February 5, 2019.
4. Dutchsinse, 8/24/2018 –Large M7.1 Earthquake strikes South America – Deep EQ at 610km – NEW UNREST POSSIBLE, https://youtube/NMq-3ipwpE4
5. André van Belkum, Earthquakes in Bible Prophecy, https://lifehopeandtruth.com/prophecy/end-times/Earthquakes-in-prophecy/
6. Inuit Elders tell NASA Earth Axis Shifted, https://watchers.news/2015/03/08/inuit-elders-tell-nasa-Earth-axis-shifted/, March 8, 2015,
7. Mike Adams, Behold, a pale horse? How the "green" environmental movement may be the Biblical Fourth Seal of an End Times global death cult, https://www.naturalnews.com/2019-02-10-behold-a-pale-horse-how-the-green-environmental-movement-may-be-the-biblical-fourth-seal-end-times.html, February 10, 2019
8. Ben Gelber, Earth's North Magnetic Pole moving east faster – NBC4's Ben Gelber explains what that means, February 15, 2019, https://www.nbc4i.com/news/local-news/Earth-s-north-magnetic-pole-rapidly-moving-east-nbc4-s-ben-gelber-explains-what-that-means/1785001457

CHAPTER TWENTY-TWO – TIME-TRAVEL IN THE BIBLE
1. Ashby, Neil (2003). "Relativity in the Global Positioning System" (PDF). Living Reviews in Relativity. 6 (1): 16. Bibcode:2003LRR....6....1A. doi:10.12942/lrr-2003-1. PMC 5253894. PMID 28163638.
2. Miller, Arthur I. (1981). *Albert Einstein's Special Theory of Relativity: Emergence* (1905) and Early Interpretation (1905–1911). Reading, Massachusetts: Addison–Wesley. ISBN 978-0-201-04679-3.
3. Darrigol, Olivier (2005). *The Genesis of the Theory of Relativity*. Einstein, 1905–2005 (PDF). Séminaire Poincaré. 1. Pp. 1–22. Doi:10.1007/3-7643-7436-5_1. ISBN 978-3-7643-7435-8.
4. Penny Bradley and Elena Ka from the Private Secret Space Program Facebook Group
5. Jamie Carter, Is 'Planet Nine' Actually A Black Hole In The Solar System? There's Only One Way To Find Out, https://www.forbes.com/sites/jamiecartereurope/2020/06/03/is-planet-nine-actually-a-black-hole-in-the-solar-system-theres-only-one-way-to-find-out/#7f924493ef40

CHAPTER TWENTY-THREE – CALENDAR SHENANIGANS
1. A New Theory On Time Indicates Present, And Future Simultaneously Exist, http://www.thescinewsreporter.com/2018/09/a-new-theory-on-time-indicates-present.html
2. Peter Dizikes | MIT News Office, Does time pass? Philosopher Brad Skow's new book says it does — but not in the way you may think. January 28, 2015
3. Brad Skow, PhD., *Objective Becoming*, Oxford University Press, Oxford, UK, 2015.
4. Ella LeBain, *Who Are the Angels* Book Three, Skypath Books, 2016.
5. Jorian Polis Schutz, Publisher, Deuteronomy Press, 5779, 5781.
6. Philip S. Berg, PhD., *Time Zones: Your Key to Control*, Research Centre of Kabbalah, Jerusalem, Israel, 1990.

CHAPTER TWENTY-FOUR – THE BATTLE FOR TIME
1. Andrew D. Basiago, *Project Pegasus*. https://www.projectpegasus.net/
2. Ella LeBain, *Who's Who in the Cosmic Zoo?* – Book One – Third Edition, Chapter: Life on Mars, pp. 310-313, Skypath Books, 2013.
3. Andrew D. Basiago, A lawyer who claims he traveled in time and even got into an old photo, https://infinityexplorers.com/andrew-basiago-a-lawyer-who-claims-he-has-traveled-in-time-and-even-got-into-an-old-photo, September 11, 2020
4. Chuck Pierce, *Reordering Your Day: Understanding and Embracing the Four Prayer Watches*, Glory of Zion International, (November 10, 2011).

CHAPTER TWENTY-FIVE – FROM SECRET SPACE PROGRAM TO SPACE FORCE
1. Trump Orders Pentagon to 'Immediately' Establish 'Space Force', https://www.thedailybeast.com/trump-orders-pentagon-to-immediately-establish-space-force, June 2018.
2. Whistleblower Jason Rice Interview, Zohar Stargate TV aka UFO & Ancient Mysteries Networking TV, (UAMN TV), YouTube Channel, March 25, 2019.
3. Jason Rice on Fade to Black with Jimmy Church, October 3, 2018

CHAPTER TWENTY-SIX – WHO OWNS SPACE?
1. "What Is Space Law?". Legal Career Path.
2. "Space Law". United Nations Office for Outer Space Affairs.
3. Gabrynowicz, Joanne Irene. "Space law: Its Cold War origins and challenges in the era of globalization". Suffolk University Law Review – via Law Journal Library.
4. Robert Wickramatunga, The Outer Space Treaty. http://www.unoosa.org.
5. Henri, Yvon. "Orbit Spectrum Allocation Procedures ITU Registration Mechanism" (PDF). International Telecommunications Union.
6. "Committee on the Peaceful Uses of Outer Space and its Subcommittees". United Nations Office for Outer Space Affairs.
7. Jacques Arnould (2011-09-15). *Icarus' Second Chance: The Basis and Perspectives of Space Ethics*. ISBN 9783709107126.
8. Stephen Gorove (1979). *"The Geostationary Orbit: Issues of Law and Policy"*. The American Journal of International Law. 73 (3): 444–461. doi:10.2307/2201144. JSTOR 2201144.
9. Yasmin Ali, Who Owns Outer Space? September 25, 2015, https://www.bbc.com/news/science-environment-34324443

10. Joseph Trevithick, Space Force Boss Says One Of Russia's Killer Satellites Fired A Projectile In Orbit, posted in THE WAR ZONE, https://www.thedrive.com/the-war-zone/35057/space-force-boss-says-russia-has-been-testing-its-killer-satellites-in-orbit
11. Joseph Trevithick, Space Force Just Received its First New Offensive Weapon, posted in THE WAR ZONE, https://www.thedrive.com/the-war-zone/32570/space-force-just-received-its-first-new-offensive-weapon
12. Joseph Trevithick, USAF Secretary Gives Ominous Warning that Show of Force Needed to Deter Space Attacks, posted in THE WAR ZONE, https://www.thedrive.com/the-war-zone/27396/usaf-secretary-gives-ominous-warning-that-show-of-force-needed-to-deter-space-attacks
13. Sandra Erwin, SpaceX President Gwynne Shotwell: 'We would launch a weapon to defend the U.S.'https://spacenews.com/spacex-president-gwynne-shotwell-we-would-launch-a-weapon-to-defend-the-u-s/, September 17, 2018.

CHAPTER TWENTY-SEVEN – THE NEW SPACE RACE
1. SpaceX – FALCON 9 – FALCON HEAVY – DRAGON – STARSHIP – HUMAN SPACEFLIGHT – STARLINK MISSION, https://www.spacex.com/launches/
2. Lara Logan, CBS 60 Minutes, Bigelow Aerospace founder says commercial world will lead in space: NASA and Las Vegas entrepreneur have partnered on a technology that could change how humans live and work in space, May 28, 2017, https://www.cbsnews.com/news/bigelow-aerospace-founder-says-commercial-world-will-lead-in-space/
3. Eric Berger, The U.S. Space Force has now actually gone to space: So far, Covid-19 has yet to substantially slow activity at the military's spaceports, https://arstechnica.com/science/2020/03/the-first-launch-for-the-us-space-force-is-set-for-today-from-florida/, 3/26/2020.

CHAPTER TWENTY-EIGHT – NASA's BOMBSHELL ON THE MOON
1. Ella LeBain, *Who's Who in the Cosmic Zoo?* Book One – Third Edition, Chapter: Solarians, pp.403. Skypath Books, 2013.
2. Janice Friedman, Ancient Aliens Blog, NASA Dropped A 2-Ton Kinetic Missile on the Moon, https://www.ancient-code.com/nasa-dropped-a-2-ton-kinetic-missile-on-the-moon-what-did-they-destroy/
3. Col Phillip S. Meilinger, USAF, Editor, The Paths of Heaven: The Evolution of Airpower Theory – by The School of Advanced Airpower Studies, Air University Press, Maxwell Air Force Base, Alabama, 1997.
4. William Elliott Butler, Perestroika and International Law, Springer; 1990 edition (January 31, 1990)
5. 'NASA Dropped Nuclear Bomb on Alien Moon Base On October 9, 2009': UFO Blogger Claims, 'Smoking Gun' Evidence, : https://www.inquisitr.com/2988781/nasa-dropped-nuclear-bomb-on-alien-moon-base-on-october-9-2009-ufo-blogger-claims-smoking-gun-evidence-video/#ixzz6IzmnxHO3, April 12, 2016.
6. Fox News Live Report of NASA Bombing the Moon, https://youtu.be/avmDnwsOGDw, October 9, 2009.
7. Scott C. Waring, CONFIRMED! NASA Did Drop the Nuke In Oct 2009 On An Alien Base, https://www.ufosightingsdaily.com/2016/04/confirmed-nasa-did-drop-nuke-in-oct.html, April 2016, Video, UFO Sighting News.

8. Ella LeBain, *Who's Who in the Cosmic Zoo?* Book One – Third Edition, Chapter: Solarians, pp.403. Skypath Books, 2013.
9. Mike Wall, Trump signs executive order to support moon mining, tap asteroid resources, The U.S. sees a clear path to the use of moon and asteroid resources, https://www.space.com/trump-moon-mining-space-resources-executive-order.html, April 6, 2020
10. whitehouse.gov/presidential-actions/executive-order-encouraging-international-support-recovery-use-space-resources/
11. whitehouse.gov/wp-content/uploads/2020/04/Fact-Sheet-on-EO-Encouraging-International-Support-for-the-Recovery-and-Use-of-Space-Resources.pdf

CHAPTER TWENTY-NINE – PRESIDENTIAL SPACE POLICIES
1. Mike Wall, Presidential Visions for Space Exploration: From Ike to Trump, https://www.space.com/11751-nasa-american-presidential-visions-space-exploration.html, February 05, 2020.

CHAPTER THIRTY – THE GRAYS & THE U.S. GOVERNMENT
1. Ella LeBain, *Who's Who in the Cosmic Zoo?* Book One – Third Edition, Chapter: The Grays, pp. 248-289, Tate, 2013.
2. Daily Mail, President Eisenhower had Three Secret Meetings with Aliens, former Pentagon Consultant claims, https://www.dailymail.co.uk/news/article-2100947/Eisenhower-secret-meetings-aliens-pentagon-consultant-claims.html, February 13, 2012.
3. Ana Ionita, TAU IX Treaty—The US Government Has a Secret Pact with the Greys, 3/26/16, http://ufoholic.com/tau-ix-treaty-the-us-government-has-a-secret-pact-with-the-greys/
4. William Cooper, "Origin, Identity, and Purpose of MJ-12," http://www.geocities.com/Area51/Shadowlands/6583/maji007.html William Cooper, A Covenant With Death, http://www.alienshift.com/id40.html taken from his book, Behold a Pale Horse, Light Technology Publishing, 1991, p. 202-203.
5. John Lear, *Disclosure Briefing, Coast to Coast Radio with Art Bell*, November 2003 http://www.coasttocoastam.com/shows/2003/11/02.html
6. Nick Redfern, *A President, a Comedian, and Pickled Aliens*, https://mysteriousuniverse.org/2017/03/a-president-a-comedian-and-pickled-aliens/, March 3, 2017.

CHAPTER THIRTY-ONE – SECRET GOVERNMENT SPACE PROGRAMS
1. Joe Pappalardo, *Popular Mechanics*: Declassified: America's Secret Flying Saucer, https://www.popularmechanics.com/military/a8699/declassified-americas-secret-flying-saucer-15075926/, February 11, 2013
2. US Air Force Tests 'Flying Saucer', https://www.military.com/video/space-technology/spacecraft/us-air-force-tests-flying-saucer/3943489006001
3. Ivana Cardinale, There Is More Than Just One 'Area 51' & They Exist In Places You Won't Believe Is Possible, https://www.websitesbb.com/2017/caribflame_dev.com/2016/07/there-is-more-than-just-one-area-51-they-exist-in-places-you-wont-believe-is-possible/, July 1, 2016
4. Ella LeBain, *Who's Who in the Cosmic Zoo?* Book One – Third Edition, See, Chapter: Dulce Underground, pp. 185-202, Skypath Books, 2013.

5. James & Lance Morcan, *Underground Bases, Subterranean Military Facilities and the Cities Beneath Our Feet* (The Underground Knowledge Series Book 7) Sterling Gate Books (November 16, 2015), https://www.goodreads.com/topic/show/1961070-undersea-bases
6. Lloyd A. Duscha, *Underground Facilities for Defense – Experience and Lessons, in Tunneling and Underground Transport: Future Developments in Technology. Economics and Policy*, ed. F.P. Davidson, pp. 109-113, New York: Elsevier Science Publishing Company, Inc., 1987.
7. Richard Sauder, Ph.D., *Hidden In Plain: Sight Underwater & Underground Bases: Surprising Facts the Government Does Not Want You to Know*, Adventures Unlimited Press (May 23, 2014)
8. Gary McKinnon, http://freegary.org.uk/
9. Profile: Gary McKinnon, https://www.bbc.co.uk/news/uk-19946902, December 14, 2012
10. Kerry Cassidy, Project Camelot interviews Gary McKinnon, Nov 20, 2007, https://www.youtube.com/watch?v=_fNsah-0vpY
11. Brett Tingley, Eyebrow raising claims about advanced space technology made by recently retired US General, The Drive, https://www.thedrive.com/the-war-zone/31445/recently-retired-usaf-general-makes-eyebrow-raising-claims-about-advanced-space-technology, December 11, 2019
12. Steven Kwast, The Urgent Need for a U.S. Space Force, https://youtu.be/KsPLmb6gAdw, December 5, 2019
13. George Filer, Filers File#28, July 9, 2008, We Have Lost Many Men and Planes Trying To Intercept UFOs, http://nationalufocenter.com/artman/publish/article_232.php
14. https://missingmoneysolari.com/dod-and-hud-missing-money-supporting-documentation/
15. 9/10/2001 Def. Secy Donald Rumsfeld, $2.3 Trillion Missing from Pentagon, https://www.youtube.com/watch?v=y7ywpfOOn7k
16. John L. Guerra, *Strange Craft: The True Story of An Air Force Intelligence Officer's Life with UFOs* Paperback, https://amzn.to/2SK3SIQ, December 19, 2018
17. Darren Perks, *Solar Warden – The Secret Space Program*, https://www.huffingtonpost.co.uk/darren-perks/solar-warden-the-secret-space-program_b_1659192.html?guccounter=1, 10.9.2012
18. Joseph Trevithick, THE WAR ZONE, https://nationalufocenter.com/?email_id=418&user_id=18945&urlpassed=aI IR0c HM6Ly91Zm8uZmFuZG9tLmNvbS93aWtpL1NvbGFyX1dhcmRlbg&controller=stats&action=analyse&wysija-page=1&wysijap=subscriptions
19. William Tompkins, *Selected by Extraterrestrials, My life in the top-secret world of UFOs., Think-Tanks and Nordic Secretaries*, July 2016, p.417, https://www.amazon.com/Selected-Extraterrestrials-secret-think-tanks-secretaries/dp/1515217469
20. *Secret Space Program*, http://conspiracy.wikia.com/wiki/Secret_Space_Program
21. Thomas Schrøder Jensen, https://nationalufocenter.com/?email_id=418&user_id=18945&urlpassed=aHR0c DovL2NvbnNwaXJhY3kud2lraWEuY29tL3dpa2kvU2VjcmV0X1NwYWNlX1Byb 2dyYW0&controller=stats&action=analyse&wysija-page=1&wysijap=subscriptions
22. Laura Valkovic, Narrated News, August 05, 2018, https://www.libertynation.com/why-are-ufos-making-it-into-the-mainstream-

media/?fbclid=IwAR25yA4o9n_ERbwNoLXGeWq3xBr_6tFRlMzewRB98lp2p1B
QyWe9yxIRE54,

CHAPTER THIRTY-TWO – REPTILIANS HELPED NAZIES BUILD A SECRET SPACE PROGRAM IN ANTARCTICA

1. Ella LeBain, *COVENANTS* – Book Four, Skypath Books, 2018, https://www.amazon.com/COVENANTS-Times-Guide-Aliens-Angels/dp/0692988637/
2. Dr. Michael Salla, William Tompkins Interview Transcript – Reptilian Aliens Helped Germans Establish Space Program in Antarctica, https://www.exopolitics.org/interview-transcript-us-navy-spies-learned-of-nazi-alliance-with-reptilian-extraterrestrials/ April 4, 2016
3. Dr. Michael Salla, William Tompkins Interview Transcript – US Navy Spies Learned of Nazi Alliance with Reptilian Extraterrestrials; https://michaelsalla.com/2016/04/14/interview-transcript-reptilian-aliens-helped-germans-establish-space-program-in-antarctica/, April 14, 2016.
4. Ella LeBain, *Who's Who in the Cosmic Zoo?* Book One – Third Edition, Tate, 2013. pp. 96-97
5. History, Feb. 15, 2020, Reptilian Aliens Helped Nazi Germany Build Secret Space Program in Antarctica, https://wokehub.com/history/reptilian-aliens-helped-nazi-germany-build-secret-space-program-in-antarctica/
6. William Mills Tompkins autobiography, Selected by Extraterrestrials: My life in the Top-Secret World of UFOs, Think-Tanks and Nordic Secretaries, CreateSpace, Dec.9, 2015.
7. William Tompkins Amazon Bio: https://www.amazon.com/Selected-Extraterrestrials-secret-think-tanks-secretaries/dp/1515217469

CHAPTER THIRTY-THREE – SPACE FORCE

1. David Axe, Pentagon Officials Are Preparing for an All-Out Space War, https://www.thedailybeast.com/pentagon-officials-are-preparing-for-an-all-out-space-war, April 11, 2019
2. NASA Astronaut Shockingly Hints at Aliens in Tweet About 'Life Forms', https://sputniknews.com/science/201912141077575518-nasa-astronaut-shockingly-hints-at-aliens-in-tweet-about-life-forms/
3. George Filer, Major USAF ret. Filer's Files, January 1, 2020 New Space Force, https://nationalufocenter.com/blog/2019/12/30/filers-files-1-2020-new-space-force/
4. https://exonews.org/space-force-what-will-the-newest-military-branch-actually-do/
5. Leonard David, Space Force: What will the new military branch actually do? February 09, 2020, https://www.space.com/united-states-space-force-next-steps.html
6. At Least 17 Billion Earth-like Planets Across the Milky Way Study, http://news.nationalpost.com/2013/01/07/a-least-17-billion-earth-like-planets-across-the-milky-way-study/, January 7, 2013.
7. Michael Moyer, Earth-Like Planets Fill the Galaxy, http://blogs.scientificamerican.com/observations/2013/01/08/earth-like-planets-fill-the-galaxy/, January 8, 2013.
8. Ker Than, National Geographic News, Every Black Hole Contains Another Universe?, http://news.nationalgeographic.com/news/2010/04/100409-black-holes-alternate-universe-multiverse-einstein-wormholes/, April 12, 2010.

9. Chris Burns, Why the Space Force logo looks like Star Trek, and Star Trek looks like NASA, https://www.slashgear.com/why-the-space-force-logo-looks-like-star-trek-and-star-trek-looks-like-nasa-24607686/, Jan 24, 2020
10. U.S. Space Force Next Steps, https://www.space.com/united-states-space-force-next-steps.html
11. Joseph Trevithick, New Space Command's Flag Sure Looks A Lot Like Old Space Command's Flag: Although the idea of an independent Space Force has made headlines for years now, Space Command isn't new. Even its flag hearkens back to the 1980s, https://www.thedrive.com/the-war-zone/29625/the-flag-for-new-space-command-sure-looks-a-lot-like-the-one-for-old-space-command, August 29, 2019.
12. Gil Carlson, The Book of Alien Races, Secret Russian KGB Book of Alien Species https://www.amazon.com.au/Book-Alien-Race-Russian-Species-ebook/dp/B07BTF6WLL/, Blue Planet Project, March 29, 2018.
13. Gil Carlson, Secret Space Program, Blue Planet Project, https://www.amazon.com.au/Air-Force-Secret-Space-Program/dp/1513660276, January 1, 2020.
14. Joseph Trevithick, USAF Secretary Gives Ominous Warning That Show Of Force Needed To Deter Space Attacks: But serious questions remain about what an all-out war in space would even look like and what hostile actions would automatically demand a response, https://www.thedrive.com/the-war-zone/27396/usaf-secretary-gives-ominous-warning-that-show-of-force-needed-to-deter-space-attacks, April 11, 2019.
15. Robin McKie, https://www.msn.com/en-us/news/technology/the-moon-mars-and-beyond…-the-space-race-in-2020/ar-BBYCSSC?ocid=spartanntp
16. *Armageddon*, 1998 Film, Director: Michael Bay, Screenplay Writers: Jonathan Hensleigh, J.J. Abrams, https://www.imdb.com/title/tt0120591/
17. Lockheed Martin, Space Exploration, https://www.lockheedmartin.com/en-us/capabilities/space/human-exploration.html?
18. *Greenland* (2020 movie), https://en.wikipedia.org/wiki/Greenland_(film)
19. Killer Asteroid Hitting Earth In 2022? Here Is What NASA Says, https://www.republicworld.com/technology-news/science/killer-asteroid-approaching-earth-in-2022-asteroid-hitting-earth.html
20. Apophis in Depth, https://solarsystem.nasa.gov/asteroids-comets-and-meteors/asteroids/apophis/in-depth/

CHAPTER THIRTY-FOUR – BIBLE DNA & TODAY'S GENETICISTS

1. Ella LeBain, *Who's Who in the Cosmic Zoo? A Spiritual Guide to ETs, Aliens, Gods & Angels*, Book One – Third Edition, Concluding Words, pp.415-444, Skypath Books, 2013.
2. Ella LeBain, *Who is God?* Book Two, Chapter: Who is Allah?, p.359, Skypath Books, 2015.
3. Zecharia Sitchin, *Genesis Revisited, Is Modern Science Catching Up with Ancient Knowledge?* (Earth Chronicles) Avon Paperback – October 1, 1990.
4. Ella LeBain, *Who's Who in the Cosmic Zoo? A Spiritual Guide to ETs, Aliens, Gods & Angels*, Book One – Third Edition, Concluding Words, pp.415-444, Skypath Books, 2013.
5. Francis Crick, *Life Itself: Its Origin and Nature*, https://www.amazon.com/Life-Itself-Its-Origin-Nature/dp/0671255622/, January 1, 1981
6. Carl Sagan, I.S. Shklovskii, *Intelligent Life in the Universe*, p.456, 463; Dell Publishing, 1966.

7. Erich Von Daniken, *The Eyes of the Sphinx: The Newest Evidence of Extraterrestrial Contact in Ancient Egypt*, https://www.amazon.com/Eyes-Sphinx-Evidence-Extraterrestial-Contact/dp/0425151301, Berkley, 1st Edition, March 1, 1996.
8. Theodor H. Gaster, *The Dead Sea Scriptures*, New York: Doubleday Anchor Books.; Ex-Monastery Library edition – January 1, 1956
9. Lee R. Berger, *In the Footsteps of Eve: The Mystery of Human Origins*, https://www.amazon.com/Footsteps-Eve-Mystery-Origins-Adventure/dp/0792276825, (Adventure Press) June 1, 2000
10. Heather Lynn PhD, *The Anunnaki Connection: Sumerian Gods, Alien DNA, and the Fate of Humanity* (From Eden to Armageddon) on www.amazon.com/Anunnaki-Connection-Sumerian, New Page Books, March 1, 2020.
11. George Filer, Filer's Files 3, 2020 Science Catches Up to the Bible, NUFOC weekly-updates@nationalufocenter.com
12. Gil Carlson, *NIBIRU PLANET X TODAY: Anunnaki Aliens, UFOs, A Blue Planet Project*, Blue Planet Project (April 8, 2018).
13. Michael Baxter, Space Force to Survey Nibiru, https://www.twistedtruth.net/featured/space-force-to-survey-nibiru/, January 15, 2020.
14. Michael Baxter, Putin Addresses Cabinet of Minister on Anunnaki in Iran; Full Transcript, https://www.twistedtruth.net/featured/putin-addressed-cabinet-of-minister-on-anunnaki-in-iran-full-transcript/, January 15, 2020.

CHAPTER THIRTY-FIVE – CLOUDSHIPS, CHARIOTS & UFOs
1. Ella LeBain, *Who is God?* Book Two in Chapter Two: Ancient Technology and Biblical Astronauts, pp.21-46, Skypath Books, 2015.
2. Ella LeBain, *Who Are the Angels?* Book Three in Chapter Three: The Celestial Hierarchy of Angels, pp.48-60, Skypath Books, 2016
3. CLOUDSHIPS – UFO CLOUDS -Federation of Light - UFO Cloud Ships?, https://youtu.be/jwpsTZIkEiA, September 16, 2008
4. UFO cloud mystery worldwide sightings - Secret experiments Alien spaceships weather, https://youtu.be/i8c87harZCw, January 27, 2010
5. Ella LeBain, *Who's Who in the Cosmic Zoo?* Book One – Third Edition, Skypath Books, 2013.
6. Order of Melchizedek, UFO, ALIENS & THE BIBLE, http://www.atam.org/UFO.html, January 22, 2011
7. Order Of Melchizedek, Celestial Chariots of God: Aliens & UFO's http://atam.org/celestial-chariots-of-god/, April 11, 2011.
8. UFO cloud mystery worldwide sightings - Secret experiments Alien spaceships weatherhttps://youtu.be/i8c87harZCw, January 27, 2010

CHAPTER THIRTY-SIX – THE CASE AGAINST FLAT EARTHERS
1. *Society of Flat Earth Debunkers, Flat Earth Theory Fabricated By TPTB To Distract From Planet X*, http://themillenniumreport.com/2016/01/flat-earth-theory-fabricated-by-tptb-to-distract-from-planet-x/, January 20, 2006
2. *Society of Flat Earth Debunkers*, Submitted: 1/20/06, http://m.beforeitsnews.com/alternative/2016/01/busted-flat-earth-theory-used-to-wage-war-against-the-truth-movement-3284388.html

3. Ella LeBain, *COVENANTS* - Book Four, Skypath Books, 2018.
4. Brown, F. et al., Brown-Driver-Briggs *Hebrew and English Lexicon: With an Appendix Containing the Biblical Aramaic*, Hendrikson Publishers, USA, p. 295, reprinted January 1999 from the 1906 edition; biblehub.com/hebrew/2329.htm.
5. Ella LeBain, *Book One of Who's Who in The Cosmic Zoo?* (See, pp. 23,50-62) Skypath Books, 2012.
6. Charles Q. Choi, Strange but True: Earth Is Not Round: It may seem round when viewed from space, but our planet is actually a bumpy spheroid, https://www.scientificamerican.com/article/earth-is-not-round/, April 12, 2007.
7. Ella LeBain, *Who is God?* Book Two, Skypath Books, 2015.
8. Ella LeBain, *Who Are the Angels?* Book Three, Skypath Books, 2016.
9. David Read, Raqia: 'Expanse or Vault?' http://advindicate.com/articles/1494, June 26, 2012
10. Crystallinity in the Atmosphere, https://en.wikipedia.org/wiki/Crystallinity
11. *Vine's Expository Dictionary of Old and New Testament Words*, 1981, p. 67.
12. *John Gill's Exposition of the Bible*, footnote to Isaiah 40:22; biblestudytools.com.
13. Dominic Statham, Isaiah 40:22 and the shape of the Earth, http://creation.mobi/isaiah-40-22-circle-sphere, August 11, 2016.
14. J. Johansen, *Flat Earthers Make Jews and Christians Look Stupid*, March 18, 2016.
15. Mike Heiser, *Who Believes the Earth is Really Flat? Does it get any dumber than this?* http://drmsh.com/christians-who-believe-the-earth-is-really-flat-does-it-get-any-dumber-than-this/
16. The Pre-flood Atmosphere, http://www.genesispark.com/exhibits/early-earth/atmosphere/
17. Henry M. Morris, *Scientific Creationism: Study Real Evidence of Origins*, Discover Scientific Flaws in Evolution, p. 211, Master Books (August 1, 1985), https://www.amazon.com/dp/B006E8ZNJA/
18. Ella LeBain, *COVENANTS* - Book Four, Chapter 31: Implants and Spiritual Limitation Devices, pp.479-521, Skypath Books, 2018.
19. Ben-Yehuda, E. and Ben-Yehuda, D, *Hebrew Dictionary*, Pocket Books (Simon & Schuster), USA, p. 252, 1961.
20. Levin, S., *Semitic and Indo-European: The Principal Etymologies*, vol. 1, John Benjamins, USA, 1995.
21. Buck, C.D., *A Dictionary of Selected Synonyms in the Principal Indo-European Languages*, University of Chicago Press, Chicago, pp. 907–8, 1949.
22. López-Menchero, F., *Proto-Indo-European Etymological Dictionary*, indo-european.info/indo-european-lexicon.pdf, 2012. Asociación Cultural Dnghu; dnghu.org.

CHAPTER THIRTY-SEVEN – THE REAL STAR WARS

1. Adam Eliyahu Berkowitz, *NASA Discovery of 7-Planet System Conforms Exactly to Zohar's Description of Pre-Messiah "Nibiru"* February 28, 2017, https://www.breakingisraelnews.com/wp-content/uploads/2017/02/TRAPPIST-1-planets
2. Yuval Ovadia, *NASA Discovery 7 Planet System Conforms Exactly Zohar's Description Pre-Messiah Nibiru*, https://www.breakingisraelnews.com/84337/nasa-discovery-7-planet-system-conforms-exactly-zohars-description-pre-messiah-nibiru/#UOLRLaZxISKXoowb.99

3. Daniel Perez, *From Jediism to Judaism: Star Wars as Jewish Allegory*, https://www.aish.com/j/as/From-Jediism-to-Judaism-Star-Wars-as-Jewish-Allegory.html, December 5, 2015.
4. Rachel Sharansky Danziger, *Star Wars and the Hanukkah Story Have More in Common Than You Think*, https://www.kveller.com/star-wars-and-the-hanukkah-story-have-more-in-common-than-you-think/, December 10, 2015.
5. Keren Keet, *Confessions of a Jewish Jedi*, https://www.aish.com/j/as/Confessions-of-a-Jewish-Jedi.html, December 12, 2015, Tatz, Worldmask, Targum Press, 1995, p34, *used with permission.*
6. Keren Keet, *The Last Jedi and the Force of the Family*, https://www.aish.com/j/as/The-Last-Jedi-and-the-Force-of-Family.html, December 10, 2017, *used with permission.*
7. Jonathan S. Tobin, *Star Trek and God, Sci-fi fun often ignores the tough questions about human existence*, https://www.aish.com/j/as/48970691.html, May 11, 2009.
8. Tony McMahon, *The Templar Knight, Mysteries of the Knights Templar* https://business.facebook.com/QuestForTheTrueCross/
9. Brian Vastag, From 'Star Wars' to reality: Astronomers discover Tatooine world with two Suns, https://www.washingtonpost.com/national/health-science/from-star-wars-to-reality-astronomers-discover-tatooine-world-with-two-suns/2011/09/15/gIQAChBzUK_story.html, September 15, 2011
10. Tony McMahon, *The Templars and the Kabbalah*, May 29, 2011 Facebook Post
11. Tony McMahon, *The Knights Templar today - who and where are they? In "Middle Ages"* 2011 Facebook Post.

CHAPTER THIRTY-EIGHT – SIGNS AND WONDERS IN THE HEAVENS
1. MUFON, Mutual UFO Network, https://www.mufon.com/
2. *Who's Who in the Cosmic Zoo?* Facebook Fan Page, http://www.Facebook.com/whoswhointhecosmiczoo
3. Ella LeBain, *Who's Who in the Cosmic Zoo?* Book One – Third Edition, Skypath Books, 2012.
4. Marshall Masters, *Surviving the Planet X Tribulation: A Faith-Based Leadership Guide* 2nd ed. Edition, Your Own Worlds Books, 2016.
5. Carl Sagan, I.S. Shklovskii, *Intelligent Life in the Universe*, p.456, 463; Dell Publishing, 1966.

CHAPTER THIRTY-NINE – SUN OR SON?
1. Ella LeBain, The Christmas Gods, Chapter 27, *Who is God?* Book Two, Skypath Books, 2015. http://www.whoswhointhecosmiczoo.com
2. Ella LeBain, *Should Christians Celebrate Christmas?*, https://spirituallydiscerning.com/f/should-christians-celebrate-christmas
3. Rabbi Arthur O. Waskow, *Seasons of Our Joy - A Modern Guide to the Jewish Holidays*, The Jewish Publication Society, 1985.
4. *Solar Mythology and the Jesus Story*, http://www.solarmythology.com/lessons/christ2002.htm
5. *Solar Mythology and the Jesus Story Explained*, http://www.solarmythology.com/david16.htm
6. Ella LeBain, *Who is God?* Book Two - Chapter 28: What Happens When You Die: Jesus Taught Reincarnation, pp. 455-479, Skypath Books, 2015.
7. Ella LeBain, *Who's Who in the Cosmic Zoo?* Book One, Chapter Two: Exopolitics & Divine Jurisprudence, Spiritual Legal Ground. [pp.43-69]

8. Dr. Eitan Bar, Isaiah 53 – The Forbidden Chapter, https://www.oneforisrael.org/bible-based-teaching-from-israel/inescapable-truth-isaiah-53/, Article later added to his books, Refuting Rabbinic Objections to Christianity & Messianic Prophecies, ONE FOR ISRAEL Ministry (January 5, 2019).
9. Isaiah Chapter 53 verses 1-12, King James Version translation from the Hebrew Masoretic Text.
10. Got Questions Ministries: What does it mean that "the Lord said to my Lord"? https://www.gotquestions.org/Lord-said-to-my-Lord.html, January 2, 2020.
11. Jon W. Quinn, *The Nature of God*, From Expository Files, 3.7; July 1996

CONCLUDING WORDS
1. Ella LeBain, *Who's Who in the Cosmic Zoo?* Book One – Third Edition, Concluding Words, pp. 415-444. Skypath Books, 2013.
2. Ella LeBain, *COVENANTS* - Book Four - An End Times Guide to ETs Aliens gods and Angels, Skypath Books, 2018. https://www.amazon.com/gp/aw/d/0692988637/
3. Ella LeBain, *Who is God?* Book Two, Chapter Four: Motherships of the Lord, p.71-85, Skypath Books, 2015.
4. Ella LeBain, *Who Are the Angels?* Book Three, Skypath Books, 2016.
5. John Pipe, *Battling the Unbelief of a Haughty Spirit*, http://www.desiringgod.org/messages/battling-the-unbelief-of-a-haughty-spirit, December 18, 1988
6. Swami Tapasyananda, Sunkarakandam of Srimad Valmiki Ramayan, *The Ramayana*, Published by The President, Sri Ramakrishna Math Printing Press, Mylapori, Chennai, India, 2006. pp.46-48

ABOUT THE AUTHOR

WHO IS ELLA LEBAIN?

ET Experiencer, UFO Researcher, Ella LeBain is the author of the Book Series, *Who's Who in The Cosmic Zoo? A Guide to ETs, Aliens, Gods & Angels*.

Ella LeBain lives in Westminster, Colorado with her husband, daughter and three cats. Ella is originally from New York City and was educated in Israel. She received a Social Sciences Degree from the Biological Research Center of the Negev in 1979 where she was trained in Biblical Hebrew. She then went on to receive an Astronomy Degree from the Hayden Planetarium in New York City in 1982. Ella has spent forty-one years in the field of UFO research, investigating alien abductions, paranormal activities along with her deep love and research of Astronomy/Astrology during which time she had many supernatural experiences of her own along the way, many of which shaped the writing of her books.

Ella is a Conference Panel Speaker on **The Truth being Stranger than Fiction!** Her panel discussions are focused on the real *Star Wars*, the newly discovered Exoplanets and how it all fits into the Ancient Prophesies of the End of this Age. SETI Astronomers nick-named their discovery of our second Sun, "Tatooine" inspired from *Star Wars*. Ella is an expert in her field of connecting the dots from science fiction to present scientific and historic realities. Ella also connects the dots to the real *Star Trek* with today's *Secret Space Program*, proving the **Truth is Stranger than Fiction**, and how these Sci-Fi films were designed to prepare the public for the Truth, that we are not alone in the Universe!

Disclosure of extraterrestrial and alien life is the truth that is stranger than fiction! Ella shares Proof that Star Wars and Star Trek are based on fact, not fiction. In her books and Panel Discussions, people learn the Truth about the real Star Wars program that has been covered up for 70+ years. Ella identifies and discerns the aliens, intraterrestrials and interdimensional beings who share our planet, who are based "inside" the Earth and inside our Solar System. In her unique way, Ella connects the dots, to the ancient prophesies concerning ETs, Aliens, gods, and Angel at the end of this age, and how the presence of recently discovered Exoplanets and a pass from Nibiru is changing our world as we know it. Welcome to the Space Age! Ella's books reveal the real Star Wars over humanity and planet Earth, which was prophesied by the Ancient Scriptures.

Ella's Book Series offers a plethora of information from a variety of sources, in addition to her own supernatural experiences incorporated into this Book set. Book One is a type of Encyclopedia, covering *Who's Who in the Cosmic Zoo of ETs and Aliens* in an A-Z Compendium.

Book Two - *Who Is God?* focuses on the Cosmic Drama, that identifies and discerns who are the ET gods of ancient history based on both biblical and exobiblical scriptures.

Book Three - *Who Are the Angels?* focuses on the hierarchy of angels (extraterrestrial messengers) including the fallen ET angels as well as those who have remained faithful to the Creator and are engaged in an ongoing galactic war with the fallen angels over this Solar System. She reveals how they have been interacting with humankind for millennia and connects the dots to the important starring roles they play at the end of this age.

Book Four - *Covenants*, explores, and discerns the roots of racism, gene wars on planet Earth and real *Star Wars* between ETs, Aliens, gods, and Angels.

Book Five - *The Heavens*, Discerning Disclosure of UFOs, SSP, Nibiru, Bible Prophecies and the second coming of Christ.

Book Six – *Heavens Witness*, discerns the Mazzaroth as Signs in the Heavens that have long been prophesied to show up as harbingers of the end of this present age, heralding the Age to Come.

Ella LeBain is available for interviews, lectures, and book signings.

Website: http://whoswhointhecosmiczoo.com

Buy the Books: http://www.whoswhointhecosmiczoo.com/buy-the-book.html

Contact Page: http://whoswhointhecosmiczoo.com/contact.html
Email: ellalebain@whoswhointhecosmiczoo.com
Facebook: https://www.Facebook.com/whoswhointhecosmiczoo

www.ingramcontent.com/pod-product-compliance
Lightning Source LLC
Chambersburg PA
CBHW082102230426
43671CB00015B/2587